Advanced Theory of Constraint and Motion Analysis for Robot Mechanisms

Advanced Theory of Constraint and Motion Analysis for Robot Mechanisms

Jingshan Zhao
Zhijing Feng
Fulei Chu
Ning Ma

ELSEVIER

AMSTERDAM • BOSTON • HEIDELBERG • LONDON
NEW YORK • OXFORD • PARIS • SAN DIEGO
SAN FRANCISCO • SINGAPORE • SYDNEY • TOKYO

Academic Press is an Imprint of Elsevier

Academic Press is an imprint of Elsevier
The Boulevard, Langford Lane, Kidlington, Oxford, OX5 1GB
225 Wyman Street, Waltham, MA 02451, USA

First published 2014

Notices
Knowledge and best practice in this field are constantly changing. As new research and experience broaden our understanding, changes in research methods, professional practices, or medical treatment may become necessary.

Practitioners and researchers must always rely on their own experience and knowledge in evaluating and using any information, methods, compounds, or experiments described herein. In using such information or methods they should be mindful of their own safety and the safety of others, including parties for whom they have a professional responsibility.

To the fullest extent of the law, neither the Publisher nor the authors, contributors, or editors, assume any liability for any injury and/or damage to persons or property as a matter of products liability, negligence or otherwise, or from any use or operation of any methods, products, instructions, or ideas contained in the material herein.

British Library Cataloguing in Publication Data
A catalogue record for this book is available from the British Library

Library of Congress Cataloguing in Publication Data
A catalog record for this book is available from the Library of Congress

ISBN: 978-0-12-420162-0

For information on all Academic Press publications
visit our website at store.elsevier.com

Printed and bound by CPI Group (UK) Ltd, Croydon, CR0 4YY

Contents

Preface

All machines are made of mechanisms. Before design, the degree of freedom of the end effector of a robot mechanism should first be determined, and then the actuation scheme can be investigated. Thus, it is the fundamental basis and prerequisite for structural analysis and synthesis to analyze the degree of freedom of the end effector of a mechanism. Structure analysis of a mechanism is focusing on proposing the structure principle of mechanisms, the classification of mechanisms, and their free motions. Meanwhile, structure analysis of mechanisms is an important aspect of kinematics, dynamics analysis and mechanism synthesis. The rapid development of science and technology calls for the creation and design of better mechanisms. Therefore, it is essential to conduct further theoretical research into the structure, type, and kinematic characteristics of mechanisms. Structure analysis presents a basis to determine the number of the actuators and carry out kinematics and dynamics research. Therefore, it is one of the most important concepts in mechanism. It offers a guarantee for innovation and structure synthesis, and lays a foundation for further analyzing the driving, and kinematics and dynamics performances of mechanisms.

This book studies the constraint and motion analysis for spatial robot mechanisms based on the unified mathematical representation of its kinematic geometry constraints. Chapter 1 roughly surveys the development and problem in the structure analysis of mechanisms in the past centuries. In order to limit the mathematical prerequisites to the knowledge of elementary screw algebra and matrix calculus, it was felt necessary to present in Chapter 2 an introduction to screw theory. The material covered is limited to that necessary for an understanding of the book, and no attempt has been made to study thoroughly such topics. It describes the primary knowledge about the screw algebra from aspects such as the Plücker column vector of a line, motion representation of rigid bodies, the screw description of the motion and force, reciprocal product of the screw and its geometry significance, the linear combination of screws and the principal screws in a screw system.

In Chapter 3, the motion screw system of kinematical chains is established through the screw representation of the free motions of kinematic pairs; the terminal constraints of the kinematic chain are derived according to the reciprocal relationship between twists and wrenches. The equivalent substitution and synthesis of kinematic chains under the conditions of the same terminal constraints are studied. The investigation into the mechanical equivalent substitution theorem lays a necessary foundation for the selection and optimization of the kinematic pairs of the spatial mechanism which has purely serial kinematic chains, the research into the idle degrees of freedom of kinematic chains, especially the complex spatial kinematic chains, and the free motion analysis of the end effector in a mechanism and the analysis of the actuation scheme. According to reciprocal screw theory, terminal constraints of the kinematic chain depend entirely on the largest linearly independent set of the kinematic screws. Therefore, the structural optimization problem of the kinematic chain can be addressed through the algebraic analysis to the specified terminal constraints; and it can also be resolved through algebra analysis of the screws. In fact, the free motions of the end effector are the results of the combination of branch kinematic chains. Therefore, the study of spatial serial mechanisms will focus on the constraints of kinematic chains to the end effector. How to determine the types of various kinematic chains and how to complete the design with the simplest and economical structure arrangement have become key issues in the synthesis of kinematic chains. This is also one of the key issues in this chapter and subsequent chapters.

In Chapter 4, terminal constraints of robot end effectors from kinematic chains and the free motions under these constraints are discussed. The analytical theory of constraint space of the end effector of parallel robot mechanism is studied according to the terminal constraint space of its kinematic chains. The degree of freedom of the end effector of a parallel robot should have the three attributes of *quantity*, *type*, and *direction*. The degree of freedom of the end effector of a robot is different from the actuations to control the mechanism. However, the traditional methods did not make any distinction between the end effector and the entire mechanical system. This chapter is dedicated to establishing an analytical theory for the constrained space of the end effectors. On this basis, the equivalent description of the complex kinematic chain is discussed, which provides a theoretical support for the equivalent description of the complex kinematic chains to pure serial kinematic chains. It allows us to transform the parallel kinematic chains into serial chains with the equivalent substitution for the compound kinematic chains. This lays a foundation for the study of actuation of the prescribed end effector.

In Chapter 5, the search for reachable workspace of the spatial robot mechanism and its symmetric homomorphism with the mechanism itself is discussed. It is proved that the structural symmetry of the mechanism itself will inevitably lead to the symmetry of the reachable workspace and the dexterous workspace of its end effector. The corresponding relation between the reachable workspace and the dexterous workspace is homomorphism mapping. A quantitative description of such symmetry relations provides certain theoretical basis for the analysis, synthesis, and structure innovative design of robot mechanism. Based on this, the coordinate transformation method for resolving the workspace based on free motion analysis of the end effector of spatial parallel mechanism is also covered in this chapter, and the synthesis of the dexterous workspace is discussed with the Schoenflies-type spatial parallel mechanism as an example. By analyzing the end effectors' degrees of freedom, a minimum of variables can be chosen to establish the kinematic constraint equations and thus the objective function can be built up. By searching the desired dexterous workspace, the optimal length of linkages in kinematic chains can be obtained. Application examples show that the method can be widely applied to the synthesis of the dexterous workspace of various parallel robot mechanisms.

In Chapter 6, the singularity of the spatial robot mechanism in the workspace is studied and it is proved that the structural symmetry of the mechanisms themselves will inevitably lead, not only to the symmetry of reachable workspace and dexterous workspace of its end effector, but also to that of singular workspace. A quantitative description of such a symmetric relation provides some theoretical basis for the analysis and synthesis of workspace of symmetrical mechanisms. With screw theory, a group of balance equations which can be used to analyze the singularity of mechanisms have been obtained and the conditions for singular mechanism are derived. The order of the coefficient matrix of final equations is identical to the number of degrees of freedom of the end effector, which constitutes great advantages for the singularity analysis of the end effector of the mechanism with less than six degrees of freedom. The fewer degrees of freedom the end effector has, the lower the order of the finally converted coefficient matrix will be.

In Chapter 7, kinematics modeling of the spatial parallel robot mechanism is discussed. On the basis of the natural coordinate method, the sufficiency and necessity have been proved that a type of kinematics model spatial parallel mechanism can be completely established with only four non-collinear points with Cartesian coordinates. This method has a significant advantage in dealing with kinematics issues of spatial parallel mechanisms with more than three degrees of freedom. And the established kinematics model shares the same advantages of general natural coordinate method and can also minimize the number of independent variables. As this model only introduces the Cartesian coordinates, and the elements of the derivative matrix merely contain the linear items or even the constant items, while the elements of the derivative matrix obtained through the modeling method based on the rotation matrix transformation are often non-linear, and even contain transcendental functions. Therefore, these advantages of the four-point coordinate method have brought great convenience to kinematics and dynamics analysis.

In Chapter 8, the kinematics and static forces of robot mechanisms are discussed. Based on the analysis of terminal constraints of the kinematic chain in Chapter 4, detailed analysis is given to the motion between the end effector and actuation and the rigid body statics. Through the analysis of the velocity of mechanism end effector, the basic conditions for the feasible motion of robot end effector have been discussed and the forward and inverse solutions of velocity are also investigated. On this basis, robot static equilibrium conditions and the forward and inverse solutions of statics are derived according to virtual power principle.

In Chapter 9, the basic way to analyze the degree of freedom is studied from the theoretical basis for the analysis model. Through investigating the theoretical basis of the degree of freedom of the end effector and the configuration degree of freedom of a mechanism, the key for analyzing the mechanism is revealed: it is a necessary condition to distinguish the degree of freedom of the end effector and the independent actuators required to control the end effector. Application examples of spatial mechanisms demonstrate that it is the necessary requirement for correctly analyzing a mechanism, and it satisfies the current needs of engineering as well.

In Chapter 10, by analyzing the calculation of configuration degrees of freedom of spatial robot mechanism under a single Cartesian coordinate system, the basic motion characteristics of the robot mechanism in the workspace are further investigated. Based on the analysis of free motion and terminal constraint of the kinematic chain in Chapter 4, specific studies have been performed to analyze the degree of freedom of the end effectors within a single Cartesian coordinate system; the configuration degrees of freedom describing the dynamic drive control of mechanisms are investigated by studying the dynamic degree of freedom of the end effector under different actuation schemes. On the basis of the unified mathematical description of kinematic geometry constraints, the analysis theory for the degree of freedom of spatial robot mechanisms has also been discussed. The theory consists of two main parts: the first is the analysis theory targeted at the end effector; the second is the analysis theory for the configuration degree of freedom mechanism with the designated end effectors. The concept of configuration degrees of freedom means that singularity analysis of complex parallel mechanisms are naturally included in the analysis of the configuration degrees of freedom; at the same time, the approach for analyzing the structural stability of mechanisms is also provided: only when the mechanism's motion inputs remain consistent with its configuration degree of freedom in the space at any time, can the entire mechanism system be stably controlled.

In Chapter 11, the mechanism theory of spatial deployable structure based on the scissor-like-element (SLE) is dealt with. To begin with, the motion analysis of the simple planar linkage, the characteristics of the deployable structure and its elements are studied step-by-step. Through in-depth analysis of the double rotating hinge connected by two SLEs, the mechanism theory in the constitution of elements of the deployable structure has been revealed. Through the analysis of elements of the planar deployable structure, the cylindrical deployable structure, and the spherical deployable structure, the constitution mechanism of spatial deployable structures based on the SLE has been investigated. This has important practical value for the promotion of deployable structures in mechanical engineering, construction engineering, and robot engineering.

In Chapter 12, application of the analytical theory in structure synthesis and innovation of mechanisms is studied. Through presenting general steps in mechanism synthesis, the basic method for the innovative design of mechanism with the universal significance is also discussed. With a comprehensive case study for the innovative design of the lower limb of a biped humanoid robot and the independent suspension of the vehicle, applications of the theory in the innovative design of mechanisms are investigated.

In Chapter 13, a new algorithm is proposed to optimize the design of a spatial parallel manipulator for the purpose of obtaining a desired dexterous workspace rather than the whole reachable workspace. With the analysis of the degree of freedom of a manipulator, we can select the least number of variables to depict the kinematic constraints of each leg of a manipulator. The optimum parameters can be obtained by searching the extreme values of the objective functions with the specified adroit workspace. An example is utilized to demonstrate the significant advantages of this method in the dexterous workspace synthesis. In applications, this method can be widely used to synthesize, optimize, and create all kinds of new spatial parallel manipulators with the desired dexterous workspace.

Regarding kinematics, many works have addressed this problem in the past. In Chapter 14, the natural coordinates are adopted to deal with the kinematics model for different spatial mechanisms. With these constraint equations in natural coordinates, we can conduct the kinematics analysis of the mechanism and then build the dynamics model with the Newton–Euler formulation.

I would like to express my gratitude for the contributions of my coauthors Zhijing Feng, Fulei Chu, and Ning Ma. In addition, I am grateful for the inspiration of Prof. Yunqing Zhang, Prof. Jian S. Dai, Prof. Yuefa Fang, Prof. J. Michael McCarthy, Dr. Ketao Zhang, Dr. Guowu Wei, Prof. Ligang Yao, Prof. Daniel Martins, and Dr. Wenxiu Lu. Last, but not least, I gratefully acknowledge the support of the Natural Science Foundation of China, the Natural Science Foundation of Beijing, the Foundation for the Author of National Excellent Doctoral Dissertation of China, and the Program for New Century Excellent Talents in the University of Education Ministry of China, and the support of the State Key Laboratory of Tribology in Tsinghua University.

Jingshan Zhao
Tsinghua University, Beijing, China

Chapter 1

Introduction

Mechanism is a branch of mechanical engineering science focusing on the study of the motion and the transmission law of mechanisms, which consists of structure analysis and synthesis [1–3]. Structure analysis of mechanism mainly investigates the mobility of the structure, and explores the trajectory, displacement, velocity, and acceleration of the mechanism [2,3]. Structure synthesis of mechanism focuses on the design of mechanisms so that they meet the kinematic requirements according to the prescribed motions. Research results of assembly principles of structure, kinematics, dynamics, and precision analysis of mechanisms provide the theories and methods for various movement generations, the kinematics and dynamics design of mechanical systems [4,5].

1.1 REVIEW OF MECHANISM

With the invention and improvement of steam engines and textile machines at the beginning of the 18th century, calculus, theoretical mechanics, and other disciplines were set up and becoming more and more sophisticated, which provided a theoretical basis and means for the design and invention of various mechanisms. At that time, mechanism was still included in the field of applied mechanics. The rapid development of mechanical industry at the second half of the 19th century proposed specific requirements for the application of mechanisms and the corresponding analysis and synthesis [1]. In the United Kingdom, scholars of mechanism, represented by R. Willis, classified mechanisms according to the characteristics of motion transformation and the speed ratio relationship, and mainly studied the gear and planetary transmission [1,2]. In Germany, scholars of mechanism, represented by F. Reuleaux and L. Burmester, introduced concepts such as kinematic pair, kinematic chain, and schematic diagram of mechanism, discussed the possible types of mechanisms from the structure principle of mechanism, and further extracted the common characteristics of mechanisms, and with graphical methods, conducted the analysis and synthesis of mechanisms according to the motion geometry principle [2,3]. In Russia, scholars, represented by П. Л. Чебышева, addressed the approximate synthesis of dimension by using algebraic methods like approximation theory of functions [4,5]. Based on these results, mechanism was developed into an independent discipline [1,2,6].

Advanced Theory of Constraint and Motion Analysis for Robot Mechanisms.
http://dx.doi.org/10.1016/B978-0-12-420162-0.00001-1

Mechanism is one of the most important fundamental disciplines on which machine design relies. The invention, innovation, and improvement of mechanisms are important tasks in machine design. The basic problems of research in mechanism can be roughly divided into two major categories: structure analysis and structure synthesis. Structure analysis focuses on investigating the mobility and constraint, kinematics and dynamics of mechanisms, so as to reveal the structure constitution, the kinematics and dynamics laws, and their inter-relationships in mechanisms, and it can provide a theoretical basis for understanding the performance and mechanical synthesis of existing mechanical systems. Mechanism synthesis is defined as the theory and method of designing new mechanisms satisfying the engineering requirements in aspects such as structure, motion, kinematics, and dynamics. It includes structure synthesis, kinematics synthesis, and dynamics synthesis. Structure synthesis of mechanism is one of the most important areas in the field of mechanism. The entire mechanism is often proposed as a multi-rigid-body system, in order to reveal the assembly laws of the structure, characteristics of the structure topology, and their intrinsic relationships with kinematics, dynamics in mechanisms, so as to provide a basis for the establishment of structure synthesis of mechanism, kinematics, and dynamics theory, and to provide a theory to design new mechanisms [7].

1.1.1 Development of Mechanism

Germany was most active in the early study of mechanism. After the 1940s, and especially after the 1960s, the United States achieved rapid development in the research of mechanism. A lot of research work has been done in areas such as structure analysis and synthesis of planar and spatial linkages, cam dynamics, dynamics of elastomeric mechanical systems, as well as mechanical design optimization. The United States leads the world in the research of many aspects of mechanism. After the 1960s, the United Kingdom accelerated its pace of research in mechanism, and achieved substantial results in cam mechanism, mechanical system dynamics, as well as areas such as spatial linkages and industry robots. Ever since the establishment of the Institute of Mechanics of the Academy of Sciences in Soviet Union during the 1930s, mechanism became one of the main directions of development in the former Soviet Union. After World War II, the Soviet Union's research in mechanism developed rapidly. The Soviet Union made important contributions in aspects like mechanical structure theory, precision analysis, lower flat pair mechanical analysis, higher pair envelope theory, spatial mechanism, and system dynamics of multiple degrees of freedom (DOFs) mechanisms. Especially from the 1950s to the1970s, the Soviet Union remained the leader in the main areas of study in mechanism in the world. Similarly, Japan studied the cam mechanism dynamics and its computer-aided design and manufacturing and the analysis and synthesis of planar linkages and so on. After the 1970s, its research in robots and manipulators developed especially fast. Significant progress had been made in the classification and performance identification of grasping mechanisms, the motion possibility, kinematic analysis, and motion stability of multi-DOF multi-joint kinematic chains. Some researchers studied the functions of fingers, arms and legs in the field of bionics in order to build artificial muscles, artificial bones, and artificial feet, and gradually formed a new discipline, biomechanical engineering, and set up the institute of bionic mechanism [8] in Japan. In recent years, there has been great progress in kinematics, ZMP and dynamics, biped walking, the generation of whole-body motion patterns, and dynamic simulation of biped humanoid robots. In 1935, Professor Xianzhou Liu, China's well-known expert in mechanical engineering at the National Tsinghua University, published *Mechanisms and Machine Theory*, the first book focused on a systematic introduction of mechanical principles. This book marked the entry of Chinese scholars into mechanism fields.

In order to strengthen the international exchange and cooperation in mechanism research, the International Federation for the Theory of Machines and Mechanisms (the IFToMM) was established in the autumn of 1969. Well-known experts in mechanism around the world formed the Executive Board and various technology development committees under the IFToMM and founded the journal *Mechanism and Machine Theory*, in which a number of important research papers on mechanism are published.

Although the research of spatial mechanism began in the late 1920s, its progress remained slow until the end of 1950s. Due to its advantages (compact structure, occupies small space and can realize various flexible and complex motions with only a small number of links), fruitful results were achieved in spatial mechanism between the 1950s and the 1990s. The emerging robot mechanisms and complex spatial mechanisms like the virtual axis machine tools all fall under the research area of spatial mechanism.

From the perspective of the development process, parallel mechanism has become one of the hot research topics in mechanism after the 1990s. In the theoretical research, various parallel mechanisms, series-parallel mechanisms, the design theory of bionic mechanisms, and performance evaluation index systems of kinematics and dynamics were all important issues in the research. In applications, people conducted research on flight simulators, virtual axis machine tools, micro-actuators, sensor frameworks, error compensators, walking robots, a variety of bio-robots, etc. Developed in the last 30 years and based on the principle of Stewart platform, the parallel mechanisms [9,10] have attracted scholars' attention both at home and abroad. It has brought about new processing equipment, a virtual axis machine tool which wins much attention from all over the world. With more than 20 years' development, it has already been profoundly understood, and it has been widely applied in many areas of mechanical engineering.

Although its prototype, Stewart platform [9], came out as early as 1965, and the Soviet Union Novosibirsk Electro Technical Institute developed a pilot machine [11] based on this new concept of virtual axis machine tool, what attracted the world's attention was the parallel machine tool VARIAX which two American companies, Ingersoll and Giddingls & Lewis, displayed for the first time at the Chicago International Machine Tool Exhibition (IMTS94) in 1994. They brought about the sensational effect of international R&D boom. After that, at the 11th European Machine Tool Exhibition in 1995, the Swiss Geodetic company exhibited GPM1000-20S-type machine tools, and the Italian Comau company exhibited the Tricept deburring robot. Parallel structures were all applied in their actuators. At the 1997 Hanover International Machine Tool Exhibition (EMO97) in Germany, more than 10 prototype parallel machine tools were displayed, and the first milling metal work-piece performances were conducted. Research on the application of virtual axis machine tools has been of general concern to the academic community around the world. For instance, Switzerland has developed a parallel machine tool in which Hexaglide parallel mechanisms are used, and Pierrot has put forward a Hexa type of parallel mechanism. The Delta mechanism is now very widely used in the three-DOFs parallel mechanisms [12]. For the latest development of the virtual axis machine tools at home and abroad, refer to the literature [13,14]. China has also made important progress in the theoretical research into the structure of planar and spatial linkages, and has reached world-class level in the inverse kinematics of serial robots, workspace analysis, kinematic analysis of the parallel multi-loop spatial linkages, as well as the synthesis of linkages.

In the theoretical study of mechanism, the use of vectors to describe the rigid body motion is one of the most basic traditional analysis methods. In spatial mechanism, the right-handed Cartesian coordinate of vector is often used. Along with that coordinate, regular expressions such as vector inner product and outer product, and vector differential method are involved in studying the mechanical kinematics and dynamics problems such as the speed and acceleration. In 1959, Shigley [15] carried out systematic research on the vector. Hartenberg and Denavit [16] first proposed the idea of using the matrix transformation of four structural size parameters and position parameters to describe projection relations of the point coordinates in the absolute coordinate system, i.e. using matrix operations of kinematic pairs to describe the position and orientation of mechanisms. Hartenberg and Denavit also conducted analysis into and synthesis of several commonly used spatial four-bar linkages, now known as the DH transformation matrix method. In the kinematic analysis of spatial mechanism, the dual number, screw vector, and dual vector are also often used. As early as the end of 1940s, ДимАитьирг, a Soviet scholar, conducted kinematic analysis and local freedom degree analysis using the dual number and a sub-screw affine. At the same time, the majority of mechanism scholars in the Soviet Union, led by the first President of IFToMM, Academician И. И. Артоболевский, creatively built systematic geometry and algebra methods for the structure constitution, kinematic analysis, dynamic analysis, and synthesis, and they made tremendous contributions to the development of mechanism. In 1967, Hunt [17] studied the problem of kinematic analysis of spatial mechanisms using screw-axis and line geometry. In 1970, Woo and Freudenstein [18] derived the algebraic formula for the kinematic analysis of mechanisms using the principles of line geometry and developed a numerical algorithm of the screw vector. Scholars like Duffy [19,20] derived the input and output displacement equations of dual numbers for the spatial five-bar linkage, using spherical trigonometry. Duffy and Keen [21] again applied the same method to RRERR mechanisms. They viewed the plane pair (E) as a combination of 2P and 1R for discussion. In 1968, Roth [22] studied the general design principles of kinematic pairs containing revolute pair (R), prismatic pair (P), and cylindrical pair (C). In 1972, Tsai and Roth [23] discussed the synthesis of limited or unlimited separated locations containing revolute pair (R), cylinder pairs (C), spherical pair (S), prismatic pair (P) and helical pair (H).

In 1978, Hunt [24] published *Kinematic Geometry of Mechanisms*, systematically introduced kinematic geometry, an effective tool for studying mechanism, and further developed Ball's screw theory [25]. The book discussed the kinematic pairs, links, planar and spherical mechanisms, the exchange of kinematic pairs, synthesis of types, number, and sizes of mechanisms, mobility and constraints; it studied the kinematic geometry representation of the planar and spatial mechanisms; it developed the theory of dual number and screw vectors; and applied these tools to the analysis of spatial mechanisms. Phillips published *Freedom in Machinery, Volume 1: Introducing Screw Theory* [26] and *Freedom in Machinery, Volume 2: Screw Theory Exemplified* [27] in 1984 and 1990, respectively, and systematically discussed the free motions and constraints of mechanisms. He directly discussed the kinematic and static constraints between rigid bodies in the three-dimensional space. As a result, he developed the screw theory. Volume II focuses on the screw theory, and its applications in the analysis and synthesis of general mechanisms, as well as the actual design of mechanisms. Huang published *Spatial Mechanism* [28] in 1991 and systematically introduced the spherical analysis theory and the screw theory in China.

Therefore, mechanism, as a basic discipline in mechanical engineering, was formed in the 19th century step by step, and developed rapidly in the 20th century. The continuing development of the mechanism theory promotes the development of robot mechanism. The analysis and synthesis methods of robot mechanism are focusing in the direction of high-precision, high-speed, and visualization. Research on the types of mechanisms focuses on the overall structure of the model, the mechanism's mobility, singularity, trajectory, rigid-body guidance, drive, workspace design and planning, as well as the design optimization of multi-dimensional constraints. In the research on the mechanism structure theory, many scholars conducted systematic analysis into the structure types and the DOF of motion by using a variety of mathematical tools such as vector, tensor, matrix, quaternion, graph theory, linear algebra, analytical geometry, dual number, screw theory, etc.

The key point of study on the DOF of kinematics is how to analyze the multi-loop spatial linkages with different general constraints in different loops and how to analyze these spatial linkages with overconstraints and local DOF; and how to explore and establish structure principle of mechanisms. Structure theory of mechanism focuses on the structure principle of mechanisms, possible types, and the calculation of kinematic DOF. Structure theory of mechanism came up very early, and initially focused on the planar mechanisms. The mechanisms in the practical engineering application at that time were generally of single DOF and of closed-loop, and had only a few small links. In recent years, mechanism theory has focused on parallel robot mechanisms, dexterous hand operation, walking robots, as well as parallel robot machine tools and so on. Mechanism scholars around the world have recently conducted research on issues such as the structure design and assembly process design of the parallel mechanism [29,30], robot position analysis [31–42] kinematics [35,43–45] and dynamic analysis [46–49], the singularity [50–55] as well as workspace [56–74], the analysis and design of flexible linkages [75–77], and have made significant progress. However, how to correctly determine the mechanisms' general constraints has always been a difficult problem in the structure analysis of complex spatial mechanisms. Many scholars [78–83] have carried out extensive research in determining the general constraints in parallel mechanisms with multiple closed-loops.

The modern complex spatial mechanisms have already been applied in practice. It can be expected that in the near future, extensive applications of modern mechanisms will contribute significantly to the invention of modern new machines, promoting the rapid development of machine industry. The modern mechanism has opened an effective way for the design of micro-machines. The development of micro-machines requires the relative research on design theories and methods. In addition, the investigation into the types of bionic mechanisms will render creations of all sorts of bionic machines, so that the application of modern mechanism will be much wider and more effective. As a result, the designs and development of robots regarding mechanism type, dimension synthesis, workspace, dynamic stiffness, and control techniques should all be guided by the modern theory and method of mechanism. This has been of broad consensus among mechanism scholars around the world. In 2003, the International Federation for the Theory of Mechanism and Machine (IFToMM) published a third edition of terminology in which many nouns such as "mechatronics," "micro-electro mechanical systems (MEMS)," and "bionic mechanisms" appeared. At the same time, IFToMM changed its name from the International Federation for Theory of Mechanism and Machine to International Federation for Promotion of Mechanism and Machine Science in 2003. New theories and methods of mechanism are being systematized, and the new theoretical system relating to the overall characteristics and new inventions in mechanism are being improved step-by-step. There are inherent inter-relationships between different level units and the overall characteristics of the system. This has led to the integration of structural synthesis of mechanism, kinematics, and dynamics. The crossover of mechanism and computer science brings about computing mechanism; the crossover of mechanism and bionics has contributed to the formation of mechanism bionics; and the crossover of mechanism, biomedicine, and automatics develops into rehabilitation engineering and robotics. Undoubtedly, the innovation in designing robots, especially robots for special purposes, relies on the development of robot mechanism. Huang and other scholars have studied quick pick and place robots [84] and the parameter identification method for 6-DOFs robotic machines [85]. It can be expected that research on the type of robot mechanism, working spatial, scale integration, and work stability has broad application prospects.

For theoretical research of mechanism, its development trend can be summarized as follows. Metamorphic mechanism is a complex spatial mechanism whose topology and mobility is changing within its workspace [86, 87]. Therefore, the mobility of the metamorphic mechanism as one of the most important topics of mechanical engineering should be investigated. The theory that could describe this phenomenon and guide the design of such mechanisms is of the first importance. For the future development of mechanism, there will be two major issues to be further addressed to establish a perfect and efficient structure synthesis theory of mechanisms, as well as correspondent presentations and means of classification; and to establish systematic methods for assessing spatial mechanisms based on topology, qualitative selection optimum and quantitative analysis so as to select the mechanism with the most rational structure for the specified working requirements.

1.1.2 Brief History of Mobility Analysis of Mechanism

All machines are made of mechanisms. Before design, the mobility of a mechanism should first be determined, and then the actuation scheme and control can be prescribed. Thus, the correct analysis of free motion of a mechanism is the basis and prerequisite for structural design and analysis. More specifically, structural design of mechanisms can also be divided into free motion design, type synthesis, and linkage synthesis. Type synthesis studies how many mechanisms with different motions can be composed, given a certain number of linkages and kinematic pairs; linkage synthesis studies the number of links and kinematic pairs composing mechanisms so as to meet the requirements of a certain motions. Although there are some differences between these two issues, they are closely connected to each other in the motion analysis of the mechanism.

Structure analysis of a mechanism focuses on proposing the structure principle of mechanisms, the classification of mechanisms, and their free motions. Mobility is the main content in the structure analysis of mechanisms. Meanwhile, structure analysis of mechanisms is an important aspect of kinematics, dynamics analysis, and mechanism synthesis. The rapid development of science and technology calls for the creation and design of better mechanisms. Therefore, it is essential to conduct further theoretical research into the structure, type, and kinematic characteristics of mechanisms. Type synthesis provides a selection method for innovative design of mechanism, while the mobility analysis presents a basis to determine the number of the original moving parts and carry out kinematics and dynamics research. Thus, mobility is one of the most basic and important concepts in mechanism. Mobility analysis plays an important role in machine design and analysis. It offers a guarantee for innovation, type synthesis, and linkage synthesis, and also lays a foundation for further analyzing the driving control and kinematics and dynamics performances of mechanisms [4,5].

Scholars of mechanism from Germany and the former Soviet Union have made an outstanding contribution to the modern mechanism structure analysis. As early as the 19th century, F. Reuleaux first made a precise definition for mechanism [1,4,5]. He systematically studied the kinematic pairs, and divided them into two categories, namely, the higher pairs and lower pairs. He defined the mechanism which was formed by links through the connection of kinematic pairs. As a link, the kinematic pair not only constrains some relative motions of two interconnected links but also allows certain relative motions between links. If we use a simple equation to represent the number of relative free motions of two links connected through a kinematic pair, we get

$$F = 6 - S \quad \text{or} \quad S = 6 - F \tag{1.1}$$

where F denotes the number of free motions of one link compared to another, and S denotes the number of constraints provided by the pair.

Equation (1.1) represents that a free rigid body has six DOFs (i.e. three rotations and three translations).

Kinematic pairs are grouped into five classes according to the number of constraints introduced by the kinematic pairs. For example, we call the kinematic pair Class I pair if $S = 1$. It means that there are five relative DOFs and one constraint (i.e. one DOF is lost) between the two interconnected links; similarly, we call the kinematic pair Class II pair if $S = 2$, which means that there are four relative DOFs and two constraints. The rest may be deduced by analogy.

Based on F. Reuleaux's research, П. Л. Чебышев (a Russian scholar), and M. Grüble (a German scholar), put forward the formula for calculating the mobility of planar mechanisms [1,2,4,5]:

$$F = 3n - 2p_5 \quad (\text{Чебышев}) \quad \text{or} \quad 2p_5 - 3n + F = 0 \quad (\text{Grüble}). \tag{1.2}$$

These two equations appeared about between the 1870 and 1883 [88]. Soon after, K. Kutzbach from Germany brought up a mobility equation for spatial kinematic chains [1,2,4,5]:

$$F = 6K - (5p_5 + 4p_4 + 3p_3 + 2p_2 + p_1) \tag{1.3}$$

where K denotes the number of active links in the kinematic chain and p_1, p_2, p_3, p_4, and p_5 are the numbers of Class I, Class II, Class III, Class IV, and Class V kinematic pairs, respectively.

Around 1930, the German scholar K. Kutzbach gave another form of equation for calculating the mobility of a mechanism,

$$F = b(n - j - 1) + \sum_{i=1}^{j} f_i \tag{1.4}$$

where b equals 6 or 3 (generally, b equals 6 for spatial mechanisms, and 3 for planar and spherical mechanisms), n denotes the number of links, j is the number of kinematic pairs, and f_i denotes the number of possible relative motions derived from kinematic pair i.

It has been found that the above equations could not help to get the correct results when applied in the calculation of DOF of some mechanisms. А. П. Малышевым et al. from the Soviet Union amended Kutzbach's equation, and in 1943 came up with a structure formula with the general constraints taken into account:

$$F = (6 - m)n - \sum_{S=m+1}^{5} (S - m)p_s \tag{1.5}$$

where m denotes the number of general constraints of the links in mechanisms ($m = 0, 1, \Lambda, 4$), S denotes the class-serial number of kinematic pairs, i.e. p_s represents the number of class-S kinematic pairs, n equals the number of movable links in mechanisms.

Despite its consideration of general constraints, the above equation did not have general applicability when it was used for checking the mobility of a mechanism. Because apart from the above-mentioned general constraints, mechanisms are restricted by the duplicate constraints, local DOF, compound hinges, etc., which will all affect the calculation result.

B. Paul, from Brown University put forward in 1960 the analytical method to use the number of independent loops to determine the mobility of planar mechanisms. But it has not reached a perfect level for calculation of mobility of multi-loop spatial mechanisms of which the general constraint of each loop is in unequal terms [1,4,5].

Crossley, from Yale University, has analyzed the planar linkage on the basis of Grübler's formula. He used a simple operator to add to different member bars. In that way, different types of articulated mechanisms composed of different links can be obtained to meet the expected DOF [1–5].

As for planar linkage mechanisms containing only the rotary pairs, the Crossley theory has equations to show the relationship between kinematic pairs and the number of links.

It is necessary to have N links and J kinematic pairs, i.e.

$$\left.\begin{aligned} N &= n_2 + n_3 + \cdots = \sum_{i=2} n_i \\ 2J &= 2n_2 + 3n_3 + 4n_4 + \cdots = \sum_{i=2} i n_i \end{aligned}\right\} \tag{1.6}$$

where n_2, n_3, and n_4 are the number of binary linkages (i.e. linkages with two kinematic pairs each), triple linkages, and quadruple linkages, respectively.

Following the Чебышев-Grübler equation for planar mechanisms, we get

$$F = 3(N - 1) - 2J$$

$F = 1$ for most engineering applications, thus

$$J = \frac{3}{2}N - 2. \tag{1.7}$$

It can be seen from the above equation that because J is the number of planar kinematic pairs, it should be an integer, and thus N should be an even number. Based on this, the simplest planar four-bar linkage only contains two-pair links, then we will see $N = n_2 = 4$. Considering equations (1.6) and (1.7), we will get $n_2 = 4$ and $J = 4$. If we further consider an eight-link mechanism with four-pair links, we have to adopt the method of trial and error to calculate the number of linkages. It is not only inconvenient, but is also easy to make mistakes. Note that formula (1.7) does not apply to the spatial mechanisms.

One of the important elements of the structure analysis of mechanisms is to identify the DOF. However, it is difficult to determine the general constraints in the calculation of DOF for spatial mechanisms. N. I. Manolescu and his collaborator V. Manafu, from Bucharest technical institutes, Romania, made a contribution to the study of determining the mechanism's "family" [1,2,4,5]. In their published works, they gave the formula for calculating the DOF of mechanisms

$$F = F_f + \sum F_{gf'} \tag{1.8}$$

where F_f denotes the DOF of mechanisms belonging to the f-family; $F_{gf'}$ represents the DOF of the Assur group or non-Assur group.

V. Manafu uses the following equation:

$$F = \sum F_f^j - F_l \tag{1.9}$$

where F_f^j denotes the mobility of mechanism j in family f, and F_l represents the overall mobility of the mechanism.

In addition, Voinea, Atanasiu, and Maroskine, from Romania, have studied the mobility of mechanisms using the matrix method [1,3,5]. Manolescu and Ozol (from the fomer Soviet Union) established a formula for calculating the DOF in accordance with the largest number of deformable closed-shaped mechanisms [1,2].

In 1971, Bagci, from the University of Tennessee, put forward a general formula for calculating the DOF of mechanisms with n links and K loops in his paper [1,2,4].

$$F = F_i + F_r = F_o + M + F_c - F_p \tag{1.10}$$

where F denotes the mobility of the mechanism; F_i is the required number of original moving parts, $F_r = \sum_{j=1}^{n-1} f_{rj}$ are the mechanism's overall redundant DOF), f_{rj} represents the redundant DOF in the equation $F_o = b(n-1) - \sum_{i=1}^{r} (b-i)N_i$, b denotes the independent parameter for the rigid body in spatial motion, r is the largest number of DOF of kinematic pairs in the mechanism, M represents the total number of general constraints, F_c denotes the total number of overconstraints, and F_p gives the total number of local DOF.

Bagci has explained and illustrated some issues such as the redundant DOF, local DOF, and general constraints and over constraints. His formula can be applied to calculate the DOF for both planar and spatial mechanisms. In 1973, Bagci gave a detailed introduction to the mobility calculation for transmission mechanisms containing cams, gears, and belts in another paper. But he also pointed out that it was not easy to correctly judge the value of the variables in the formula. In addition to the type and configuration of kinematic pairs, local DOF is also subjective to the impact of the relative size and geometry relationship, which Bagci did not take into account in his formula.

At almost the same time as when Bagci published his paper, the American scholars Gupia and Radcliffe published articles using geometry to study the mobility of planar and spatial mechanisms [1–3]. In the articles, they pointed out that general formulas did not take into account the sizes of various links. For example, in the planar four-bar linkage, the link size has a direct impact on whether the crank exists; while for the spatial mechanism, its DOFs are related with the geometric form of the link connection and link size. The article used the maximum and minimum length of the linkage as the parameters and made a mobility chart for planar and spatial mechanisms, which showed the scope of motion of mechanisms. However, their method did not have a unified description for geometry parameters and the structural constraints, and it is therefore not applicable to all mechanisms.

Although the study of mechanism has a long history [88], and has achieved many results, it has not yet been perfected. The calculation of mobility of mechanisms and the structure synthesis of type, number, and size are still urgent problems of current mechanism.

1.2 CONTRADICTION BETWEEN CALCULATION AND PRACTICE OF MOBILITY OF SPATIAL MECHANISM

It can be said that the emergence and wide applications of complex spatial parallel mechanisms embody significant achievements of research of spatial mechanism in recent decades. According to the number of DOFs of the end effector, spatial parallel mechanisms can be divided into four basic categories of which the numbers of DOFs are six, five, four, and three, respectively. Of course, the two- and one-DOF spatial parallel mechanisms are also available in engineering practice. For instance, the simple spatial four-bar linkage and its basic variations have been applied in aircraft landing gear. According to the different types of driving pairs, mechanisms can be divided into internal pair drive, external pair drive, and internal and external combined drives, of which connecting rods with constant length are more and more popular.

With the emergence of complex spatial mechanisms, many researchers [17,24,26,78,79,88–90] have found that the former formulas for calculating the mobility of mechanisms no longer apply to some spatial mechanisms, such as complex parallel mechanisms, metamorphic mechanisms, and so on. Although many scholars have proposed a number of remedial programs, there are still some limitations, and they can not completely address the problem. Due to the limitations of the formulas for counting the DOFs or mobility of mechanisms, there are a certain invaluable mechanisms with superior performance could not be synthesized. For example, on page 120 of one of his works [2], Professor

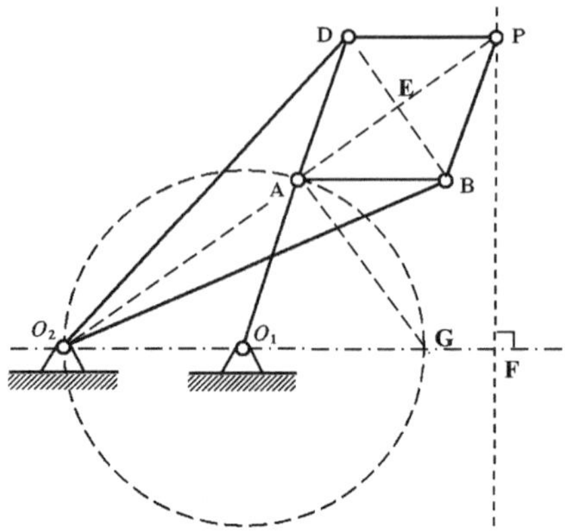

FIGURE 1.1 An eight-link and 10-pair Peaucellier–Lipkin mechanism.

Norton wrote, To generate an exact straight line with only pin joints requires more than four links. At least six links and seven pin joints are needed to generate an exact straight line with a pure revolute-jointed linkage, i.e. a Watt's or Stephenson's six bar.

His statement was obtained completely based on the traditional Kutzbach–Grübler formula. In fact, six links and six revolute pairs are enough to output a straight line motion. This book will present detailed analysis and application examples in Chapter 12. In the history of mechanism, the most well-known straight-line guidance mechanism should be the Peaucellier–Lipkin eight-link and 10-pair mechanism [91,92] as is illustrated in Figure 1.1. In this mechanism, if the quadrilateral *ABPD* is a rhombus, and O_2B and O_2D, as well as O_1O_2 and O_1A are equal in terms of the length, joint P will trace a rectilinear motion along the dotted line which is perpendicular to line O_1O_2.

For clarity, here we give a simple proof of plane geometry. Connect B and D to get BD. Based on the known conditions, we can see that ΔO_2BD is an isosceles triangle with BD as the bottom edge. Then with the same operation, we can get line O_2A, and $\Delta O_2BA \cong \Delta O_2DA$, and therefore the line O_2A bisects $\angle BO_2D$ and thus $O_2A \perp BD$. Assume that AP intersects with BD at E. Because *ABPD* is a rhombus, the two diagonal lines will satisfy that $AP \perp BD$. In this way, O_2A and AP are two line segments which go through the same point A and are perpendicular to the same line BD. Therefore, O_2A and AP will always be collinear. Put differently, the hinge point A will always be on line O_2P. For convenience, if $\|AE\| = \|EP\| = a$, $\|BE\| = \|ED\| = b$, then

$$\begin{aligned}
\|o_2B\|^2 &= \|o_2E\|^2 + \|BE\|^2 \\
&= (\|o_2A\| + e_1)(\|o_2P\| - a) + b^2 \\
&= \|o_2A\|\|o_2P\| + (\|o_2P\| - \|o_2A\|)a - a^2 + b^2 \\
&= \|o_2A\|\|o_2P\| + a^2 + b^2 \\
&= \|o_2A\|\|o_2P\| + \|AB\|^2.
\end{aligned}$$

Draw a vertical line of O_2O_1 and let its intersection with O_2O_1 be F. Let G be the intersection of the line O_2O_1 and the circle going through O_2 with the center O_1. Connect A and G to get AG. Obviously, the right-angled triangle O_2FP is similar to O_2AG. Then we have

$$\frac{\|O_2G\|}{\|O_2P\|} = \frac{\|O_2A\|}{\|O_2F\|}$$

and then

$$\|O_2F\| = \frac{\|O_2A\|\|O_2P\|}{\|O_2G\|}.$$

Because $\|O_2G\| = 2r$, $\|o_2A\|\|o_2P\| = \|o_2B\|^2 - \|AB\|^2$, we can get to

$$\|o_2F\| = \frac{\|o_2B\|^2 - \|AB\|^2}{2r}$$

which is a fixed value. That is to say, the locus of P is the fixed line PF.

As shown in Figure 1.1, the Peaucellier–Lipkin mechanism strictly follows the Kutzbach–Grübler formula. Let us look at the Sarrus mechanism as shown in Figure 1.2 [93]; although this mechanism does not satisfy Kutzbach–Grübler formula, the link C_1C_2 is able to output rectilinear motion in the direction of the double arrows, as shown in Figure 1.2. Just from the perspective of the structural realization, the Sarrus mechanism in Figure 1.2 is clearly simpler than the eight-link and 10-pair Peaucellier–Lipkin mechanism.

In view of the above analysis, we can see that it is unreasonable to use the Kutzbach–Grübler conditions to synthesize mechanism. The shackles of Kutzbach–Grübler conditions must be broken down so that it can be possible to synthesize a simpler mechanism like Sarrus. Two cases are given in the next section to illustrate the contradictions existing between calculation results derived from the traditional formulas for calculating the DOF of the end effector and the number of actuations to control it.

For example, in the analysis of the three different spatial parallel mechanisms in Figure 1.3, it is very difficult to obtain realistic results using the traditional

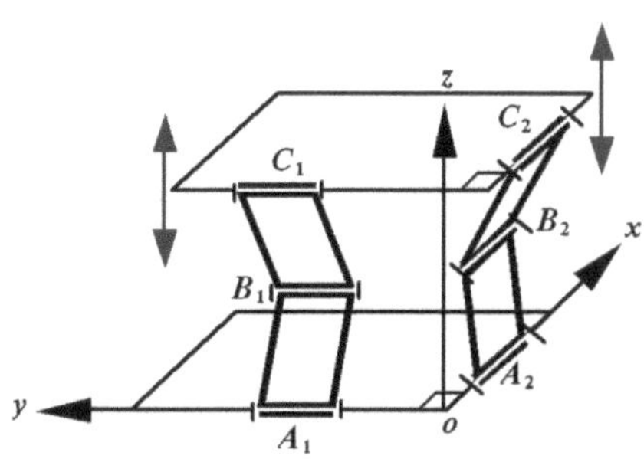

FIGURE 1.2 Sarrus mechanism.

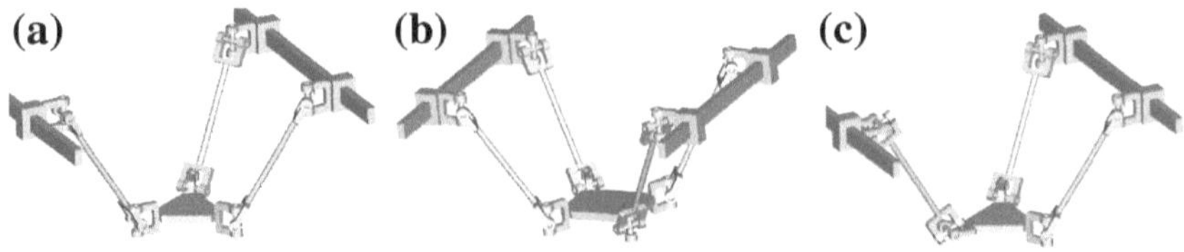

FIGURE 1.3 Three types of spatial parallel mechanisms composed of *PUU* kinematic chains.

formulas and methods based on the simple algebraic summation of the number of links, the number of kinematic pairs, and the number of constraints introduced by kinematic pairs, whether it is an end effector of the parallel mechanism or the whole mechanism. In Figure 1.3, mechanisms (a) and (c) have the same number of links, the number of kinematic pairs and the number of constraints introduced by kinematic pairs. Therefore, according to the widely used Kutzbach–Grübler formula, the number of DOF (in the general position) of mechanism (a) should be consistent with that of (c) in Figure 1.3; however, they are essentially different; on the contrary, the end effectors of the parallel mechanisms as shown in (a) and (b) have the same DOF (i.e. the number, type and direction). For more detailed analysis, see Chapter 4.

In another example, for the two different types of spatial parallel mechanisms as shown in Figure 1.4, links and kinematic pairs are equal in terms of the number. The only difference is that there is a change of 90° in the relative position of kinematic pairs of the three-branch kinematic chains of the end effector and the base. This has caused the number of DOFs of the end effector shown in (a) to be four, while that of (b) to be three. For detailed analysis of the DOF of the end effector shown in (b), see Chapter 4.

Using other traditional methods or formulas to analyze mechanisms, as shown in Figures 1.3 and 1.4, will also have similar problems. Objectively speaking, there is a sharp contradiction between the theoretical formula and practical applications. There is an urgent need to resolve this contradiction in order to ensure the continuous invention and application promotion of new functional mechanisms.

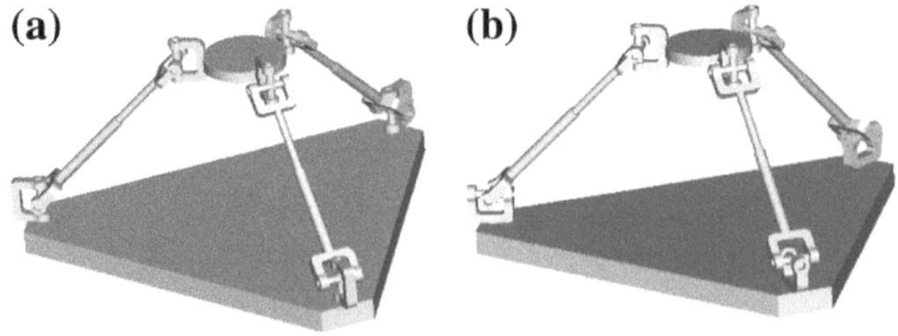

FIGURE 1.4 Two types of spatial parallel mechanisms composed of *UPU* kinematic pairs.

1.3 POSSIBLE CAUSES FOR CONTRADICTION BETWEEN THE CALCULATED DOF AND THE ACTUAL ONE

1.3.1 Cause I

In general, a mechanism consists of four parts, namely the actuator, the end effector, transmission links, and the controller. Therefore, the number of actuators and the DOF of the end effector are priorities in analysis and synthesis of a mechanism. The number of actuators and the DOF of the end effector are two different concepts, which should be treated separately in the kinematic analysis. However, in the development history of the mobility formula of mechanisms, these two concepts have not been distinguished strictly. It is generally believed that the calculation result of the DOF equals the number of the actuators. But in fact, the number of actuators of the mechanism is related to the selection of the end effector. For example, as shown in Figure 1.5, it is a special four-bar mechanism [94], whose projection in the z-direction is shown in Figure 1.6.

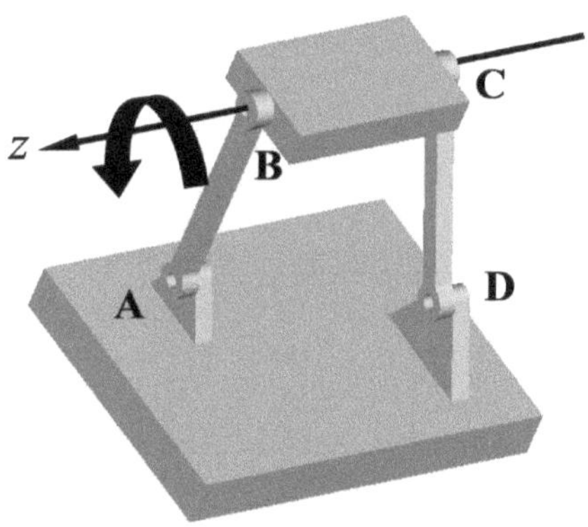

FIGURE 1.5 A special spatial four-bar mechanism.

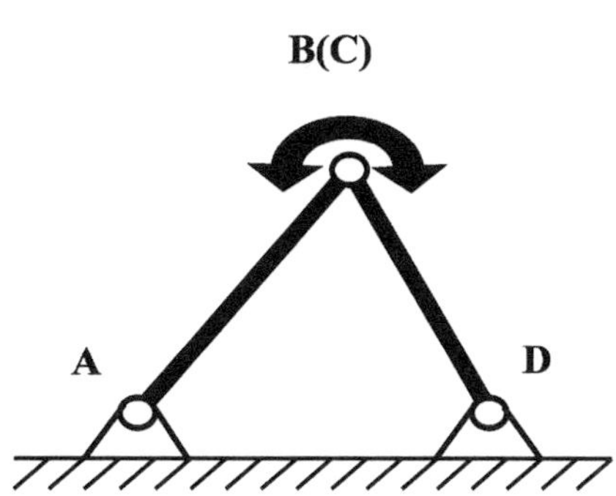

FIGURE 1.6 A special spatial four-bar mechanism.

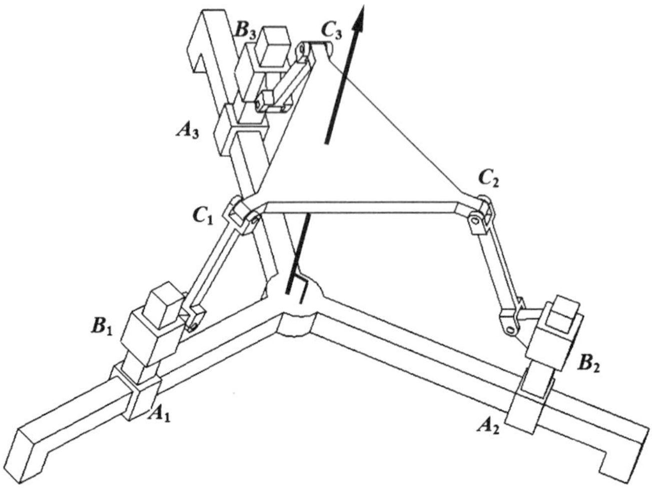

FIGURE 1.7 3-PPRR mechanism.

If the linkage *CD* is selected to be the end effector, then according to Figure 1.6, the mobility should be zero; however, if *BC* is selected as the end effector, then it will have one rotational DOF around the *z*-axis. Obviously its DOF is not zero. Two different results appear in a same mechanism, which presents difficulty for the analysis and the practical application.

In the 3-PPRR spatial mechanism shown in Figure 1.7 [95], it can be found through the analysis that its end effector $C_1C_2C_3$ has only one DOF which can move up and down, while the slider B_i ($i = 1,2,3$) has two DOFs in two orthogonal directions. As for the link B_iC_i, its DOFs are different from that of the above mechanisms. Using the traditional calculation formula (1.4), it can be found that all the links of the mechanism share the same rotational constraint (as shown in Figure 1.7 in the direction of the arrow) which is perpendicular to the bottom surface. The number of the general constraint of this mechanism is one, and $b = 6-1 = 5$, in turn. Substituted it into formula (1.4), the calculation result $F = 2$ can be obtained. According to conventional knowledge, it means that after the two actuations are imposed on the mechanism, they will be able to fully control the entire motion of the whole mechanism. For example, if the two actuators are added to the sliders A_1 and A_2, it can be found that the end effector $C_1C_2C_3$ and the linkage B_iC_i can still freely move up and down; but if these two actuators are assigned to A_i and B_i ($i = 1,2,3$), respectively, the end effector $C_1C_2C_3$ could be completely controlled. And each of the RRPP planar closed chain mechanisms formed by the other two kinematic chains in their planes has one DOF, individually. In this actuation scheme, the motion of the other two kinematic chains is still uncertain. This issue will be analyzed in detail in the subsequent chapter and particularly in Chapter 9.

1.3.2 Cause II

In structure analysis of a mechanism, we all wish to accomplish the task using a simple algebraic summation of the number of links, the number of kinematic pairs, and the number of constraints introduced by kinematic pairs. In order to achieve this goal, some scholars of mechanism have been committed to the study of counting issues such as general constraints, local DOF, duplicate constraints, and compound pairs. In the course of this research, many methodologies have appeared through the establishment of concrete analytical models to complete the calculation of some parameters. However, different selection of end effectors will lead to different general constraints imposed on the end effector from the kinematic chains. Therefore, without pointing out the end effector, analyzing the general constraints directly would inevitably lead to calculation errors.

In the spatial mechanisms in Figures 1.3 and 1.4, the general constraints of all the links are zero. However, when investigating constraints of the end effector imposed by kinematic chains, duplicate constraints often occur. This will be discussed in detail in the subsequent chapters. Again in the special spatial four-bar mechanism as shown in Figure 1.5, if CD is the end effector, the number of general constraints is four; if BC is the end effector, however, the number of general constraints is three. Therefore, different selection of the end effector will lead to a different number of general constraints of the end effector imposed by kinematic chains.

1.3.3 Cause III

The number of actuators required to control the mechanism has something to do with the number of DOFs of the end effector and is also influenced by the actuation schemes. In the 3-PPS parallel mechanism shown in Figure 1.8, according to formula (1.3), it can be obtained that $F = 3$. From the viewpoint of

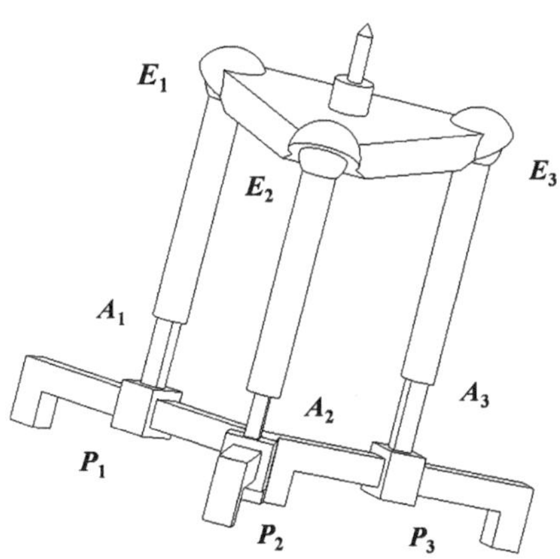

FIGURE 1.8 3-PPS parallel mechanism.

the traditional formula, $F = 3$ means that, as long as there are three actuators, the motion of the entire mechanism should be controlled. As for the mechanism in Figure 1.8, these three actuators can be set on any three sliding pairs of P_i and $A_i (i = 1,2,3)$. Because the mechanism has six prismatic pairs, the number of feasible actuation schemes can be $C_6^3 = 20$. Taking into account the spatial symmetry of the mechanism, these 20 different actuation schemes contain six essentially different ones. For simplicity, here we only give the analysis of the numbers of actuators required in two actuation schemes.

When the three actuators are set on the three prismatic pairs (i.e. P_1, P_2, and P_3), respectively, it can be known through analysis that the end effector $E_1E_2E_3$ still has one DOF which allows it to move up and down. Therefore, the three actuators do not completely control the motion of the end effector. However, if these three actuators are set on another three prismatic pairs (i.e., A_1, A_2, and A_3), they can completely control the motion of the end effector. Thus, if different actuating schemes are selected, the DOF of the end effector will differ in spite of the same mechanism and the same end effector.

1.3.4 Cause IV

The terminal constraints exerted on the end effector by the kinematic chains are functions of spatial position of mechanisms. They change as the position and orientation of the end effector change in its reachable workspace. Therefore, it is unreasonable to attempt to define the general constraint of kinematic chains applied to the end effector in a general sense. The DOF of the end effectors is a function of the pose and position of the mechanism, which dynamically describes the type, quantity, and direction of the free motions of the end effector.

With the development of modern mechanisms, this problem has become increasingly prominent. In complex spatial mechanisms, especially in metamorphic mechanisms, the mobility is no longer constant [79,86,87,94–99] and is instead subjected to the position of mechanisms in the workspace.

In one-and-a-half centuries' history of mechanism, researchers have always been trying to find a fast calculation method for the mobility through a simple algebraic sum of the number of the links, kinematic pairs, and the constraints introduced by the pairs. However, the changing mechanical positions and configurations make it impossible to calculate constant common constraints between the kinematic pair and the links.

As the mechanical structure in general consists of four parts, namely, the actuation part, execution part, transmission part and control part, the analysis of mobility can be divided into the analysis of DOFs of the end effector and the analysis of actuation schemes. Based on the above analysis we can find that, in order to completely address the analysis of mobility, we must first solve the following problems:

(1) clearly identify the end effector from a mechanism;

(2) analyze the actuation schemes to control the end effector;
(3) dynamically calculate the DOF of the end effector of the mechanism.

It is unattainable to solve the above problems with a simple algebraic sum of the number of the links, the number of kinematic pairs, and the number of constraints introduced by kinematic pairs. Therefore, analytical tools are indispensable to completely address the problem of mobility.

1.4 CONTENTS OF THE BOOK

Considering the existing problems in the mobility analysis of spatial mechanisms, this book studies the constraint and motion analysis for spatial robot mechanisms based on the unified mathematical description of its kinematic geometry constraints. The main contents are as follows.

In Chapter 2, the screw theory is briefly introduced. It describes the primary knowledge about the screw theory from aspects such as the Plücker column vector of a line, motion description of the rigid body, the screw description of the motion and force, the reciprocal product of the screw and its geometric significance, the linear combination of screw and the principal screws in a screw system [100,101].

In Chapter 3, the motion screw system of kinematical chains is established through the screw description of the free motions of kinematic pairs; the terminal constraints of the kinematic chain are derived according to reciprocal relationship between the kinematic screw and constraint screw. The equivalent substitution [102] and synthesis [103] of kinematic chains under the condition of the same terminal constraints are studied. The investigation into the mechanical equivalent replacement theorem lays a necessary foundation for the selection and optimization of the kinematic pair of the spatial mechanism which has purely serial kinematic chains, the research on the idle DOF of kinematic chains, especially the complex spatial kinematic chains, and the free motion analysis of the end effector in a mechanism and the analysis of actuation scheme. According to reciprocal screw theory, terminal constraints of the kinematic chain depend entirely on the largest linearly independent set of the kinematic screws. Therefore, the structural optimization problem of the kinematic chain can be addressed through the algebraic analysis to the specified terminal constraints; and it can also be resolved through algebraic treatment to the largest linearly independent set of screws. In fact, the free motions of the end effector are the results of the combination of branch kinematic chains. Therefore, the study of spatial serial mechanisms will focus on the motion constraints of kinematic chains to the end effector. How to determine the types of various kinematic chains and how to complete the design with the simplest and economical structure arrangement have become key issues in the synthesis of kinematic chains. This is also one of the key issues in Chapter 3 and subsequent chapters.

In Chapter 4, terminal constraints of robot end effectors from kinematic chains and the free motions under these constraints are discussed. The analytical theory of constraint space of the end effector of parallel robot mechanism is studied according to the terminal constraint space of its kinematic chains [96–98,104]. The DOF of the end effector of a parallel robot should have the three attributes of the quantity, type, and direction. The DOF of robot end effector is one of the core issues in the free motion analysis of mechanisms. However, the traditional analysis and calculation methods did not make any distinction between the end effector in the mechanism and the entire mechanical system. Therefore, this chapter establishes analytical theory for the constrained space of the end effectors. On this basis, the equivalent description of the complex kinematic chain is discussed, which provides a theoretical support for the equivalent description of the complex kinematic chains to pure serial kinematic chains. It changes the parallel kinematic chains to serial kinematic chains with the equivalent substitution, through its equivalent description of the output motion of the compound kinematic chains. This lays a foundation for the study of actuation of the prescribed end effector.

In Chapter 5, the search for reachable workspace of the spatial robot mechanism [67] and its symmetric homomorphism with the mechanism itself [68,74] is discussed. It is proved that the structural symmetry of the mechanism itself will inevitably lead to the symmetry of the reachable workspace and the flexible workspace of its end effector. The corresponding relation between the reachable workspace and the flexible workspace is homomorphism mapping [74]. A quantitative description of such symmetry relations provides certain theoretical basis for the analysis, synthesis, and structure innovative design of robot mechanism. Based on this, the coordinate transformation method for the resolution of workspace based on free motion analysis of the end effector of spatial parallel mechanism is also covered in this chapter and the synthesis of the flexible workspace is also discussed with the Schoenflies-type spatial parallel mechanism as an example [104]. By analyzing the end effectors' DOF, a minimum of variables can be chosen to establish the kinematic constraint equations and thus the objective function can be established. By searching the desired dexterous workspace, the optimal length of linkages in kinematic chains can be obtained. Application examples show that the method can be widely applied to the synthesis of the flexible workspace of various parallel robot mechanisms [105].

In Chapter 6, the singularity of the spatial robot mechanism in the workspace is studied [106–108] and it is proved that the structural symmetry of the mechanisms themselves will inevitably lead not only to the symmetry of reachable workspace and flexible workspace of its end effector, but also to that of singular workspace. A quantitative description of such a symmetric relation provides some theoretical basis for the analysis and synthesis of workspace of symmetrical mechanisms. With screw theory, a group of balance equations which can be used to analyze the singularity of mechanisms have been obtained and the conditions for singular mechanism have been derived. The order of the coefficient matrix of final equations equals the number of DOFs of the end effector, which constitutes great advantages for the singularity analysis of the end effector of the mechanism with less than six DOFs. The fewer the DOFs the end effector has, the lower the order of the finally converted coefficient matrix will be. According to equivalent substitution Theorems 3.1 and 3.2 of the kinematic chain, it is also available to use this method for dealing with the singularity of compound mechanisms [109]. The analysis of spatial robot mechanism singularity is in fact an aspect of the mobility analysis.

In Chapter 7, kinematics modeling of the spatial parallel robot mechanism is discussed [110]. On the basis of the natural coordinate method, the sufficiency and necessity have been proved that a type of kinematics model spatial parallel mechanism can be completely established with only four non-collinear points with Cartesian coordinates. This method has a significant advantage in dealing with kinematics issues of spatial parallel mechanisms with more than three DOFs. And the established kinematics model shares the same advantages [46] of the general natural coordinate method and can also minimize the number of independent variables. This can particularly apply to solving the kinematics modeling of the multi-DOF spatial parallel robot mechanisms. As this model introduces only the Cartesian coordinate, the elements of the derivative matrix contain only the linear term or even the constant term, while the elements of the derivative matrix obtained through the modeling method of the rotation matrix transformation are often non-linear, or even contain transcendental functions. Therefore, these advantages of the four-point coordinate method have brought great convenience to kinematic and dynamic analysis.

In Chapter 8, the kinematics and static force of robot mechanisms are studied. Based on the analysis of terminal constraints of the kinematic chain in the Chapter 4, detailed analysis is given to the motion between the end effector and drive joint input and the rigid body statics. Through the analysis of the velocity of the end effector of robot mechanism, the basic conditions for the feasible motion of a robot's end effector have been discussed and the forward and inverse solutions of velocity are also investigated [111,112]. On this basis, robot static equilibrium conditions and the forward and inverse solutions of statics are derived according to the virtual power principle.

In Chapter 9, the basic way to correctly analyze the mobility of a mechanism is studied through investigating the theoretical basis of the end effector DOF and the configuration DOF of mechanisms, the key for analyzing the mechanism mobility, is revealed: it is a necessary condition to correctly analyze the mechanism mobility to distinguish the end effector DOF and the independent actuators required to control the end effector. Application examples of spatial mechanisms show that it is not only a necessary requirement for correctly analyzing the mechanism DOF but also meets the needs of engineering applications to distinguish the end effector DOF from the independent actuators required to control the end effector.

In Chapter 10, by analyzing the robot mechanism configuration under a single Cartesian coordinate system [98,99], the basic motion characteristics of the robot mechanism in the workspace is further investigated. Based on the analysis of free motion and terminal constraint of the kinematic chain in Chapter 4, specific studies have been given to the DOF analysis of the end effectors within a single Cartesian coordinate system; the configuration DOF describing the dynamic drive control of mechanisms is investigated by studying the dynamic DOF of the end effector under different actuation schemes. Through analyzing the motion characteristics of the underdetermined, determined, and overdetermined input systems, it is revealed that these three mechanisms are essentially among the deterministic input system. On the basis of the unified mathematical description of kinematic geometry constraints, the analysis theory for the DOF of spatial robot mechanisms has also been discussed. The analysis theory consists of two main parts, the first is the DOF analysis theory targeted at the end effector [96–98, 113]; the second is the configuration DOF analysis theory for the mechanism with the designated end effectors [94,99]. The concept of configuration DOF makes singularity analysis of complex parallel mechanisms be naturally included in the analysis of the configuration DOF; at the same time, the approach for analyzing the structural stability of mechanisms is also provided: only when the mechanism's motion inputs remain consistent with its configuration DOF in the space at any time, can the entire mechanism system be stabilized.

In Chapter 11, the mechanism theory of spatial deployable structure based on the Scissor-Like-Element (SLE) [114] is dealt with. Beginning with the motion analysis of the simple planar linkage, the motion characteristics of the deployable structure and its elements are studied step by step. Through in-depth analysis of the double rotating hinge connected by two SLEs, the mechanism theory in the constitution of elements of the deployable structure has been revealed. Through the analysis of elements of the planar deployable structure, the cylindrical deployable structure, and the spherical deployable structure, the constitution mechanism of spatial deployable structures based on the SLE has been investigated. This has important practical value for the promotion of deployable structures in mechanical engineering, structure engineering, and robot engineering.

In Chapter 12, the application of DOF analysis theory in structure synthesis and innovation of mechanisms has been studied. Through presenting general steps in mechanism synthesis, the basic method for the innovative design of mechanism with the universal significance is also studied. With a comprehensive case study for the innovative design of the lower limb machine of a biped humanoid robot and the independent suspension of the vehicle [103], applications of DOF analysis theory in the innovative design of mechanisms are investigated.

In Chapter 13, a new algorithm is proposed to optimize the design of a spatial parallel manipulator for the purpose of obtaining a desired dexterous workspace rather than the whole reachable workspace. With the analysis of the DOF of a manipulator, we can select the least number of variables to depict the kinematic constraints of each leg of a manipulator. The optimum parameters can be obtained by searching the extreme values of the objective functions with the specified adroit workspace. An example is utilized to demonstrate the significant advantages of this method in the dexterous workspace synthesis. In applications, this method can be widely used to synthesize, optimize, and create all kinds of new spatial parallel manipulator with the desired dexterous workspace.

In Chapter 14, the natural coordinates are adopted to deal with the kinematics model for different spatial mechanisms. After obtaining the constraint equations in natural coordinates, we can conduct the kinematics analysis of the mechanism and then build the dynamics model with Newton–Euler's formulation.

REFERENCES

[1] J.E. Shigley, J.J. Ucker, Theory of Machines and Mechanisms, McGraw–Hill, New York, 1980.

[2] R.L. Norton, Design of Machinery (3rd edition): An Introduction to the Synthesis and Analysis of Mechanisms and Machines, Tata McGraw–Hill, 2004.

[3] J.M. McCarthy, S.G. Song, Geometric Design of Linkages, Springer, New York, 2011.

[4] H.H. Mahie, C.F. Reiholtz, Mechanisms and Dynamics of Machinery, John Wiley & Sons, New York, 1987.

[5] R.G. Budynas, J.K. Nisbett, Mechanical Engineering Design, McGraw–Hill Science, 2010.

[6] J.S. Dai, An historical review of the theoretical development of rigid body displacements from rodrigues parameters to the finite twist, Mechanism and Machine Theory 41 (1) (2006) 41–52.

[7] S. Wen, M. Li, Strategy Study on Mechanology Development, Tsinghua University Press, Beijing, 2003, January.

[8] S. Kajita, Humanoid Robot (Translator to Chinese: Guan, Y.), University Press, Beijing, 2007, March.

[9] D.A. Stewart, Platform with six degrees of freedom, Proceedings of the Institute of Mechanical Engineers Part I 180 (15) (1965) 371–386.

[10] K. Zhou, D.Z. Mao, Z. Tao, Kinematic analysis and application research on a high-speed travelling fouble gour-rod spatial parallel mechanism, International Journal of Advanced Manufacturing Technology 19 (12) (2002) 873–878.

[11] K. Zhou, B.C. Wang, Drive and control of virtual axis NC machine tools, Chinese Journal of Mechanical Engineering 12 (2) (1999) 106–111.

[12] M.A. Laribi, L. Romdhane, S. Zeghloul, Analysis and dimensional synthesis of the DELTA robot for a prescribed workspace, Mechanism and Machine Theory 42 (7) (2007) 859–870.

[13] O. Company, F. Pierrot, Modelling and design issues of a 3-axis parallel machine-tool, Mechanism and Machine Theory 37 (11) (2002) 1325–1345.

[14] R. Whalley, A.A. Abdul-Ameer, K.M. Ebrahimi, The axes response and resonance identification for a machine tool, Mechanism and Machine Theory 46 (8) (2011) 1171–1192.

[15] J.E. Shigley, Kinematic Analysis of Mechanisms, McGraw–Hill Book Company Inc., New York, 1959.

[16] R.S. Hartenberg, J. Denavit, Kinematic Synthesis of Linkages, McGraw–Hill Book Company Inc., New York, 1964.

[17] K.H. Hunt, Screw axes and mobility in spatial mechanisms via the linear complex, Journal of Mechanisms 2 (3) (1967) 307–327.

[18] L. Woo, F. Freudenstein, Application of line geometry to theoretical kinematics and the kinematic analysis of mechanical systems, Journal of Mechanisms 5 (3) (1970) 417–460.

[19] J. Duffy, H.Y. Habib-Olahi, A displacement analysis of spatial five-link 3R–2C mechanisms Part II: analysis of the RRCRC mechanism, Journal of Mechanisms 6 (3) (1971) 463–473.

[20] J. Duffy, H.Y. Habib-Olahi, A displacement analysis of spatial five-link 3R–2C mechanisms Part III: analysis of the RCRRC mechanism, Mechanism and Machine Theory 7 (1) (1972) 71–84.

[21] J. Duffy, D.A. Keen, A displacement analysis of spatial five-link RERRR mechanism, Mechanism and Machine Theory 7 (2) (1972) 211–230.
[22] B. Roth, The design of binary cranks with revolute cylindric and prismatic joints, Journal of Mechanisms 3 (2) (1968) 61–72.
[23] L.W. Tsai, B. Roth, Design of dyads with helical, cylindrical, spherical, revolute and prismatic joints, Mechanism and Machine Theory 7 (1) (1972) 85–102.
[24] K.H. Hunt, Kinematic Geometry of Mechanisms, Oxford University Press, Oxford, 1978.
[25] R.S. Ball, A Treatise on the Theory of Screws, Cambridge University Press, Cambridge, 1900.
[26] J. Phillips, Freedom in machinery, Introducing Screw Theory, vol. 1, Cambridge University Press, Cambridge, 1984.
[27] J. Phillips, Freedom in machinery, Screw Theory Exemplified, vol. 2, Cambridge University Press, Cambridge, 1990.
[28] Z. Huang, Spatial Mechanism, Machine Industry Press, Beijing, 1991, May.
[29] H.S. Kim, L.W. Tsai, Design optimization of a cartesian parallel manipulator, Journal of Mechanical Design, Transactions of ASME 125 (2003) 43–51.
[30] M. Ceccarelli, C.A. Lanni, Multi-objective optimum design of general 3R manipulators for prescribed workspace limits, Mechanism and Machine Theory 39 (2) (2004) 119–132.
[31] M. Griffis, J.A. Duffy, Forward displacement analysis of a class of Stewart platforms, Journal of Robotic Systems 6 (6) (1989) 703–720.
[32] V. Parenti-Castelli, C. Innocenti, Forward displacement analysis of parallel mechanisms: closed form solution of PRR-3S and PPR-3S structures, Journal of Mechanical Design, Transactions of the ASME 114 (1992) 68–73.
[33] W. Lin, M. Griffis, J. Duffy, Forward displacement analysis of the 2–4 Stewart platforms, Journal of Mechanical Design, Transactions of the ASME 114 (1992) 442–450.
[34] M. Raghavan, The Stewart platform of general geometry has 40 configurations, Journal of Mechanical Design, Transactions of the ASME 115 (1993) 115–277.
[35] M. Husain, K.J. Waldron, Position kinematics of a two limbed mixed mechanism, Mechanism and Machine Theory 28 (6) (1993) 763–775.
[36] J.P. Yin, C.G. Liang, The forward displacement analysis of a kind of special platform manipulator mechanisms, Mechanism and Machine Theory 29 (1) (1994) 1–9.
[37] F. Wen, C. Liang, Displacement analysis of the 6–6 Stewart platform mechanisms, Mechanism and Machine Theory 29 (4) (1994) 547–557.
[38] C.M. Gosselin, J. Merlet, Determination of the workspace of planar parallel manipulators with joint limits, Robotics and Autonomous Systems 17 (3) (1996) 129–138.
[39] N.X. Chen, S.M. Song, Direct position analysis of the 2–6 Stewart platforms, Journal of Mechanical Design, Transactions of the ASME 116 (1994) 61–66.
[40] Y.K. Byun, H.S. Cho, Analysis of a novel 6-DOF, 3-PPSP parallel manipulator, The International Journal of Robotics Research 16 (6) (1997) 859–872.
[41] J.S. Wang, X.J. Liu, Analysis of a novel cylindrical 3-DOF parallel robot, Robotics and Autonomous Systems 42 (1) (2003) 31–46.
[42] V.P. Hernández, Position analysis of planar mechanisms with R-pairs using a geometrical–iterative method, Mechanism and Machine Theory 39 (2) (2004) 133–152.
[43] J.P. Merlet, Forward kinematics of nonpolyhedral parallel manipulators, Journal of Mechanical Design 115 (1993) 938–940.
[44] H.Y. Lee, C. Woernie, M. Hiller, A complete solution for the inverse kinematic problem of the general 6R robot manipulator, Journal of Mechanical Design, Transactions of the ASME 113 (1991) 481–486.
[45] V. Murthy, K.J. Waldron, Position kinematics of the generalized lobster arm and its series–parallel dual, Journal of Mechanical Design, Transactions of the ASME 114 (1992) 406–413.
[46] J.G. Jalón, E. Bayo, Kinematic and Dynamic Simulation of Multibody Systems: The Real-Time Challenge, Springer-Verlag, NewYork, 1994.
[47] F.C. Park, J. Choi, S.R. Ploen, Symbolic formulation of closed chain dynamics in independent coordinates, Mechanism and Machine Theory 34 (5) (1999) 731–751.
[48] J. Gallardo, J.M. Rico, A. Frisoli, Dynamics of parallel manipulators by means of screw theory, Mechanism and Machine Theory 38 (11) (2003) 1113–1131.

[49] K. Koser, A slider crank mechanism based robot arm performance and dynamic analysis, Mechanism and Machine Theory 39 (2) (2004) 169–182.
[50] C. Gosselin, J. Angeles, Singularity analysis of closed-loop kinematic chains, IEEE Transactions on Robotics and Automation 6 (3) (1990) 281–290.
[51] C.L. Collins, J.M. McCarthy, The quadratic singularity surfaces of planar platforms in the Clifford algebra of the projective plane, Mechanism and Machine Theory 33 (7) (1998) 931–944.
[52] P. Choudhury, A. Ghosal, Singularity and controllability analysis of parallel manipulators and closed-loop mechanisms, Mechanism and Machine Theory 35 (10) (2000) 1455–1479.
[53] L. Romdhane, Z. Affi, M. Fayet, Design and singularity analysis of a 3-translational-DOF in-parallel manipulator, Journal of Mechanical Design, Transactions of the ASME 124 (2002) 419–426.
[54] K. Zhou, Z. Tao, J.S. Zhao, Singularity loci research on high-speed travelling type of double four-rod spatial parallel mechanism, Mechanism and Machine Theory 38 (3) (2003) 195–211.
[55] A. Wolf, E. Ottaviano, M. Shoham, Application of line geometry and linear complex approximation to singularity analysis of the 3-DOF CaPaMan Parallel Manipulator, Mechanism and Machine Theory 39 (1) (2004) 75–95.
[56] A. Kumar, K.J. Waldron, The workspaces of a mechanical mechanism, Journal of Mechanical Design 103 (1981) 665–672.
[57] Z.C. Lai, C.H. Menq, The dexterous workspace of simple manipulators, IEEE Transactions on Robotics 4 (1) (1988) 99–103.
[58] C. Gosselin, Determination of the workspace of 6-DOF parallel manipulators, Journal of Mechanical Design 112 (3) (1990) 331–336.
[59] J.P. Merlet, Designing a parallel manipulator for a specific workspace, International Journal of Robotics Research 16 (4) (1997) 545–556.
[60] H.T. Liu, T. Huang, et al., Kinematic design of a 5-DOF hybrid robot with large workspace/limb-stroke ratio, Journal of Mechanical Design 129 (5) (2007) 530–537.
[61] J.P. Merlet, C. Gosselin, N. Mouly, Workspaces of planar parallel manipulators, Mechanism and Machine Theory 33 (1) (1998) 7–20.
[62] T. Huang, J.S. Wang, D.J. Whitehouse, Closed form solution to workspace of hexapod-based virtual axis machine tools, Journal of Mechanical Design 121 (1) (1999) 26–31.
[63] J.P. Merlet, Determination of 6D workspaces of gough-type parallel manipulator and comparison between different geometries, International Journal of Robotics Research 18 (9) (1999) 902–916.
[64] I.A. Bonev, J.A. Ryu, New approach to orientation workspace analysis of 6-DOF parallel mechanisms, Mechanism and Machine Theory 36 (1) (2001) 15–28.
[65] F. Gao, X.J. Liu, X. Chen, The relationships between the shapes of the workspaces and the link lengths of 3-DOF symmetrical planar parallel mechanisms, Mechanism and Machine Theory 36 (2) (2001) 205–220.
[66] Y. Wang, G.S. Chirikjian, Workspace generation of hyper-redundant manipulators as a diffusion process on SE(N), IEEE Transactions on Robotics 20 (3) (2004) 399–408.
[67] J.S. Zhao, Z.J. Feng, K. Zhou, On the workspace of spatial parallel manipulator with multi-translational degrees of freedom, The International Journal of Advanced Manufacturing Technology 27 (1–2) (2005) 112–118.
[68] J.S. Zhao, M. Chen, K. Zhou, J.X. Dong, Z.J. Feng, Workspace of mechanisms with symmetry identical kinematic chains, Mechanism and Machine Theory 41 (6) (2006) 632–645.
[69] H. Li, C.M. Gosselin, M.J. Richard, Determination of maximal singularity–free zones in the workspace of planar three-degree-of-freedom parallel mechanisms, Mechanism and Machine Theory 41 (10) (2006) 1157–1167.
[70] I.A. Bonev, C.M. Gosselin, Analytical determination of the workspace of symmetrical spherical parallel mechanisms, IEEE Transactions on Robotics 22 (5) (2006) 1011–1017.
[71] M. Gouttefarde, C.M. Gosselin, Analysis of the Wrench-Closure workspace of planar parallel cable-driven mechanisms, IEEE Transactions on Robotics 22 (3) (2006) 434–445.

[72] H. Li, C.M. Gosselin, M.J. Richard, Determination of maximal singularity-free zones in the six-dimensional workspace of the general Gough–Stewart platform, Mechanism and Machine Theory 42 (4) (2007) 497–511.

[73] J.Z. Yang, K. Abdel-Malek, Y.Q. Zhang, On the workspace boundary determination of serial manipulators with non-unilateral constraints, Robotics and Computer Integrated Manufacturing 24 (1) (2007) 60–76.

[74] J.-S. Zhao, F.-L. Chu, Z.-J. Feng, Symmetrical characteristics of the workspace shape for spatial parallel mechanisms with symmetric structure, Mechanism and Machine Theory 43 (4) (2008) 427–444.

[75] M. Giovagnoni, A. Rossi, Transient analysis of a flexible crank, Mechanism and Machine Theory 24 (4) (1989) 231–243.

[76] Y.-X. Wang, Multifrequency resonances of flexible linkages, Mechanism and Machine Theory 33 (3) (1998) 255–271.

[77] X. Wang, J.K. Kills, Dynamic modeling of a flexible-link planar parallel platform using a substructuring approach, Mechanism and Machine Theory 41 (6) (2006) 671–687.

[78] J.R. Phillips, K.H. Hunt, On the theorem of three axes in the spatial motion of three bodies, Australian Journal of Exerted Science 15 (1964) 267–287.

[79] K.J. Waldron, The constraint analysis of mechanisms, Journal of Mechanisms 1 (1966) 101–114.

[80] J.A. Mirth, T.R. Chase, Circuit analysis of watt chain six-bar mechanisms, Journal of Mechanical Design, Transactions of the ASME 115 (1993) 212–222.

[81] T.R. Chase, J.A. Mirth, Circuits and branches of single-degree-of-freedom planar linkages, Journal of Mechanical Design, Transactions of the ASME 115 (1993) 223–230.

[82] R. Srivatsan, D.A. Hoeltzel, A. Das, An analytical method for the identification of redundant constraints in multiple-loop spatial mechanisms, Journal of Mechanical Design, Transactions of the ASME 115 (1993) 322–330.

[83] M.J. Tsai, H.W. Lee, The transmissivity and manipulability of spatial mechanisms, Journal of Mechanical Design, Transactions of the ASME 116 (1994) 37–143.

[84] T. Huang, D.G. Chetwynd, D.J. Whitehouse, J.S. Wang, A general and novel approach for parameter identification of 6-DOF parallel kinematic machines, Mechanism and Machine Theory 40 (2) (2005) 219–239.

[85] T. Huang, J.P. Mei, Z.X. Li, X.M. Zhao, A method for estimating servomotor parameters of a parallel robot for rapid pick-and-place operations, Journal of Mechanical Design 127 (3) (2005) 596–601.

[86] J.S. Dai, J.R. Jones, Mobility in metamorphic mechanisms of foldable/erectable kinds, ASME Journal of Mechanical Design 121 (3) (1999) 375–382.

[87] J.S. Dai, J.R. Jones, Matrix representation of topological changes in metamorphic mechanisms, ASME Journal of Mechanical Design 127 (4) (2005) 837–840.

[88] G. Gogu, Mobility of mechanisms: a critical review, Mechanism and Machine Theory 40 (9) (2005) 1068–1097.

[89] J.M. McCarthy, Geometric Design of Linkages, Springer-Verlag, New York, 2000.

[90] G. Gogu, Mobility and spatiality of parallel robots revisited via theory of linear transformations, European Journal of Mechanics A/Solids 24 (2005) 690–711.

[91] A.B. Kempe, How to Draw a Straight Line, Macmillan, London, (2013). http://kmoddl.slibrary.cornell.edu/tutorials/11/.html , Cited by D.W. Henderson, and D. Taimina, 27 August.

[92] <http://kmoddl.library.cornell.edu/model.php?m=244.html>, 27 August 2013.

[93] <http://mathworld.wolfram.com/SarrusLinkage.html.html>, 27 August 2013.

[94] J.-S. Zhao, K. Zhou, Z.J. Feng, A theory of degrees of freedom for mechanisms, Mechanism and Machine Theory 39 (6) (2004) 621–643.

[95] J.S. Zhao, F. Chu, Z.J. Feng, Jian S. Dai, Actuation schemes of a spatial 3-PPRR parallel mechanism, Journal of Mechanical Engineering Science, Proc. IMechE, Part C 226 (1) (2012) 228–241.

[96] J.S. Zhao, K. Zhou, D.Z. Mao, Y.F. Gao, Y. Fang, A new method to study the degree of freedom of spatial parallel mechanisms, The International Journal of Advanced Manufacturing Technology 23 (3–4) (2004) 288–294.

[97] J.S. Zhao, Z.J. Feng, K. Zhou, D.W. Jin, Re-analysis of the degree-of-freedom configuration of the platforms in spatial parallel mechanisms with constraints spaces, The International Journal of Advanced Manufacturing Technology 28 (1–2) (2006) 190–196.
[98] J.S. Zhao, Z.J. Feng, L.P. Wang, J.X. Dong, The free mobility of a parallel manipulator, Robotica 24 (5) (2006) 635–641.
[99] J.S. Zhao, Z.J. Feng, J.X. Dong, Computation of the configuration degree of freedom of a spatial parallel mechanism by using reciprocal screw theory, Mechanism and Machine Theory 41 (12) (2006) 1486–1504.
[100] J.S. Zhao, F. Chu, Z.J. Feng, Identification of principal screws of two- and three-screw systems, Proceedings of IMechE, Journal of Mechanical Engineering Science, Part C 221 (12) (2007) 1701–1715.
[101] J.S. Zhao, H.X. Zhou, Z.J. Feng, Jian S. Dai, An algebraic methodology to identify the principal screws and their pitches of screw systems, Proceedings of IMechE, Journal of Mechanical Engineering Science, Part C 223 (8) (2009) 1931–1941.
[102] J.S. Zhao, Z.J. Feng, The design of kinematic chain based on the principle of invariable terminal constraints, The International Journal of Advanced Manufacturing Technology 26 (1–2) (2005) 138–143.
[103] J.S. Zhao, F.L. Chu, Z.J. Feng, Synthesis of rectilinear motion generating spatial mechanism with application to automotive suspension, Journal of Mechanical Design 130 (6) (2008) 065001-1–065001-8.
[104] J.S. Zhao, Y.Z. Fu, K. Zhou, Z.J. Feng, Mobility properties of a schoenflies type parallel manipulator, Robotics and Computer-Integrated Manufacturing 22 (2) (2006) 124–133.
[105] J.S. Zhao, S.L. Zhang, et al., Optimizing the kinematic chains for a spatial parallel manipulator via searching the desired dexterous workspace, Robotics and Computer-Integrated Manufacturing 23 (1) (2007) 38–46.
[106] J.S. Zhao, Z.J. Feng, et al., Analysis of the singularity of spatial parallel manipulator with terminal constraints, Mechanism and Machine Theory 40 (3) (2005) 275–284.
[107] J.S. Zhao, K. Zhou, et al., A novel methodology to study the singularity of spatial parallel mechanism, The International Journal of Advanced Manufacturing Technology 23 (9–10) (2004) 750–754.
[108] J.S. Zhao, F.L. Chu, Z.J. Feng, Singularities within the workspace of spatial parallel mechanisms with symmetric structure, Proceedings of IMechE, Journal of Mechanical Engineering Science, Part C 224 (2) (2010) 459–472.
[109] J.S. Zhao, K. Zhou, Z.J. Feng, Z.Y. Tan, The singularity study of spatial hybrid mechanism based on screw theory, The International Journal of Advanced Manufacturing Technology 25 (9–10) (2005) 1053–1061.
[110] K. Zhou, J.S. Zhao, et al., The kinematics study of a class of spatial parallel mechanism with fewer degrees of freedom, The International Journal of Advanced Manufacturing Technology 25 (9–10) (2005) 972–978.
[111] J.S. Zhao, Y. Yun, L.P. Wang, et al., Investigate the forward kinematics of Gough–Stewart manipulator with natural coordinates, The International Journal of Advanced Manufacturing Technology 30 (7–8) (2006) 700–716.
[112] J.S. Zhao, X. Yang, L. Zhu, et al., On the forward and inverse displacement of spatial parallel manipulators, The International Journal of Advanced Manufacturing Technology 29 (11–12) (2006) 1284–1294.
[113] J.-S. Zhao, Z.J. Feng, K. Zhou, D.W. Jin, Re–analysis on the degree of freedom of the platforms in spatial parallel mechanisms with constraints spaces, The International Journal of Advanced Manufacturing Technology 28 (1–2) (2006) 190–196.
[114] J.S. Zhao, F.L. Chu, Z.J. Feng, The mechanism theory and application of deployable structures based on SLE, Mechanism and Machine Theory 44 (3) (2009) 252–263.

Chapter 2

A Brief Introduction to Screw Theory

Screw theory started at the second half of the 19th century from the study on line geometry by Plücker [1–4]. In 1900, the publication of the classic work, *A Treatise on the Theory of Screws* [4], marked that screw theory is relatively mature. A rigid body's instantaneous velocity and the force that it is subjected to can be expressed as screws which are named twist and torque, respectively. Therefore, screws can be used to represent the constraints that a spatial rigid mechanism subjected to and its free motions under the constraints. In addition, a screw is determined by its axis and pitch in geometry, which provides a great convenience for the application and promotion of screw theory.

2.1 PLÜCKER VECTOR

Suppose that there are two different points which are expressed as $\boldsymbol{r}_1 = (x_1 \quad y_1 \quad z_1)^T$ and $\boldsymbol{r}_2 = (x_2 \quad y_2 \quad z_2)^T$ in a Cartesian coordinate system, as shown in Figure 2.1. These two points determine one line uniquely.

The direction of the line can be expressed by $\boldsymbol{S}$

$$\boldsymbol{S} = \begin{bmatrix} x_2 - x_1 & y_2 - y_1 & z_2 - z_1 \end{bmatrix}^T \tag{2.1}$$

Suppose that $\boldsymbol{r}$ denotes any point on line l_{AB}. Obviously, we have

$$\left(\boldsymbol{r} - \boldsymbol{r}_1\right) \times \boldsymbol{S} = 0. \tag{2.2}$$

Rearranging equation (2.2) presents

$$\boldsymbol{r} \times \boldsymbol{S} = \boldsymbol{S}_0 \tag{2.3}$$

where

$$\boldsymbol{S}_0 = \boldsymbol{r}_1 \times \boldsymbol{S}. \tag{2.4}$$

Advanced Theory of Constraint and Motion Analysis for Robot Mechanisms.
http://dx.doi.org/10.1016/B978-0-12-420162-0.00002-3

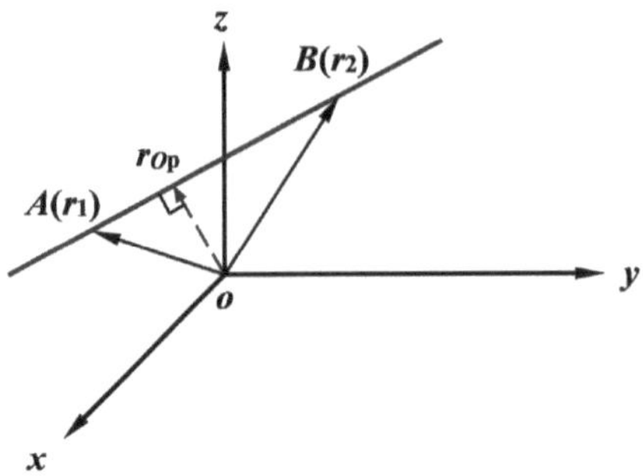

FIGURE 2.1 A straight line in Cartesian coordinates.

Therefore, the vector $\boldsymbol{S}$ and any known point r_1 on it can completely and solely determine line l_{AB}. In other words, line l_{AB} depends entirely on the vector pair

$$\$ = \begin{bmatrix} \boldsymbol{S} \\ \boldsymbol{S}_0 \end{bmatrix}. \tag{2.5}$$

Equation (2.5) is called the Plücker homogeneous coordinates of l_{AB}. Assume that

$$\begin{cases} x_2 - x_1 = L \\ y_2 - y_1 = M \\ z_2 - z_1 = N. \end{cases} \tag{2.6}$$

Then equation (2.1) can be expressed as

$$\boldsymbol{S} = \begin{bmatrix} L & M & N \end{bmatrix}^T. \tag{2.7}$$

Suppose

$$\|\boldsymbol{S}\| = \sqrt{L^2 + M^2 + N^2} \tag{2.8}$$

where $\|\cdot\|$ denotes the norm of a three-dimensional vector "·". Therefore, the unit direction vector of the line can be expressed as

$$s = \frac{\boldsymbol{S}}{\|\boldsymbol{S}\|} = \begin{bmatrix} l \\ m \\ n \end{bmatrix} \tag{2.9}$$

where

$$\begin{cases} l = \frac{L}{\sqrt{L^2+M^2+N^2}} \\ m = \frac{M}{\sqrt{L^2+M^2+N^2}} \\ n = \frac{N}{\sqrt{L^2+M^2+N^2}}. \end{cases} \tag{2.10}$$

Obviously,

$$\|s\| = l^2 + m^2 + n^2 = 1. \tag{2.11}$$

Suppose that

$$s_0 = \frac{S_0}{\|S\|}.$$

From equation (2.5), the Plücker homogeneous coordinates of l_{AB} can also be expressed as

$$\$ = \begin{bmatrix} s^T & s_0^T \end{bmatrix}^T \tag{2.12}$$

where

$$r \times s = s_0. \tag{2.13}$$

Screws expressed by equation (2.12) which satisfies equation (2.11) are called unit screws. Therefore, equation (2.12) is called the unit screw of a line, where s_0 indicates the distance from the origin of coordinate system to the line. It is noteworthy that S_0 is generally expressed as a function of S

$$S_0 = g(S)$$

According to the definition of screws, we have

$$g(S) = r \times S + hS$$

where r represents the position vector of any point on a screw axis and h is the pitch of the screw.

Suppose that $s = \frac{S}{\|S\|}$, thus s represents a unit three-dimension vector and $S = \|S\|\, s$.

Substituting $S = \|S\|\, s$ and $g(S) = r \times S + hS$ into S_0 yields

$$S_0 = \|S\| \left(r \times s + hs\right).$$

Substituting $S = \|S\|\, s$ and S_0 into equation (2.5) yields

$$\$ = \|S\| \begin{bmatrix} s \\ s_0 \end{bmatrix}$$

where

$$s_0 = r \times s + hs.$$

Assume that

$$\$ = \|S\| \$_u$$

where

$$\$_u = \begin{bmatrix} s \\ s_0 \end{bmatrix}.$$

$\$_u$ is a unit screw. We know that the unit screw of a specified screw is exclusive if one does not consider the minus form of $\boldsymbol{s}$. Obviously, the screw represented by $\$_u$ is the same line represented by equation (2.5). As a matter of fact, any form of $k\$_u$ (k is a scalar and $k \neq 0$) represents the same screw as $\$_u$ in the name of motion or constraint. In other words, any form of $k\$_u$ can be characterized entirely by $\$_u$. Therefore, it is not difficult to find the following theorem.

Theorem 2.1. *For any specified spatial line, its unit screw expression (2.12) is exclusive.*

Proof. As shown in Figure 2.1, suppose that $\boldsymbol{r}_{OP}$ denotes the projection from the origin of coordinate system o on line l_{AB}; then we can find that $\boldsymbol{r}_{OP} \perp \boldsymbol{s}$ and $\boldsymbol{r}_{OP}$ is exclusive.

Therefore

$$\boldsymbol{r}_{OP} \cdot \boldsymbol{s} = 0. \tag{2.14}$$

According to equations (2.3) and (2.4), we obtain that

$$\boldsymbol{r}_{OP} \times \boldsymbol{s} = \boldsymbol{s}_0. \tag{2.15}$$

Because the projection $\boldsymbol{r}_{OP}$ from the origin on the specified line and the unit direction vector $\boldsymbol{s}$ are both exclusive, the unit screw expression (2.12) is exclusive.

Left cross multiplying $\boldsymbol{s}$ at both sides of equation (2.15) yields

$$\boldsymbol{s} \times \left(\boldsymbol{r}_{OP} \times \boldsymbol{s}\right) = \boldsymbol{s} \times \boldsymbol{s}_0. \tag{2.16}$$

Expanding the left side of equation (2.16) and associating equation (2.14), we have

$$\boldsymbol{s} \times \left(\boldsymbol{r}_{OP} \times \boldsymbol{s}\right) = \left(\boldsymbol{s} \cdot \boldsymbol{s}\right) \boldsymbol{r}_{OP} - \left(\boldsymbol{r}_{OP} \cdot \boldsymbol{s}\right) \boldsymbol{s} = \left(\boldsymbol{s} \cdot \boldsymbol{s}\right) \boldsymbol{r}_{OP}$$

Therefore, we have

$$\boldsymbol{r}_{OP} = \frac{\boldsymbol{s} \times \boldsymbol{s}_0}{\|\boldsymbol{s}\|^2} = \boldsymbol{s} \times \boldsymbol{s}_0. \tag{2.17}$$

This means that the projection of the origin of coordinate system on the line is just the cross product of the pair of vectors $\boldsymbol{s}$ and $\boldsymbol{s}_0$ in the screw. If the line passes through the origin of the coordinate system, $\boldsymbol{s}_0 = 0$ in line with equation (2.15) and $\boldsymbol{r}_{OP} = 0$ from equation (2.17), which is the same as the fact.

Therefore, the unit screw expression (2.12) is exclusive for any prescribed line. Dot multiplying $\boldsymbol{s}$ at both sides of equation (2.13) yields

$$\boldsymbol{s} \cdot \left(\boldsymbol{r} \times \boldsymbol{s}\right) = \boldsymbol{s} \cdot \boldsymbol{s}_0 = 0. \tag{2.18}$$

Suppose that

$$\boldsymbol{s}_0 = \begin{bmatrix} p & q & r \end{bmatrix}^T. \tag{2.19}$$

Then the Plücker homogeneous coordinates of line l_{AB} can be expressed as

$$\$ = \begin{bmatrix} l & m & n & p & q & r \end{bmatrix}^T. \tag{2.20}$$

Expanding equation (2.18) yields

$$lp + mq + nr = 0. \tag{2.21}$$

It is not difficult to find that the unit screw of a specified straight line contains six Plücker coordinates with two associated constraints equations (2.11) and (2.21). Therefore, Plücker homogeneous coordinates of a line is determined by four independent variables.

2.2 RIGID BODY'S MOTION EXPRESSION

The distance between any two points of a moving rigid body will always remain constant. The typical motions of rigid bodies include rotation and translation.

2.2.1 Rotation of a Rigid Body

For the sake of convenience, suppose that the fixed coordinate system $o_0x_0y_0z_0$ and the moving coordinate system $o_1x_1y_1z_1$ attached to the rigid body have a common origin, and z_1-axis coincides with z_0-axis. As shown in Figure 2.2, $z_1\left(z_0\right)$-axis is the instantaneous axis of the rigid body. Assume that in the fixed coordinate system $o_0x_0y_0z_0$, the coordinates of a vector $\boldsymbol{p}$ that starts from the origin is $\boldsymbol{p}_0$, while it is $\boldsymbol{p}_1$ in the coordinate system $o_1x_1y_1z_1$. When the rigid body rotates with an angle of θ around the instantaneous axis, assume that $\mathbf{R}_{01}$ denotes the coordinate transformation matrix from $o_1x_1y_1z_1$ to $o_0x_0y_0z_0$, therefore we have

$$\boldsymbol{p}_0 = \mathbf{R}_{01}\boldsymbol{p}_1 \tag{2.22}$$

where $\mathbf{R}_{01}$ denotes the coordinate transformation matrix from $o_1x_1y_1z_1$ to $o_0x_0y_0z_0$.

Before and after rotation, the squares of the lengths of vectors $\boldsymbol{p}_1$ attached to the rigid body and $\boldsymbol{p}_0$ in the absolute coordinate system are

$$\left\|\boldsymbol{p}_1\right\|^2 = \boldsymbol{p}_1^T\boldsymbol{p}_1 \tag{2.23}$$

$$\left\|\boldsymbol{p}_0\right\|^2 = \left(\mathbf{R}_{01}\boldsymbol{p}_1\right)^T\left(\mathbf{R}_{01}\boldsymbol{p}_1\right) = \boldsymbol{p}_1^T\left(\mathbf{R}_{01}^T\mathbf{R}_{01}\right)\boldsymbol{p}_1. \tag{2.24}$$

For any vector attached to the rigid body, its length remains constant, so

$$\left\|\boldsymbol{p}_0\right\|^2 - \left\|\boldsymbol{p}_1\right\|^2 = 0. \tag{2.25}$$

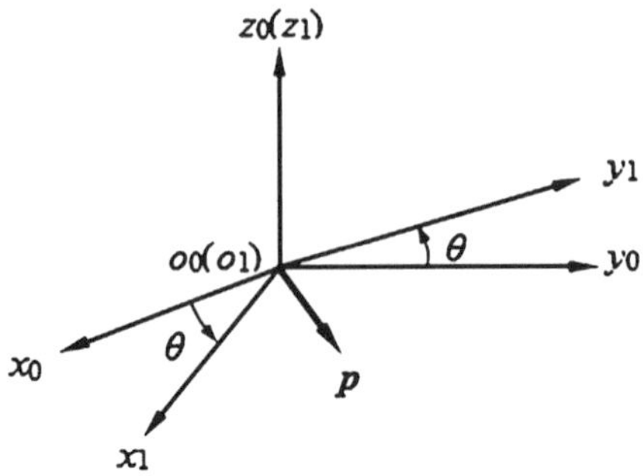

FIGURE 2.2 Rotation of a rigid body.

Substituting equations (2.23) and (2.24) into equation (2.25) yields

$$p_1^T \left(\mathbf{R}_{01}^T \mathbf{R}_{01} - \mathbf{I}\right) p_1 = 0 \tag{2.26}$$

where **I** is a 3 × 3 unit screw matrix.

Considering p_1 is an arbitrary vector, the necessary and sufficient condition where equation (2.26) holds is

$$\mathbf{R}_{01}^T \mathbf{R}_{01} = \mathbf{I}. \tag{2.27}$$

Equation (2.27) indicates that the rotational transformation matrix of a coordinate system is an orthogonal matrix. Obviously, one also has

$$\mathbf{R}_{01} \mathbf{R}_{01}^T = \mathbf{I}. \tag{2.28}$$

Differentiating both sides of equation (2.22) with respect to time yields

$$\dot{p}_0 = \dot{R}_{01} p_1 + R_{01} \dot{p}_1 = \dot{R}_{01} p_1 \tag{2.29}$$

where p_1 is constant in the coordinate system $o_1x_1y_1z_1$ attached to the rigid body. Therefore, the derivative of p_1 with respect to time is zero. According to equation (2.22), we obtain

$$p_1 = \mathbf{R}_{01}^{-1} p_0 = \mathbf{R}_{01}^T p_0. \tag{2.30}$$

Substituting equation (2.30) into equation (2.29) yields [5]

$$\dot{p}_0 = \dot{\mathbf{R}}_{01} p_1 = \left(\dot{\mathbf{R}}_{01} \mathbf{R}_{01}^T\right) p_0. \tag{2.31}$$

For a rigid body that rotates around a fixed axis, the velocity of any point on it can be expressed as

$$\dot{p}_0 = \hat{\omega} p_0 \tag{2.32}$$

where $\boldsymbol{\omega}$ indicates the rotational angular velocity vector around the fixed axis, and $\widehat{\boldsymbol{\omega}} = \boldsymbol{\omega}\times$ which represents the cross product of vector $\boldsymbol{\omega}$.

Associating equation (2.31) and equation (2.32), we get

$$\widehat{\boldsymbol{\omega}} = \dot{\mathbf{R}}_{01}\mathbf{R}_{01}^T. \tag{2.33}$$

Differentiating both sides of equation (2.28) with respect to time and rearranging it yields

$$\dot{\mathbf{R}}_{01}\mathbf{R}_{01}^T = -\left(\dot{\mathbf{R}}_{01}\mathbf{R}_{01}^T\right)^T. \tag{2.34}$$

Therefore, $\widehat{\boldsymbol{\omega}}$ is a skew matrix which is also called antisymmetric matrix in Chinese.

$$\widehat{\boldsymbol{\omega}} = \begin{bmatrix} 0 & -c_3 & -c_2 \\ c_3 & 0 & -c_1 \\ c_2 & c_1 & 0 \end{bmatrix} \tag{2.35}$$

Suppose $\boldsymbol{\omega} = \begin{bmatrix} \omega_1 & \omega_2 & \omega_3 \end{bmatrix}^T$ and $\boldsymbol{r} = \begin{bmatrix} x & y & z \end{bmatrix}^T$. For $\boldsymbol{v} = \boldsymbol{\omega} \times \boldsymbol{r}$, there is

$$\boldsymbol{v} = \begin{vmatrix} \boldsymbol{i} & \boldsymbol{j} & \boldsymbol{k} \\ \omega_1 & \omega_2 & \omega_3 \\ x & y & z \end{vmatrix} = \left(\omega_2 z - \omega_3 y\right)\boldsymbol{i} + \left(\omega_3 x - \omega_1 z\right)\boldsymbol{j} + \left(\omega_1 y - \omega_2 x\right)\boldsymbol{k}. \tag{2.36}$$

Substituting equation (2.35) into equation (2.32) yields

$$\boldsymbol{v} = \boldsymbol{\omega} \times \boldsymbol{r} = \widehat{\boldsymbol{\omega}} r = \begin{bmatrix} 0 & -c_3 & -c_2 \\ c_3 & 0 & -c_1 \\ c_2 & c_1 & 0 \end{bmatrix} \begin{bmatrix} x \\ y \\ z \end{bmatrix} = \begin{bmatrix} -c_3 y - c_2 z \\ c_3 x - c_1 z \\ c_2 x + c_1 y \end{bmatrix}. \tag{2.37}$$

Equation (2.36) minus equation (2.37) presents

$$\begin{bmatrix} \left(c_3 - \omega_3\right) y + \left(\omega_2 + c_2\right) z \\ \left(\omega_3 - c_3\right) x + \left(c_1 - \omega_1\right) z \\ -\left(\omega_2 + c_2\right) x + \left(\omega_1 - c_1\right) y \end{bmatrix} = 0. \tag{2.38}$$

Considering the arbitrariness of $\boldsymbol{r}$, the necessary and sufficient condition where equation (2.38) holds is

$$\begin{cases} c_1 = \omega_1 \\ c_2 = -\omega_2 \\ c_3 = \omega_3 \end{cases} \tag{2.39}$$

Substituting equation (2.39) into equation (2.35) yields

$$\widehat{\boldsymbol{\omega}} = \begin{bmatrix} 0 & -\omega_3 & \omega_2 \\ \omega_3 & 0 & -\omega_1 \\ -\omega_2 & \omega_1 & 0 \end{bmatrix}. \tag{2.40}$$

Furthermore, the following two equations hold simultaneously:

$$\widehat{\boldsymbol{\omega}}^2 = \begin{bmatrix} -\left(\omega_2^2+\omega_3^2\right) & \omega_1\omega_2 & \omega_1\omega_3 \\ \omega_1\omega_2 & -\left(\omega_1^2+\omega_3^2\right) & \omega_2\omega_3 \\ \omega_1\omega_3 & \omega_2 c_3 & -\left(\omega_1^2+\omega_2^2\right) \end{bmatrix} = \boldsymbol{\omega}\boldsymbol{\omega}^T - \|\boldsymbol{\omega}\|^2 \mathbf{I} \text{ and} \tag{2.41}$$

$$\widehat{\boldsymbol{\omega}}^3 = -\|\boldsymbol{\omega}\|^2 \widehat{\boldsymbol{\omega}}. \tag{2.42}$$

Separating variables at both sides of the differential equation (2.32) and integrating it by time yields

$$\boldsymbol{p}_0(t) = \mathbf{R}_{01}\boldsymbol{p}_0(0) \tag{2.43}$$

where $\boldsymbol{p}_0(0)$ indicates the initial position of the proscribed point and $\mathbf{R}_{01} = \exp\left(\widehat{\boldsymbol{\omega}}t\right)$ and $\exp\left(\widehat{\boldsymbol{\omega}}t\right)$ is an exponential matrix

$$\exp\left(\widehat{\boldsymbol{\omega}}t\right) = \mathbf{I} + \widehat{\boldsymbol{\omega}}t + \frac{1}{2!}\widehat{\boldsymbol{\omega}}^2 t^2 + \frac{1}{3!}\widehat{\boldsymbol{\omega}}^3 t^3 + \cdots \tag{2.44}$$

Substituting equations (2.41) and (2.42) into the infinite series (2.44) yields

$$\begin{aligned} \exp\left(\widehat{\boldsymbol{\omega}}t\right) = \mathbf{I} &+ \frac{\widehat{\boldsymbol{\omega}}}{\|\boldsymbol{\omega}\|}\left(\|\boldsymbol{\omega}\|t - \frac{1}{3!}\|\boldsymbol{\omega}\|^3 t^3 + \frac{1}{5!}\|\boldsymbol{\omega}\|^5 t^5 + \cdots\right) \\ &+ \frac{\widehat{\boldsymbol{\omega}}^2}{\|\boldsymbol{\omega}\|^2}\left[1 - \left(1 - \frac{1}{2!}\|\boldsymbol{\omega}\|^2 t^2 + \frac{1}{4!}\|\boldsymbol{\omega}\|^4 t^4 + \cdots\right)\right]. \end{aligned} \tag{2.45}$$

Transforming equation (2.45) with trigonometric series yields

$$\exp\left(\widehat{\boldsymbol{\omega}}t\right) = \mathbf{I} + \frac{\widehat{\boldsymbol{\omega}}}{\|\boldsymbol{\omega}\|}\sin\left(\|\boldsymbol{\omega}\|t\right) + \frac{\widehat{\boldsymbol{\omega}}^2}{\|\boldsymbol{\omega}\|^2}\left[1 - \cos\left(\|\boldsymbol{\omega}\|t\right)\right]. \tag{2.46}$$

Equation (2.46) is called the Rodrigues formula [6]. Assume that

$$\theta = \|\boldsymbol{\omega}\|t \tag{2.47}$$

where θ denotes the angle with which the vector $\boldsymbol{p}$ rotates around the axis $\boldsymbol{\omega}$. Equation (2.46) can be expressed as

$$\exp\left(\widehat{\boldsymbol{\omega}}t\right) = \mathbf{I} + \frac{\widehat{\boldsymbol{\omega}}}{\|\boldsymbol{\omega}\|}\sin\theta + \frac{\widehat{\boldsymbol{\omega}}^2}{\|\boldsymbol{\omega}\|^2}\left(1 - \cos\theta\right). \tag{2.48}$$

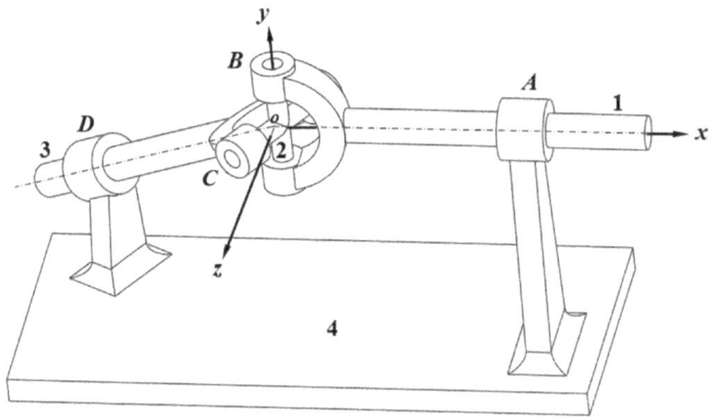

FIGURE 2.3 Single universal coupling.

When $\|\boldsymbol{\omega}\| = 1$, equation (2.48) can be further simplified as

$$\exp\left(\widehat{\boldsymbol{\omega}}t\right) = \mathbf{I} + \widehat{\boldsymbol{\omega}}\sin\theta + \widehat{\boldsymbol{\omega}}^2\left(1 - \cos\theta\right). \quad (2.49)$$

Therefore, any coordinate transformation matrix can be determined entirely by equation (2.48) or (2.49) when the direction of rotational axis, $\boldsymbol{\omega}$, and the rotated angle of θ are known. We can find that this transformation matrix is only associated with the axis's unit direction vector $\frac{\boldsymbol{\omega}}{\|\boldsymbol{\omega}\|}$ and angle θ. Therefore, $\boldsymbol{\omega}$ will be a unit vector without special instructions in this book. As a result, it is available to establish the spatial motion equations for a rigid body. In order to verify the methodology, an example is presented below.

Example 2.1. Universal couplings can be used to deliver rotation between two spatial intersecting axes. Figure 2.3 shows the structure of a single universal coupling. The driving shaft 1 and driven shaft 3 both have planar forks on their ends. These two forks and the crosshead 2 are composed of two perpendicular revolute pairs B and C. The axis 1, axis 3, and the base 4 are composed of revolute pairs A and D. The axes of A and B are perpendicular and so are the axes of C and D. Obviously, the axes of B and C are perpendicular, too. Their intersection is the center of the crosshead, o. We next analyze the expression of the input angle θ_1 of driving shaft 1 and the output angle θ_3 of driven shaft 3, as well as the transmission ratio $i_{31} = \frac{\omega_3}{\omega_1}$.

Solution: We can establish a right-handed Cartesian coordinate shown in Figure 2.3, where the x-axis is the straight line and where the driving shaft of the universal coupling lies, the origin of coordinate system is the center o, and the y-axis is the axis of the revolute pair B which composes of the crosshead and driving shaft 1. Assume that when the driving shaft 1 rotates around its axial unit direction vector $\boldsymbol{\omega}_1 = \begin{bmatrix} 1 & 0 & 0 \end{bmatrix}^T$ with an angle of θ_1, the driven shaft 3 rotates around its axial unit direction vector $\boldsymbol{\omega}_3 = \begin{bmatrix} \cos\alpha & \cos\beta & \cos\gamma \end{bmatrix}^T$ with an angle of θ_3. Obviously, we have

$$\cos^2\alpha + \cos^2\beta + \cos^2\gamma = 1$$

$$\boldsymbol{R}_1 = \boldsymbol{I} + \begin{bmatrix} 0 & 0 & 0 \\ 0 & 0 & -1 \\ 0 & 1 & 0 \end{bmatrix} \sin\theta_1 + \begin{bmatrix} 0 & 0 & 0 \\ 0 & 0 & -1 \\ 0 & 1 & 0 \end{bmatrix}^2 (1 - \cos\theta_1)$$
$$= \begin{bmatrix} 1 & 0 & 0 \\ 0 & \cos\theta_1 & -\sin\theta_1 \\ 0 & \sin\theta_1 & \cos\theta_1 \end{bmatrix}$$

$$\boldsymbol{R}_3 = \boldsymbol{I} + \begin{bmatrix} 0 & -\cos\gamma & \cos\beta \\ \cos\gamma & 0 & -\cos\alpha \\ -\cos\beta & \cos\alpha & 0 \end{bmatrix} \sin\theta_3 + \begin{bmatrix} 0 & -\cos\gamma & \cos\beta \\ \cos\gamma & 0 & -\cos\alpha \\ -\cos\beta & \cos\alpha & 0 \end{bmatrix}^2 (1 - \cos\theta_3)$$
$$= \begin{bmatrix} 1 - (\cos^2\beta + \cos^2\gamma)(1 - \cos\theta_3) & \cos\alpha\cos\beta(1 - \cos\theta_3) - \cos\gamma\sin\theta_3 & \cos\alpha\cos\gamma(1 - \cos\theta_3) + \cos\beta\sin\theta_3 \\ \cos\alpha\cos\beta(1 - \cos\theta_3) + \cos\gamma\sin\theta_3 & 1 - (\cos^2\alpha + \cos^2\gamma)(1 - \cos\theta_3) & \cos\beta\cos\gamma(1 - \cos\theta_3) - \cos\alpha\sin\theta_3 \\ \cos\alpha\cos\gamma(1 - \cos\theta_3) - \cos\beta\sin\theta_3 & \cos\beta\cos\gamma(1 - \cos\theta_3) + \cos\alpha\sin\theta_3 & 1 - (\cos^2\alpha + \cos^2\beta)(1 - \cos\theta_3) \end{bmatrix}.$$

Suppose that r denotes the radius of the universal coupling's crosshead. While at the initial position, the coordinates of the intersection of the crosshead and the driving shaft are denoted by $\boldsymbol{r}_{B_1} = \begin{bmatrix} 0 & r & 0 \end{bmatrix}^T$, the coordinate of the intersection of the crosshead and the driven shaft is denoted by $\boldsymbol{r}_{C_1} = \begin{bmatrix} r\cos\gamma & 0 & -r\cos\alpha \end{bmatrix}^T$. The direction vector of $\boldsymbol{r}_C$ can be determined by $y \times \boldsymbol{\omega}_o = \begin{bmatrix} \cos\gamma & 0 & -\cos\alpha \end{bmatrix}^T$. Therefore,

$$\boldsymbol{r}_{B_2} = \boldsymbol{R}_1 \begin{bmatrix} 0 \\ r \\ 0 \end{bmatrix} = \begin{bmatrix} 1 & 0 & 0 \\ 0 & \cos\theta_1 & -\sin\theta_1 \\ 0 & \sin\theta_1 & \cos\theta_1 \end{bmatrix} \begin{bmatrix} 0 \\ r \\ 0 \end{bmatrix} = r \begin{bmatrix} 0 \\ \cos\theta_1 \\ \sin\theta_1 \end{bmatrix},$$

$$\boldsymbol{r}_{C_2} = \boldsymbol{R}_3 \begin{bmatrix} r\cos\gamma \\ 0 \\ -r\cos\alpha \end{bmatrix}.$$

Expanding $\boldsymbol{r}_{C_2}$ yields

$$\boldsymbol{r}_{C_2} = r \begin{bmatrix} \cos\gamma\left[1 - (\cos^2\beta + \cos^2\gamma)(1 - \cos\theta_3)\right] - \cos\alpha\left[\cos\alpha\cos\gamma(1 - \cos\theta_3) + \cos\beta\sin\theta_3\right] \\ \cos\gamma\left[\cos\alpha\cos\beta(1 - \cos\theta_3) + \cos\gamma\sin\theta_3\right] - \cos\alpha\left[\cos\beta\cos\gamma(1 - \cos\theta_3) - \cos\alpha\sin\theta_3\right] \\ \cos\gamma\left[\cos\alpha\cos\gamma(1 - \cos\theta_3) - \cos\beta\sin\theta_3\right] - \cos\alpha\left[1 - (\cos^2\alpha + \cos^2\beta)(1 - \cos\theta_3)\right] \end{bmatrix}.$$

In fact, $\boldsymbol{r}_{B_2} \cdot \boldsymbol{r}_{C_2} = 0$ always holds. Therefore,

$$\{\cos\gamma\left[\cos\alpha\cos\beta(1 - \cos\theta_3) + \cos\gamma\sin\theta_3\right]$$
$$- \cos\alpha\left[\cos\beta\cos\gamma(1 - \cos\theta_3) - \cos\alpha\sin\theta_3\right]\}\cos\theta_1$$

$$+\left\{\cos\gamma\left[\cos\alpha\cos\gamma(1-\cos\theta_3)-\cos\beta\sin\theta_3\right]\right.$$
$$\left.-\cos\alpha\left[1-\left(\cos^2\alpha+\cos^2\beta\right)(1-\cos\theta_3)\right]\right\}\sin\theta_1=0.$$

Simplifying it yields

$$\left(\cos^2\alpha+\cos^2\gamma\right)\sin\theta_3-\left(\cos\alpha\cos\theta_3+\cos\beta\cos\gamma\sin\theta_3\right)\tan\theta_1=0.$$

Therefore, we have

$$\tan\theta_3=\frac{\cos\alpha\tan\theta_1}{\cos^2\alpha+\cos^2\gamma-\cos\beta\cos\gamma\tan\theta_1}.$$

Differentiating both sides of the equation above with respect to time and considering that α, β, and γ are structure parameters, we obtain

$$i_{31}=\frac{\omega_3}{\omega_1}=\frac{\cos\alpha\left(\cos^2\alpha+\cos^2\gamma\right)\sec^2\theta_1}{\cos^2\alpha\tan^2\theta_1+\left(\cos^2\alpha+\cos^2\gamma-\cos\beta\cos\gamma\tan\theta_1\right)^2}.$$

When $\alpha=\frac{\pi}{3}$, $\beta=\frac{\pi}{4}$ and $\gamma=\frac{\pi}{3}$, the relationship of transmission ratio $i_{31}=\frac{\omega_3}{\omega_1}$ and the driving shaft 1's angular velocity ω_1 with respect to rotational angle is shown in Figure 2.4.

When $\gamma=\frac{\pi}{2},\beta=\frac{\pi}{2}-\alpha$. The transmission ratio can be simplified as

$$i_{31}=\frac{\omega_3}{\omega_1}=\frac{\cos\alpha}{1-\sin^2\alpha\cos^2\theta_1}$$

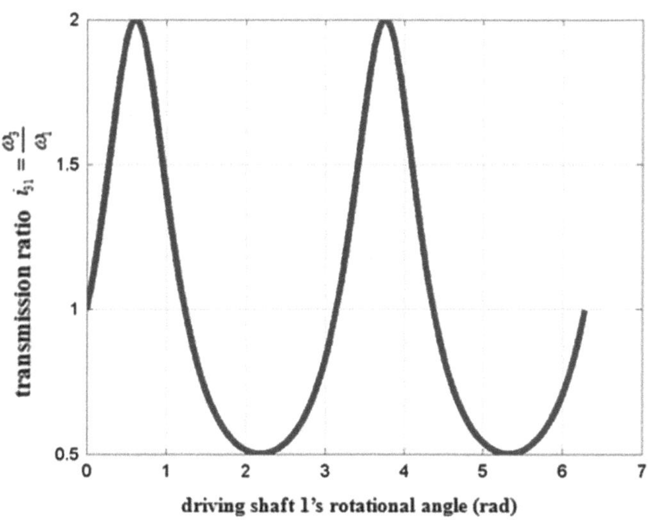

FIGURE 2.4 The relationship of transmission ratio $i_{31}=\frac{\omega_3}{\omega_1}$ and the driving shaft 1's angular velocity ω_1 with respect to rotational angle.

2.2.2 Composited Motion of a Rigid Body

The rotation of a rigid body is discussed in the previous section. This section will discuss the situation where a rigid body has both rotation and translation. Obviously, for the origin of the coordinate system attached to the rigid body, its motion relative to the fixed coordinate system can be utilized to describe the rigid body's translation. In addition, equation (2.49) indicates that the rotation of rigid body is entirely determined by the unit direction vector of the instantaneous rotation axis and the angular velocity of the rotation. Suppose $\boldsymbol{p}_{01}$ denotes the vector from the origin of the fixed coordinate system to the origin of the coordinate system connected with a rigid body, $\mathbf{R}_{01}$ denotes the coordinate transformation matrix from the system attached to the rigid body to the fixed one. Then the general motion of a rigid body is entirely determined by $(\mathbf{R}_{01}, \boldsymbol{p}_{01})$.

$$\boldsymbol{p}_0 = \mathbf{R}_{01}\boldsymbol{p}_1 + \boldsymbol{p}_{01} \tag{2.50}$$

Equation (2.50) is an affine transformation. Introducing a homogeneous coordinate $\boldsymbol{q} = \begin{bmatrix} p_{0_1} & p_{0_2} & p_{0_3} & 1 \end{bmatrix}^T$, equation (2.50) can be expressed as

$$\boldsymbol{q}_0 = \begin{bmatrix} \boldsymbol{p}_0 \\ 1 \end{bmatrix} = \begin{bmatrix} \mathbf{R}_{01} & \boldsymbol{p}_{01} \\ 0 & 1 \end{bmatrix} \begin{bmatrix} \boldsymbol{p}_1 \\ 1 \end{bmatrix}. \tag{2.51}$$

Suppose $\mathbf{T}_{01} = \begin{bmatrix} \mathbf{R}_{01} & \boldsymbol{p}_{01} \\ 0 & 1 \end{bmatrix}$ and $\boldsymbol{q}_1 = \begin{bmatrix} \boldsymbol{p}_1 \\ 1 \end{bmatrix}$, equation (2.51) can be rewritten as

$$\boldsymbol{q}_0 = \mathbf{T}_{01}\boldsymbol{q}_1. \tag{2.52}$$

Equation (2.52) indicates the position and orientation of the system attached to the rigid body relative to the fixed coordinate system. If n rigid bodies are connected with the fixed coordinate system in series through pairs and they compose a rigid body system, the position and orientation of the end effector relative to the fixed system can be expressed as

$$\mathbf{T}_{0n} = \prod_{i=0}^{n-1} \mathbf{T}_{i(i+1)} \tag{2.53}$$

where $\mathbf{T}_{i(i+1)}$ denotes the position and orientation matrix of the coordinate system attached to the $i+1^{\text{th}}$ rigid body relative to the system attached to the ith rigid body.

$$\mathbf{T}_{02} = \mathbf{T}_{01}\mathbf{T}_{12} = \begin{bmatrix} \mathbf{R}_{01}\mathbf{R}_{12} & \mathbf{R}_{01}\boldsymbol{p}_{12} + \boldsymbol{p}_{01} \\ 0 & 1 \end{bmatrix} \tag{2.54}$$

Rearranging equation (2.52) yields

$$\boldsymbol{q}_1 = \mathbf{T}_{01}^{-1}\boldsymbol{q}_0. \tag{2.55}$$

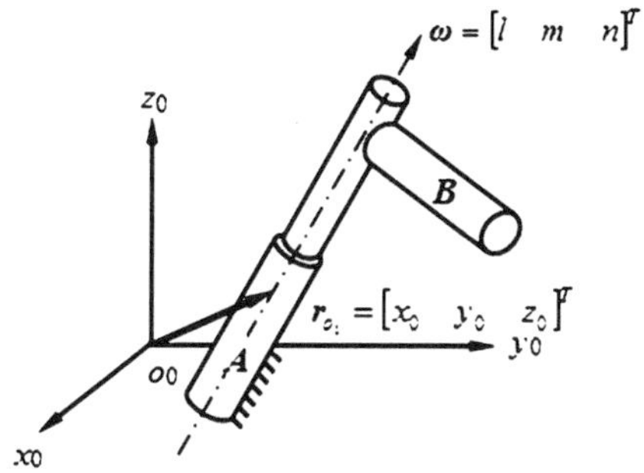

FIGURE 2.5 Relative motion of two rigid bodies connected by cylindrical pairs.

where

$$\mathbf{T}_{01}^{-1} = \begin{bmatrix} \mathbf{R}_{01}^T & -\mathbf{R}_{01}^T \boldsymbol{p}_{01} \\ 0 & 1 \end{bmatrix}. \tag{2.56}$$

Example 2.2. As shown in Figure 2.5, rigid body B is connected with the fixed coordinate system A through a cylindrical pair C. The vector of the cylindrical pair's axis in the fixed coordinate system $o_0x_0y_0z_0$ is denoted by $\boldsymbol{\omega} = \begin{bmatrix} l & m & n \end{bmatrix}^T$, and the initial position vector of a point o_1 is denoted by $\boldsymbol{r}_{o_1} = \begin{bmatrix} x_0 & y_0 & z_0 \end{bmatrix}^T$. If c_1 denotes the rotational angular velocity and c_2 denotes the translation velocity, the position and orientation matrix $\mathbf{T}_{AB}$ of the coordinate system B relative to the fixed coordinate system A can be discussed.

According to equation (2.46),

$$\begin{aligned} \mathbf{R}_{AB} &= \exp(\widehat{\boldsymbol{\omega}} t) \\ &= \mathbf{I} + \sin(c_1 t) \begin{bmatrix} 0 & -n & m \\ n & 0 & -l \\ -m & l & 0 \end{bmatrix} \\ &\quad + [1 - \cos(c_1 t)] \begin{bmatrix} -m^2 - n^2 & lm & ln \\ lm & -l^2 - n^2 & mn \\ ln & mn & -l^2 - m^2 \end{bmatrix} \\ &= \begin{bmatrix} 1 - (m^2 + n^2)[1 - \cos(c_1 t)] & lm[1 - \cos(c_1 t)] - n\sin(c_1 t) & ln[1 - \cos(c_1 t)] + m\sin(c_1 t) \\ lm[1 - \cos(c_1 t)] + n\sin(c_1 t) & 1 - (l^2 + n^2)[1 - \cos(c_1 t)] & mn[1 - \cos(c_1 t)] - l\sin(c_1 t) \\ ln[1 - \cos(c_1 t)] - m\sin(c_1 t) & mn[1 - \cos(c_1 t)] + l\sin(c_1 t) & 1 - (l^2 + m^2)[1 - \cos(c_1 t)] \end{bmatrix}. \end{aligned}$$

Because $\boldsymbol{p}_{AB} = \boldsymbol{r}_{o_1} + c_2\boldsymbol{\omega}$, the position and orientation matrix of system B relative to system A can be expressed as

$$\begin{aligned} &\mathbf{T}_{AB} \\ &= \begin{bmatrix} 1 - (m^2 + n^2)[1 - \cos(c_1 t)] & lm[1 - \cos(c_1 t)] - n\sin(c_1 t) & ln[1 - \cos(c_1 t)] + m\sin(c_1 t) & x_0 + c_2 lt \\ lm[1 - \cos(c_1 t)] + n\sin(c_1 t) & 1 - (l^2 + n^2)[1 - \cos(c_1 t)] & mn[1 - \cos(c_1 t)] - l\sin(c_1 t) & y_0 + c_2 mt \\ ln[1 - \cos(c_1 t)] - m\sin(c_1 t) & mn[1 - \cos(c_1 t)] + l\sin(c_1 t) & 1 - (l^2 + m^2)[1 - \cos(c_1 t)] & x_0 + c_2 nt \\ 0 & 0 & 0 & 1 \end{bmatrix}. \end{aligned}$$

Differentiating both sides of equation (2.52) with respect to time yields

$$\dot{\boldsymbol{q}}_0 = \dot{\mathbf{T}}_{01}\boldsymbol{q}_1. \tag{2.57}$$

Substituting equation (2.55) into equation (2.57) yields

$$\dot{\boldsymbol{q}}_0 = \left(\dot{\mathbf{T}}_{01}\mathbf{T}_{01}^{-1}\right)\boldsymbol{q}_0. \tag{2.58}$$

Similar to equation (2.33), the operator "^" is defined as

$$\widehat{\boldsymbol{\xi}} = \dot{\mathbf{T}}_{01}\mathbf{T}_{01}^{-1}. \tag{2.59}$$

Expanding equation (2.59) presents

$$\widehat{\boldsymbol{\xi}} = \begin{bmatrix} \dot{\mathbf{R}}_{01} & \dot{\boldsymbol{p}}_{01} \\ 0 & 0 \end{bmatrix}\begin{bmatrix} \mathbf{R}_{01}^T & -\mathbf{R}_{01}^T\boldsymbol{p}_{01} \\ 0 & 1 \end{bmatrix} = \begin{bmatrix} \dot{\mathbf{R}}_{01}\mathbf{R}_{01}^T & -\dot{\mathbf{R}}_{01}\mathbf{R}_{01}^T\boldsymbol{p}_{01} + \dot{\boldsymbol{p}}_{01} \\ 0 & 0 \end{bmatrix}. \tag{2.60}$$

Suppose $\dot{\mathbf{R}}_{01}\mathbf{R}_{01}^T = k\widehat{\boldsymbol{\omega}}$, where $\|\boldsymbol{\omega}\| = 1$ is a unit vector. Substituting it into equation (2.60) yields

$$\widehat{\boldsymbol{\xi}} = k\begin{bmatrix} \hat{\boldsymbol{\omega}} & -\hat{\boldsymbol{\omega}}\boldsymbol{p}_{01} + \frac{\dot{\boldsymbol{p}}_{01}}{k} \\ 0 & 0 \end{bmatrix}. \tag{2.61}$$

Suppose $\boldsymbol{v} = -\hat{\boldsymbol{\omega}}\boldsymbol{p}_{01} + \frac{\dot{\boldsymbol{p}}_{01}}{k}$ and define the operator "∨" as

$$\boldsymbol{\xi} = \left(\hat{\boldsymbol{\xi}}\right)^{\vee} = k\begin{bmatrix} \hat{\boldsymbol{\omega}} & \boldsymbol{v} \\ 0 & 0 \end{bmatrix}^{\vee} = k\begin{bmatrix} \boldsymbol{\omega} \\ \boldsymbol{v} \end{bmatrix}. \tag{2.62}$$

The twist, $\boldsymbol{\xi}$, is called the Plücker homogeneous column vector of $\hat{\boldsymbol{\xi}}$. When $k=1$, $\boldsymbol{\xi}$ is a unit screw. Substituting equation (2.59) into equation (2.58), separating variables of both sides of equation (2.58) and integrating them yields

$$\boldsymbol{q}_0(t) = \mathbf{T}_{01}\boldsymbol{q}_0(0) \tag{2.63}$$

where $\mathbf{T}_{01} = \exp\left(\hat{\boldsymbol{\xi}}t\right)$ is the matrix exponential function of the 4×4 matrix $\hat{\boldsymbol{\xi}}t$.

Equation (2.63) describes the position and orientation of a rigid body in a fixed coordinate system, where $\boldsymbol{q}_0(0)$ indicates the initial position of the rigid body. Moreover

$$\exp\left(\widehat{\boldsymbol{\xi}}t\right) = \mathbf{I}_{4\times4} + \widehat{\boldsymbol{\xi}}t + \frac{1}{2!}\left(\widehat{\boldsymbol{\xi}}t\right)^2 + \frac{1}{3!}\left(\widehat{\boldsymbol{\xi}}t\right)^3 + \cdots \tag{2.64}$$

where $\mathbf{I}_{4\times4}$ is a 4×4 unit matrix.

When $\omega = 0$, equation (2.64) can be simplified as

$$\exp\left(\widehat{\xi}t\right) = \begin{bmatrix} \mathbf{I}_{3\times3} & kvt \\ 0 & 1 \end{bmatrix}. \tag{2.65}$$

If $\omega \neq 0$, we can define the translation matrix of a rigid body as

$$\mathbf{T} = \begin{bmatrix} \mathbf{I}_{3\times3} & \omega \times v \\ 0 & 1 \end{bmatrix}. \tag{2.66}$$

Considering $\|\omega\| = 1$, we have

$$\begin{aligned} \hat{\xi}' &= \mathbf{T}^{-1}\hat{\xi}\mathbf{T} = k\begin{bmatrix} \mathbf{I}_{3\times3} & -\omega \times v \\ 0 & 1 \end{bmatrix}\begin{bmatrix} \hat{\omega} & v \\ 0 & 0 \end{bmatrix}\begin{bmatrix} \mathbf{I}_{3\times3} & \omega \times v \\ 0 & 1 \end{bmatrix} \\ &= k\begin{bmatrix} \hat{\omega} & \omega\omega^T v \\ 0 & 0 \end{bmatrix}. \end{aligned} \tag{2.67}$$

Suppose $h = \omega \cdot v = \omega^T v$, equation (2.67) can be expressed as

$$\hat{\xi}' = k\begin{bmatrix} \hat{\omega} & h\omega \\ 0 & 0 \end{bmatrix} \tag{2.68}$$

Assume that

$$\mathbf{T}_1 = \mathbf{T}^{-1}\exp\left(\hat{\xi}t\right)\mathbf{T}. \tag{2.69}$$

Substituting equation (2.64) into equation (2.69) yields

$$\mathbf{T}_1 = \mathbf{I}_{4\times4} + \left(\mathbf{T}^{-1}\hat{\xi}\mathbf{T}\right)t + \frac{1}{2!}\left(\mathbf{T}^{-1}\hat{\xi}\mathbf{T}\right)^2 t^2 + \frac{1}{3!}\left(\mathbf{T}^{-1}\hat{\xi}\mathbf{T}\right)^3 t^3 + \cdots = \exp\left(\hat{\xi}'t\right). \tag{2.70}$$

Therefore,

$$\exp\left(\hat{\xi}t\right) = \mathbf{T}\exp\left(\hat{\xi}'t\right)\mathbf{T}^{-1}. \tag{2.71}$$

According to equation (2.64), we have

$$\exp\left(\hat{\xi}'t\right) = \mathbf{I}_{4\times4} + \hat{\xi}'t + \frac{1}{2!}\left(\hat{\xi}'t\right)^2 + \frac{1}{3!}\left(\hat{\xi}'t\right)^3 + \cdots \tag{2.72}$$

and

$$\left(\hat{\xi}'\right)^2 = k^2\begin{bmatrix} \hat{\omega}^2 & 0 \\ 0 & 0 \end{bmatrix}, \quad \left(\hat{\xi}'\right)^3 = k^3\begin{bmatrix} \hat{\omega}^3 & 0 \\ 0 & 0 \end{bmatrix}, \ldots$$

Therefore,

$$\exp\left(\hat{\xi}'t\right) = \begin{bmatrix} \exp\left(k\hat{\omega}t\right) & kh\omega t \\ 0 & 1 \end{bmatrix}. \tag{2.73}$$

Substituting equation (2.73) into equation (2.71) yields

$$\exp\left(\widehat{\xi}t\right) = \begin{bmatrix} \mathbf{I}_{3\times3} & \omega \times v \\ 0 & 1 \end{bmatrix} \begin{bmatrix} \exp\left(k\widehat{\omega}t\right) & kh\omega t \\ 0 & 1 \end{bmatrix} \begin{bmatrix} \mathbf{I}_{3\times3} & -\omega \times v \\ 0 & 1 \end{bmatrix}.$$

Let $\theta = kt$ and the equation can be simplified as

$$\exp\left(\widehat{\xi}t\right) = \begin{bmatrix} \exp\left(\widehat{\omega}\theta\right) & \left[\boldsymbol{I}_{3\times3} - \exp\left(\widehat{\omega}\theta\right)\right]\omega \times v + h\omega\theta \\ 0 & 1 \end{bmatrix}. \tag{2.74}$$

Similarly, if $\mathbf{T}_0\left(0\right)$ indicates the initial position and orientation of rigid body 1 relative to the fixed coordinate system 0, the position and orientation of the end effector can be expressed as

$$\mathbf{T}_0(t) = \mathbf{T}_{01}(\theta)\mathbf{T}_0(0). \tag{2.75}$$

If n rigid bodies are connected with the fixed coordinate system in series through joints and are composed of a serial rigid-body system, we can obtain the position and orientation of the end effector.

$$\mathbf{T}_0(t) = \mathbf{T}_{0n}\mathbf{T}_0(0) \tag{2.76}$$

where $\mathbf{T}_{0n} = \prod_{i=1}^{n} \mathbf{T}_{(i-1)i}\left(\theta_i\right)$ and $\mathbf{T}_{(i-1)i}$ indicates the coordinate transform matrix from the ith rigid body to the $(i-1)^{\text{th}}$ rigid body.

2.2.3 Velocity of a Rigid Body

The general motion expressed by equation (2.76) for a rigid body in the fixed coordinate system renders that

$$\mathbf{T}_0(0) = \mathbf{T}_{0n}^{-1}\mathbf{T}_0\left(t\right). \tag{2.77}$$

Differentiating both sides of equation (2.76) with respect to time and using equation (2.77) yields

$$\dot{T}_0(t) = \left(\dot{\mathbf{T}}_{0n}\mathbf{T}_{0n}^{-1}\right)\mathbf{T}_0\left(t\right). \tag{2.78}$$

Let $\widehat{V}_{0n} = \dot{\mathbf{T}}_{0n}\mathbf{T}_{0n}^{-1}, V_{0n}$ indicates the velocity of the end effector of the rigid-body system in the fixed coordinate system.

$$\widehat{\mathbf{V}}_{0n} = \sum_{i=1}^{n}\left(\frac{\partial \mathbf{T}_{0n}}{\partial\theta_i}\mathbf{T}_{0n}^{-1}\dot{\theta}_i\right) \tag{2.79}$$

where

$$\frac{\partial \mathbf{T}_{0n}}{\partial\theta_i}\mathbf{T}_{0n}^{-1} = \left(e^{\hat{\xi}_1\theta_1}e^{\hat{\xi}_2\theta_2}\cdots e^{\hat{\xi}_i\theta_i}\hat{\xi}_1\cdots e^{\hat{\xi}_n\theta_n}\right)$$
$$\left(e^{\hat{\xi}_n\theta_n}e^{\hat{\xi}_{n-1}\theta_{n-1}}\cdots e^{\hat{\xi}_i\theta_i}\cdots e^{-\hat{\xi}_2\theta_2}e^{-\hat{\xi}_1\theta_1}\right)^{-1},$$

and

$$\frac{\partial \mathbf{T}_{0n}}{\partial \theta_i}\mathbf{T}_{0n}^{-1} = \left(e^{\hat{\xi}_1\theta_1}e^{\hat{\xi}_2\theta_2}\ldots e^{\hat{\xi}_{i-1}\theta_{i-1}}\right)\hat{\boldsymbol{\xi}}_i\left(e^{\hat{\xi}_1\theta_1}e^{\hat{\xi}_2\theta_2}\ldots e^{\hat{\xi}_{i-1}\theta_{i-1}}\right)^{-1} \tag{2.80}$$

Equation (2.80) indicates the matrix transform of $\hat{\boldsymbol{\xi}}_i$ in the fixed coordinate system. Let

$$\hat{\boldsymbol{\xi}}_i^0 = \left(e^{\hat{\xi}_1\theta_1}e^{\hat{\xi}_2\theta_2}\ldots e^{\hat{\xi}_{i-1}\theta_{i-1}}\right)\hat{\boldsymbol{\xi}}_i\left(e^{\hat{\xi}_1\theta_1}e^{\hat{\xi}_2\theta_2}\ldots e^{\hat{\xi}_{i-1}\theta_{i-1}}\right)^{-1} \tag{2.81}$$

Equation (2.79) can be rewritten as

$$\hat{V}_{0n} = \hat{\mathbf{J}}\dot{\theta} \tag{2.82}$$

where

$$\hat{J} = \begin{bmatrix}\hat{\boldsymbol{\xi}}_1^0 & \hat{\boldsymbol{\xi}}_2^0 & \cdots & \hat{\boldsymbol{\xi}}_n^0\end{bmatrix},\ \dot{\theta} - \begin{bmatrix}\dot{\theta}_1 & \dot{\theta}_2 & \cdots & \dot{\theta}_n\end{bmatrix}^T.$$

It is noteworthy that $\hat{J}$ is a $4 \times 4n$ partitioned matrix. Therefore, the velocity of the terminal actuator can be expressed as a Plücker column vector.

$$V_{0n} = \left(\hat{V}_{0n}\right)^{\vee} = \mathbf{J}\dot{\boldsymbol{\theta}} \tag{2.83}$$

where $\mathbf{J} = \begin{bmatrix}\boldsymbol{\xi}_1^0 & \boldsymbol{\xi}_2^0 & \cdots & \boldsymbol{\xi}_n^0\end{bmatrix}$. $\mathbf{J}$ is a $6 \times n$ Jacobian matrix and is also called motion screw matrix, whose column vectors are free motion screws of the pairs in the fixed coordinate system. $\boldsymbol{\xi}_i^0\ (i = 1,2,\ldots,n)$ is the Plücker vector of the ith unit motion screw in the fixed coordinate system and $\dot{\theta}_i$ denotes the ith general angular speed.

Equation (2.83) indicates that, in the fixed coordinate system, the velocity screw of the end effector, which belongs to a rigid-body system composed of n general revolute joints in series, can be expressed as a linear combination of all the motion screws in the kinematic chain. The linear coefficients describe the amplitude of the angular speed of the corresponding general rotational joints. Therefore, the motion and constraints of a rigid-body system can be obtained by analyzing the motion screw matrix of the kinematic chain.

2.3 SCREW EXPRESSION OF MOTION AND FORCE

According to Theorem 2.1, there is a one-to-one relationship between a spatial straight line and the specified unit screw. Therefore, any physical quantity relative to lines can be described as screws. A spatial rigid body's instantaneous motion and constraints on it can be expressed as screws.

2.3.1 Motion Screw and Force Screw

When a rigid body rotates around a specified fixed axis, its instantaneous motion can be described by a Plücker homogeneous coordinate equation (2.12).

$$\$_\omega = \boldsymbol{\omega}\begin{bmatrix} \boldsymbol{s} \\ \boldsymbol{s}_0 \end{bmatrix} \tag{2.84}$$

where $\boldsymbol{\omega}$ indicates the instantaneous rotational angular velocity of the rigid body, $\boldsymbol{s}$ indicates the unit direction vector or the unit angular velocity of the screw's axis, $\boldsymbol{s}_0$ indicates the line moment of the axis about the origin of the coordinate system and $\boldsymbol{s}_0 = \boldsymbol{r} \times \boldsymbol{s}$, which in physical meaning is just the linear velocity of the instantaneous intersection of the rigid body and the origin.

For the rigid body's motion in an instantaneous screw, it can be expressed as:

$$\$_\omega = \boldsymbol{\omega}\begin{bmatrix} \boldsymbol{s} \\ \boldsymbol{s}_0 + h_\omega \boldsymbol{s} \end{bmatrix} \tag{2.85}$$

where h_ω is the pitch of the instantaneous motion screw, which indicates the change of rigid body's translational velocity along the axis relative to rotational angular velocity.

Equation (2.85) is a general form of motion screw of a spatial rigid body. Similarly, the forces that a rigid body is subjected to can also be expressed as

$$\$_f = f\begin{bmatrix} \boldsymbol{s} \\ \boldsymbol{s}_0 + h_f \boldsymbol{s} \end{bmatrix} \tag{2.86}$$

where f indicates the instantaneous force that a rigid body subjected to, $\boldsymbol{s}$ indicates the unit direction vector of the force, $\boldsymbol{s}_0 = \boldsymbol{r} \times \boldsymbol{s}$ indicates the static moment of the unit force about the origin of coordinate system along $\boldsymbol{s}$-direction, and h_f is the pitch of the force screw, which indicates the changing ratio of the moment acting on the rigid body along the axis's direction relative to the force.

Equation (2.86) is a general form of force screw acting on a spatial rigid body. When the screw is a pure force, the degenerated screw is

$$\$_f = f\begin{bmatrix} \boldsymbol{s} \\ \boldsymbol{s}_0 \end{bmatrix}. \tag{2.87}$$

2.3.2 Conditions that the Rigid Bodies Keep Balance

This section will discuss a rigid body's balance motion $\$_\omega$ when it is subjected to a force screw $\$_f$. According to equation (2.85), the rigid body's instantaneous motion screw can be denoted by

$$\$_\omega = \omega\$_1 = \begin{bmatrix} \omega\boldsymbol{s}_1 \\ \omega\boldsymbol{s}_{10} + \omega h_\omega \boldsymbol{s}_1 \end{bmatrix}. \tag{2.88}$$

For a rigid body with the instantaneous motion expressed by equation (2.88), the force screw that the rigid body subjected to is

$$\$_f = f\$_2 = \begin{bmatrix} f\boldsymbol{s}_2 \\ f\boldsymbol{s}_{20} + fh_f\boldsymbol{s}_2 \end{bmatrix} \tag{2.89}$$

where $\$_1$ and $\$_2$ are unit screws.

When the rigid body is balanced, the virtual power of the force screw on the motion screw must be zero. The virtual power consists of two parts:

(1) The power of the force on the translational velocity is

$$P_1 = f\boldsymbol{s}_2 \cdot \omega(\boldsymbol{s}_{10} + h_\omega\boldsymbol{s}_1) = f\omega(\boldsymbol{s}_2 \cdot \boldsymbol{s}_{10} + h_\omega\boldsymbol{s}_2 \cdot \boldsymbol{s}_1) \tag{2.90}$$

(2) The power of the twisting moment on the rotational velocity is

$$P_2 = f(\boldsymbol{s}_{20} + h_f\boldsymbol{s}_2) \cdot \omega\boldsymbol{s}_1 = f\omega(\boldsymbol{s}_{20} \cdot \boldsymbol{s}_1 + h_f\boldsymbol{s}_2 \cdot \boldsymbol{s}_1) \tag{2.91}$$

Therefore, the virtual power of the force screw on the motion screw is

$$P = P_1 + P_2 = f\omega[(\boldsymbol{s}_1 \cdot \boldsymbol{s}_{20} + \boldsymbol{s}_2 \cdot \boldsymbol{s}_{10}) + (h_f + h_\omega)\boldsymbol{s}_1 \cdot \boldsymbol{s}_2]. \tag{2.92}$$

As $P = 0$, we have

$$(\boldsymbol{s}_1 \cdot \boldsymbol{s}_{20} + \boldsymbol{s}_2 \cdot \boldsymbol{s}_{10}) + (h_f + h_\omega)\boldsymbol{s}_1 \cdot \boldsymbol{s}_2 = 0. \tag{2.93}$$

Equation (2.93) presents the condition under which a spatial rigid body is balanced. Neither the magnitude of force f nor the angular velocity ω appears in this condition. Therefore, the condition where a rigid body keeps balance is only determined by the unit force screw and the unit motion screw. Thus, the Plücker homogeneous coordinate equations (2.88) and (2.89) can be equivalently expressed as

$$\$_\omega = \begin{bmatrix} \boldsymbol{s}_1 \\ \boldsymbol{s}_{10} + h_\omega\boldsymbol{s}_1 \end{bmatrix} \tag{2.94}$$

$$\$_f = \begin{bmatrix} \boldsymbol{s}_2 \\ \boldsymbol{s}_{20} + h_f\boldsymbol{s}_2 \end{bmatrix}. \tag{2.95}$$

Suppose that

$$\begin{aligned} M_{12} &= \boldsymbol{s}_1 \cdot \left(\boldsymbol{s}_{20} + h_f\boldsymbol{s}_2\right) + \boldsymbol{s}_2 \cdot \left(\boldsymbol{s}_{10} + h_\omega\boldsymbol{s}_1\right) \\ &= \left(\boldsymbol{s}_1 \cdot \boldsymbol{s}_{20} + \boldsymbol{s}_2 \cdot \boldsymbol{s}_{10}\right) + \left(h_\omega + h_f\right) \boldsymbol{s}_1 \cdot \boldsymbol{s}_2 \end{aligned} \tag{2.96}$$

where M_{12} is a scalar and it is called the reciprocal product of screws $\$_\omega$ and $\$_f$. Two screws, whose reciprocal product is zero, are called a pair of reciprocal screws. M_{12} can also be expressed in matrix form as [7]

$$M_{12} = \$_1^T \Delta \$_2 \tag{2.97}$$

where $\$_1 = \$_\omega, \$_2 = \$_f, \Delta = \begin{bmatrix} 0 & I \\ I & 0 \end{bmatrix}$, 0 and I are 3 × zero matrix and unit matrix, individually.

The physical meaning of equation (2.97) is the virtual power of the force screw on the motion screw, so

$$\$_1^T \Delta \$_2 = 0. \tag{2.98}$$

Two screws that satisfy equation (2.98) are called a pair of reciprocal screws. Next we will study the problem of how to simplify Plücker homogeneous coordinates when s_{10} and s_{20} are finite, h_ω and h_f tend to be infinite. According to equation (2.94), we have

$$\lim_{h_\omega\to\infty} \frac{\$_\omega}{h_\omega} = \begin{bmatrix} \lim\limits_{h_\omega\to\infty} \frac{s_1}{h_\omega} \\ \lim\limits_{h_\omega\to\infty} \frac{s_{10}}{h_\omega} + s_1 \end{bmatrix} = \begin{bmatrix} 0 \\ s_1 \end{bmatrix} \tag{2.99}$$

Therefore, when h_ω tends to be infinite, equation (2.94) is degenerated as

$$\$_\omega = \begin{bmatrix} 0 \\ s_1 \end{bmatrix}. \tag{2.100}$$

Similarly, when h_ω tends to be infinite, equation (2.95) degenerates to

$$\$_f = \begin{bmatrix} 0 \\ s_2 \end{bmatrix}. \tag{2.101}$$

Obviously, equation (2.100) indicates the translation of a rigid body along s_1-direction, equation (2.101) indicates the moment the rigid body subjected to is around the s_2-direction.

From equations (2.94), (2.95), (2.99), (2.100), and (2.101), any unit screw can be uniformly expressed by

$$\$ = \begin{bmatrix} s \\ s_0 + hs \end{bmatrix}. \tag{2.102}$$

Definition 2.1. $\$$ is called a *proper screw* when $\|s\| \neq 0$; on the other hand, $\$$ is called an *improper screw* when $\|s\| = 0$ but $\|s_0\| \neq 0$.

According to equation (2.99), no matter how proper or improper a screw is, it can be uniformly expressed with equation (2.102). When the pitch of a screw is zero, equation (2.102) can be simplified as $\$ = \begin{bmatrix} s \\ s_0 \end{bmatrix}$. When the pitch of a screw tends to be infinite, equation (2.102) can be simplified as $\$ = \begin{bmatrix} 0 \\ s \end{bmatrix}$. When the origin of the coordinate system is on the screw's axis, equation (2.102) can be simplified as $\$ = \begin{bmatrix} 0 \\ hs \end{bmatrix}$.

2.4 RECIPROCAL PRODUCT OF SCREWS AND ITS GEOMETRIC MEANING

This section will systematically investigate the reciprocal product of two screws. Assume that the pitches of two unit screws, $\$_1$ and $\$_2$, are h_1 and h_2, the subtended angle of their axes is α. Without loss of generality, we can establish a coordinate system shown in Figure 2.6 where the x-axis is along the axis of $\$_1$, the z-axis is superimposed with the common perpendicular of the axes of $\$_1$ and $\$_2$. Consequently, the y-axis can be determined in accordance with the right-hand rule. Supposing that the length of the common perpendicular of the axes of $\$_1$ and $\$_2$ is denoted by d, and the intersections of the z-axis and the axes of $\$_1$ and $\$_2$ are o_1 and o_2.

In the coordinate system shown in Figure 2.6, these two screws can be expressed as:

$$\begin{cases} \$_1 = \begin{bmatrix} s_1 \\ h_1 s_1 \end{bmatrix} \\ \$_2 = \begin{bmatrix} s_2 \\ d e_3 \times s_2 + h_2 s_2 \end{bmatrix} \end{cases} \tag{2.103}$$

where $e_3 = \begin{bmatrix} 0 & 0 & 1 \end{bmatrix}^T$.

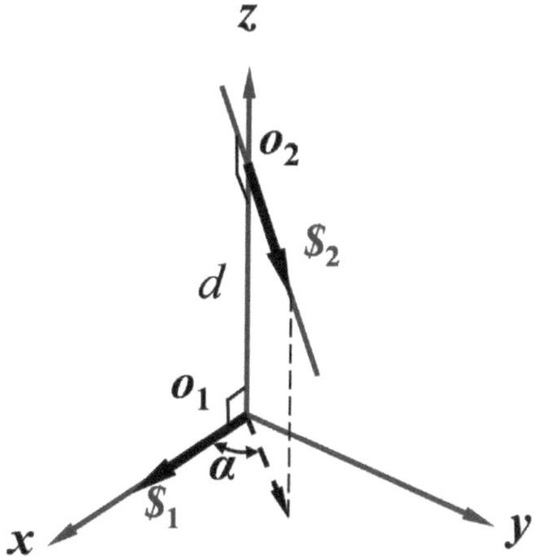

FIGURE 2.6 Reciprocal product of two screws.

From equation (2.97), the reciprocal product of screws $\$_1$ and $\$_2$ is

$$\begin{aligned} M_{12} &= \$_1^T \Delta \$_2 = \boldsymbol{s}_1 \cdot \left(d\boldsymbol{e}_3 \times \boldsymbol{s}_2 + h_2 \boldsymbol{s}_2\right) + h_1 \boldsymbol{s}_1 \cdot \boldsymbol{s}_2 \\ &= \left(h_1 + h_2\right) \boldsymbol{s}_1 \cdot \boldsymbol{s}_2 + d\boldsymbol{s}_1 \cdot \left(\boldsymbol{e}_3 \times \boldsymbol{s}_2\right). \end{aligned} \tag{2.104}$$

From Figure 2.6, the following equations hold

$$\begin{cases} \boldsymbol{s}_1 \cdot \boldsymbol{s}_2 = \cos\alpha \\ \boldsymbol{s}_1 \cdot \left(\boldsymbol{e}_3 \times \boldsymbol{s}_2\right) = -\sin\alpha. \end{cases} \tag{2.105}$$

Substituting equation (2.105) into equation (2.104) yields

$$M_{12} = \$_1^T \Delta \$_2 = \left(h_1 + h_2\right) \cos\alpha - d\sin\alpha. \tag{2.106}$$

M_{12} is called the reciprocal distance between $\$_1$ and $\$_2$. $\frac{1}{2}\left[\left(h_1 + h_2\right) \cos \alpha - d \sin \alpha\right]$ is called the *virtual coefficient* by Ball [4]. According to equation (2.98), if $\$_1$ and $\$_2$ are reciprocal screws, we have

$$\left(h_1 + h_2\right) \cos \alpha - d \sin \alpha = 0. \tag{2.107}$$

If the axes of $\$_1$ and $\$_2$ passes through a same point, then $d = 0$. The conditions under which equation (2.107) holds is

$$h_1 = -h_2 \tag{2.108}$$

or

$$\alpha = \pm\frac{\pi}{2}. \tag{2.109}$$

Now, we can investigate the geometric meaning of the reciprocal product of any two lines. Suppose the coordinates of unit screws of these two straight lines are expressed as

$$\$_1 = \begin{bmatrix} \boldsymbol{s}_1 \\ \boldsymbol{s}_{10} \end{bmatrix} = \begin{bmatrix} \boldsymbol{s}_1 \\ \boldsymbol{r}_1 \times \boldsymbol{s}_1 \end{bmatrix} \tag{2.110}$$

$$\$_2 = \begin{bmatrix} \boldsymbol{s}_2 \\ \boldsymbol{s}_{20} \end{bmatrix} = \begin{bmatrix} \boldsymbol{s}_2 \\ \boldsymbol{r}_2 \times \boldsymbol{s}_2 \end{bmatrix} \tag{2.111}$$

where $\boldsymbol{r}_1$ denotes any point on the line $\$_1$ and $\boldsymbol{r}_2$ denotes any point on the line $\$_2$.

According to equation (2.97), we obtain the reciprocal product of these two straight lines.

$$\begin{aligned} M_{12} &= \boldsymbol{s}_1 \cdot \boldsymbol{s}_{20} + \boldsymbol{s}_2 \cdot \boldsymbol{s}_{10} \\ &= \boldsymbol{s}_1 \cdot \left(\boldsymbol{r}_2 \times \boldsymbol{s}_2\right) + \boldsymbol{s}_2 \cdot \left(\boldsymbol{r}_1 \times \boldsymbol{s}_1\right) \\ &= \boldsymbol{r}_2 \cdot \left(\boldsymbol{s}_2 \times \boldsymbol{s}_1\right) - \boldsymbol{r}_1 \cdot \left(\boldsymbol{s}_2 \times \boldsymbol{s}_1\right) \\ &= \left(\boldsymbol{r}_2 - \boldsymbol{r}_1\right) \cdot \left(\boldsymbol{s}_2 \times \boldsymbol{s}_1\right) \end{aligned} \tag{2.112}$$

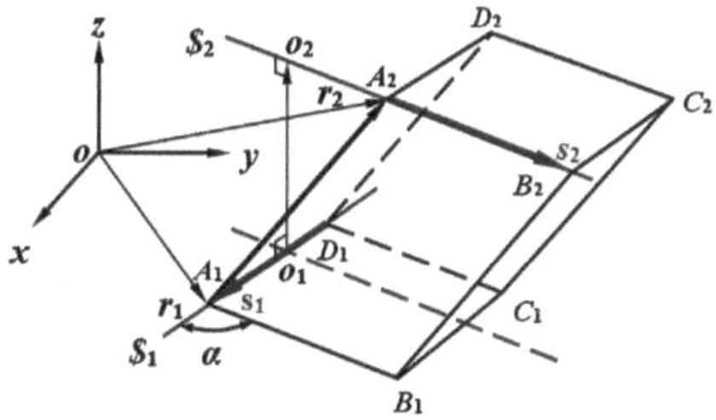

FIGURE 2.7 Geometric representation for the reciprocal product of two straight lines.

As shown in Figure 2.7, the reciprocal product of straight lines $\$_1$ and $\$_2$ indicates the opposite of the volume of hexahedron $A_1B_1C_1D_1 - A_2B_2C_2D_2$ composed of unit vectors $\boldsymbol{s}_1$,$\boldsymbol{r}_2 - \boldsymbol{r}_1$ and $\boldsymbol{s}_2$. Expanding equation (2.112) yields

$$M_{12} = -\left\|\overrightarrow{o_1o_2}\right\| \|s_2\| \|s_1\| \sin\alpha = -d\sin\alpha \tag{2.113}$$

where o_1o_2 is the common perpendicular of $\boldsymbol{s}_1$ and $\boldsymbol{s}_2$,$d = \left\|\overrightarrow{o_1o_2}\right\|$.

Suppose $M_{12} = 0$, from equation (2.113) we can find that $d = 0$ or $\sin\alpha = 0$. When $d = 0$, the straight lines $\$_1$ and $\$_2$ pass through a same point. When $\sin\alpha = 0$, straight lines $\$_1$ and $\$_2$ are parallel or superimposed with each other.

From the above analysis, we can find that if the reciprocal product of two straight lines is equal to zero, these two straight lines must be in a same plane.

2.5 LINEAR COMBINATIONS OF SCREWS AND PRINCIPAL SCREWS OF A SCREW SYSTEM

This section will investigate the linear combinations of screws and the principal screws and principal coordinate system of a screw system. According to the number of independent screws, screw systems can be divided into one-order screw systems, two-order screw systems, three-order screw systems, four-order screw systems, and five-order screw systems. According to the properties of reciprocal screws, four-order screw systems and five-order screw systems can be transformed to two-order screw systems and one-order screw systems to investigate their principal screws [1,8].

Firstly, we can investigate the linear combinations of two-order screws. Suppose that $\$_1$ and $\$_2$ are two independent unit screws, any screw in the two-order screw system can be expressed as a linear combination of $\$_1$ and $\$_2$.

$$\$ = k_1\$_1 + k_2\$_2 \tag{2.114}$$

where k_1 and k_2 are any real numbers that are non-zero simultaneously.

Theorem 2.2. *For any two independent proper screws* $\$_1 = \begin{bmatrix} s_1 \\ s_{01} + h_1 s_1 \end{bmatrix}$ *and* $\$_2 = \begin{bmatrix} s_2 \\ s_{02} + h_2 s_2 \end{bmatrix}$ *in a two-order screw system, when their linear combination equation (2.114) is a real screw, the axis of* $\$$ *must intersect with the common perpendicular of* $\$_1$ and $\$_2$ *at a right angle.*

Proof. Suppose that $\$ = k_1\$_1 + k_2\$_2$, where

$$s_1 \cdot s_\perp = s_2 \cdot s_\perp = 0. \tag{2.115}$$

From equation (2.107) we can find that

$$\$_\perp^T \Delta \$_1 = \$_\perp^T \Delta \$_2 = 0. \tag{2.116}$$

Because $\$$ is a real screw, there should be

$$k_1 s_1 + k_2 s_1 \neq 0.$$

According to equation (2.115), we have

$$\left(k_1 s_1 + k_2 s_1\right) \cdot s_\perp = k_1 s_1 \cdot s_\perp + k_2 s_1 \cdot s_\perp = 0. \tag{2.117}$$

Therefore, the axis of $\$$ is perpendicular to the axis of $\$_\perp$. According to equation (2.116), we have

$$\$_\perp^T \Delta \$ = \$_\perp^T \Delta \left(k_1 \$_1 + k_2 \$_2\right) = k_1 \$_\perp^T \Delta \$_1 + k_2 \$_\perp^T \Delta \$_2 = 0. \tag{2.118}$$

According to equations (2.107) and (2.117), the distance from the axis of $\$$ to the axis of $\$_\perp$ is $d = 0$. In other words, $\$$ and $\$_\perp$ pass through a same point. Therefore, from equations (2.117) and (2.118) we can find that $\$$ must intersect with the common perpendicular of $\$_1$ and $\$_2$ at a right angle when $\$$ is a proper screw.

Theorem 2.3. *Suppose that* $\$_1$ *and* $\$_2$ *are two proper screws which have parallel axes but opposite directions,* $\$ = \$_1 + \$_2 = \mathbf{0}$ *when the axes of* $\$_1$ *and* $\$_2$ *are on a same straight line and* $h_1 = h_2 = h$.

Proof. Because $s_1 = -s_2$,

$$s = s_1 + s_2 = 0 \tag{2.119}$$

When the axes of $\$_1$ and $\$_2$ are on the same straight line,

$$s_{01} = r \times s_1 = r \times \left(-s_2\right) = -s_{02}$$

Because $h_1 = h_2 = h$,

$$s_0 = \left(s_{01} + h_1 s_1\right) + \left(s_{02} + h_2 s_2\right) = \left(s_{01} + s_{02}\right) + h\left(s_1 + s_2\right) = 0 \quad (2.120)$$

From equations (2.119) and (2.120) we can find that

$$\$ = 0$$

Theorem 2.4. *A two-order screw system must be an improper screw system if it is generated by two linearly independent improper screws.*

Proof. Suppose that $\$_1 = \begin{bmatrix} 0 \\ s_1 \end{bmatrix}$ and $\$_2 = \begin{bmatrix} 0 \\ s_2 \end{bmatrix}$ are two linearly independent unit improper screws. Any screw $\$$ generated by $\$_1$ and $\$_2$ can be uniformly expressed as $\$ = k_1\$_1 + k_2\$_2 = \begin{bmatrix} 0 \\ k_1 s_1 + k_2 s_2 \end{bmatrix}$, where k_1,k_2 are non-zero numbers simultaneously. According to Definition 2.1, $\$$ is an improper screw. Therefore, the two-order screw system generated by $\$_1$ and $\$_2$ is improper.

Theorem 2.5. *If there are two independent proper screws whose axes are parallel in a two-order screw system, the whole proper screws in this screw system will be in the plane that is determined by these two screws, and all the improper screws will be perpendicular with the common perpendicular of these two screws.*

Proof. Assume that $\$_1 = \begin{bmatrix} s \\ s_{01} + h_1 s \end{bmatrix}$ and $\$_2 = \begin{bmatrix} s \\ s_{02} + h_2 s \end{bmatrix}$ are two parallel independent proper unit screws in the two-order screw system. Because any screw $\$$ in the screw system can be expressed as

$$\$ = k_1\$_1 + k_2\$_2 = \begin{bmatrix} \left(k_1 + k_2\right) s \\ k_1 s_{01} + k_2 s_{02} + \left(k_1 h_1 + k_2 h_2\right) s \end{bmatrix} \quad (2.121)$$

where k_1 and k_2 are any numbers but not zeros at the same time.

When $k_1 + k_2 \neq 0$, the axis of $\$$ is parallel to both of the axes of $\$_1$ and $\$_2$. According to Theorem 2.2, $\$$ must cross a same point with the common perpendicular of the axes of $\$_1$ and $\$_2$ at a right angle. Therefore, $\$$ must be in the plane that is determined by $\$_1$ and $\$_2$.

Furthermore, the pitch of $\$$ is

$$h = \frac{\left(k_1 + k_2\right) s \cdot \left[k_1 s_{01} + k_2 s_{02} + \left(k_1 h_1 + k_2 h_2\right) s\right]}{\left(k_1 + k_2\right)^2} = \frac{k_1}{k_1 + k_2} h_1 + \frac{k_2}{k_1 + k_2} h_2.$$

If $k_2 = -k_1 \neq 0, k_1 + k_2 = 0$. Equation (2.121) can be simplified as

$$\begin{aligned}\$ = k_1\$_1 + k_2\$_2 &= \begin{bmatrix} 0 \\ k_1\left(s_{01} - s_{02}\right) + k_1\left(h_1 - h_2\right)s \end{bmatrix} \\ &= k_1 \begin{bmatrix} 0 \\ \left(r_1 - r_2\right) \times s + \left(h_1 - h_2\right)s \end{bmatrix}.\end{aligned}$$

Suppose that the unit direction vector of the common perpendicular of the axes of $\$_1$ and $\$_2$ is denoted by $s_{12}^{\perp}$, then we have $s \cdot s_{12}^{\perp} = 0$. Obviously, $\$$ is an improper screw whose direction vector is represented by

$$\begin{aligned}s_0 &= \left(r_1 - r_2\right) \times s + \left(h_1 - h_2\right)s \\ &= \left(ds_{12}^{\perp} + ps\right) \times s + \left(h_1 - h_2\right)s = ds_{12}^{\perp} \times s + \left(h_1 - h_2\right)s\end{aligned}$$

where s_0 represents the direction vector of the combined screw $\$$, d indicates the vertical distance of the axes of $\$_1$ and $\$_2$, p denotes the length of the projection from $r_1 - r_2$ to the direction of the axes of $\$_1$ and $\$_2$.

$$S_0 \cdot s_{12}^{\perp} = \left[ds_{12}^{\perp} \times s + \left(h_1 - h_2\right)s\right] \cdot s_{12}^{\perp} = \left(ds_{12}^{\perp} \times s\right) \cdot s_{12}^{\perp} + \left(h_1 - h_2\right)s \cdot s_{12}^{\perp} = 0$$

Therefore, the whole improper screws in this two-order screw system are perpendicular with the common perpendicular of the axes of $\$_1$ and $\$_2$.

Theorem 2.6. *When the axes of two linearly independent proper screws are not in the same plane, any screw generated by these two screws is a proper screw.*

Proof. Suppose $\$_1 = \begin{bmatrix} s_1 \\ s_{01} + h_1 s \end{bmatrix}$ and $\$_2 = \begin{bmatrix} s_2 \\ s_{02} + h_2 s \end{bmatrix}$ are two linearly independent unit proper screws. Any screw $\$$ generated by $\$_1$ and $\$_2$ can be expressed as

$$\$ = k_1\$_1 + k_2\$_2 = \begin{bmatrix} k_1 s_1 + k_2 s_2 \\ \left(k_1 s_{01} + k_2 s_{02}\right) + \left(k_1 h_1 s_1 + k_2 h_2 s_2\right) \end{bmatrix} \tag{2.122}$$

where k_1 and k_2 are any real numbers but not zeros simultaneously.

Because the axes of $\$_1$ and $\$_2$ are not in the same plane, s_1 and s_2 are linearly independent. Therefore, $k_1 s_1 + k_2 s_2 = 0$ holds if and only if $k_1 = k_2 = 0$. As a matter of fact, k_1 and k_2 are any real numbers but not zeros at the same time, so any screw $\$$ generated by $\$_1$ and $\$_2$ is a proper screw.

Theorem 2.7. *If the axes of two linearly independent proper screws are superimposed, there is only one improper screw in the two-order screw system generated by these two screws.*

Proof. Suppose $\$_1 = \begin{bmatrix} s \\ s_{01} + h_1 s \end{bmatrix}$ and $\$_2 = \begin{bmatrix} s \\ s_{02} + h_2 s \end{bmatrix}$ are two linearly independent unit proper screws whose axes are superimposed. Any screw $\$$ generated by $\$_1$ and $\$_2$ can be expressed as

$$\$ = k_1\$_1 + k_2\$_2 = \begin{bmatrix} (k_1 + k_2)\, s \\ (k_1 + k_2)\, s_0 + (k_1 h_1 + k_2 h_2)\, s \end{bmatrix} \tag{2.123}$$

where k_1 and k_2 are any real numbers and not both zero.

Obviously, the axis of $\$$ expressed by equation (2.123) is superimposed with the axes of $\$_1$ and $\$_2$. When $k_2 = -k_1 \neq 0, k_1 + k_2 = 0$. As a result, equation (2.123) can be simplified as

$$\$ = k_1 (h_1 - h_2) \begin{bmatrix} 0 \\ s \end{bmatrix}. \tag{2.124}$$

Because $\$_1$ and $\$_2$ are linearly independent, $h_1 \neq h_2$. Equation (2.124) indicates that any improper screw generated by $\$_1$ and $\$_2$ is only determined by the unit screw $\$_u = \begin{bmatrix} 0 & s^T \end{bmatrix}^T$, which proves that there is only one improper screw in the two-order screw system generated by $\$_1$ and $\$_2$.

All of the screws in a two-order screw system are located on an irregular spatial surface, called a *cylindroid*. The cylindroid was first found by Hamilton in 1830 and named by Cayley in 1871.

If there is at least one improper screw in a two-order screw system, this screw system is degenerated. In a degenerated screw system, the geometric shape of its cylindroid is simple and the axes of improper screws are infinite in distance. But in screw systems that are not degenerated, the geometry shapes of their cylindroids are more complicated, which will be discussed next.

Theorem 2.8. *If the cylindroid of a two-order system is not degenerated to a plane, there will be only one line which perpendicularly intersects with all of the axes of screws in the cylindroids.*

Proof. Obviously, for two different screws in the cylindroid, there is only one common perpendicular of their axes. According to Theorem 2.2, any screw generated by these two screws must perpendicularly intersect with their common perpendicular. In accordance with the definition of two-order screw systems, any screw in the cylindroid can be expressed as the linear combination of these two specified screws. Consequently, the theorem is proved.

Theorem 2.9. *If the cylindroid of a two-order system is not degenerated to a plane, there must be only two screws which are perpendicular with each other.*

Proof. Suppose that $\$_1$ and $\$_2$ are two screws perpendicular with each other, whose pitches are h_1 and h_2. The length of the common perpendicular of the axes of $\$_1$ and $\$_2$ is denoted by d_{12}. Therefore, any two screws perpendicular to each other in this screw system can be expressed as

$$\begin{cases} \$_1\left(\alpha\right) = \$_1\cos\alpha + \$_2\sin\alpha \\ \$_2\left(\alpha\right) = -\$_1\sin\alpha + \$_2\cos\alpha \end{cases} \tag{2.125}$$

where α is an undetermined constant.

According to equation (2.106), the reciprocal distance of $\$_1(\alpha)$ and $\$_2(\alpha)$ is

$$M_{12}\left(\alpha\right) = -d_\alpha \tag{2.126}$$

where d_α indicates the length of the common perpendicular between $\$_1\left(\alpha\right)$ and $\$_2\left(\alpha\right)$.

From equation (2.126), if $M_{12}(\alpha) = 0$ and $d_\alpha = 0$ hold simultaneously, $\$_1\left(\alpha\right)$ and $\$_2\left(\alpha\right)$ must pass through a same point. Therefore,

$$\begin{aligned} M_{12}\left(\alpha\right) = \$_1^T\left(\alpha\right)\Delta\$_2\left(\alpha\right) = \left(\cos^2\alpha - \sin^2\alpha\right)\$_1^T\Delta\$_2 \\ +\cos\alpha\sin\alpha\left(\$_2^T\Delta\$_2 - \$_1^T\Delta\$_1\right). \end{aligned} \tag{2.127}$$

Because the axes of $\$_1$ and $\$_2$ are perpendicular, from equation (2.106) the following equation holds

$$\$_1^T\Delta\$_2 = -d_{12} \tag{2.128}$$

where d_{12} denotes the length of the common perpendicular between the axes of $\$_1$ and $\$_2$.

Because of $h_1 = \frac{1}{2}\$_1^T\Delta\$_1$ and $h_2 = \frac{1}{2}\$_2\Delta\$_2$, equation (2.127) can be also simplified as

$$M_{12}\left(k_1,k_2\right) = -d_{12}\cos 2\alpha + \left(h_2 - h_1\right)\sin 2\alpha. \tag{2.129}$$

When $M_{12}(k_1,k_2) = 0$, equation (2.129) results in

$$d_{12}\cos 2\alpha + \left(h_1 - h_2\right)\sin 2\alpha = 0. \tag{2.130}$$

When $h_2 = h_1$,

$$\alpha = \frac{\pi}{4} \quad (2.131)$$

Hence, $\$_1(\alpha)$ and $\$_2(\alpha)$ are perpendicular with each other.
When $h_2 \neq h_1, \alpha \neq \frac{\pi}{4}$, equation (2.130) can be further simplified as

$$\tan 2\alpha = \frac{d_{12}}{h_2 - h_1}. \quad (2.132)$$

Solving this trigonometric equation (2.132) yields

$$\alpha = \frac{1}{2}\arctan\frac{d_{12}}{h_2 - h_1}. \quad (2.133)$$

Equations (2.131) and (2.133) indicate that there are only two screws which are perpendicular to each other. Consequently, Theorem 2.9 is proved.

These two unique screws perpendicular with each other in the cylindroids are called the principal screws in the two-order screw system [1,8]. When the cylindroid is degenerated to a plane, any two screws perpendicular with each other in this plane can be selected as the principal screws.

For the sake of convenience, assume that the unique principal screws in the cylindroid are selected as x-axis and y-axis. The z-axis is determined by the right-hand rule. Suppose $\$_1$ and $\$_2$ are perpendicular unit principal screws, s_1 and s_2 are the unit direction vectors of $\$_1$ and $\$_2$, x-axis and y-axis are superimposed with the axes of $\$_1$ and $\$_2$. Such a right-hand coordinate system is called principal coordinate system. In the principal coordinate system, any unit screw in the two-order screw system is expressed as

$$\$ = \$_1\cos\psi + \$_2\sin\psi, \psi \in [0, 2\pi). \quad (2.134)$$

When the principal screws are known, $\$$ will be completely dependent to ψ. As is shown in Figure 2.8, ψ denotes the subtended angle of the axes of $\$_1$ and $\$_2$. Suppose that $o_1\begin{pmatrix}0 & 0 & z\end{pmatrix}$ denotes the coordinates of the intersection of $\$$ and z-axis, Point A denotes the terminal of the screw vector $\$$. Point A_P is the projection of point A on the principal plane xoy.

According to Theorem 2.9, we can construct a screw $\$'$ perpendicular to the axis of $\$$.

$$\$' = -\$_1\sin\psi + \$_2\cos\psi \quad (2.135)$$

According to Theorem 2.2, $\$'$ is also perpendicular to the z-axis. Suppose that $o_2\begin{pmatrix}0 & 0 & z'\end{pmatrix}$ denotes the coordinates of the intersection of $\$'$ and the z-axis, Point B denotes the terminal of the screw vector $\$'$. Point B_P is the projection of point B on the principal plane xoy. Therefore, in the principal

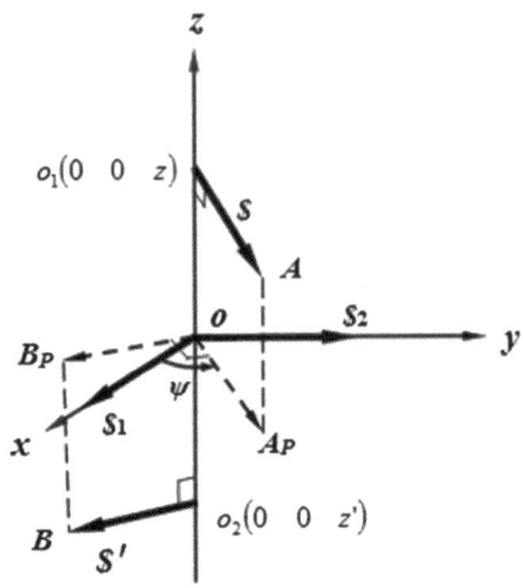

FIGURE 2.8 A screw system generated by principal screws.

coordinate system, the two principle screws are

$$\begin{cases} \$_1 = \begin{bmatrix} \boldsymbol{e}_1 \\ h_1\boldsymbol{e}_1 \end{bmatrix} \\ \$_2 = \begin{bmatrix} \boldsymbol{e}_2 \\ h_2\boldsymbol{e}_2 \end{bmatrix} \end{cases}. \tag{2.136}$$

Consequently, $\$$ and $\$'$ can be expressed as

$$\$ = \begin{bmatrix} \boldsymbol{e}_1\cos\psi + \boldsymbol{e}_2\sin\psi \\ h_1\cos\psi\boldsymbol{e}_1 + h_2\sin\psi\boldsymbol{e}_2 \end{bmatrix} \tag{2.137}$$

$$\$' = \begin{bmatrix} -\boldsymbol{e}_1\sin\psi + \boldsymbol{e}_2\cos\psi \\ -h_1\boldsymbol{e}_1\sin\psi + h_2\boldsymbol{e}_2\cos\psi \end{bmatrix}. \tag{2.138}$$

Because $\$$ intersects with z-axis at $\begin{pmatrix} 0 & 0 & z \end{pmatrix}$, $\$$ can be also expressed as

$$\$ = \begin{bmatrix} \boldsymbol{e}_1\cos\psi + \boldsymbol{e}_2\sin\psi \\ z\boldsymbol{e}_3 \times (\boldsymbol{e}_1\cos\psi + \boldsymbol{e}_2\sin\psi) + h(\boldsymbol{e}_1\cos\psi + \boldsymbol{e}_2\sin\psi) \end{bmatrix} \tag{2.139}$$

where h is the pitch of $\$$.

Similarly, $\$'$ can also be expressed as

$$\$' = \begin{bmatrix} -\boldsymbol{e}_1\sin\psi + \boldsymbol{e}_2\cos\psi \\ z'\boldsymbol{e}_3 \times (-\boldsymbol{e}_1\sin\psi + \boldsymbol{e}_2\cos\psi) + h'(-\boldsymbol{e}_1\sin\psi + \boldsymbol{e}_2\cos\psi) \end{bmatrix} \tag{2.140}$$

where h' is the pitch of $\boldsymbol{\$}'$.

Associating equations (2.137) and (2.139) presents

$$h_1\cos\psi\boldsymbol{e}_1 + h_2\sin\psi\boldsymbol{e}_2 = z\boldsymbol{e}_3 \times \left(\boldsymbol{e}_1\cos\psi + \boldsymbol{e}_2\sin\psi\right) + h\left(\boldsymbol{e}_1\cos\psi + \boldsymbol{e}_2\sin\psi\right) \tag{2.141}$$

Expanding equation (2.141) yields

$$\begin{aligned} & h_1\cos\psi\boldsymbol{e}_1 + h_2\sin\psi\boldsymbol{e}_2 \\ & = z\left(\boldsymbol{e}_3 \times \boldsymbol{e}_1\cos\psi + \boldsymbol{e}_3 \times \boldsymbol{e}_2\sin\psi\right) + h\left(\boldsymbol{e}_1\cos\psi + \boldsymbol{e}_2\sin\psi\right) \\ & = z\left(\boldsymbol{e}_2\cos\psi - \boldsymbol{e}_1\sin\psi\right) + h\left(\boldsymbol{e}_1\cos\psi + \boldsymbol{e}_2\sin\psi\right) \\ & = \boldsymbol{e}_1\left(-z\sin\psi + h\cos\psi\right) + \boldsymbol{e}_2\left(z\cos\psi + h\sin\psi\right). \end{aligned} \tag{2.142}$$

Therefore, we can obtain the following

$$\begin{cases} h_1 = h - z\tan\psi \\ h_2 = h + z\cot\psi. \end{cases} \tag{2.143}$$

According to equations (2.138) and (2.140) we have

$$\begin{aligned} -h_1\sin\psi\boldsymbol{e}_1 + h_2\cos\psi\boldsymbol{e}_2 = {} & z'\boldsymbol{e}_3 \times \left(-\boldsymbol{e}_1\sin\psi + \boldsymbol{e}_2\cos\psi\right) \\ & + h'\left(-\boldsymbol{e}_1\sin\psi + \boldsymbol{e}_2\cos\psi\right). \end{aligned} \tag{2.144}$$

Substituting equation (2.143) into equation (2.144) yields

$$\begin{aligned} & -\left(h - z\tan\psi\right)\sin\psi\boldsymbol{e}_1 + \left(h + z\cot\psi\right)\cos\psi\boldsymbol{e}_2 \\ & = z'\left(-\boldsymbol{e}_2\sin\psi - \boldsymbol{e}_1\cos\psi\right) + h'\left(-\boldsymbol{e}_1\sin\psi + \boldsymbol{e}_2\cos\psi\right). \end{aligned} \tag{2.145}$$

Dot multiplying $\boldsymbol{e}_1$ on both sides of equation (2.145) yields

$$\left(z\tan\psi - h\right)\sin\psi = -z'\cos\psi - h'\sin\psi. \tag{2.146}$$

Dot multiplying $\boldsymbol{e}_2$ on both sides of equation (2.145) yields

$$\left(h + z\cot\psi\right)\cos\psi = -z'\sin\psi + h'\cos\psi. \tag{2.147}$$

Associating equations (2.146) and (2.147) yields

$$\begin{cases} z'\cos\psi + h'\sin\psi = -\left(z\tan\psi - h\right)\sin\psi \\ z'\sin\psi - h'\cos\psi = -\left(z\cot\psi + h\right)\cos\psi. \end{cases} \tag{2.148}$$

Solving equation set (2.148) yields

$$\begin{cases} z' = -z \\ h' = 2z\cot 2\psi + h. \end{cases} \tag{2.149}$$

From equation (2.106) we have

$$\begin{cases} d = z - z' = 2z \\ \$^T \Delta \$' = (h + h')\cos\frac{\pi}{2} - 2z\sin\frac{\pi}{2} = -2z. \end{cases} \tag{2.150}$$

According to equations (2.137) and (2.138), we can also obtain

$$\$^T \Delta \$' = (\$_1\cos\psi + \$_2\sin\psi)\,\Delta\,(-\$_1\sin\psi + \$_2\cos\psi) = (h_2 - h_1)\sin 2\psi \tag{2.151}$$

where $h_1 = \frac{1}{2}\$_1^T \Delta \$_1, h_2 = \frac{1}{2}\$_2^T \Delta \$_2$ and $\$_1^T \Delta \$_2 = \$_2^T \Delta \$_1 = 0$ are utilized.

Associating equations (2.150) and (2.151) yields

$$z = \frac{1}{2}(h_1 - h_2)\sin 2\psi. \tag{2.152}$$

Similarly, we have

$$h = \frac{1}{2}\$^T \Delta \$ = h_1\cos^2\psi + h_2\sin^2\psi = \frac{1}{2}(h_1 + h_2) + \frac{1}{2}(h_1 - h_2)\cos 2\psi. \tag{2.153}$$

Associating equations (2.152) and (2.153) and replacing ψ, we obtain the circle expression of the degenerated two-order system [1,8].

$$z^2 + \left(h - \frac{h_1 + h_2}{2}\right) = \left(\frac{h_1 - h_2}{2}\right)^2 \tag{2.154}$$

As shown in Figure 2.9, equation (2.154) can be expressed as a circle in h–z plane. When $h_1 \neq h_2$, any point on the circle represents a unique unit screw in the two-order system. When $h_1 = h_2$, the circle in h–z plane is degenerated to a point. At the same time, the cylindroid is degenerated to a plane. According to the h–z circle expression of the two-order system, it is not difficult to find that

(1) The pitches of principal screws are the extremes.

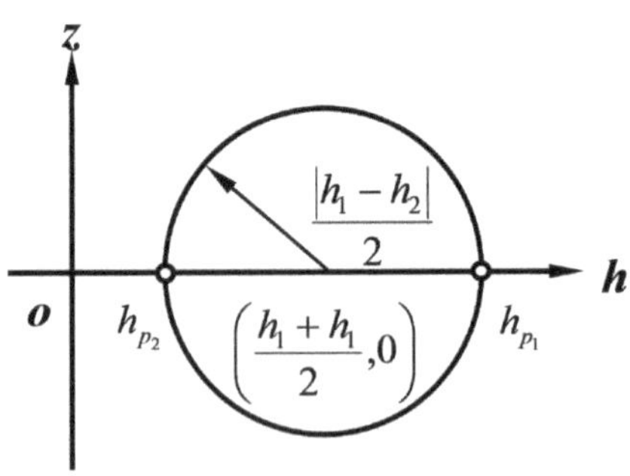

FIGURE 2.9 h–z circle expression of two-order system.

(2) The distribution of the cylindroid along the z-axis is $|h_1 - h_2|$.
(3) Plane xoy bisects the distribution of the cylindroid along the z-axis.

Equation (2.152) can be transformed into

$$z - (h_1 - h_2)\sin\psi\cos\psi = 0. \tag{2.155}$$

From Figure 2.8, ψ can be expressed by the Cartesian coordinates of the end point of the corresponding screw in the cylindroid.

$$\begin{cases} \cos\psi = \frac{x}{\sqrt{x^2+y^2}} \\ \sin\psi = \frac{y}{\sqrt{x^2+y^2}} \end{cases} \tag{2.156}$$

Substituting equation (2.156) into equation (2.155) yields

$$z\left(x^2 + y^2\right) - (h_1 - h_2)\,xy = 0. \tag{2.157}$$

When $z = 0$, equation (2.157) can be simplified as $x = 0$ or $y = 0$ which represents the two principal coordinate axes. When $z \neq 0$, equation (2.157) can be expressed as

$$\left[x - \frac{(h_1 - h_2) + \sqrt{(h_1 - h_2)^2 - 4z^2}}{2z}y\right]\left[x - \frac{(h_1 - h_2) - \sqrt{(h_1 - h_2)^2 - 4z^2}}{2z}y\right] = 0. \tag{2.158}$$

Equation (2.158) indicates that the cylindroid intersects with any plane that is perpendicular to the z-axis at two straight lines, which passes through a same point in the z-axis.

$$x - \frac{(h_1 - h_2) + \sqrt{(h_1 - h_2)^2 - 4z^2}}{2z}y = 0 \text{ and}$$
$$x - \frac{(h_1 - h_2) - \sqrt{(h_1 - h_2)^2 - 4z^2}}{2z}y = 0$$

Therefore, the cylindroid of a two-order system is a plane with straight lines, which can be illustrated by Figure 2.10 in Cartesian coordinate system. According to the shape of the cylindroid shown in Figure 2.10, we can see (a) that the lengths of the screw lines represent the magnitudes of pitches, (b) the front view of cylindroid along the minus x-direction, (c) the top view of cylindroids along the z-direction (where $h_1 < h_2$), which actually constitutes a polar coordinate figure where the pitches represent the polar radii, and (d) the left-side view of cylindroid along y-direction.

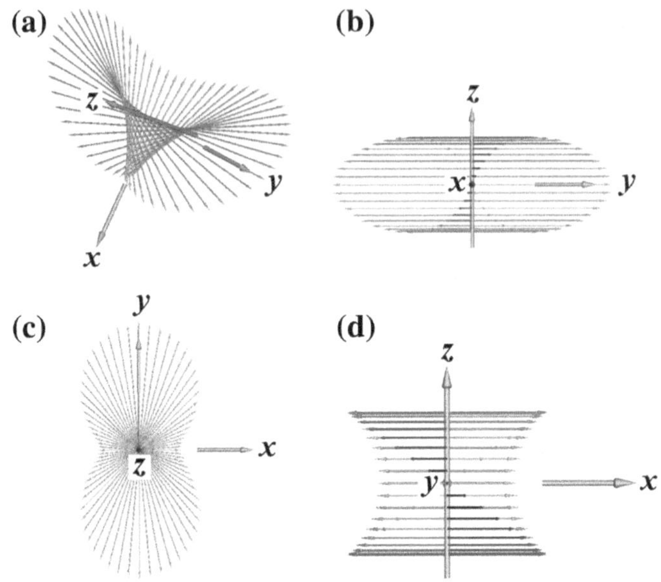

FIGURE 2.10 Three-dimensional expression of cylindroid. (a) Shape of cylindroid in Cartesian coordinate system; (b) front view of cylindroid; (c) top view of cylindroid; (d) left-side view of cylindroid.

From the discussion above we can deduce that the principal coordinates and principal coordinate system of a two-order screw system can completely determine the whole screw system. Consequently, the next section will focus on the problem of how to identify the principal coordinates and coordinate system in a screw system. After Ball [8], many scholars [1,8–15] investigated the methodology of identifying the principal screws of a screw system. This chapter provides a methodology which can identify both proper and improper screws [16,17].

2.6 IDENTIFICATION OF PRINCIPAL SCREWS OF A SCREW SYSTEM

Any screw in a screw system can be expressed as a linear combination of a set of maximal linearly independent screws of the system [1–4]. Therefore, the maximal linearly independent screws in a screw system can be utilized to completely represent the system. Theoretically, any set of such independent screws is enough to determine the screw system. For a specified screw system, the maximum and minimum pitches of the screws present the explicit characteristics of the system. Traditionally, the principal screws of a screw system are defined as the screws whose pitches are the extremes and equal the order of the system in number. As a matter of fact, when the principal screws of a system are known, the characteristics of the screw system and its principal coordinate system are also completely and solely determined. In applications, the kinematic screws of a series robot form a specified screw system which is determined by the structure parameters of the links and joints of the robot. The principal screws of the system determine the kinematic characteristics, such as the mobility, the singularity and the instantaneous motion stability within its reachable workspace, and so on. The same is true for the parallel manipulators. Consequently, in order to completely understand a screw system, we have to identify the principal screws from the screw system.

Ball first systematically investigated screw theory in his treatise initially published in 1900 and reprinted in 1998 by Lipkin and Duffy [4]. Dai [18] presented a comprehensive review of the theoretical development of rigid body displacements from Rodrigues parameters to the finite twist in an annalistic form. Hunt [1], Phillips [2,3], Gibson and Hunt [8,9], Rico and Duffy [10–12], Dai and Jones [19], *et al.*, made great contributions to the contemporary development of screw theory and its applications after Ball [4]. For the first-order screw system, any screw in the system can be selected as the principal screw. As to the second order screw system, the principal screws are the screws with minimum and maximum pitches, and the identification procedure is also relatively simple [1]. However, it is complicated to identify the principal screws of the third-order system [1,8–12,14,15]. Ball [4] expressed the third-order screw system by planar conics, and the three principal screws can be obtained when the conics are degenerated. Gibson and Hunt [8,9] classified one-order, two-order and three-order screw systems by means of projective geometry. However, Rico and Duffy [10–12] doubt the completeness of these classifications. Alternatively, they provided a consistent representation of screw systems in terms of an orthonormal basis of the orthogonal subspaces associated with the screw systems under consideration. Tsai and Lee [13] presented a method to search the principal screws of a screw system based on the reduced echelon form of the screw matrix.

As is discussed above, the simple fact is that a screw system is completely determined by any set of maximal linearly independent screws in the system. Therefore, it is reasonable to expect that the principal screws and their pitches be obtained by simple mathematical processing. This section presents an algebraic methodology to identify the principal screws and their pitches [16,17]. According to reciprocal screw theory, the identifications of principal screws and principal pitches of a system whose order is larger than three can be equivalently transformed into the identifications of those of its reciprocal screw system whose order is less than three [1]. Consequently, this section only discusses the applications to the second and third-order screw systems after the theoretical analysis of this method.

2.6.1 Representations of a Specified Screw System

The homogeneous coordinates of an improper screw can be expressed by equation (2.102) if an infinity pitch $h = H$ is introduced. Similarly, we can

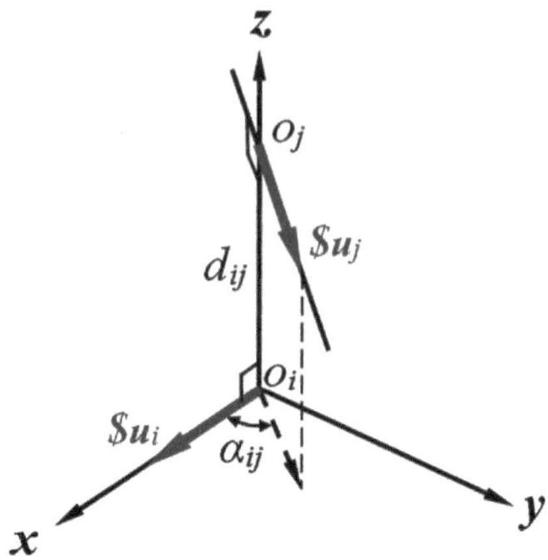

FIGURE 2.11 Coordinate system of two specified screws.

introduce two infinity pitches H_1 and H_2 for a system that contains two improper screws, and introduce three infinity pitches H_1, H_2, and H_3 that contains three improper screws. As a result, the screws that have the proper form of equation (2.102) can be uniformly discussed.

Assume that the pitches of two unit screws, $\$_{u_i}$ and $\$_{u_j}$, are h_i and h_j, the subtended angle of their axes is α_{ij}. Without loss of generality, we can establish a coordinate system shown in Figure 2.11 where the x-axis is along the axis of $\$_{u_i}$, the z-axis is superimposed with the common perpendicular of the axes of $\$_{u_i}$ and $\$_{u_j}$. Consequently, the y-axis can be determined in accordance with the right-hand rule. Supposing that the length of the common perpendicular of the axes of $\$_{u_i}$ and $\$_{u_j}$ is d_{ij}, and the intersections of the z-axis and the axes of $\$_{u_i}$ and $\$_{u_j}$ are o_i and o_j, individually, we can get the following coordinates for these two screws:

$$\begin{cases} \$_{u_i} = \begin{bmatrix} \boldsymbol{s}_i \\ h_i \boldsymbol{s}_i \end{bmatrix} \\ \$_{u_j} = \begin{bmatrix} \boldsymbol{s}_j \\ d_{ij} \boldsymbol{e}_3 \times \boldsymbol{s}_j + h_j \boldsymbol{s}_j \end{bmatrix} \end{cases} \tag{2.159}$$

where $\boldsymbol{e}_3 = \begin{bmatrix} 0 & 0 & 1 \end{bmatrix}^T$.

The reciprocal product of screws $\$_{u_i}$ and $\$_{u_j}$ is

$$\begin{aligned} M_{ij} &= \$_{u_i}^T \Delta \$_{u_j} = \boldsymbol{s}_i \cdot \left(d_{ij} \boldsymbol{e}_3 \times \boldsymbol{s}_j + h_j \boldsymbol{s}_j \right) + h_i \boldsymbol{s}_i \cdot \boldsymbol{s}_j \\ &= \left(h_i + h_j \right) \boldsymbol{s}_i \cdot \boldsymbol{s}_j + d_{ij} \boldsymbol{s}_i \cdot \left(\boldsymbol{e}_3 \times \boldsymbol{s}_j \right). \end{aligned} \tag{2.160}$$

From Figure 2.11, the following equations hold

$$\begin{cases} \boldsymbol{s}_i \cdot \boldsymbol{s}_j = \cos\alpha_{ij} \\ \boldsymbol{s}_i \cdot \left(\boldsymbol{e}_3 \times \boldsymbol{s}_j \right) = -\sin\alpha_{ij} \end{cases}. \tag{2.161}$$

Substituting equation (2.161) into equation (2.160) yields

$$M_{ij} = \left(h_i + h_j\right)\cos\alpha_{ij} - d_{ij}\sin\alpha_{ij}. \tag{2.162}$$

When $i = j$, the subtended angle, α_{ij}, will be zero. Therefore, equation (2.162) can be simplified as

$$M_{ii} = 2h_i \tag{2.163}$$

where $i = 1,2,\ldots$ Equation (2.162) can be simplified as the following form when $\alpha_{ij} = -\frac{\pi}{2}$:

$$M_{ij} = d_{ij}. \tag{2.164}$$

Therefore, it is not difficult to find from equation (2.163) that the pitch of a unit screw, $\$_{u_i}$ $\left(i = 1,2,\ldots\right)$, can also be expressed as

$$h_i = \frac{1}{2}M_{ii} = \frac{1}{2}\$_{u_i}^T \Delta \$_{u_i}. \tag{2.165}$$

For simplicity in what follows, the screws all mean unit ones if there are not special notes. Suppose a screw matrix, $\mathbf{A}_{6\times n}$, consists of n independent screws, that is, $A_{6\times n} = \begin{bmatrix} \$_1 & \$_2 & \cdots & \$_n \end{bmatrix}, n = 1,2,\ldots,6$. Then, Rank $(\mathbf{A}) = n$. Therefore, we can define the reciprocal product matrix, $\mathbf{M}_A$, to be:

$$\mathbf{M}_A = \mathbf{A}_{6\times n}^T \boldsymbol{\Delta}\mathbf{A}_{6\times n} = \begin{bmatrix} M_{11} & M_{12} & \cdots & M_{1n} \\ M_{21} & M_{22} & \cdots & M_{2n} \\ & & \cdots & \\ M_{n1} & M_{n2} & \cdots & M_{nn} \end{bmatrix} \tag{2.166}$$

where $n = 1,2,\ldots,6$, but $n = 1,2,3$ in this chapter. Obviously, $\mathbf{M_A}$ is a real symmetric matrix.

2.6.2 Matrix Representation of the Pitch of a Screw

Assume that a screw matrix, $\mathbf{A} = \begin{bmatrix} \$_1 & \cdots & \$_n \end{bmatrix}$, consists of n independent unit screws. The column of the screw matrix is denoted by $\$_i = \begin{bmatrix} \mathbf{s}_i \\ \mathbf{s}_{0i} \end{bmatrix}, i = 1,2,\ldots,n$ where $\mathbf{s}_i$ is a unit vector, n represents the order of the screw system and $1 \le n \le 3$. Obviously, the screw matrix, $\mathbf{A}$, completely determines an n-screw system.

Definition 2.2. If the axes of $n(n = 2,3)$ screws denoted by the columns of matrix, $\mathbf{A}$, are mutually perpendicular; $\mathbf{A}$ is called axis-perpendicular screw matrix.

Theorem 2.10. *For any specified n-proper-screw system, we can always find at least one set of n screws whose axes are mutually perpendicular.*

Proof. For a prescribed n-proper-screw system, any set of n screws can be utilized to determine the screw system. According to the definition of a unit screw, we can suppose without loss of generality that there are n independent unit screws pertaining to the specified n-proper-screw system:

$$\left\{ \$_1 \quad \cdots \quad \$_n \right\}$$

Therefore, any screw of this n-system can be expressed by the linear combinations of $\$_1 \quad \cdots \quad \$_n$. Specifically, a Gram–Schmidt orthogonalization process on the set of screws does not alter the properties that the combined screws still belong to the same n-system. As a result, we can always find a set of axis-perpendicular screws for any specified proper screw system.

For a three-proper-screw system, suppose that the generating screw matrix of the system is $\mathbf{A} = \left[\$_1 \quad \$_2 \quad \$_3 \right]$, we can have the following orthogonalization process:

Step 1: Let $\$_1'' = \$_1$.
Step 2: Let $\$_2' = \$_2 - \left(s_2 \cdot s_1'' \right) \$_1''$.
Step 3: Let $\$_2'' = \frac{\$_2'}{\|s_2'\|}$.
Step 4: Let $\$_3' = \$_3 - \left(s_3 \cdot s_1'' \right) \$_1'' - \left(s_3 \cdot s_2'' \right) \$_2''$.
Step 5: Let $\$_3'' = \frac{\$_3'}{\|s_3'\|}$.

Obviously, $\$_1''$,$\$_2''$, and $\$_3''$ are three independent unit screws that belong to the screw system determined by $\mathbf{A} = \left[\$_1 \quad \$_2 \quad \$_3 \right]$. What is more, $\$_1''$,$\$_2''$, and $\$_3''$ form an axis-perpendicular screw matrix which is denoted by $\mathbf{A}_\perp = \left[\$_1'' \quad \$_2'' \quad \$_3'' \right]$.

The screw system determined by $\mathbf{A} = \left[\$_1 \quad \$_2 \quad \$_3 \right]$ can be equivalently established by $\mathbf{A}_\perp = \left[\$_1'' \quad \$_2'' \quad \$_3'' \right]$. Therefore, it is explicit that we can always find a set of axis-perpendicular unit screws which are equivalent in generating the screw system for any prescribed set of independent screws. As a result, in the following part of this chapter, a screw matrix always means one whose columns are all axis-perpendicular unit screws. The reciprocal product matrix of $\mathbf{A}_\perp$ is easily obtained with equations (2.164) and (2.166)

$$\mathbf{M}_A = \mathbf{A}_\perp^T \Delta \mathbf{A}_\perp = \begin{bmatrix} M_{11} & -d_{12}\sin\alpha_{12} & -d_{13}\sin\alpha_{13} \\ -d_{12}\sin\alpha_{12} & M_{22} & -d_{23}\sin\alpha_{23} \\ -d_{13}\sin\alpha_{13} & -d_{23}\sin\alpha_{23} & M_{33} \end{bmatrix} \tag{2.167}$$

where d_{ij} represents the length of common perpendicular of the ith and j^{th} screw axes in $\mathbf{A}_\perp$, and α_{ij} represents the subtended angle between the ith and j^{th} screw axes in $\mathbf{A}_\perp$.

Theorem 2.11. *If the columns of a screw matrix,* $\mathbf{A}$*, are all axis-perpendicular unit screws, for a set of real numbers* $-1 \le k_1, k_2, \ldots, k_n \le 1$ *that form an n-dimensional vector* $\mathbf{k} = \begin{pmatrix} k_1 & k_2 & \cdots & k_n \end{pmatrix}^T$*, the screw*

$$\$_A = \mathbf{A}\mathbf{k} \tag{2.168}$$

must be a unit screw so long as $\sum_{i=1}^{n} k_i^2 = 1$.

Proof. Because the columns of the screw matrix, $\mathbf{A}$, are all axis-perpendicular unit screws, we can rewrite $\mathbf{A}$ in the following form

$$\mathbf{A} = \begin{bmatrix} \mathbf{A}_1 \\ \mathbf{A}_2 \end{bmatrix} \tag{2.169}$$

where $\mathbf{A}_1 = \begin{bmatrix} \boldsymbol{s}_1 & \cdots & \boldsymbol{s}_n \end{bmatrix}, \mathbf{A}_2 = \begin{bmatrix} \boldsymbol{s}_{10} & \cdots & \boldsymbol{s}_{n0} \end{bmatrix}$ and $\boldsymbol{s}_i \cdot \boldsymbol{s}_j = \begin{cases} 0, i \neq j \\ 1, i - j \end{cases}, n = 2,3$.

$$\boldsymbol{s}_\mathbf{A} \cdot \boldsymbol{s}_\mathbf{A} = (A_1\boldsymbol{k})^T \mathbf{A}_1\boldsymbol{k} = \boldsymbol{k}^T \left(\mathbf{A}_1^T \mathbf{A}_1\right)\boldsymbol{k} = \boldsymbol{k}^T \mathbf{I}_{n\times n}\boldsymbol{k} = \sum_{i=1}^{n} k_i^2 = 1$$

where $\mathbf{I}_{n\times n}$ is an n-dimension identity matrix.

Therefore, $\$_A = \mathbf{A}\boldsymbol{k}$ is also a unit screw. Generally, the pitch of a unit screw expressed by equation (2.167) can be expressed in the form of equation (2.165).

$$h = \frac{1}{2}\$_A^T \Delta \$_A = \frac{1}{2}\boldsymbol{k}^T \left(\mathbf{A}^T \Delta \mathbf{A}\right)\boldsymbol{k} \tag{2.170}$$

where h is called a principal pitch when equation (2.169) gets to the extremum according to the definition of the principal screws of a screw system [1–4].

Substituting equation (2.166) into equation (2.169) yields

$$h = \frac{1}{2}\boldsymbol{k}^T \mathbf{M}_\mathbf{A}\boldsymbol{k}.$$

Consequently, the pitch of any screw of a screw system can be represented by a homogeneous quadratic form of a real vector $\boldsymbol{k} = \begin{pmatrix} k_1 & k_2 & \cdots & k_n \end{pmatrix}^T$ for $n \le 3$, the quadratic matrix of which is the reciprocal product matrix, $\mathbf{M}_\mathbf{A}$.

2.6.3 Identifications of the Principal Pitches and Principal Screws

Theorem 2.12. *The reciprocal product matrix,* $\mathbf{M}_\mathbf{A}$*, can be diagonalized and there is an orthogonal transformation matrix,* $\mathbf{Q}$*, such that*

$$\mathbf{Q}^T \mathbf{M}_\mathbf{A} \mathbf{Q} = \boldsymbol{\Lambda}$$

where $\mathbf{\Lambda} = \mathrm{diag}\left(2h_{p1}, 2h_{p2}, \ldots, 2h_{pn}\right)$, *and* h_{p_i} $\left(i = 1,2,\ldots,n\right)$ *denotes the* i^{th} *principal pitch of the screw system. The principal screw matrix of the system can be expressed as*

$$\mathbf{A}_p = \mathbf{AQ} \tag{2.171}$$

where $\mathbf{A}_P$ *is a matrix composed of the principal screws,* $\$_{p1}, \$_{p2}, \ldots, \$_{pn}$, *corresponding to the principal pitches,* $h_{p1}, h_{p2}, \ldots, h_{pn}$.

Proof. From equation (2.166), we can find that

$$\mathbf{M}_{\mathbf{A}}^T = \mathbf{M}_{\mathbf{A}}$$

and

$$\mathrm{Rank}\left(\mathbf{M}_{\mathbf{A}}\right) = \mathrm{Rank}(\mathbf{A}) = n.$$

Therefore, $\mathbf{M}_{\mathbf{A}}$ has $n(n \leq 3)$ non-zero eigenvalues and $\mathbf{M}_{\mathbf{A}}$ can be diagonalized according to the knowledge of linear algebra. Assume that the ith $\left(i = 1,2,\ldots,n\right)$ eigenvalue of $\mathbf{M}_{\mathbf{A}}$ is denoted by $2h_{p_i}$ and the eigenvector corresponding to h_{p_i} is represented by $\boldsymbol{\xi}_i$. The following equations must hold:

$$\begin{cases} M_A \boldsymbol{\xi}_i = 2h_{p_i} \boldsymbol{\xi}_i \\ \boldsymbol{\xi}_i \cdot \boldsymbol{\xi}_j = \delta_{ij} \end{cases}$$

where $\delta_{ij} = \begin{cases} 0, i \neq j \\ 1, i = j \end{cases}, \quad i,j = 1,2,\ldots,n$ for $n \leq 3$.

Therefore, the following equation holds:

$$\mathbf{M}_{\mathbf{A}}\mathbf{Q} = \mathbf{Q}\mathbf{\Lambda} \tag{2.172}$$

where $\mathbf{Q} = \begin{bmatrix} \boldsymbol{\xi}_1 & \boldsymbol{\xi}_2 & \cdots & \boldsymbol{\xi}_n \end{bmatrix}, \mathbf{\Lambda} = \mathrm{diag}\left(2h_{p1}, 2h_{p2}, \ldots, 2h_{pn}\right)$.

Because $\mathbf{Q}^T\mathbf{Q} = \mathbf{I}$, then $\mathbf{Q}^{-1} = \mathbf{Q}^T$. Equation (2.172) can be equivalently denoted by

$$\mathbf{Q}^T\mathbf{M}_{\mathbf{A}}\mathbf{Q} = \mathbf{\Lambda}. \tag{2.173}$$

Suppose

$$\boldsymbol{k} = \mathbf{Qc} \tag{2.174}$$

where $\mathbf{c} = \begin{pmatrix} c_1 & c_2 & \cdots & c_n \end{pmatrix}, c_i \in \mathrm{R}$, and $\sum_{i=1}^{n} c_i^2 = 1$.

Substituting equation (2.174) into equation (2.170) and applying the result of equation (2.173) yield

$$h = \frac{1}{2}\boldsymbol{c}^T\left(\mathbf{Q}^T\mathbf{M}_{\mathbf{A}}\mathbf{Q}\right)\boldsymbol{c} = \frac{1}{2}\boldsymbol{c}^T\mathbf{\Lambda}\boldsymbol{c} = \sum_{i=1}^{n} h_{p_i}c_i^2. \tag{2.175}$$

Of course, from equation (2.175), we can find that the pitches of the screw system that satisfy

$$\min\left\{h_{p_1} \quad h_{p_2} \quad \cdots \quad h_{p_n}\right\} \leq h_A \leq \max\left\{h_{p_1} \quad h_{p_2} \quad \cdots \quad h_{p_n}\right\}$$

where $\max\left\{h_{p_1} \quad h_{p_2} \quad \cdots \quad h_{p_n}\right\}$ represents the maximum value of the pitch, and $\min\left\{h_{p_1} \quad h_{p_2} \quad \cdots \quad h_{p_n}\right\}$ represents the minimum value of the pitch. According to the definition of the principal screws of a screw system [1,2], $\left\{h_{p_1} \quad h_{p_2} \quad \cdots \quad h_{p_n}\right\}$ are the principal pitches of the screw system for $n \leq 3$.

Replacing $\mathbf{M}_A$ with $A^T \Delta A$ in accordance with equation (2.166), we can transform equation (2.173) into

$$\mathbf{Q}^T \mathbf{A}^T \boldsymbol{\Delta} \mathbf{A} \mathbf{Q} = \boldsymbol{\Lambda}. \tag{2.176}$$

Let

$$\mathbf{A}_p = \left[\$_{p_1} \quad \cdots \quad \$_{p_n}\right] = \mathbf{AQ}.$$

Equation (2.176) can be transformed to

$$\mathbf{A}_p^T \boldsymbol{\Delta} \mathbf{A}_p = \boldsymbol{\Lambda}. \tag{2.177}$$

According to the definition of the principal screws of a screw system, the principal screws of a proper screw system should perpendicularly pass through one point. Equation (2.177) indicates that $\mathbf{A}_p$ is a matrix that consists of the principal screws because each of the non-diagonal elements, denoted by $\boldsymbol{\Lambda}(ij)$, of the reciprocal matrix $\boldsymbol{\Lambda}$, which indicates that the right distance between the ith and the jth screw, is zero from equation (2.164).

2.6.4 Principal Screws of Two-system

The identification of principal screws and their pitches for the two-system is relatively simple. However, in order to verify the methodology proposed in this chapter, two cases of two-systems will still be discussed in this section.

Suppose that there are two independent unit proper screws, $\$_1$ and $\$_2$, shown in Figure 2.12. $\$_1$ passes through the origin of the Cartesian coordinate system and $\$_2$ passes through point $A\left(-4\sqrt{3} \quad 4 \quad 0\right)$ and is perpendicular to both the axes of $\$_1$ and the dashed line oA. The pitches of $\$_1$ and $\$_2$ are $h_1 = 3$ and $h_2 = 4$, respectively. Assume $\alpha = \frac{\pi}{3}, \beta = \gamma = \frac{\pi}{4}$. In the Cartesian coordinate system $oxyz$ shown in Figure 2.12, the Plücker coordinates of the unit screw, $\$_1$, can be expressed as:

$$\$_1 = \left[\tfrac{\sqrt{2}}{4} \ \tfrac{\sqrt{6}}{4} \ \tfrac{\sqrt{2}}{2} \ \tfrac{3\sqrt{2}}{4} \ \tfrac{3\sqrt{6}}{4} \ \tfrac{3\sqrt{2}}{2}\right]^T.$$

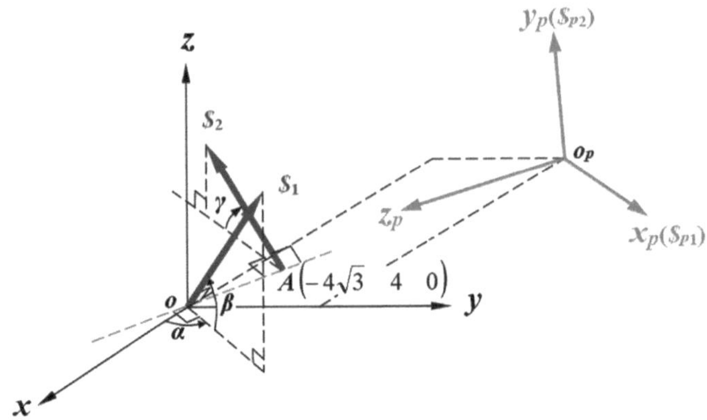

FIGURE 2.12 Principal screws of a two-system.

The Plücker coordinates of the unit screw, $\$_2$, can be expressed as:

$$\$_2 = \left[-\tfrac{\sqrt{2}}{4}\ -\tfrac{\sqrt{6}}{4}\ \tfrac{\sqrt{2}}{2}\ \sqrt{2}\ \sqrt{6}\ 6\sqrt{2} \right]^T.$$

Because

$$s_1 \cdot s_2 = \frac{\sqrt{2}}{4} \times \left(-\frac{\sqrt{2}}{4}\right) + \frac{\sqrt{6}}{4} \times \left(-\frac{\sqrt{6}}{4}\right) + \frac{\sqrt{2}}{2} \times \frac{\sqrt{2}}{2} = 0,$$

the axes of $\$_1$ and $\$_2$ are perpendicular. Therefore, one need not apply the Gram–Schmidt orthogonal process and can directly obtain an axis-perpendicular matrix, $\mathbf{A} = \left[\$_1\ \ \$_2 \right]$. The reciprocal product matrix of $\mathbf{A}$ is easily obtained with equation (2.166):

$$\mathbf{M_A} = \mathbf{A}^T \mathbf{\Delta} \mathbf{A} = \begin{bmatrix} 6 & 8 \\ 8 & 8 \end{bmatrix}. \tag{2.178}$$

The eigenvalues of $\mathbf{M_A}$ can be gained by solving the eigenvalue polynomial equation below:

$$\left| \mathbf{M_A} - 2h_p \mathbf{I}_{2\times 2} \right| = 0 \tag{2.179}$$

where $\mathbf{I}_{2\times 2}$ is a 2×2 identity matrix. Expanding equation (2.179) yields

$$h_p^2 - 7h_p - 4 = 0. \tag{2.180}$$

The two roots of equation (2.180) are $h_{p_1} = \frac{1}{2}\left(7 - \sqrt{65}\right)$ and $h_{p_2} = \frac{1}{2}\left(7 + \sqrt{65}\right)$. Suppose that the eigenvectors of matrix $\mathbf{M_A}$ corresponding to eigenvalues h_{p_1} and h_{p_2} are x_1 and x_2, there will be

$$\mathbf{M_A} \boldsymbol{x}_1 = 2h_{p_1} \boldsymbol{x}_1.$$

Therefore, we have

$$\left(\mathbf{M_A} - 2h_{p1} I\right)\boldsymbol{x}_1 = 0. \tag{2.181}$$

Substituting $h_{p_1} = \frac{1}{2}\left(7 - \sqrt{65}\right)$ into equation (2.181) yields

$$\begin{bmatrix} -1+\sqrt{65} & 8 \\ 8 & 1+\sqrt{65} \end{bmatrix}\begin{bmatrix} x_{11} \\ x_{12} \end{bmatrix} = 0 \tag{2.182}$$

The basic solution of equation (2.182) is

$$\boldsymbol{x}_1 = \begin{bmatrix} x_{11} \\ x_{12} \end{bmatrix} = \begin{bmatrix} \frac{8}{\sqrt{130-2\sqrt{65}}} \\ \frac{1-\sqrt{65}}{\sqrt{130-2\sqrt{65}}} \end{bmatrix}.$$

Similarly, we can find out the eigenvector x_2 corresponding to $h_{p_2} = 7+\sqrt{65}$:

$$\boldsymbol{x}_2 = \begin{bmatrix} x_{21} \\ x_{22} \end{bmatrix} = \begin{bmatrix} \frac{8}{\sqrt{130+2\sqrt{65}}} \\ \frac{1+\sqrt{65}}{\sqrt{130+2\sqrt{65}}} \end{bmatrix}.$$

Therefore, we obtain $\mathbf{Q}$ from equations (2.181) and (2.182):

$$Q = \begin{bmatrix} \boldsymbol{x}_1 & \boldsymbol{x}_2 \end{bmatrix} = \begin{bmatrix} \frac{8}{\sqrt{130-2\sqrt{65}}} & \frac{8}{\sqrt{130+2\sqrt{65}}} \\ \frac{1-\sqrt{65}}{\sqrt{130-2\sqrt{65}}} & \frac{1+\sqrt{65}}{\sqrt{130+2\sqrt{65}}} \end{bmatrix}. \tag{2.183}$$

Consequently, the principal screws of the system determined by $\$_1$ and $\$_2$ can be obtained by substituting equation (2.183) into equation (2.171):

$$\begin{bmatrix} \$_{p1} & \$_{p2} \end{bmatrix} = \mathbf{A}_{6\times n}\mathbf{Q}_{n\times n} = \begin{bmatrix} \frac{\sqrt{2}}{4} & -\frac{\sqrt{2}}{4} \\ \frac{\sqrt{6}}{4} & -\frac{\sqrt{6}}{4} \\ \frac{\sqrt{2}}{2} & \frac{\sqrt{2}}{2} \\ \frac{3\sqrt{2}}{4} & \sqrt{2} \\ \frac{3\sqrt{6}}{4} & \sqrt{6} \\ \frac{3\sqrt{2}}{2} & 6\sqrt{2} \end{bmatrix} \begin{bmatrix} \frac{8}{\sqrt{130-2\sqrt{65}}} & \frac{8}{\sqrt{130+2\sqrt{65}}} \\ \frac{1-\sqrt{65}}{\sqrt{130-2\sqrt{65}}} & \frac{1+\sqrt{65}}{\sqrt{130+2\sqrt{65}}} \end{bmatrix}$$

$$= \begin{bmatrix} 0.499034 & -0.031068 \\ 0.864352 & -0.053813 \\ 0.0621374 & 0.998068 \\ -0.140776 & 1.762153 \\ -0.243832 & 3.052138 \\ -4.025273 & 7.765126 \end{bmatrix}.$$

$\boldsymbol{\$}_{p_1}$ and $\boldsymbol{\$}_{p_2}$ can be denoted by their unit forms:

$$\boldsymbol{\$}_{p_1} = \begin{bmatrix} \boldsymbol{s}_{p_1} \\ \boldsymbol{r}_1 \times \boldsymbol{s}_{p_1} + h_{p_1}\boldsymbol{s}_{p_1} \end{bmatrix}, \boldsymbol{\$}_{p_2} = \begin{bmatrix} \boldsymbol{s}_{p_2} \\ \boldsymbol{r}_2 \times \boldsymbol{s}_{p_2} + h_{p_2}\boldsymbol{s}_{p_2} \end{bmatrix}.$$

Because $\boldsymbol{s}_{p_1} \cdot (\boldsymbol{r}_1 \times \boldsymbol{s}_{p_1} + h_{p_1}\boldsymbol{s}_{p_1}) = \boldsymbol{s}_{p_1} \cdot (\boldsymbol{r}_1 \times \boldsymbol{s}_{p_1}) + h_{p_1}(\boldsymbol{s}_{p_1} \cdot \boldsymbol{s}_{p_1}) = h_{p_1}$, it is not difficult to find that $h_{p_1} = \boldsymbol{s}_{p_1} \cdot (\boldsymbol{r}_1 \times \boldsymbol{s}_{p_1} + h_1\boldsymbol{s}_{p_1}) = \frac{1}{2}\left(7 - \sqrt{65}\right) = -0.531129$. Similarly, we can find that $h_{p_2} = \boldsymbol{s}_{p_2} \cdot (\boldsymbol{r}_2 \times \boldsymbol{s}_{p_2} + h_{p_2}\boldsymbol{s}_{p_2}) = \frac{1}{2}\left(7 + \sqrt{65}\right) = 7.531129$ and $\boldsymbol{s}_{p_1} \cdot \boldsymbol{s}_{p_2} = 0$. Therefore, the axes of the two principal screws cross a same point at a right angle. This common cross point can be solved:

$$\begin{pmatrix} x_{OP} \\ y_{OP} \\ z_{OP} \end{pmatrix} = \begin{pmatrix} -20.099934 \\ 11.604702 \\ 0.000000 \end{pmatrix}. \tag{2.184}$$

Equation (2.184) indicates that the two principal screws of the system really cross orthogonally at the same point. Substituting equation (2.178) into equation (2.170) yields

$$h = \frac{1}{2}\begin{pmatrix} k_1 & k_2 \end{pmatrix} \begin{bmatrix} 6 & 8 \\ 8 & 8 \end{bmatrix} \begin{pmatrix} k_1 \\ k_2 \end{pmatrix} = 3k_1^2 + 8k_1k_2 + 4k_2^2. \tag{2.185}$$

Equation (2.185) represents a quadratic surface in the $ok_1k_2h_A$-space, which is shown in Figure 2.13. The intersections of any h-plane and the surface form conic curves, which are roughly shown by the contours in the k_1ok_2-plane.

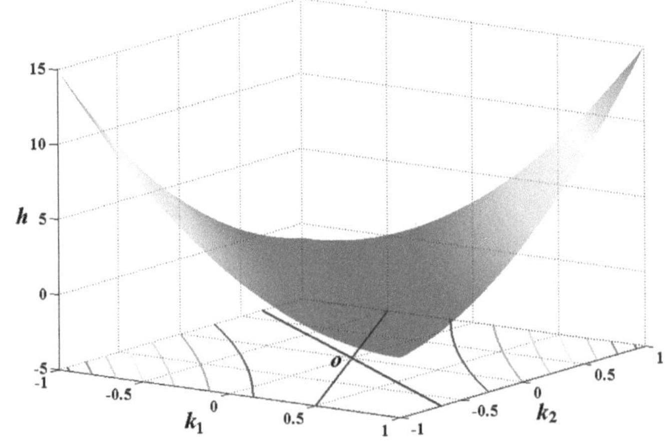

FIGURE 2.13 The h distribution with respect to k_1,k_2 and the contours in k_1ok_2-plane.

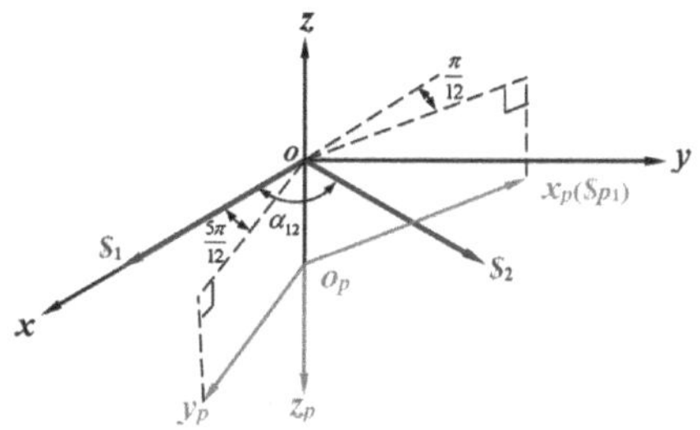

FIGURE 2.14 Principal screws of another two-system.

However, the principal directions are all along the axes of the two principal screws. The extremum values of h under the condition $k_1^2 + k_2^2 = 1$ are the principal pitches.

Now, we can investigate another special two-system. Suppose that there are two independent unit proper screws, $\$_1$ and $\$_2$, which pass through one point and the subtended angle of their axes is $\alpha_{12} = \frac{\pi}{3}$. Their pitches are $h_1 = 5$ and $h_2 = 6$, respectively. For convenience, we can let the origin of the Cartesian coordinate system superimpose with the cross-point of these two screws, let the axis of one screw be along the x-direction, and let the z-axis be perpendicular to the plane determined by these two screws, which is shown in Figure 2.14; the Plücker coordinates of the unit screws, $\$_1$ and $\$_2$, can be expressed as:

$$\begin{cases} \$_1 = \begin{bmatrix} 1 & 0 & 0 & 5 & 0 & 0 \end{bmatrix}^T \\ \$_2 = \begin{bmatrix} \cos\frac{\pi}{3} & \sin\frac{\pi}{3} & 0 & 6\cos\frac{\pi}{3} & 6\sin\frac{\pi}{3} & 0 \end{bmatrix}^T \end{cases}.$$

Because $\boldsymbol{s}_1 \cdot \boldsymbol{s}_2 = \cos\alpha_{12} = \cos\frac{\pi}{3} = \frac{1}{2} \neq 0$, the axes of $\$_1$ and $\$_2$ are not perpendicular. One needs to apply the Gram–Schmidt orthogonal process to obtain an axis-perpendicular screw matrix.

Step 1: Let $\$_1'' = \$_1 = \begin{bmatrix} 1\,0\,0\,5\,0\,0 \end{bmatrix}^T$.

Step 2: Let $\$_2' = \$_2 - \left(\boldsymbol{s}_2 \cdot \boldsymbol{s}_1''\right)\$_1'' = \begin{bmatrix} 0\ \frac{\sqrt{3}}{2}\ 0\ \frac{1}{2}\ 3\sqrt{3}\ 0 \end{bmatrix}^T$.

Step 3: Let $\$_2'' = \frac{\$_2'}{\|\boldsymbol{s}_2'\|} = \begin{bmatrix} 0\,1\,0\ \frac{\sqrt{3}}{3}\ 6\,0 \end{bmatrix}^T$.

Obviously, $\$_1''$ and $\$_2''$ are two independent unit screws belonging to the screw system determined by $\$_1$ and $\$_2$.

Let

$$\mathbf{A} = \begin{bmatrix} \$_1'' & \$_2'' \end{bmatrix} = \begin{bmatrix} 1 & 0 & 0 & 5 & 0 & 0 \\ 0 & 1 & 0 & \frac{\sqrt{3}}{3} & 6 & 0 \end{bmatrix}^T.$$

The reciprocal product matrix of **A** is easily obtained with equation (2.166)

$$\mathbf{M_A} = \mathbf{A}^T \boldsymbol{\Delta} \mathbf{A} = \begin{bmatrix} 10 & \frac{\sqrt{3}}{3} \\ \frac{\sqrt{3}}{3} & 12 \end{bmatrix}. \tag{2.186}$$

The eigenvalues of $\mathbf{M_A}$ can be solved by the following eigenvalue polynomial:

$$\left| \mathbf{M_A} - 2h_p \mathbf{I}_{2\times 2} \right| = 0 \tag{2.187}$$

where $\mathbf{I}_{2\times 2}$ is a 2×2 identity matrix. Expanding equation (2.187) yields

$$3h_p^2 - 66h_p + 359 = 0. \tag{2.188}$$

The two roots of equation (2.188) are $h_{p_1} = \frac{1}{6}\left(33 - 2\sqrt{3}\right) = 4.922650$ and $h_{p_2} = \frac{1}{6}\left(33 + 2\sqrt{3}\right) = 6.077350$. Therefore, we obtain the orthogonal transformation matrix **Q**:

$$\mathbf{Q} = \begin{bmatrix} -0.965926 & 0.258819 \\ 0.258819 & 0.965926 \end{bmatrix}.$$

Therefore, the principal screws of the system determined by $\$_1$ and $\$_2$ can be expressed with equation (2.171):

$$\begin{bmatrix} \$_{p_1} & \$_{p_2} \end{bmatrix} = \mathbf{AQ} = \begin{bmatrix} -0.965926 & 0.258819 \\ 0.258819 & 0.965926 \\ 0 & 0 \\ -4.680200 & 1.851773 \\ 1.552914 & 5.795555 \\ 0 & 0 \end{bmatrix}.$$

It is not difficult to verify that $h_{p1} = 4.922650$. Similarly, we can also verify that $h_{p2} = 6.077350$ and $\mathbf{s}_{p_1} \cdot \mathbf{s}_{p_2} = 0$. The axes of the two principal screws cross a same point at right angle. This common cross point can be obtained:

$$\begin{pmatrix} x_{OP} \\ y_{OP} \\ z_{OP} \end{pmatrix} = \begin{pmatrix} 0 \\ 0 \\ -0.288675 \end{pmatrix}.$$

Substituting equation (2.186) into equation (2.170) yields

$$h_A = \frac{1}{2}\begin{pmatrix} k_1 & k_2 \end{pmatrix} \begin{bmatrix} 10 & \frac{\sqrt{3}}{3} \\ \frac{\sqrt{3}}{3} & 12 \end{bmatrix} \begin{pmatrix} k_1 \\ k_2 \end{pmatrix} = 5k_1^2 + \frac{\sqrt{3}}{3}k_1 k_2 + 6k_2^2. \tag{2.189}$$

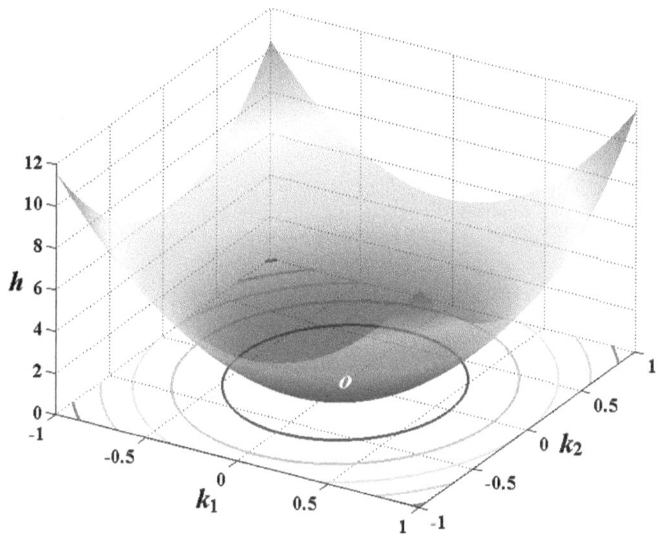

FIGURE 2.15 The h distribution with respect to k_1,k_2 and the contours in k_1ok_2-plane.

Equation (2.189) represents a quadratic surface in the $ok_1k_2h_A$ space, which is shown in Figure 2.15. The intersections of each h-plane and the surface form ellipse curves, which are shown by the contours in the k_1ok_2-plane.

However, the principal directions are all along the axes of the two principal screws. The extremum values of h under the condition $k_1^2 + k_2^2 = 1$ are the principal pitches.

2.6.5 Principal Screws of Three-system

The third screw systems are the most complicated. To illustrate their particular properties, Hunt [1] classified them into 10 special cases. This chapter discusses the identification problems of the third-order screw systems with the reciprocal screw matrix method proposed in this chapter.

Suppose that a third-order screw system consists of three independent unit proper screws:

$$\begin{cases} \$_1 = \left[\frac{\sqrt{2}}{2} \ 0 \ \frac{\sqrt{2}}{2} \ \sqrt{2} \ 0 \ \sqrt{2}\right]^T \\ \$_2 = \left[1\,0\,0\,3\,0\,0\right]^T \\ \$_3 = \left[0 \ \frac{\sqrt{3}}{2} \ \frac{1}{2} \ 0 \ 2\sqrt{3} \ 2\right]^T. \end{cases}$$

Because the axes of the three screws are not mutually perpendicular, we can follow the following orthogonalization process:

Step 1: Let $\$_1'' = \$_1 = \left[\frac{\sqrt{2}}{2} \ 0 \ \frac{\sqrt{2}}{2} \ \sqrt{2} \ 0 \ \sqrt{2}\right]^T$.

Step 2: Let $\$_2' = \$_2 - \left(s_2 \cdot s_1''\right)\$_1'' = \left[\frac{1}{2}\ 0\ -\frac{1}{2}\ 2\ 0\ -1\right]^T$.

Step 3: Let $\$_2'' = \frac{\$_2'}{\|s_2'\|} = \left[\frac{\sqrt{2}}{2}\ 0\ -\frac{\sqrt{2}}{2}\ 2\sqrt{2}\ 0\ -\sqrt{2}\right]^T$.

Step 4: Let $\$_3' = \$_3 - \left(s_3 \cdot s_1''\right)\$_1'' - \left(s_3 \cdot s_2''\right)\$_2'' = \left[0\ \frac{\sqrt{3}}{2}\ 0\ \frac{1}{2}\ 2\sqrt{3}\ 1\right]^T$.

Step 5: Let $\$_3'' = \frac{\$_3'}{\|s_3'\|} = \left[0\ 1\ 0\ \frac{\sqrt{3}}{3}\ 4\ \frac{2\sqrt{3}}{3}\right]^T$.

Obviously, $\$_1''$,$\$_2''$ and $\$_3''$ are three independent unit screws that pertain to the screw system determined by $\$_1$,$\$_2$ and $\$_3$.

Let

$$\mathbf{A} = \left[\$_1''\ \ \$_2''\ \ \$_3''\right] = \begin{bmatrix} \frac{\sqrt{2}}{2} & \frac{\sqrt{2}}{2} & 0 \\ 0 & 0 & 1 \\ \frac{\sqrt{2}}{2} & -\frac{\sqrt{2}}{2} & 0 \\ \sqrt{2} & 2\sqrt{2} & \frac{\sqrt{3}}{3} \\ 0 & 0 & 4 \\ \sqrt{2} & -\sqrt{2} & \frac{2\sqrt{3}}{3} \end{bmatrix}.$$

$\mathbf{M_A}$ can be obtained in accordance with equation (2.166)

$$\mathbf{M_A} = \mathbf{A}^T \boldsymbol{\Delta} \mathbf{A} = \begin{bmatrix} 4 & 1 & \frac{\sqrt{6}}{2} \\ 1 & 6 & -\frac{\sqrt{6}}{6} \\ \frac{\sqrt{6}}{2} & -\frac{\sqrt{6}}{6} & 8 \end{bmatrix}. \tag{2.190}$$

The eigenvalues of $\mathbf{M_A}$ can be gained by solving the following eigenvalue polynomial equation:

$$\left|\mathbf{M_A} - 2h_p \mathbf{I}_{3\times3}\right| = 0 \tag{2.191}$$

where $\mathbf{I}_{3\times3}$ is a 3×3 identity matrix. Expanding equation (2.191) yields

$$3h_p^3 - 27h_p^2 + 76h_p - 65 = 0. \tag{2.192}$$

The three roots of equation (2.192) are $h_{P1} = 1.618702$, $h_{P2} = 3.205183$ and $h_{P3} = 4.176116$, individually. The orthogonal transformation matrix, $\mathbf{Q}$, made up of the eigenvectors corresponding to these three eigenvalues is

$$\mathbf{Q} = \begin{bmatrix} -0.894673 & 0.364586 & 0.258142 \\ 0.362443 & 0.930222 & -0.057633 \\ 0.261142 & -0.041999 & 0.964386 \end{bmatrix}.$$

Therefore, the principal screws of the system determined by $\$_1$,$\$_2$, and $\$_3$ can be expressed with equation (2.171)

$$\$_p = \mathbf{A}_{6\times3}\mathbf{Q}_{3\times3}$$
$$= \begin{bmatrix} \frac{\sqrt{2}}{2} & \frac{\sqrt{2}}{2} & 0 \\ 0 & 0 & 1 \\ \frac{\sqrt{2}}{2} & -\frac{\sqrt{2}}{2} & 0 \\ \sqrt{2} & 2\sqrt{2} & \frac{\sqrt{3}}{3} \\ 0 & 0 & 4 \\ \sqrt{2} & -\sqrt{2} & \frac{2\sqrt{3}}{3} \end{bmatrix} \begin{bmatrix} -0.894673 & 0.364586 & 0.258142 \\ 0.362443 & 0.930222 & -0.057633 \\ 0.261142 & -0.041999 & 0.964386 \end{bmatrix}.$$

The ultimate result of $\$_p$ is

$$\$_p = \mathbf{A}_{6\times3}\mathbf{Q}_{3\times3} = \begin{bmatrix} 0.141781 & 0.915567 & -0.376344 \\ 0.964386 & -0.041999 & 0.261142 \\ 0.223287 & -0.399966 & -0.888915 \\ 0.758845 & 3.122420 & -0.089345 \\ 3.857545 & -0.167995 & 1.044567 \\ 1.560151 & -0.848427 & -1.476290 \end{bmatrix}.$$

It is not difficult to verify that the axes of the three principal screws really cross the same point at right angles. This common cross-point can be expressed as:

$$\begin{pmatrix} x_{OP} \\ y_{OP} \\ z_{OP} \end{pmatrix} = \begin{pmatrix} 0.577350 \\ -0.500000 \\ -0.288675 \end{pmatrix}.$$

Substituting equation (2.190) into equation (2.170) yields

$$h = \frac{1}{2}\begin{pmatrix} k_1 & k_2 & k_3 \end{pmatrix} \begin{bmatrix} 4 & 1 & \frac{\sqrt{6}}{2} \\ 1 & 6 & -\frac{\sqrt{6}}{6} \\ \frac{\sqrt{6}}{2} & -\frac{\sqrt{6}}{6} & 8 \end{bmatrix} \begin{pmatrix} k_1 \\ k_2 \\ k_3 \end{pmatrix}$$
$$= 2k_1^2 + 3k_2^2 + 4k_3^2 + 2k_1k_2 + \frac{\sqrt{6}}{2}k_1k_3 - \frac{\sqrt{6}}{6}k_2k_3.$$

The principal directions are clearly represented by Figure 2.16. The extremum values of h under the condition $k_1^2 + k_2^2 + k_3^2 = 1$ are just the principal pitches.

2.7 CONCLUSIONS

This chapter briefly introduces screw algebra. The definitions and primary concepts of screws in constraint and motion analysis for rigid body systems are discussed. The screw matrix of a kinematic chain represents the free

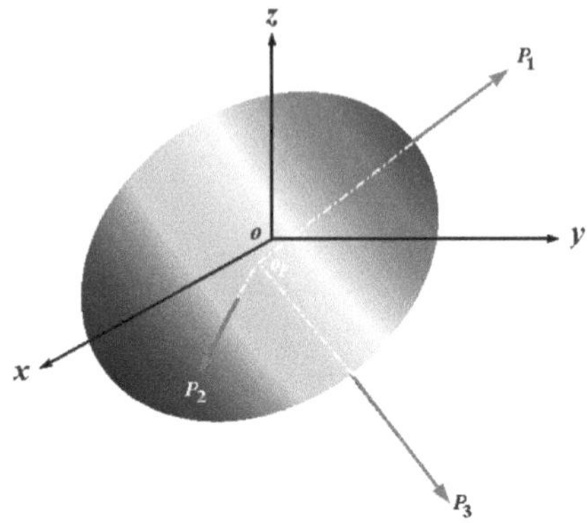

FIGURE 2.16 The rough representations of the principal screws by pitch ellipsoid.

motions of the chain and its reciprocal screws give a full expression of the terminal constraints. The reciprocal relationship between the motion screw and constraint screws provides a good means to analyze and synthesize a spatial mechanism. For a screw system, the principal screws determine the mechanical characteristics of the system. Therefore, identification of principal screws and their pitches is also investigated with linear algebra.

REFERENCES

[1] K.H. Hunt, Kinematic Geometry of Mechanisms, Oxford University Press, Oxford, 1978.
[2] J. Phillips, Freedom in machinery, Introducing Screw Theory, vol. 1, Cambridge University Press, Cambridge, 1984.
[3] J. Phillips, Freedom in machinery, Screw Theory Exemplified, vol. 2, Cambridge University Press, Cambridge, 1990.
[4] R.S. Ball, A treatise on the theory of screws, Cambridge University Press, 1900.
[5] O. Bottema, B. Roth, Theoretical Kinematics, North-Holland Publishing Company, New York, 1979.
[6] R.M. Murray, Z.X. Li, S.S. Sastry, A Mathematical Introduction to Robotic Manipulation, CRC Press, Boca Raton, 1994.
[7] V. Kumar, Instantaneous kinematics of parallel-chain robotic mechanisms, ASME Journal of Mechanical Design 114 (1992) 349–358.
[8] C.G. Gibson, K.H. Hunt, Geometry of screw systems:1 Screws: genesis and geometry, Mechanism and Machine Theory 25 (1990) 1–10.
[9] C.G. Gibson, K.H. Hunt, Geometry of screw systems:2 Classification of screw systems, Mechanism and Machine Theory 25 (1) (1990) 1–27.
[10] Martínez J.M. Rico, J. Duffy, Orthogonal spaces and screw systems, Mechanism and Machine Theory 27 (4) (1992) 451–458.
[11] Martínez J.M. Rico, J. Duffy, Classification of screw systems:i. One- and two-systems, Mechanism and Machine Theory 27 (4) (1992) 459–470.
[12] Martínez J.M. Rico, J. Duffy, Classification of screw SYStems: II Three-systems, Mechanism and Machine Theory 27 (1992) 471–490.
[13] M.J. Tsai, H.W. Lee, On the special bases of two- and three-screw systems, Journal of Mechanical Design 115 (1993) 540–546.
[14] W.X. Zhang, Z.C. Xu, Algebraic construction of the three-system of screws, Mechanism and Machine Theory 33 (7) (1998) 925–930.

[15] Y.F. Fang, Z. Huang, Analytical identification of the principal screws of the third order screw system, Mechanism and Machine Theory 33 (7) (1998) 987–992.
[16] J.S. Zhao, F.L. Chu, Z.J. Feng, Identification of principal screws of two- and three-screw systems, Proceedings of the Institution of Mechanical Engineers, Part C, Journal of Mechanical Engineering Science 221 (2007) 1701–1715.
[17] J.-S. Zhao, H.-X. Zhou, Z.-J. Feng, J.S. Dai, An algebraic methodology to identify the principal screws and pitches of screw systems, Journal of Mechanical Engineering Science, Proceedings of the Institution of Mechanical Engineers, Part C 223 (2009) 1931–1941.
[18] J.S. Dai, An historical review of the theoretical development of rigid body displacements from rodrigues parameters to the finite twist, Mechanism and Machine Theory 41 (2006) 41–52.
[19] J.S. Dai, J.R. Jones, Null-space construction using cofactors from a screw-algebra context, Proceedings of the Royal Society of London, Series A 458 (2002) 1845–1866.

Chapter 3

Twists and Wrenches of a Kinematic Chain

As is discussed in Chapter 2, the twists and wrenches are uniformly expressed with Plücker homogeneous coordinates in one Cartesian coordinate system. In the following, we will investigate the free relative motions between two members and the constraints induced by their kinematic pairs in 10 primary cases.

3.1 FREE MOTIONS AND THE CONSTRAINTS OF A KINEMATIC PAIR

Generally, the homogeneous coordinates of a wrench $\$^r$ have the following form:

$$\$^r = \begin{bmatrix} \boldsymbol{s} \\ \boldsymbol{s}_0 + h\boldsymbol{s} \end{bmatrix} \tag{3.1}$$

where $\boldsymbol{s} = [\, l^r \quad m^r \quad n^r \,]^T$ is a unit vector pointing to the direction of the screw axis, $\boldsymbol{s}_0 = \boldsymbol{r} \times \boldsymbol{s} = [\, p^r \quad q^r \quad r^r \,]^T$. Then, the twist constrained by $\$^r$ may be obtained:

$$\$_M^C = \begin{bmatrix} \boldsymbol{s} \\ \boldsymbol{s}_0 + \frac{\boldsymbol{s}}{h} \end{bmatrix}. \tag{3.2}$$

3.1.1 Helical Pair

In a Cartesian coordinate system shown in Figure 3.1, the screw pair O jointed by a bolt and a nut has one helical degree of freedom (DOF) about the x-axis. Therefore, the Plücker homogeneous coordinates for the relative motion is

$$\$_H = \begin{bmatrix} 1 & 0 & 0 & h & 0 & 0 \end{bmatrix}^T \tag{3.3}$$

where h denotes the pitch of the screw [1–3].

The wrenches applied to the nut by the fixed bolt may be obtained with equation (2.98):

$$\$_H^r = \begin{bmatrix} \$_H^{r1} & \$_H^{r2} & \$_H^{r3} & \$_H^{r4} & \$_H^{r5} \end{bmatrix} \tag{3.4}$$

Advanced Theory of Constraint and Motion Analysis for Robot Mechanisms.
http://dx.doi.org/10.1016/B978-0-12-420162-0.00003-5

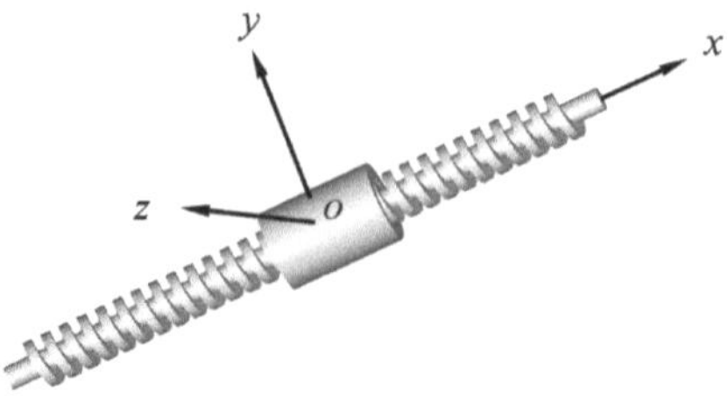

FIGURE 3.1 Cartesian coordinate system for a helical pair.

where

$\$_H^{r1} = \begin{bmatrix} 1 & 0 & 0 & -h & 0 & 0 \end{bmatrix}^T$, which denotes a force screw constraint along the x-axis with a negative pitch $-h$;
$\$_H^{r2} = \begin{bmatrix} 0 & 1 & 0 & 0 & 0 & 0 \end{bmatrix}^T$, which denotes a force constraint along the y-axis;
$\$_H^{r3} = \begin{bmatrix} 0 & 0 & 1 & 0 & 0 & 0 \end{bmatrix}$, which denotes a force constraint along the z-axis;
$\$_H^{r4} = \begin{bmatrix} 0 & 0 & 0 & 0 & 1 & 0 \end{bmatrix}$, which denotes a moment constraint about the y-axis;
$\$_H^{r5} = \begin{bmatrix} 0 & 0 & 0 & 0 & 0 & 1 \end{bmatrix}$, which denotes a moment constraint about the z-axis.

With equation (2.98), we can find that the twists constrained by wrenches (3.4) are

$$\$_M^C = \begin{bmatrix} \$_M^{C_1} & \$_M^{C_2} & \$_M^{C_3} & \$_M^{C_4} & \$_M^{C_5} \end{bmatrix} \tag{3.5}$$

where

$\$_M^{C_1} = \begin{bmatrix} 1 & 0 & 0 & 0 & -\frac{1}{h} & 0 \end{bmatrix}$, which denotes a screw about the x-axis;
$\$_M^{C_2} = \begin{bmatrix} 0 & 0 & 0 & 0 & 1 & 0 \end{bmatrix}$, which denotes a translation along the y-axis;
$\$_M^{C_3} = \begin{bmatrix} 0 & 0 & 0 & 0 & 0 & 1 \end{bmatrix}$, which denotes a translation along the z-axis;
$\$_M^{C_4} = \begin{bmatrix} 0 & 1 & 0 & 0 & 0 & 0 \end{bmatrix}$, which denotes a rotation about the y-axis;
$\$_M^{C_5} = \begin{bmatrix} 0 & 0 & 1 & 0 & 0 & 0 \end{bmatrix}$, which denotes a rotation about the z-axis.

3.1.2 Revolute Pair

If the pitch of a screw is zero, the kinematic pair expressed by the screw will be reduced to a revolute one. To investigate the relative free motion between the two members jointed by a revolute pair (R), we establish a Cartesian coordinate system as Figure 3.2 shows. Supposing x-axis is superimposed with the axis of the R pair, we have the Plücker homogeneous coordinates of this pair

$$\$_R = \begin{bmatrix} 1 & 0 & 0 & 0 & 0 & 0 \end{bmatrix}^T. \tag{3.6}$$

The wrenches applied to one member by the other through a revolute pair are obtained with equation (2.98):

$$\$_R^r = \begin{bmatrix} \$_R^{r1} & \$_R^{r2} & \$_R^{r3} & \$_R^{r4} & \$_R^{r5} \end{bmatrix} \tag{3.7}$$

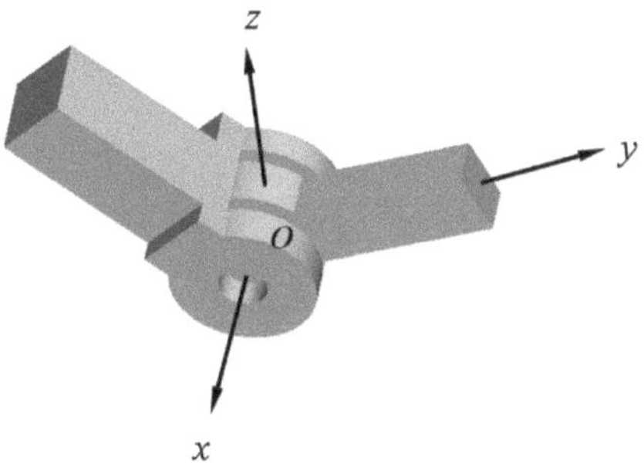

FIGURE 3.2 Cartesian coordinate system for a revolute pair.

where

$\$_R^{r1} = [1\ 0\ 0\ 0\ 0\ 0]^T$, which denotes a force constraint along the x-axis;
$\$_R^{r2} = [0\ 1\ 0\ 0\ 0\ 0]^T$, which denotes a force constraint along the y-axis;
$\$_R^{r3} = [0\ 0\ 1\ 0\ 0\ 0]^T$, which denotes a force constraint along the z-axis;
$\$_R^{r4} = [0\ 0\ 0\ 0\ 1\ 0]^T$, which denotes a moment constraint about the y-axis;
$\$_R^{r5} = [0\ 0\ 0\ 0\ 0\ 1]^T$, which denotes a moment constraint about the z-axis.

With equation (2.98), we can find that the twists constrained by the wrenches expressed by equation (3.7) are

$$\$_M^C = \begin{bmatrix} \$_M^{C_1} & \$_M^{C_2} & \$_M^{C_3} & \$_M^{C_4} & \$_M^{C_5} \end{bmatrix} \tag{3.8}$$

where

$\$_M^{C_1} = [0\ \ 0\ \ 0\ \ 1\ \ 0\ \ 0]^T$, which denotes a translation along the x-axis;
$\$_M^{C_2} = [0\ \ 0\ \ 0\ \ 0\ \ 1\ \ 0]^T$, which denotes a translation along the y-axis;
$\$_M^{C_3} = [0\ \ 0\ \ 0\ \ 0\ \ 0\ \ 1]^T$, which denotes a translation along the z-axis;
$\$_M^{C_4} = [0\ \ 1\ \ 0\ \ 0\ \ 0\ \ 0]^T$, which denotes a rotation about the y-axis;
$\$_M^{C_5} = [0\ \ 0\ \ 1\ \ 0\ \ 0\ \ 0]^T$, which denotes a rotation about the z-axis.

3.1.3 Prismatic Pair

If the pitch of a screw is infinite, which is characterized with equation (2.94) in formula form, the screw pair will degenerate to a prismatic one [3]. In the Cartesian coordinate system as Figure 3.3 shows, the slider P has one translational DOF along the x-axis. Therefore, the Plücker homogeneous coordinates for the relative motion are defined by (2.99), namely

$$\$_P = [0\ \ 0\ \ 0\ \ 1\ \ 0\ \ 0]^T. \tag{3.9}$$

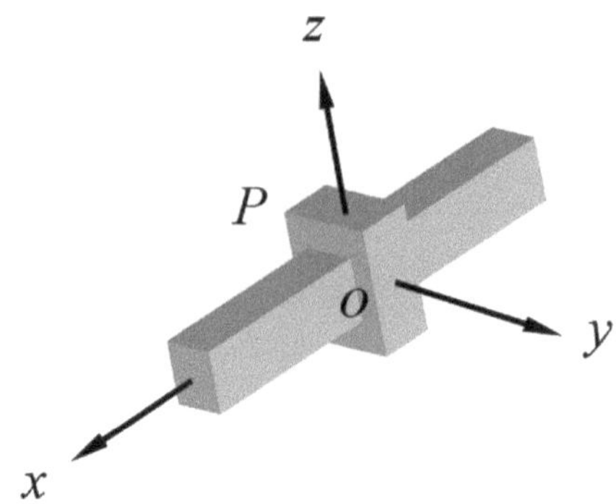

FIGURE 3.3 Cartesian coordinate system for a prismatic pair.

The wrenches applied to the slider by the fixed guide may be obtained with equation (2.98):

$$\$_R^r = \begin{bmatrix} \$_R^{r1} & \$_R^{r2} & \$_R^{r3} & \$_R^{r4} & \$_R^{r5} \end{bmatrix} \tag{3.10}$$

where

$\$_R^{r1} = \begin{bmatrix} 0 & 1 & 0 & 0 & 0 & 0 \end{bmatrix}^T$, which denotes a force constraint along the y-axis;
$\$_R^{r2} = \begin{bmatrix} 0 & 0 & 1 & 0 & 0 & 0 \end{bmatrix}^T$, which denotes a force constraint along the z-axis;
$\$_R^{r3} = \begin{bmatrix} 0 & 0 & 0 & 1 & 0 & 0 \end{bmatrix}^T$, which denotes a moment constraint about the x-axis;
$\$_R^{r4} = \begin{bmatrix} 0 & 0 & 0 & 0 & 1 & 0 \end{bmatrix}^T$, which denotes a moment constraint about the y-axis;
$\$_R^{r5} = \begin{bmatrix} 0 & 0 & 0 & 0 & 0 & 1 \end{bmatrix}^T$, which denotes a moment constraint about the z-axis.

With equation (3.2), we can find that the twists constrained by wrenches (3.10) are

$$\$_M^C = \begin{bmatrix} \$_M^{C_1} & \$_M^{C_2} & \$_M^{C_3} & \$_M^{C_4} & \$_M^{C_5} \end{bmatrix} \tag{3.11}$$

where

$\$_M^{C_1} = \begin{bmatrix} 0 & 0 & 0 & 0 & 1 & 0 \end{bmatrix}^T$, which denotes a translation along the y-axis;
$\$_M^{C_2} = \begin{bmatrix} 0 & 0 & 0 & 0 & 0 & 1 \end{bmatrix}^T$, which denotes a translation along the z-axis;
$\$_M^{C_3} = \begin{bmatrix} 1 & 0 & 0 & 0 & 0 & 0 \end{bmatrix}^T$, which denotes a rotation about the x-axis;
$\$_M^{C_4} = \begin{bmatrix} 0 & 1 & 0 & 0 & 0 & 0 \end{bmatrix}^T$, which denotes a rotation about the y-axis;
$\$_M^{C_5} = \begin{bmatrix} 0 & 0 & 1 & 0 & 0 & 0 \end{bmatrix}^T$, which denotes a rotation about the z-axis.

3.1.4 Universal Pair

To investigate the relative free motion between members A and B jointed via a universal pair (U), we establish a Cartesian coordinate system as Figure 3.4

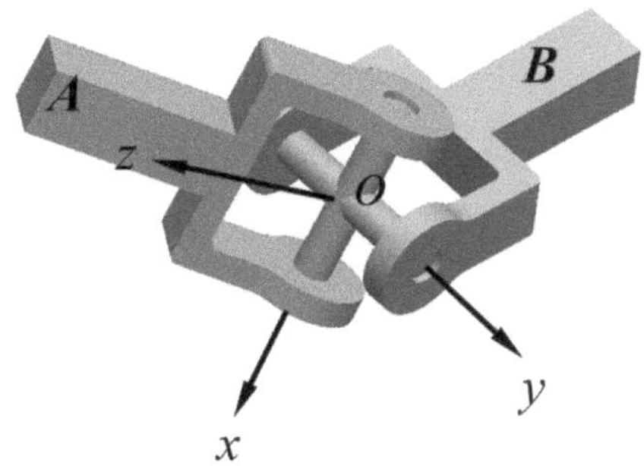

FIGURE 3.4 Cartesian coordinate system for a universal pair.

shows. Supposing the x-axis is superimposed with one axis of the U pair, and the y-axis is superimposed with the other, we obtain the Plücker homogeneous coordinates of the U pair

$$\$_U = \begin{bmatrix} 1 & 0 & 0 & 0 & 0 & 0 \\ 0 & 1 & 0 & 0 & 0 & 0 \end{bmatrix}^T. \tag{3.12}$$

The wrenches applied to one member by the other through the universal pair may be obtained with equation (2.98):

$$\$_U^r = \begin{bmatrix} \$_U^{r1} & \$_U^{r2} & \$_U^{r3} & \$_U^{r4} \end{bmatrix} \tag{3.13}$$

where

$\$_U^{r1} = \begin{bmatrix} 1 & 0 & 0 & 0 & 0 & 0 \end{bmatrix}^T$, which denotes a force constraint along the x-axis;
$\$_U^{r2} = \begin{bmatrix} 0 & 1 & 0 & 0 & 0 & 0 \end{bmatrix}^T$, which denotes a force constraint along the y-axis;
$\$_U^{r3} = \begin{bmatrix} 0 & 0 & 1 & 0 & 0 & 0 \end{bmatrix}^T$, which denotes a force constraint along the z-axis;
$\$_U^{r4} = \begin{bmatrix} 0 & 0 & 0 & 0 & 0 & 1 \end{bmatrix}^T$, which denotes a moment constraint about the z-axis.

With equation (3.2), we can find that the twists constrained by wrenches (3.13) are

$$\$_M^C = \begin{bmatrix} \$_M^{C_1} & \$_M^{C_2} & \$_M^{C_3} & \$_M^{C_4} \end{bmatrix} \tag{3.14}$$

where

$\$_M^{C_1} = \begin{bmatrix} 0 & 0 & 0 & 1 & 0 & 0 \end{bmatrix}^T$, which denotes a translation along the x-axis;
$\$_M^{C_2} = \begin{bmatrix} 0 & 0 & 0 & 0 & 1 & 0 \end{bmatrix}^T$, which denotes a translation along the y-axis;
$\$_M^{C_3} = \begin{bmatrix} 0 & 0 & 0 & 0 & 0 & 1 \end{bmatrix}^T$, which denotes a translation along the z-axis;
$\$_M^{C_4} = \begin{bmatrix} 0 & 0 & 1 & 0 & 0 & 0 \end{bmatrix}^T$, which denotes a rotation about the z-axis.

What must be pointed out here is that the relative motion of such a universal pair at any instant is a linear combination of the two vectors of the basis (3.12). That is, the relative motion of one member of the pair with respect to the other can be expressed

$$\$_{Cr} = \begin{bmatrix} \lambda_1 & \lambda_2 & 0 & 0 & 0 & 0 \end{bmatrix}^T \tag{3.15}$$

where λ_1 and λ_2 are two parameters independent to the basis of the Plücker homogeneous coordinates of the U pair.

3.1.5 Cylindrical Pair

To investigate the relative free motions between members A and B jointed by a cylindrical pair (C), we establish a Cartesian coordinate system as Figure 3.5 shows.

Supposing that the y-axis is superimposed with the axis of a C pair, we obtain a basis of the Plücker homogeneous coordinates of the C pair

$$\$_C = \begin{bmatrix} 1 & 0 & 0 & 0 & 0 & 0 \\ 0 & 0 & 0 & 1 & 0 & 0 \end{bmatrix}^T. \tag{3.16}$$

The wrenches applied to the one member by the other through the cylindrical pair can be obtained with equation (2.98):

$$\$_C^r = \begin{bmatrix} \$_C^{r1} & \$_C^{r2} & \$_C^{r3} & \$_C^{r4} \end{bmatrix} \tag{3.17}$$

where

$\$_C^{r1} = [0\ 1\ 0\ 0\ 0\ 0]^T$, which denotes a force constraint along the y-axis;
$\$_C^{r2} = [0\ 0\ 1\ 0\ 0\ 0]^T$, which denotes a force constraint along the z-axis;
$\$_C^{r3} = [0\ 0\ 0\ 0\ 1\ 0]^T$, which denotes a moment constraint about the y-axis;
$\$_C^{r4} = [0\ 0\ 0\ 0\ 0\ 1]^T$, which denotes a moment constraint about the z-axis.

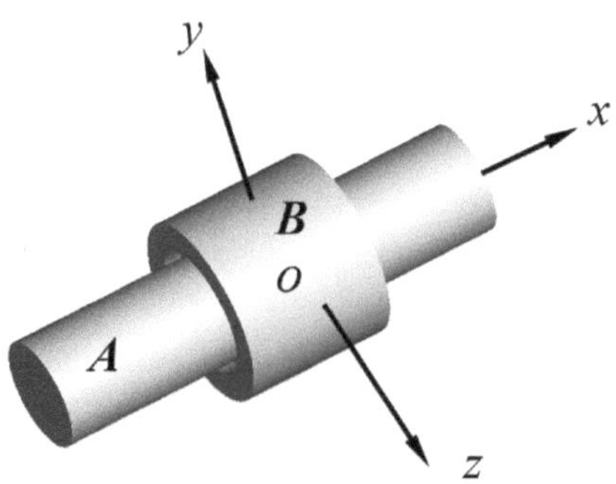

FIGURE 3.5 Cartesian coordinate system for a cylindrical pair.

With equation (3.2), we can find that the twists constrained by wrenches (3.17) are

$$\$_C^C = \begin{bmatrix} \$_C^{C_1} & \$_C^{C_2} & \$_C^{C_3} & \$_C^{C_4} \end{bmatrix} \tag{3.18}$$

where

$\$_C^{C_1} = \begin{bmatrix} 0 & 0 & 0 & 0 & 1 & 0 \end{bmatrix}^T$, which denotes a translation along the y-axis;
$\$_C^{C_2} = \begin{bmatrix} 0 & 0 & 0 & 0 & 0 & 1 \end{bmatrix}^T$, which denotes a translation along the z-axis;
$\$_C^{C_3} = \begin{bmatrix} 0 & 1 & 0 & 0 & 0 & 0 \end{bmatrix}^T$, which denotes a rotation about the y-axis;
$\$_C^{C_4} = \begin{bmatrix} 0 & 0 & 1 & 0 & 0 & 0 \end{bmatrix}^T$, which denotes a rotation about the z-axis.

What must be pointed out here is that the relative motion of such a C pair at any instant is a linear combination of the two vectors of the basis (3.16). That is, the relative motion of one member of the pair with respect to the other can be expressed

$$\$_{Cr} = \begin{bmatrix} \lambda_1 & 0 & 0 & \lambda_2 & 0 & 0 \end{bmatrix}^T \tag{3.19}$$

where λ_1 and λ_2 are two parameters independent to the basis of the Plücker homogeneous coordinates of the C pair.

Generally, if the basis of the Plücker homogeneous coordinates of the C pair is selected as

$$\$_C = \begin{bmatrix} l & m & n & p & q & r \\ 0 & 0 & 0 & l & m & n \end{bmatrix}^T \tag{3.20}$$

where $[l \quad m \quad n]^T$ is the unit direction vector of the axis of the C pair and $[l \quad m \quad n \quad p \quad q \quad r]^T$ denotes the axis screw of the pair. Therefore, the relative motion of one member of the pair with respect to the other at any instant can be expressed as

$$\$_{Cr} = \begin{bmatrix} \lambda_1 l & \lambda_1 m & \lambda_1 n & p + \lambda_2 l & q + \lambda_2 m & r + \lambda_2 n \end{bmatrix}^T \tag{3.21}$$

where λ_1 and λ_2 are real numbers and should not equal zero simultaneously.

The bases, (3.16) and (3.20), of such a C pair are dependent on the setting of the Cartesian coordinate system.

3.1.6 Gear Pair

A Cartesian coordinate system for the gear pair is shown in Figure 3.6. Supposing the x-axis is superimposed with the contact line of two teeth of the gears and the z-axis is along the sliding line of these two teeth, we obtain a basis of the Plücker homogeneous coordinates of the gear pair

$$\$_G = \begin{bmatrix} 1 & 0 & 0 & 0 & 0 & 0 \\ 0 & 0 & 0 & 0 & 0 & 1 \end{bmatrix}^T. \tag{3.22}$$

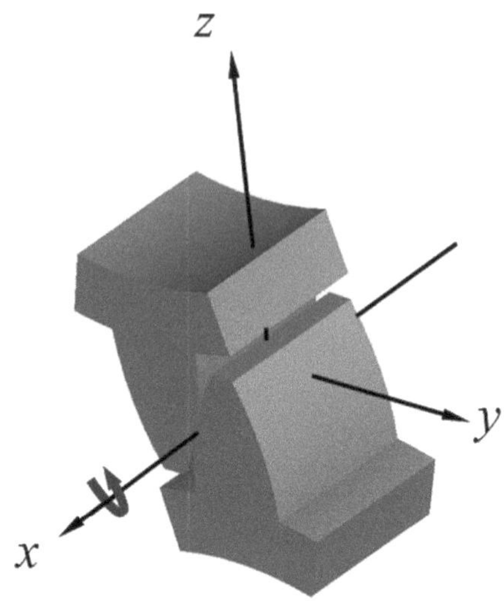

FIGURE 3.6 Cartesian coordinate system for a gear pair.

The wrenches applied to one member by the other through the gear pair may be obtained with equation (2.98):

$$\$_G^r = \begin{bmatrix} \$_G^{r1} & \$_G^{r2} & \$_G^{r3} & \$_G^{r4} \end{bmatrix} \tag{3.23}$$

where

$\$_G^{r1} = [1\ 0\ 0\ 0\ 0\ 0]^T$, which denotes a force constraint along the x-axis;
$\$_G^{r2} = [0\ 1\ 0\ 0\ 0\ 0]^T$, which denotes a force constraint along the y-axis;
$\$_G^{r3} = [0\ 0\ 0\ 0\ 1\ 0]^T$, which denotes a moment constraint about the y-axis;
$\$_G^{r4} = [0\ 0\ 0\ 0\ 0\ 1]^T$, which denotes a moment constraint about the z-axis.

With equation (3.2), we can find that the twists constrained by wrenches (3.23) are

$$\$_G^C = \begin{bmatrix} \$_G^{C_1} & \$_G^{C_2} & \$_G^{C_3} & \$_G^{C_4} \end{bmatrix} \tag{3.24}$$

where

$\$_G^{C_1} = [0\ \ 0\ \ 0\ \ 1\ \ 0\ \ 0]^T$, which denotes a translation along the x-axis;
$\$_G^{C_2} = [0\ \ 0\ \ 0\ \ 0\ \ 1\ \ 0]^T$, which denotes a translation along the y-axis;
$\$_G^{C_3} = [0\ \ 1\ \ 0\ \ 0\ \ 0\ \ 0]^T$, which denotes a rotation about the y-axis;
$\$_G^{C_4} = [0\ \ 0\ \ 1\ \ 0\ \ 0\ \ 0]^T$, which denotes a rotation about the z-axis.

What must be pointed out here is that the relative motion of such a G pair at any instant is a linear combination of the two vectors of the basis(3.22). That is, the relative motions of one member of the pair with respect to the other can be expressed as

$$\$_{Cr} = \begin{bmatrix} \lambda_1 & 0 & 0 & 0 & 0 & \lambda_2 \end{bmatrix}^T \tag{3.25}$$

where λ_1 and λ_2 are two parameters independent to the basis of the Plücker homogeneous coordinates of the G pair and are not equal to zero simultaneously.

Generally, if the basis of the Plücker homogeneous coordinates of the gear pair is selected as

$$\$_C = \begin{bmatrix} l & m & n & p & q & r \\ 0 & 0 & 0 & l & m & n \end{bmatrix}^T \tag{3.26}$$

where $[l \quad m \quad n]^T$ is the unit direction vector of the axis of the G pair and $[l \quad m \quad n \quad p \quad q \quad r]^T$ denotes the axis of the pair. Therefore, the relative motion of one member of the pair with respect to the other can be expressed as

$$\$_{Cr} = \begin{bmatrix} \lambda_1 l & \lambda_1 m & \lambda_1 n & \lambda_1 p + \lambda_2 l & \lambda_1 q + \lambda_2 m & \lambda_1 r + \lambda_2 n \end{bmatrix}^T. \tag{3.27}$$

The form of bases, (3.22) and (3.26), of such a gear pair is dependent on the setting of the Cartesian coordinate system.

3.1.7 Spherical Pair

To investigate the free relative motion between members A and B jointed by a spherical pair (S), we establish a Cartesian coordinate system, as Figure 3.7

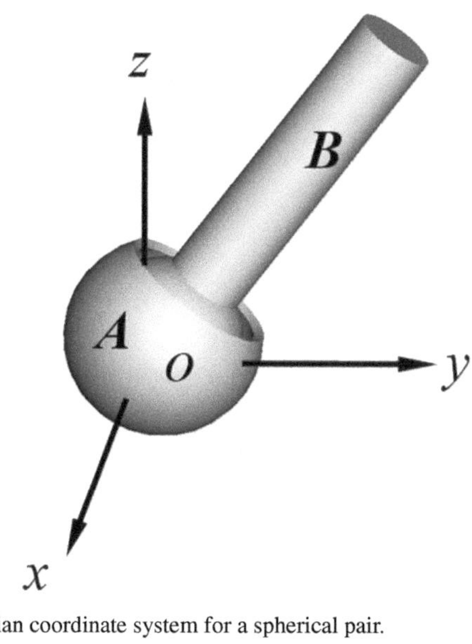

FIGURE 3.7 Cartesian coordinate system for a spherical pair.

shows. Supposing that the origin of the system is superimposed with the center of the sphere, we obtain a basis of the Plücker homogeneous coordinates of the S pair:

$$\$_S = \begin{bmatrix} 1 & 0 & 0 & 0 & 0 & 0 \\ 0 & 1 & 0 & 0 & 0 & 0 \\ 0 & 0 & 1 & 0 & 0 & 0 \end{bmatrix}^T . \tag{3.28}$$

The wrenches to the one member by the other through the universal pair may be obtained with equation (2.98):

$$\$_S^r = \begin{bmatrix} \$_S^{r1} & \$_S^{r2} & \$_S^{r3} \end{bmatrix} \tag{3.29}$$

where

$\$_S^{r1} = \begin{bmatrix} 1 & 0 & 0 & 0 & 0 & 0 \end{bmatrix}^T$, which denotes a force constraint along the x-axis;
$\$_S^{r2} = \begin{bmatrix} 0 & 1 & 0 & 0 & 0 & 0 \end{bmatrix}^T$, which denotes a force constraint along the y-axis;
$\$_S^{r3} = \begin{bmatrix} 0 & 0 & 1 & 0 & 0 & 0 \end{bmatrix}^T$, which denotes a force constraint along the z-axis;

With equation (3.2), we can find that the twists constrained by wrenches (3.29) are

$$\$_M^C = \begin{bmatrix} \$_M^{C_1} & \$_M^{C_2} & \$_M^{C_3} \end{bmatrix} \tag{3.30}$$

where

$\$_M^{C_1} = \begin{bmatrix} 0 & 0 & 0 & 1 & 0 & 0 \end{bmatrix}^T$, which denotes a translation along x-axis;
$\$_M^{C_2} = \begin{bmatrix} 0 & 0 & 0 & 0 & 1 & 0 \end{bmatrix}^T$, which denotes a translation along the y-axis;
$\$_M^{C_3} = \begin{bmatrix} 0 & 0 & 0 & 0 & 0 & 1 \end{bmatrix}^T$, which denotes a translation along the z-axis.

Through a similar deduction with the cylindrical pair, we know that the relative motions of the two members of the S pair are the linear combinations of the basis (3.28):

$$\$_{Cr} = \begin{bmatrix} \lambda_1 & \lambda_2 & \lambda_3 & 0 & 0 & 0 \end{bmatrix}^T \tag{3.31}$$

where λ_1, λ_2, and λ_3 are three parameters independent of the basis of the Plücker homogeneous coordinates of the S pair and are not equal to zero simultaneously.

Generally, if the Cartesian coordinates of the center of the spherical pair are $(x_s \ \ y_s \ \ z_s)$, the three independent free motions can be denoted as:

$$\$_S = \begin{bmatrix} 1 & 0 & 0 & 0 & z_s & y_s \\ 0 & 1 & 0 & -z_s & 0 & x_s \\ 0 & 0 & 1 & y_s & -x_s & 0 \end{bmatrix}^T . \tag{3.32}$$

Therefore, the instantaneous relative motion of the two members of S pair is represented by:

$$\$_{Cr} = \begin{bmatrix} \lambda_1 & \lambda_2 & \lambda_3 & \lambda_3 y_s - \lambda_2 z_s & -\lambda_3 x_s + \lambda_1 z_s & \lambda_2 x_s + \lambda_1 y_s \end{bmatrix}^T. \quad (3.33)$$

3.1.8 Planar Pair

To investigate the relative free motions between two members jointed by a planar pair, which are often called an E pair after the German word for a plane, *Ebene*, we establish the Cartesian coordinate system shown in Figure 3.8. Supposing that the E pair is perpendicular to the x-axis of the system, we obtain the Plücker homogeneous coordinates of the E pair

$$\$_E = \begin{bmatrix} 1 & 0 & 0 & 0 & 0 & 0 \\ 0 & 0 & 0 & 0 & 1 & 0 \\ 0 & 0 & 0 & 0 & 0 & 1 \end{bmatrix}^T. \quad (3.34)$$

The wrenches applied to the one member by the other through a universal pair may be obtained with equation (2.98):

$$\$_E^r = \begin{bmatrix} \$_E^{r1} & \$_E^{r2} & \$_E^{r3} \end{bmatrix} \quad (3.35)$$

where

$\$_E^{r1} = \begin{bmatrix} 1 & 0 & 0 & 0 & 0 & 0 \end{bmatrix}^T$, which denotes a force constraint along the x-axis;
$\$_E^{r2} = \begin{bmatrix} 0 & 0 & 0 & 0 & 1 & 0 \end{bmatrix}^T$, which denotes a moment constraint about the y-axis;
$\$_E^{r3} = \begin{bmatrix} 0 & 0 & 0 & 0 & 0 & 1 \end{bmatrix}^T$, which denotes a moment constraint about the the z-axis;

With equation (3.2), we can find that the twists constrained by wrenches (3.35) are

$$\$_E^C = \begin{bmatrix} \$_E^{C_1} & \$_E^{C_2} & \$_E^{C_3} \end{bmatrix} \quad (3.36)$$

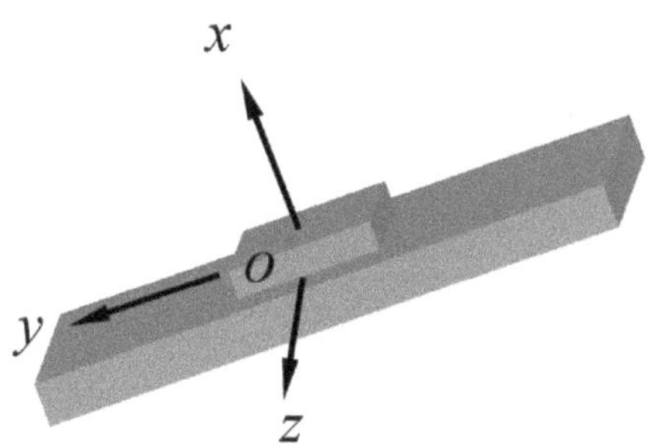

FIGURE 3.8 Cartesian coordinate system for an E-pair.

where

$\$_E^{C_1} = \begin{bmatrix} 0 & 0 & 0 & 1 & 0 & 0 \end{bmatrix}^T$, which denotes a translation along the x-axis;
$\$_E^{C_2} = \begin{bmatrix} 0 & 1 & 0 & 0 & 0 & 0 \end{bmatrix}^T$, which denotes a rotation about the y-axis;
$\$_E^{C_3} = \begin{bmatrix} 0 & 0 & 1 & 0 & 0 & 0 \end{bmatrix}^T$, which denotes a rotation about the z-axis.

The relative motions of the jointed members can be analyzed with a similar process to the above.

3.1.9 Sphere-Cylindrical Pair

To investigate the relative free motion between two members jointed by a sphere–cylindrical pair, we establish a Cartesian coordinate system, as Figure 3.9 shows. Supposing the origin of the system is superimposed with the center of the sphere, we obtain a basis of the Plücker homogeneous coordinates of the S pair:

$$\$_{SC} = \begin{bmatrix} 1 & 0 & 0 & 0 \\ 0 & 1 & 0 & 0 \\ 0 & 0 & 1 & 0 \\ 0 & 0 & 0 & 1 \\ 0 & 0 & 0 & 0 \\ 0 & 0 & 0 & 0 \end{bmatrix}. \tag{3.37}$$

The wrenches applied to one member by the other through a universal pair can be obtained with equation (2.98):

$$\$_{SC}^r = \begin{bmatrix} \$_{SC}^{r1} & \$_{SC}^{r2} \end{bmatrix} \tag{3.38}$$

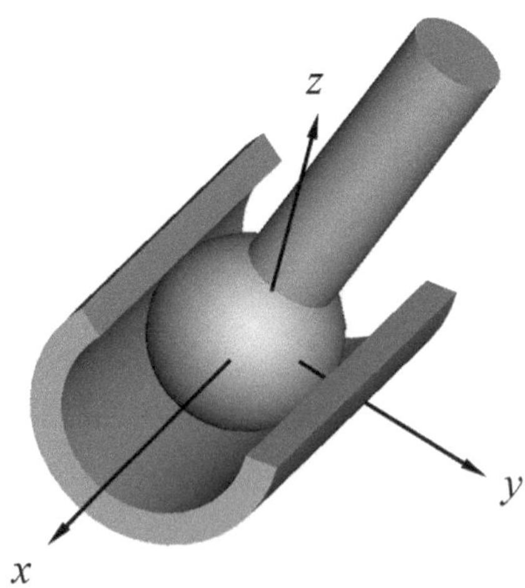

FIGURE 3.9 Cartesian coordinate system for the sphere–cylindrical pair.

where

$\$^{r_1}_{SC} = \begin{bmatrix} 0 & 1 & 0 & 0 & 0 & 0 \end{bmatrix}^T$, which denotes a force constraint along the y-axis;
$\$^{r_2}_{SC} = \begin{bmatrix} 0 & 0 & 1 & 0 & 0 & 0 \end{bmatrix}^T$, which denotes a force constraint along the z-axis.

With equation (3.2), we can directly find that the twists constrained by wrenches (3.38) are

$$\$^{C}_{SC} = \begin{bmatrix} \$^{C_1}_{SC} & \$^{C_2}_{SC} \end{bmatrix} \tag{3.39}$$

where

$\$^{C_1}_{SC} = \begin{bmatrix} 0 & 0 & 0 & 0 & 1 & 0 \end{bmatrix}^T$, which denotes a translation along the y-axis;
$\$^{C_2}_{SC} = \begin{bmatrix} 0 & 0 & 0 & 0 & 0 & 1 \end{bmatrix}^T$, which denotes a translation along the z-axis.

Through a similar deduction of the above process, we know that the transitory relative motion of the two members of SC pair is the linear combination of the basis (3.37):

$$\$_{Cr} = \begin{bmatrix} \lambda_1 & \lambda_2 & \lambda_3 & \lambda_4 & 0 & 0 \end{bmatrix}^T \tag{3.40}$$

where λ_1,λ_2,λ_3, and λ_4 are four parameters independent of the basis of the Plücker homogeneous coordinates of the SC pair and are not equal to zero simultaneously.

Generally, when the Cartesian coordinates of the center of the spherical pair are $(x_S \quad y_S \quad z_S)$, the 3 DOFs independent motions can be denoted as:

$$\$_{SC} = \begin{bmatrix} 1 & 0 & 0 & 0 \\ 0 & 1 & 0 & 0 \\ 0 & 0 & 1 & 0 \\ 0 & -z_s & y_s & a \\ z_s & 0 & -x_s & b \\ y_s & x_s & 0 & c \end{bmatrix} \tag{3.41}$$

where $a^2 + b^2 + c^2 = 1$. Therefore, the transitory relative motion of the two members of the S pair is:

$$\$_{Cr} = \begin{bmatrix} \lambda_1 & \lambda_2 & \lambda_3 & \lambda_3 y_s - \lambda_2 z_s + a\lambda_4 & -\lambda_3 x_s + \lambda_1 z_s + b\lambda_4 & \lambda_2 x_s + \lambda_1 y_s + c\lambda_4 \end{bmatrix}^T. \tag{3.42}$$

3.1.10 Sphere-Planar Pair

To investigate the free relative motion between two members, A and B, jointed by a sphere–planar pair, which is denoted "SP pair," we establish a Cartesian coordinate system as Figure 3.10 shows. Supposing the xoy-plane is parallel to

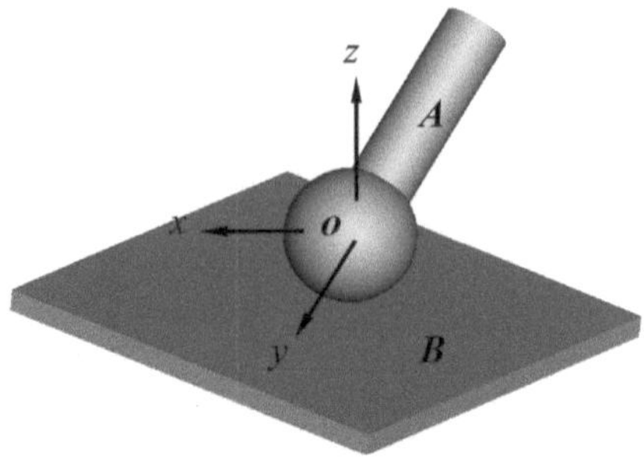

FIGURE 3.10 Cartesian coordinate system for the sphere–planar pair.

the plane B, we obtain the Plücker homogeneous coordinates of the SP pair

$$\$_{SP} = \begin{bmatrix} 1 & 0 & 0 & 0 & 0 \\ 0 & 1 & 0 & 0 & 0 \\ 0 & 0 & 1 & 0 & 0 \\ 0 & 0 & 0 & 1 & 0 \\ 0 & 0 & 0 & 0 & 1 \\ 0 & 0 & 0 & 0 & 0 \end{bmatrix}. \tag{3.43}$$

The wrenches applied to the one member by the other through the sphere-planar pair may be obtained with equation (2.98):

$$\$^{r}_{SE} = \begin{bmatrix} 0 & 0 & 0 & 0 & 0 & 1 \end{bmatrix}^{T} \tag{3.44}$$

where $\$^{r3}_{SE}$ denotes a moment constraint about z-axis.

With equation (3.2), we can find that the twists constrained by wrenches (3.44) are

$$\$^{C}_{SE} = \begin{bmatrix} 0 & 0 & 1 & 0 & 0 & 0 \end{bmatrix}^{T} \tag{3.45}$$

where $\$^{C_3}_{SE}$ denotes a rotation about z-axis.

The relative motions of the paired members can be analyzed with similar processes to the sphere–planar pair.

3.2 TWISTS OF KINEMATIC CHAINS

Because the parallel and the hybrid kinematic chains can be decomposed to a number of simple serial ones, we only discuss the series kinematic chain here for the sake of simplicity. A simple kinematic chain is made up of a number of members jointed by kinematic pairs in series. Assume that the free motion of a rigid body, connected with the fixed base through a series kinematic chain, consists of $n(n > 0)$ DOFs, and the axes of the general twists of every DOF are given in Plücker homogeneous coordinates in one Cartesian coordinate system. Then, the twist of the rigid body should be the sum of these relative twists:

$$\$_B = \sum_{i=1}^{n} \omega_i \$_i \tag{3.46}$$

where ω_i is the amplitude of the ith twist, and $\$_i$ denotes the unit Plücker homogeneous coordinates of the ith twist.

Suppose the Plücker homogeneous coordinates of the ith twist has the equation form:

$$\$_i = \begin{bmatrix} l_i & m_i & n_i & p_i & q_i & r_i \end{bmatrix}^T. \tag{3.47}$$

Therefore, equation (3.46) can be equivalently expressed in matrix multiplication form

$$\$_B = \$\boldsymbol{\omega} \tag{3.48}$$

where

$$\$ = \begin{bmatrix} l_1 & l_2 & & l_n \\ m_1 & m_2 & & m_n \\ n_1 & n_2 & \cdots & n_n \\ p_1 & p_2 & & p_n \\ q_1 & q_2 & & q_n \\ r_1 & r_2 & & r_n \end{bmatrix} \tag{3.49}$$

$$\boldsymbol{\omega} = \begin{bmatrix} \omega_1 & \omega_2 & \cdots & \omega_n \end{bmatrix}^T \tag{3.50}$$

where $\$$ is called the screw matrix of the kinematic chain. According to the theory of reciprocal screws, the terminal constraint of the chain is the reciprocal screw(s) of $f\$^r$. With equation (2.98), we have

$$(\$\boldsymbol{\omega})^T \boldsymbol{\Delta}(f\$^r) = 0 \tag{3.51}$$

where scale f denotes the intensity of $\$^r$. Therefore, equation (3.51) can be transformed into

$$(f\boldsymbol{\omega}^T)\$^T \boldsymbol{\Delta}\$^r = 0. \tag{3.52}$$

Now, let's consider the physical meanings of f and $\boldsymbol{\omega}$. f interprets the intensity of the reciprocal screw, and $\boldsymbol{\omega}$ represents the magnitudes of the actuations. Therefore, these two parameters may be any value. As a result, the necessary and sufficient criteria for equation (3.52) are simplified to

$$\$^T \boldsymbol{\Delta}\$^r = 0. \tag{3.53}$$

That is to say, if we get a set of reciprocal screws/wrenches exerted on a rigid body, the free motion(s) can also be solved with equation (3.53).

Theorem 3.1. *The spaces spanned by the screws/twists of a set of kinematic chains should be equivalent if the terminal constraints/wrenches of these kinematic chains are equivalent.*

Proof. Suppose that a basis of the $6 \times m$-dimension subspace spanned by the twists of a kinematic chain is denoted by

$$\mathbf{S}_1 = \begin{bmatrix} \boldsymbol{a}_1 & \boldsymbol{a}_2 & \cdots & \boldsymbol{a}_m \end{bmatrix} \tag{3.54}$$

And a basis of the $6 \times n$-dimension subspace spanned by the twists of any other kinematic chain is

$$\mathbf{S}_2 = \begin{bmatrix} \boldsymbol{b}_1 & \boldsymbol{b}_2 & \cdots & \boldsymbol{b}_n \end{bmatrix} \tag{3.55}$$

where $m,n \leq 6$ are natural numbers and $\boldsymbol{a}_i$ and $\boldsymbol{b}_i$ are unit vectors in a 6-dimension space.

Because the subspace $\mathbf{S}_1$ is spanned by the twists, $\$_1$, of the first kinematic chain, we can immediately formulate $\$_1$ with $\mathbf{S}_1$ as a linear transformation of the basis of the $6 \times m$-dimension subspace:

$$\$_1 = \mathbf{S}_1\mathbf{C} \tag{3.56}$$

where

$$\mathbf{C} = \begin{bmatrix} c_{11} & c_{12} & \cdots & c_{1k_1} \\ c_{21} & c_{22} & \cdots & c_{2k_1} \\ & & \vdots & \\ c_{m1} & c_{m2} & \cdots & c_{mk_1} \end{bmatrix} \tag{3.57}$$

and $k_1 \geq m$ and Rank $(\mathbf{C}) = m$. And therefore,

$$\left|\mathbf{C}\mathbf{C}^T\right| \neq 0 \tag{3.58}$$

where $|\cdot|$ denotes the determinate of "·".

From equations (3.56) and (3.58) we have

$$\mathbf{S}_1 = \$_1\mathbf{C}^T\left(\mathbf{C}\mathbf{C}^T\right)^{-1}. \tag{3.59}$$

Similarly, the twists of any other kinematic chain can also be formulated as a linear combination of the basis of the $6 \times n$-dimension subspace spanned by its twists:

$$\$_2 = \mathbf{S}_2\mathbf{D} \tag{3.60}$$

where

$$\mathbf{D} = \begin{bmatrix} d_{11} & d_{12} & \cdots & d_{1k_2} \\ d_{21} & d_{22} & \cdots & d_{2k_2} \\ & & \vdots & \\ d_{n1} & d_{n2} & \cdots & d_{nk_2} \end{bmatrix} \tag{3.61}$$

and $k_2 \geq n$ and Rank $(\mathbf{D}) = n$. Therefore,

$$\left|\mathbf{D}\mathbf{D}^T\right| \neq 0. \tag{3.62}$$

Further, from equations (3.60) and (3.62) we have

$$\mathbf{S}_2 = \$_2\mathbf{D}^T\left(\mathbf{D}\mathbf{D}^T\right)^{-1}. \tag{3.63}$$

Because the terminal constraints of the kinematic chains are equivalent, we know from equation (2.98) that $\$_1^T\boldsymbol{\Delta}\$^r = 0$ and $\$_2^T\boldsymbol{\Delta}\$^r = 0$ should have the same solutions.

Therefore, there are $\mathbf{K}_1$ and $\mathbf{K}_2$ such that

$$\$_1 = \$_2\mathbf{K}_1 \tag{3.64}$$

and

$$\$_2 = \$_1\mathbf{K}_2 \tag{3.65}$$

where $\mathbf{K}_1$ is a $k_2 \times k_1$ coefficient matrix and $\mathbf{K}_2$ is a $k_1 \times k_2$ coefficient matrix.

Therefore, with equations (3.59), (3.60), and (3.64), we have

$$\mathbf{S}_1 = \mathbf{S}_2\mathbf{D}\mathbf{K}_1\mathbf{C}^T\left(\mathbf{C}\mathbf{C}^T\right)^{-1} \tag{3.66}$$

and with equations (3.56), (3.63), and (3.65), we have

$$\mathbf{S}_2 = \mathbf{S}_1\mathbf{C}\mathbf{K}_2\mathbf{D}^T\left(\mathbf{D}\mathbf{D}^T\right)^{-1}. \tag{3.67}$$

Therefore, the bases $\mathbf{S}_1$ and $\mathbf{S}_2$ can be mutually linear formulated. So

$$\dim\left(\mathbf{S}_1\right) = \dim\left(\mathbf{S}_2\right). \tag{3.68}$$

That is,

$$m = n. \tag{3.69}$$

Therefore, $\mathbf{S}_1$ and $\mathbf{S}_2$ are equivalent. Alternatively, we have the following Theorem 3.2.

Theorem 3.2. *The terminal constraints/wrenches of a set of kinematic chains should be equivalent if the spaces spanned by the screws/twists of the kinematic chains are equivalent.*

Proof. Suppose the spaces spanned by the screws of the kinematic chains are denoted with $\mathbf{S}_1$ and $\mathbf{S}_2$. Because $\mathbf{S}_1$ and $\mathbf{S}_2$ are equivalent, we have

$$\dim\left(\mathbf{S}_1\right) = \dim\left(\mathbf{S}_2\right) \tag{3.70}$$

and

$$\mathbf{S}_2 = \mathbf{S}_1\mathbf{K} \tag{3.71}$$

where $|\mathbf{K}| \neq 0$.

Suppose $\dim(\mathbf{S}_1) = \dim(\mathbf{S}_2) = m$, the kinematic screws of the two chains can be expressed

$$\$_1 = \mathbf{S}_1\mathbf{C} \tag{3.72}$$

and

$$\$_2 = \mathbf{S}_2\mathbf{D} \tag{3.73}$$

where $\text{Rank}(\mathbf{C}) = \text{Rank}(\mathbf{D}) = m$. Therefore, $|\mathbf{C}\mathbf{C}^T| \neq 0$ and $|\mathbf{D}\mathbf{D}^T| \neq 0$.

Substituting equation (3.71) into equation (3.73) yields

$$\$_2 = \mathbf{S}_1\mathbf{K}\mathbf{D}. \tag{3.74}$$

From equation (3.59), we know

$$\mathbf{S}_1 = \$_1\mathbf{C}^T\left(\mathbf{C}\mathbf{C}^T\right)^{-1}. \tag{3.75}$$

Substituting equation (3.59) into equation (3.75) yields

$$\$_2 = \$_1\mathbf{C}^T\left(\mathbf{C}\mathbf{C}^T\right)^{-1}\mathbf{K}\mathbf{D}. \tag{3.76}$$

Similarly, we can deduce that

$$\$_1 = \$_2\mathbf{D}^T\left(\mathbf{D}\mathbf{D}^T\right)^{-1}\mathbf{K}^{-1}\mathbf{C}. \tag{3.77}$$

Therefore, from equations (3.76) and (3.77), we know that $\$_1$ and $\$_2$ can be linearly expressed with each other. And from equation (2.98) or equation (3.53), we know that the terminal constraints of the two kinematic chains are equivalent.

Consequently, if the terminal constraints of a pure series kinematic chain are known, the maximal linearly independent twists are determined; and *vice versa*, if the maximal linearly independent twists are specified, the terminal constraints of the kinematic chain can also be obtained. In other words, any kinematic chains with the same terminal constraints should at least consist of the maximal linearly independent twists. Therefore, we can design different kinematic chains, which have the same terminal constraints. In addition, the best kinematic chain can also be gained from all of the chains that have the same terminal constraints, which is especially useful in engineering applications when we need to select the best kinematic chain to carry out the same work.

Suppose the Plücker homogeneous coordinates of the reciprocal screw $\$^r$ has the following form:

$$\$^r = \begin{bmatrix} l^r & m^r & n^r & p^r & q^r & r^r \end{bmatrix}^T. \tag{3.78}$$

Substituting equation (3.78) into equation (3.53) yields

$$\begin{bmatrix} p_1 & q_1 & r_1 & l_1 & m_1 & n_1 \\ p_2 & q_2 & r_2 & l_2 & m_2 & n_2 \\ & & & \vdots & & \\ p_n & q_n & r_n & l_n & m_n & n_n \end{bmatrix} \begin{bmatrix} l^r \\ m^r \\ n^r \\ p^r \\ q^r \\ r^r \end{bmatrix} = 0. \tag{3.79}$$

Therefore, the Plücker homogeneous coordinates for the terminal constraint of any series chain can be gained with equation (3.79). According to linear algebra, the reciprocal screws of the kinematic chain, $\$^r$, are determined by the maximum independent set of the kinematic screws, $\$$.

3.3 THEORY OF RECIPROCAL SCREWS

As is discussed above, the procedure to create and optimally synthesize a serial kinematic chain can be given as follows:

1. Establish the coordinate system and express the Plücker coordinates of every pair of the kinematic chain in a coordinate system.
2. Form the kinematic screw matrix of the chain, denoted by $\$_{chain}$, and then find the maximum linearly independent set of $\$_{chain}$.
3. Optimize and synthesize the kinematic chain by combining the maximum linearly independent set of screws, $\$_{chain}$, based on the engineering requirements.

In step 3, there are many techniques available to composite the kinematic pairs, which will be illustrated through examples.

3.3.1 Kinematic Chain of Zero Terminal Constraint

A kind of kinematic chain of zero terminal constraint is shown in Figure 3.11, which is made up of *SCCS* pairs. The coordinate system can be set with the origin superposing with one of the spherical joints and the z-axis is along their common line of the kinematic pairs. Assume the distance between A and D is $l (l > 0)$.

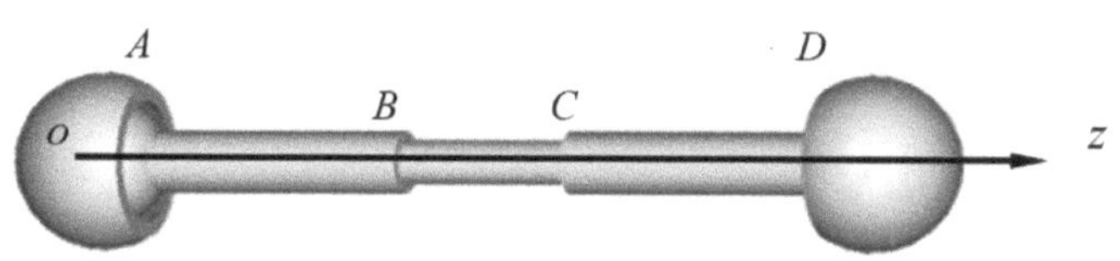

FIGURE 3.11 Kinematic chain of zero terminal constraint.

The kinematic screw set, $\$_{AD}$, includes:

1. The kinematic screws of spherical joint A are:

$$\$_1 = [1\ \ 0\ \ 0\ \ 0\ \ 0\ \ 0]^T, \$_2 = [0\ \ 1\ \ 0\ \ 0\ \ 0\ \ 0]^T,$$
$$\$_3 = [0\ \ 0\ \ 1\ \ 0\ \ 0\ \ 0]^T.$$

2. The kinematic screws of cylindrical joint B are:

$$\$_4 = [0\ \ 0\ \ 1\ \ 0\ \ 0\ \ 0]^T, \$_5 = [0\ \ 0\ \ 0\ \ 0\ \ 0\ \ 1]^T.$$

3. The kinematic screws of cylindrical joint C are:

$$\$_6 = [0\ \ 0\ \ 1\ \ 0\ \ 0\ \ 0]^T, \$_7 = [0\ \ 0\ \ 0\ \ 0\ \ 0\ \ 1]^T.$$

4. The kinematic screws of spherical joint D are:

$$\$_8 = [1\ \ 0\ \ 0\ \ 0\ \ l\ \ 0]^T, \$_9 = [0\ \ 1\ \ 0\ \ -l\ \ 0\ \ 0]^T,$$
$$\$_{10} = [0\ \ 0\ \ 1\ \ 0\ \ 0\ \ 0]^T.$$

Therefore, $\text{Rank}(\$_{AD}) = 6$. The maximal independent set can be selected as

$$\$_{AD}^M = \{\$_1\ \ \$_2\ \ \$_3\ \ \$_4\ \ \$_5\ \ \$_6\ \ \$_7\ \ \$_8\ \ \$_9\}. \tag{3.80}$$

If we select $\$_{AD}^M$ as the screws of the kinematic chain, we can design the kinematic chain APG in Figure 3.12 to replace $ABCD$ in Figure 3.11. The elements of screw set $\$_{AD}^M$ can be divided as:

1. The kinematic screws of universal joint A are:

$$\$_1 = [1\ \ 0\ \ 0\ \ 0\ \ 0\ \ 0]^T, \$_2 = [0\ \ 1\ \ 0\ \ 0\ \ 0\ \ 0]^T.$$

2. The kinematic screws of prismatic joint P is:

$$\$_5 = [0\ \ 0\ \ 0\ \ 0\ \ 0\ \ 1]^T.$$

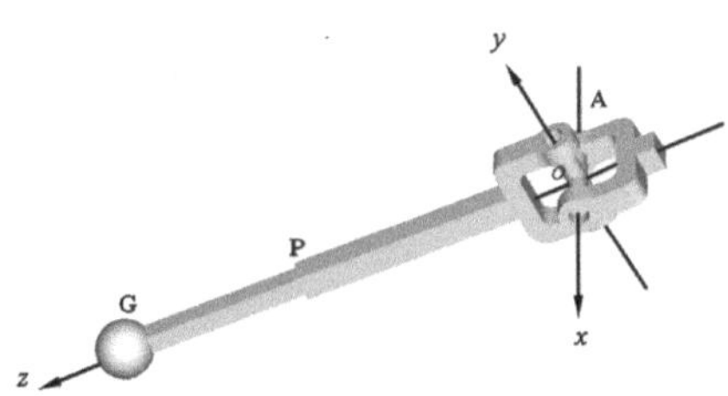

FIGURE 3.12 Kinematic chain made up of SPT.

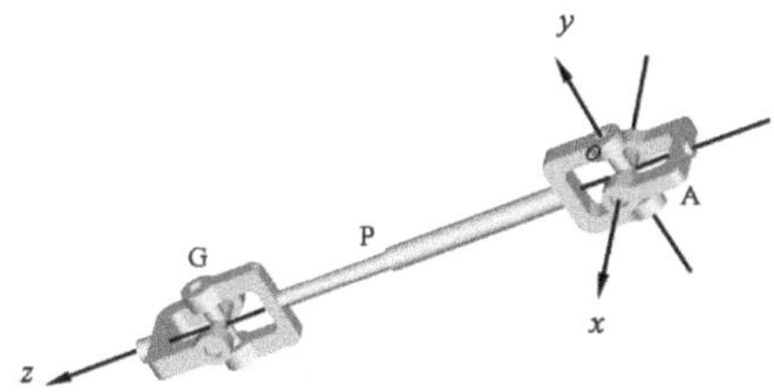

FIGURE 3.13 Kinematic chain made up of TCT.

3. The kinematic screws of spherical joint G are:

$$\$_8 = [1 \;\; 0 \;\; 0 \;\; 0 \;\; l \;\; 0]^T, \$_9 = [0 \;\; 1 \;\; 0 \;\; -l \;\; 0 \;\; 0]^T,$$
$$\$_{10} = [0 \;\; 0 \;\; 1 \;\; 0 \;\; 0 \;\; 0]^T = \$_3.$$

Similarly, we can design the kinematic chain APG with $\mathbf{\$}_{AD}^M$ in Figure 3.13 to replace $ABCD$ in Figure 3.11. The elements of screw set $\mathbf{\$}_{AD}^M$ can be divided as:

1. The kinematic screws of universal joint A are:

$$\$_1 = [1 \;\; 0 \;\; 0 \;\; 0 \;\; 0 \;\; 0]^T, \$_2 = [0 \;\; 1 \;\; 0 \;\; 0 \;\; 0 \;\; 0]^T.$$

2. The kinematic screws of cylindrical joint P are:

$$\$_3 = [0 \;\; 0 \;\; 1 \;\; 0 \;\; 0 \;\; 0]^T, \$_5 = [0 \;\; 0 \;\; 0 \;\; 0 \;\; 0 \;\; 1]^T.$$

3. The kinematic screws of universal joint G are:

$$\$_8 = [1 \;\; 0 \;\; 0 \;\; 0 \;\; l \;\; 0]^T, \$_9 = [0 \;\; 1 \;\; 0 \;\; -l \;\; 0 \;\; 0]^T.$$

If we select $\mathbf{\$} = \{\$_1 \;\; \$_2 \;\; \$_3 \;\; \$_4 \;\; \$_5 \;\; \$_6 \;\; \$_7 \;\; \$_8 \;\; \$_9\}$ as the screw set of the kinematic chain, we can realize it as Figure 3.14 shows.

In a Stewart platform shown in Figure 3.15, the platform $ABCDEF$connects the base $GHIJKL$ through six SCS kinematic chains. Because of the symmetry

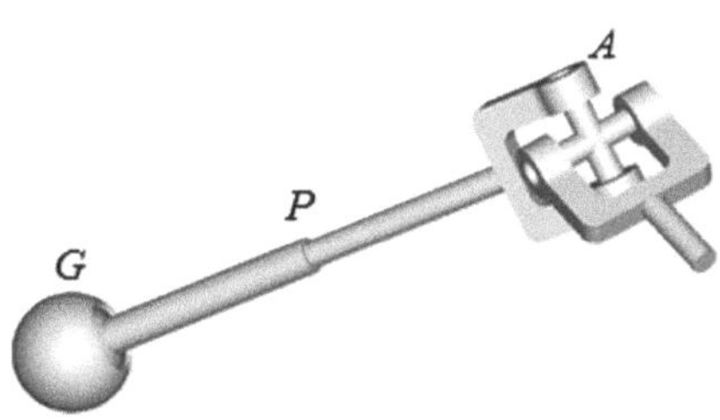

FIGURE 3.14 Kinematic chain made up of SCT.

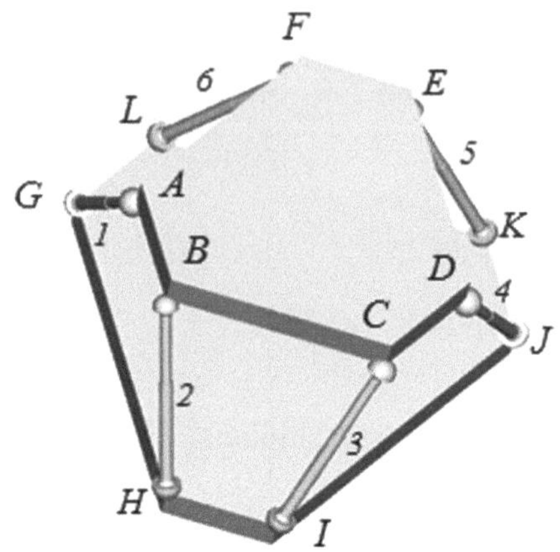

FIGURE 3.15 A Stewart platform.

of the mechanism, we can select one of them (say *AG*) to analyze the kinematic screws of the branch. The local coordinate system is shown in Figures 3.12 and 3.13.

The maximal independent set for the kinematic screws of *GPA* is equation (3.80). Therefore, the limb of the Stewart platform can be equivalently substituted by either of *SPT* in Figure 3.12, or *TCT* in Figure 3.13 or *SCT* in Figure 3.14, whose maximal linearly independent kinematic screws are all equal to equation (3.80). The limbs of the Stewart platform can be equivalently replaced by any one of the limbs in Figures 3.12, 3.13 without changing the kinematic characteristics of the platform.

3.3.2 Kinematic Chain of One Terminal Constraint

3.3.2.1 A Force as the Terminal Constraint

Assume that a rigid shaft is coupled with other rigid bodies by spherical pairs (*S*) at both ends. We create a local coordinate system *oxyz* as Figure 3.16 shows, in which the origin of the coordinate system is superposed with one

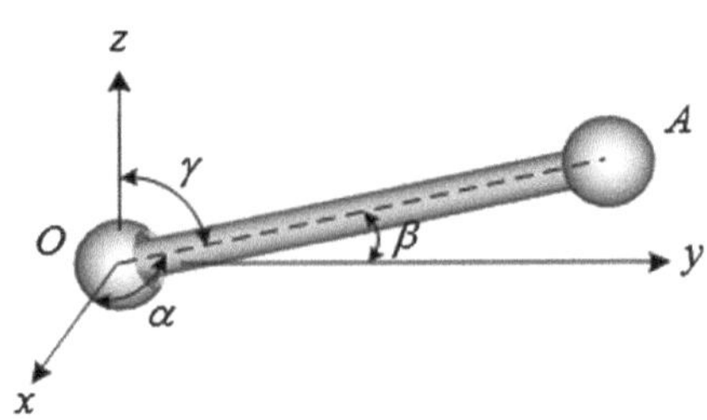

FIGURE 3.16 SS kinematic chain.

of the spherical pairs. The length of the shaft is denoted by l. The angular coordinates of the shaft are denoted by α,β, and γ. So, the coordinates of point A are$(l\cos\alpha \quad l\cos\beta \quad l\cos\gamma)$. The cross product matrix of vector **r** is:

$$\hat{\boldsymbol{r}} = \begin{bmatrix} 0 & -l\cos\gamma & l\cos\beta \\ l\cos\gamma & 0 & -l\cos\alpha \\ -l\cos\beta & l\cos\alpha & 0 \end{bmatrix}. \tag{3.81}$$

The free motions of the spherical joint O can be decomposed as three revolutions around the orthogonal coordinate axes. The pivots' directional vectors can be written as follows:

$$\boldsymbol{s}_1 = [1 \quad 0 \quad 0]^T, \boldsymbol{s}_2 = [0 \quad 1 \quad 0]^T, \boldsymbol{s}_3 = [0 \quad 0 \quad 1]^T.$$

The free motions of the spherical joint A can also be decomposed as three revolutions around the orthogonal coordinate axes. In this case, the pivots' directional vectors can be written as:

$$\boldsymbol{s}_4 = [1 \quad 0 \quad 0]^T, \boldsymbol{s}_5 = [0 \quad 1 \quad 0]^T, \boldsymbol{s}_6 = [0 \quad 0 \quad 1]^T.$$

As

$$\boldsymbol{s}_{0i} = \boldsymbol{r} \times \boldsymbol{s}_i \quad (i = 1,2,\ldots,6)$$

O is the original point of the system, so we obtain:

$$\boldsymbol{s}_{01} = \boldsymbol{s}_{02} = \boldsymbol{s}_{03} = [0 \quad 0 \quad 0]^T.$$

So the Plücker coordinates of O and A can be written as follows:

$$\begin{aligned}
&\$_1 = [1 \quad 0 \quad 0 \quad 0 \quad 0 \quad 0]^T, \$_2 = [0 \quad 1 \quad 0 \quad 0 \quad 0 \quad 0]^T,\\
&\$_3 = [0 \quad 0 \quad 1 \quad 0 \quad 0 \quad 0]^T,\\
&\$_4 = [1 \quad 0 \quad 0 \quad 0 \quad l\cos\gamma \quad -l\cos\beta]^T,\\
&\$_5 = [0 \quad 1 \quad 0 \quad -l\cos\gamma \quad 0 \quad l\cos\alpha]^T,\\
&\$_6 = [0 \quad 0 \quad 1 \quad l\cos\beta \quad -l\cos\alpha \quad 0]^T.
\end{aligned}$$

Suppose the kinematic screw set as $\$_{SS} = \{\$_1 \quad \$_2 \quad \$_3 \quad \$_4 \quad \$_5 \quad \$_6\}$. According to reciprocal screw theory, we derive the following equation:

$$\$_{SS}^T \boldsymbol{\Delta} \$_{SS}^{\mathrm{r}} = 0.$$

The rank of the kinematic screw set $\$_{SS}$ is:

$$\mathrm{Rank}(\$_{SS}) = 5.$$

The reciprocal screw can be obtained:

$$\$^r_{SS} = k[\cos\alpha \quad \cos\beta \quad \cos\gamma \quad 0 \quad 0 \quad 0]^T \tag{3.82}$$

where $k \in R, k \neq 0$.

The reciprocal screw merely denotes a linear force vector, which constrains the displacement in both directions along the axis of the limb. When one of the above joints is replaced by a universal joint, the kinematic chain is changed into *ST* or *TS*. Assume the origin of the local coordinates system is superposed with the universal joint, and one of the orthogonal axes of the universal joints lies along the *z*-axis shown in Figure 3.17. The angle from the shaft to the *xoy*-plane is denoted by ψ, and the angle from the projective line of the shaft to *x*-axis is denoted by φ. As a result, the parameters of α, β, γ, ψ and φ are not independent, and have the following relationships:

$$\begin{cases} l\cos\psi\cos\varphi = l\cos\alpha \\ l\cos\psi\sin\varphi = l\cos\beta \\ l\sin\psi = l\cos\gamma. \end{cases} \tag{3.83}$$

So we have

$$\cos\phi = \frac{\cos\alpha}{\sqrt{\cos^2\alpha + \cos^2\beta}}, \quad \sin\varphi = \frac{\cos\beta}{\sqrt{\cos^2\alpha + \cos^2\beta}}.$$

According to the coordinate system established above, the Plücker coordinates of *O* and *A* are as follows:

$$\$_1 = \left[-\frac{\cos\beta}{\sqrt{\cos^2\alpha + \cos^2\beta}} \quad \frac{\cos\alpha}{\sqrt{\cos^2\alpha + \cos^2\beta}} \quad 0 \quad 0 \quad 0 \quad 0 \right]^T,$$

$$\$_2 = [0 \quad 0 \quad 1 \quad 0 \quad 0 \quad 0]^T, \$_3 = [1 \quad 0 \quad 0 \quad 0 \quad l\cos\gamma \quad -l\cos\beta]^T,$$

$$\$_4 = [0 \quad 1 \quad 0 \quad -l\cos\gamma \quad 0 \quad l\cos\alpha]^T,$$

$$\$_5 = [0 \quad 0 \quad 1 \quad l\cos\beta \quad -l\cos\alpha \quad 0]^T.$$

The rank of the screw set $\$_{ST}$ is:

$$\text{Rank}(\$_{ST}) = 5.$$

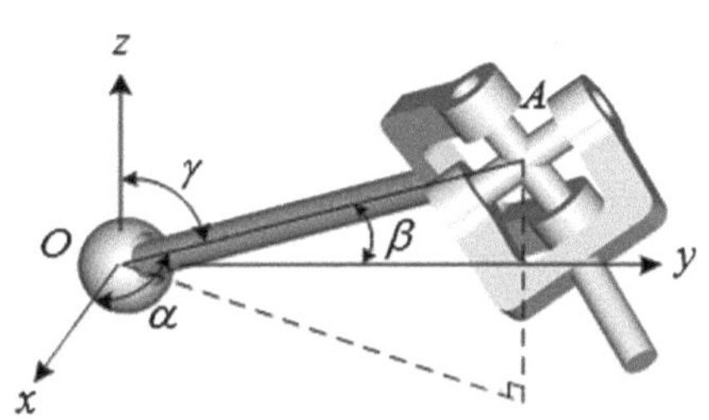

FIGURE 3.17 ST kinematic chains.

The reciprocal screw of $\$_{ST}$ is:

$$\$^r_{ST} = [\cos\alpha \quad \cos\beta \quad \cos\gamma \quad 0 \quad 0 \quad 0]^T. \tag{3.84}$$

Seen from the rank of the screw set, the *SS* kinematic chain and *TS* or *ST* kinematic chains have the same rank value. That is, Rank($\$_{SS}$) = Rank($\$_{ST}$) = 5. The reciprocal screw sets of the two are also equal. So, the kinematic chains of *TS* (or *ST*) and *SS* have the same terminal constraints. In fact, $\$_{ST}$ is one of the maximal independent sets of kinematic screws $\$_{SS}$.

From the above analysis, we can also deduce that the rank of the screws and their reciprocal screws are not influenced by the different selections of coordinate systems.

3.3.2.2 A Moment of Couple as the Terminal Constraint [4]

In the PUU kinematic chain shown in Figure 3.18, we assume that the end *O* consists of one prismatic joint and one universal joint, which are all denoted by point *O*.

The Plücker coordinates of slider *O* are:

$$\$_1 = \begin{bmatrix} 0 & 0 & 0 & 1 & 0 & 0 \end{bmatrix}^T.$$

The universal pair *O* can be decomposed as two orthogonal rotational pairs, the direction vectors of which are denoted by $\boldsymbol{e}_u^1$ and $\boldsymbol{e}_u^2$, individually, where $\boldsymbol{e}_u^1$ is parallel to the *z*-axis. As a result, $\boldsymbol{e}_u^1$ has the form of $\boldsymbol{e}_u^1 = \begin{bmatrix} 0 & 0 & 1 \end{bmatrix}^T$, $\boldsymbol{e}_u^2$ can be found by the following fomula:

$$\boldsymbol{e}_u^2 = \frac{\overrightarrow{PB_P}}{\left\|\overrightarrow{PB_P}\right\|} \times \overrightarrow{oz} = \begin{bmatrix} 0 & 0 & \sin\alpha \\ 0 & 0 & -\cos\alpha \\ -\sin\alpha & \cos\alpha & 0 \end{bmatrix} \begin{bmatrix} 0 \\ 0 \\ 1 \end{bmatrix} = \begin{bmatrix} \sin\alpha \\ -\cos\alpha \\ 0 \end{bmatrix}.$$

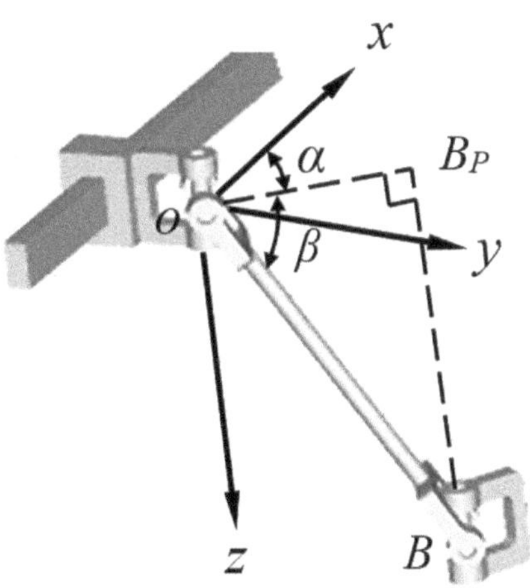

FIGURE 3.18 PUU kinematic chain.

Therefore, the individual Plücker coordinates corresponding to universal pair O are:

$$\$_2 = \begin{bmatrix} 0 & 0 & 1 & 0 & 0 & 0 \end{bmatrix}^T, \$_3 = \begin{bmatrix} \sin\alpha & -\cos\alpha & 0 & 0 & 0 & 0 \end{bmatrix}^T.$$

And the same are true for the universal pair B. The coordinates of B are:

$$B\ (l\cos\alpha\cos\beta \quad l\sin\alpha\cos\beta \quad l\sin\beta).$$

The Plücker coordinates can be denoted as:

$$\$_4 = \begin{bmatrix} s_4 \\ s_{40} \end{bmatrix}, \$_5 = \begin{bmatrix} s_5 \\ s_{50} \end{bmatrix}$$

where

$$\begin{aligned} s_4 &= \begin{bmatrix} \sin\alpha & -\cos\alpha & 0 \end{bmatrix}^T, s_5 = \begin{bmatrix} 0 & 0 & 1 \end{bmatrix}^T, \\ s_{40} &= \boldsymbol{r}_B \times s_{40} = \begin{bmatrix} l\cos\alpha\sin\beta & l\sin\alpha\sin\beta & -l\cos\beta \end{bmatrix}^T, \\ s_{50} &= \boldsymbol{r}_B \times s_{50} = \begin{bmatrix} l\sin\alpha\cos\beta & -l\cos\alpha\cos\beta & 0 \end{bmatrix}^T. \end{aligned}$$

Therefore,

$$\begin{aligned} \$_4 &= \begin{bmatrix} \sin\alpha & -\cos\alpha & 0 & l\cos\alpha\sin\beta & l\sin\alpha\sin\beta & -l\cos\beta \end{bmatrix}^T, \\ \$_5 &= \begin{bmatrix} 0 & 0 & 1 & l\sin\alpha\cos\beta & -l\cos\alpha\cos\beta & 0 \end{bmatrix}^T. \end{aligned}$$

As a result, the kinematic screws of the kinematic chain OB can be expressed as:

$$\$_{OB} = \begin{bmatrix} \$_1 & \$_2 & \$_3 & \$_4 & \$_5 \end{bmatrix}. \tag{3.85}$$

The terminal constraint of kinematic chain OB can be gained:

$$\$^r_{OB} = \begin{bmatrix} 0 & 0 & 0 & \cos\alpha & \sin\alpha & 0 \end{bmatrix}^T. \tag{3.86}$$

According to the physical meaning of the reciprocal screws, $\$^r_{OB}$ denotes a pure moment constraint that is perpendicular to the universal pair plane. And as a result, this terminal constraint will restrict the rotation about the norm direction of the universal pair plane.

3.3.3 Kinematic Chain of Other Terminal Constraints

3.3.3.1 A Spatial Six-Link Mechanism

Figure 3.19 shows a spatial six link mechanism *ABCDEF* [5]. We create a local coordinate system *oxyz* as Figure 3.19 shows, in which the x-axis is perpendicular to the plane of kinematic chain *ABC*. The y'-axis is the axis of the

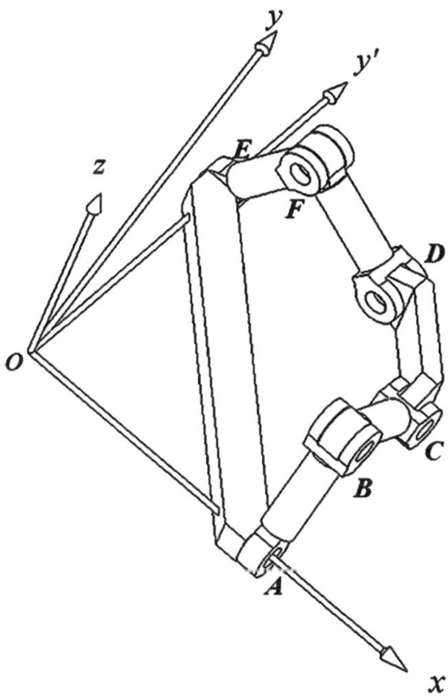

FIGURE 3.19 Spatial six-link mechanism with a straight-line translational end effector.

revolute joint *F* which is perpendicular to the plane of kinematic chain *DEF*. It is not difficult to find that α is the angle from the *x*-coordinate to the *y*'-axis and $\boldsymbol{n}_{FED} = (\cos\alpha \quad \sin\alpha \quad 0)^T$. Then, we get the coordinates of different joints: $A(a,0,0)$, $B(a,y_B,z_B)$, $C(a,y_C,z_C)$, $D(x_D,y_D,z_D)$, $E(x_E,y_E,z_E)$, $F(x_F,y_F,z_F)$.

Then, the twists of the first kinematic chain *ABC* will be:

$$\$_{ABC} = [\$^1_{ABC} \quad \$^2_{ABC} \quad \$^3_{ABC}] \tag{3.87}$$

where $\$^1_{ABC} = [1 \quad 0 \quad 0 \quad 0 \quad 0 \quad 0]^T$, $\$^2_{ABC} = [1 \quad 0 \quad 0 \quad 0 \quad z_B \quad -y_B]^T$, $\$^3_{ABC} = [1 \quad 0 \quad 0 \quad 0 \quad z_C \quad -y_C]^T$.

According to equation (3.53), we have the terminal constraint screw matrix of kinematic chain *ABC*

$$\$^r_{ABC} = \begin{bmatrix} 1 & 0 & 0 \\ 0 & 0 & 0 \\ 0 & 0 & 0 \\ 0 & 0 & 0 \\ 0 & 1 & 0 \\ 0 & 0 & 1 \end{bmatrix}. \tag{3.88}$$

Similarly, the twists of the kinematic chain *FDE* will be:

$$\$_{FED} = [\$^1_{FED} \quad \$^2_{FED} \quad \$^3_{FED}] \tag{3.89}$$

where

$$\begin{aligned}
\$^1_{FED} &= [\cos\alpha \quad \sin\alpha \quad 0 \quad -z_F\sin\alpha \quad z_F\cos\alpha \quad x_F\sin\alpha - y_F\cos\alpha]^T \\
\$^2_{FED} &= [\cos\alpha \quad \sin\alpha \quad 0 \quad -z_E\sin\alpha \quad z_E\cos\alpha \quad x_E\sin\alpha - y_E\cos\alpha]^T \\
\$^3_{FED} &= [\cos\alpha \quad \sin\alpha \quad 0 \quad -z_D\sin\alpha \quad z_D\cos\alpha \quad x_D\sin\alpha - y_D\cos\alpha]^T.
\end{aligned}$$

According to equation (2.98), we have the terminal constraint screw matrix of kinematic chain *FED*

$$\$^r_{FED} = \begin{bmatrix} \cos\alpha & 0 & 0 \\ \sin\alpha & 0 & 0 \\ 0 & 0 & 0 \\ 0 & -\sin\alpha & 0 \\ 0 & \cos\alpha & 0 \\ 0 & 0 & 1 \end{bmatrix}. \tag{3.90}$$

From the above analysis, it is not difficult to find that the two kinematic chains *ABC* and *FED* can surely safeguard the pure rectilinear motion of the end effector *CD* so long as $\$_{ABC}$ and $\$_{FED}$ do not degenerate.

$$\$_{CD}(\alpha) = \begin{bmatrix} 1 & \cos\alpha & 0 & 0 & 0 \\ 0 & \sin\alpha & 0 & 0 & 0 \\ 0 & 0 & 0 & 0 & 0 \\ 0 & 0 & \sin\alpha & 0 & 0 \\ 0 & 0 & -\cos\alpha & 1 & 0 \\ 0 & 0 & 0 & 0 & 1 \end{bmatrix} \tag{3.91}$$

Let $\det\left(\$^r_{CD}(\alpha)\right) = 0$, we can deduce that $\alpha = 0$ or $\alpha = \pi$. Therefore, in order to keep the end effector *CD* having one straight-line translation along the *z*-axis, there will be $\alpha \neq 0$ and $\alpha \neq \pi$. According to equation (3.53), we have the terminal constraint screw of kinematic chain *CD*:

$$\$^r_{CD} = [0 \quad 0 \quad 0 \quad 0 \quad 0 \quad 1]^T. \tag{3.92}$$

Let $\mathbf{A} = \$^r_{CD}(\alpha)$ and

$$\text{Cond}(\mathbf{A})_2 = \|\mathbf{A}\|_2 \left\|\mathbf{A}^{-1}\right\|_2 = \sqrt{\frac{\lambda_{\max}\left(\mathbf{A}^T\mathbf{A}\right)}{\lambda_{\min}\left(\mathbf{A}^T\mathbf{A}\right)}} = 1 \tag{3.93}$$

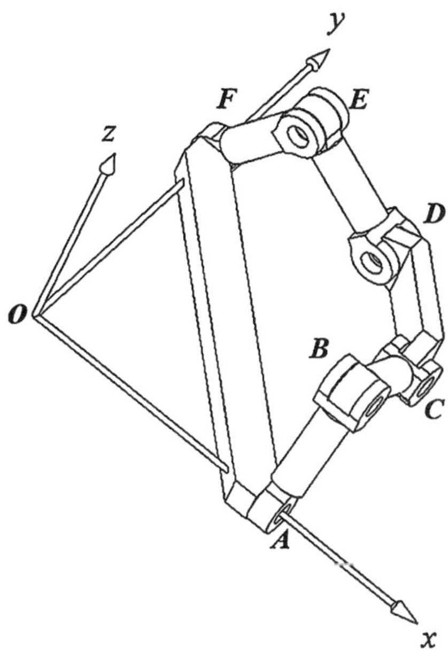

FIGURE 3.20 Spatial six-link mechanism with the best structure stability.

where Cond $(\mathbf{A})_2$ indicate the condition number of matrix $\mathbf{A}$, $\|\mathbf{A}\|_2$ indicates the 2-norm of matrix $\mathbf{A}$, and $\lambda\left(\mathbf{A}^T\mathbf{A}\right)$ indicates the eigenvalue of matrix $\mathbf{A}^T\mathbf{A}$.

The solution of equation (3.93) is:

$$\alpha = \frac{\pi}{2}. \tag{3.94}$$

Therefore, when $\alpha = \frac{\pi}{2}$, the mechanism will have the best structure stability which is shown in Figure 3.20.

3.3.3.2 A Bionic Robot Mechanism

A bionic mechanism for a tortoise is shown in Figure 3.21, the kinematic chain of its leg can be assumed to be that shown in Figure 3.22, which consists of one prismatic pair and two spherical pairs.

In what follows, we will investigate the synthesis problem of the kinematic chain shown in Figure 3.22.

The coordinate system can be set with the origin superposing with one of the spherical pair A, and the z-axis being along the center line of the kinematic pairs A and B. Assume the distance between A and B is $l (l > 0)$, and the angle from the prismatic guide C to the z-axis is α. As a result, the coordinates of pair B are $(0 \quad 0 \quad l)$, the kinematic screw set, $\$_{ABC}$, includes:

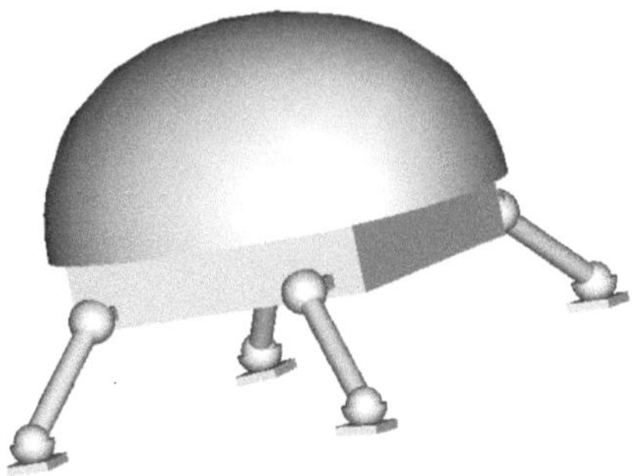

FIGURE 3.21 A bionic mechanism for tortoise.

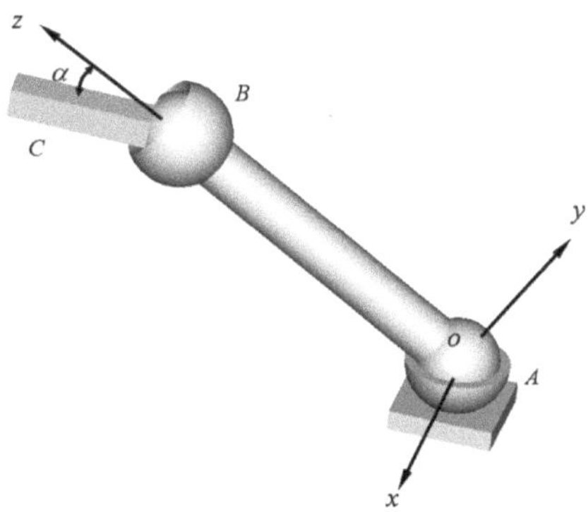

FIGURE 3.22 One leg of the bionic mechanism for tortoise.

1. The kinematic screws of spherical pair A are

$$\$_1 = [1\ \ 0\ \ 0\ \ 0\ \ 0\ \ 0]^T, \$_2 = [0\ \ 1\ \ 0\ \ 0\ \ 0\ \ 0]^T,$$
$$\$_3 = [0\ \ 0\ \ 1\ \ 0\ \ 0\ \ 0]^T;$$

2. The kinematic screws of spherical pair B are

$$\$_4 = [1\ \ 0\ \ 0\ \ 0\ \ l\ \ 0]^T, \$_5 = [0\ \ 1\ \ 0\ \ -l\ \ 0\ \ 0]^T,$$
$$\$_6 = [0\ \ 0\ \ 1\ \ 0\ \ 0\ \ 0]^T;$$

3. The kinematic screws of prismatic pair C are

$$\$_7 = [0\ \ 0\ \ 0\ \ 0\ \ -\sin\alpha\ \ \cos\alpha]^T.$$

Therefore,

$$\$_{ABC} = [\$_1\ \ \$_2\ \ \$_3\ \ \$_4\ \ \$_5\ \ \$_6\ \ \$_7] \tag{3.95}$$

and

$$\text{Rank}(\$_{ABC}) = 6. \tag{3.96}$$

The maximal linearly independent set can be selected as:

$$\$^{M}_{ABC} = \begin{bmatrix} 1 & 0 & 0 & 1 & 0 & 0 \\ 0 & 1 & 0 & 0 & 1 & 0 \\ 0 & 0 & 1 & 0 & 0 & 0 \\ 0 & 0 & 0 & 0 & -l & 0 \\ 0 & 0 & 0 & l & 0 & -\sin\alpha \\ 0 & 0 & 0 & 0 & 0 & \cos\alpha \end{bmatrix}. \tag{3.97}$$

When we select $\$ = [\$_1 \ \ \$_2 \ \ \$_3 \ \ \$_4 \ \ \$_5 \ \ \$_6 \ \ \$_7]$ as the screws of the kinematic chain, we can synthesize the kinematic chain ABC in Figure 3.23 to replace that in Figure 3.22. In the case of Figure 3.23, the elements of screw set $\$$ can be divided as:

1. The kinematic screws of spherical pair A are

$$\$_1 = [1 \ \ 0 \ \ 0 \ \ 0 \ \ 0 \ \ 0]^T, \$_2 = [0 \ \ 1 \ \ 0 \ \ 0 \ \ 0 \ \ 0]^T,$$
$$\$_3 = [0 \ \ 0 \ \ 1 \ \ 0 \ \ 0 \ \ 0]^T;$$

2. The kinematic screws of universal pair B are

$$\$_4 = [1 \ \ 0 \ \ 0 \ \ 0 \ \ l \ \ 0]^T, \$_5 = [0 \ \ 1 \ \ 0 \ \ -l \ \ 0 \ \ 0]^T;$$

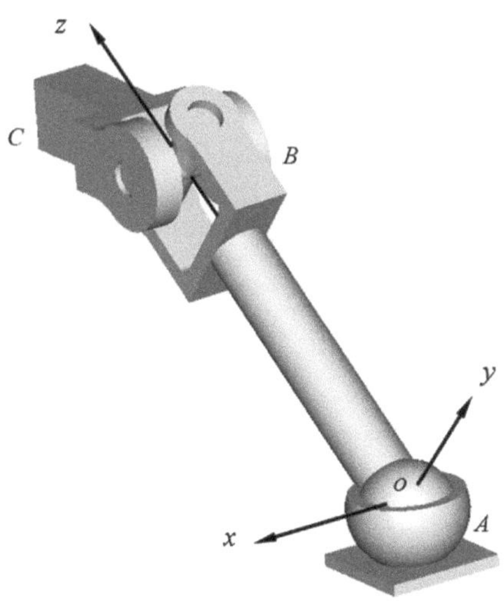

FIGURE 3.23 Simplification of the bionic mechanism for a Tortoise.

3. The kinematic screws of spherical pair C are

$$\$_7 = [0 \quad 0 \quad 0 \quad 0 \quad -\sin\alpha \quad \cos\alpha]^T.$$

From the above two examples, we can see that, in the general cases, the maximum linearly independent sets of the kinematic chains *GPA* and *ABC* can all be spanned into the same six-dimensional space and have the same terminal constraints. Therefore, each serial kinematic chain of the bionic mechanism shown in Figure 3.22 can also be simplified by the *SUP* kinematic chain shown in Figure 3.23.

3.4 CONCLUSIONS

This chapter focuses on the equivalent substitution of kinematic chains in engineering applications. Through proving that the maximum linearly independent set of kinematic screws of a serial kinematic chain determines the terminal constraint of the chain, this chapter proposes a synthesis theorem and presents an innovation methodology for serial kinematic chains. It provides an innovative methodology for kinematic chains with invariable principles of terminal constraint. According to the reciprocal screw theory, the terminal constraint of a kinematic chain is decided by the maximal independent set of kinematic screws and the simplest screws for the branch kinematic chains can be obtained through algebraic analysis. So, the best kinematic chain, whose terminal constraints will not change, can be synthesized by combining the kinematic screws of the maximal independent set. This allows the engineers to possibly use the most economic and simplest kinematic chains to design the mechanisms with some specified requirements.

REFERENCES

[1] K.H. Hunt, Kinematic Geometry of Mechanisms, Oxford University Press, Oxford, 1978.
[2] R.S. Ball, A Treatise on the Theory of Screws, Cambridge University Press, 1900.
[3] J.-S. Zhao, F.-L. Chu, Z.-J. Feng, Identification of principal screws of two- and three-screw systems, Proceedings of the Institution of Mechanical Engineers, Part C. Journal of Mechanical Engineering Science 221 (2007) 1701–1715.
[4] J.-S. Zhao, K. Zhou, Z.J. Feng, A theory of degrees of freedom for mechanisms, Mechanism and Machine Theory 39 (2004) 621–643.
[5] J.S. Zhao, F.L. Chu, Z.J. Feng, Synthesis of rectilinear motion generating spatial mechanism with application to automotive suspension, Journal of Mechanical Design 130 (2008) 065001-1–065001-18.

Chapter 4

Free Motion of the End Effector of a Robot Mechanism

In Chapter 3, we discussed the reciprocity of twists and wrenches, and the free relative motion between two paired members and the constraint induced by their pairs. Next, we will investigate the terminal constraints of a kinematic chain and free motions induced by the constraints [1–4].

4.1 FREE MOTION SPACE AND CONSTRAINT SPACE OF KINEMATIC CHAIN

As proven in Chapter 2, the necessary and sufficient criterion for equation (2.98) is:

$$\left(\$^{T}\boldsymbol{\Delta}\right)\$^{r} = 0 \tag{4.1}$$

where $\$$ is a $6 \times m$ matrix made up of m unit twists of the chain, and $\$^{r}$ represents any one of the unit terminal constraint(s) of the kinematic chain.

From the analysis in Chapter 3, it is not difficult to find that the reciprocity of unit motion(s) and the terminal constraint(s) provide a perfect means to analyze the instantaneous constraint(s) exerted on the end effector by its every kinematic chain.

Assume

$$\mathbf{A} = \$^{T}\boldsymbol{\Delta} \tag{4.2}$$

and

$$\boldsymbol{x} = \$^{r}. \tag{4.3}$$

Equation (4.1) can be rewritten as

$$\mathbf{A}\boldsymbol{x} = 0. \tag{4.4}$$

Next, the solution spaces of the homogeneous linear equation set (4.4) will be discussed. Within equation (4.4), $\mathbf{A}$ is a matrix of $m \times 6(m = 1,2,3,\ldots)$, and therefore, the solution space $N(\mathbf{A})$ should be $6 \times n(n = 1,2,3,\ldots,6)$. Although, there are perfect interpretations of the abstract relationship between $\mathbf{A}$ and $N(\mathbf{A})$

Advanced Theory of Constraint and Motion Analysis for Robot Mechanisms.
http://dx.doi.org/10.1016/B978-0-12-420162-0.00004-7

in any advanced algebra, this chapter mainly explores their relationships from physical meanings.

First, let us consider the singular value decomposition (SVD) of matrix **A**. Assume **A** has the following form of SVD [4]:

$$\mathbf{A} = \mathbf{US}\boldsymbol{Q}^T \tag{4.5}$$

where **U** is an $m \times m$ matrix; $\mathbf{S} = \begin{bmatrix} \mathbf{M}_r & 0 \\ 0 & 0 \end{bmatrix}_{m\times 6}$, $\mathbf{M}_r = \mathrm{diag}\left(\sigma_1,\sigma_2,\ldots,\sigma_r\right)$, $\sigma_i > 0 (i > 1,2\cdots r)$ and $r = \mathrm{Rank}(\mathbf{A})$ which means that the rank of matrix **A**, $\boldsymbol{Q}$ is a 6×6 matrix; **U** and $\boldsymbol{Q}$ are all orthogonal matrices.

Substituting equation (4.5) into equation (4.4) yields:

$$\mathbf{US}\boldsymbol{Q}^T \boldsymbol{x} = 0. \tag{4.6}$$

Presuming

$$\boldsymbol{y} = \boldsymbol{Q}^T \boldsymbol{x} \tag{4.7}$$

and multiplying $\mathbf{U}^T$ at both sides of equation (4.6) yields:

$$\mathbf{S}\boldsymbol{y} = 0. \tag{4.8}$$

Assume

$$N(\mathbf{S}) = \{\boldsymbol{y} \in \boldsymbol{R}^6 \,|\mathbf{S}\boldsymbol{y} = 0\}. \tag{4.9}$$

Obviously, $N(\mathbf{S})$ denotes the solution space of equation (4.8). Therefore, according to equation (4.7), we can presume

$$N(\mathbf{A}) = \{\boldsymbol{x} \,|\boldsymbol{x} = \boldsymbol{Q}y, \boldsymbol{y} \in N(\mathbf{S}\,)\}. \tag{4.10}$$

Therefore, $N(\mathbf{A})$ denotes the solution space of equation (4.4). In engineering, $N(\mathbf{A})$ represents the spaces spanned by the terminal constraint(s) of the kinematic chain *\$* and denotes the twists of this chain. Reciprocally, $N(\mathbf{A})$ represents the spaces spanned by the free motion(s) of the end effector and *\$* denotes the constraints exerted on the end effector.

Equation (4.7) indicates that the solution space of a homogeneous linear equation set can be obtained through SVD while the basis of the solution space $N(\mathbf{S})$ can be directly gained. The dimension of $N(\mathbf{A})$ is

$$\dim N(\mathbf{A}) = \dim N(\mathbf{S}) = 6 - r. \tag{4.11}$$

Bearing in mind the above general theory, we can find both the terminal constraints and the free motions of the end effector of a spatial manipulator.

4.2 GENERAL STEPS TO ANALYZE THE DEGREE OF FREEDOM OF THE END EFFECTOR

The following are the general steps:

(1) **Decompose the System**

In general, we can decompose the system into n limbs connected with the end effector. The rule of decomposition is as follows:

We can assume that each limb connects once and only once with the fixed base, otherwise the limb contains sub-parallel chains which can be replaced with the screw systems of the sub-parallel chains.

(2) **Study the $ith(1 \leq i \leq n)$ Limb and Write out the Plücker Coordinates of the Kinematic Pairs of the Limb**

To facilitate the description, a local coordinate system often needs to be created for every limb. The rule of the establishment of the coordinate system is as follows:

Select as many lines along, parallel, or perpendicular to the kinematic pairs as possible as coordinate axes, such as the prismatic pairs or revolute pairs, which can facilitate the Plücker coordinates.

After the establishment of the coordinate system, the Plücker coordinates of every kinematic pairs can be written out.

(3) **Simplify the Kinematic Screws of the $ith(1 \leq i \leq n)$ Limb**

Make up the kinematic screws $\mathbf{\$}_i$ of the $ith(1 \leq i \leq n)$ limb.

(4) **Solve the Reciprocal Screws $\$_i^r$ of the $ith(1 \leq i \leq n)$ Limb**

Assume $\mathbf{A} = (\mathbf{\$}_i)^T\mathbf{\Delta}$ and substitute it into equation (4.2). According to equation (4.1), we could work out the unit terminal constraint of the ith kinematic chain $\$_i^r$.

(5) **Solve the Reciprocal Screws of the $(i+1)th(1 \leq i \leq n-1)$ Limb in Turn**

When the kinematic screws of the $(i+1)th(1 \leq i \leq n-1)$ limb are all the same as those of any of the branches solved, then move on to the next step; otherwise, return to execute steps 2, 3, and 4, until all of the reciprocal screws of every limb of the mechanism have been solved.

(6) **Decide the Dimension of the Reciprocal Screws That Can Be Spanned in the Normal Linear Spaces and Compute the Degree of Freedom of the End Effector, F, of the System**

Suppose $\mathbf{\$}^r = \{\$_{i^r} \,|\, i = 1,2,\ldots n\}$ and $\mathbf{A} = (\mathbf{\$}^r)^T\mathbf{\Delta}$. Substituting A into equation (4.4), we could calculate the spaces spanned by the terminal constraint(s) of the kinematic chain $N(\mathbf{A})$ according to equation (4.11).

$$\dim N(\mathbf{A}) = 6 - r \tag{4.12}$$

4.3 APPLICATION OF THE ANALYTICAL THEORY OF THE DEGREE OF FREEDOM OF THE END EFFECTOR

4.3.1 A Planar Four-Bar Mechanism

A planar four-bar mechanism and the absolute coordinate system are shown in Figure 4.1. Because link BC is connected with the fixed base AD through crank AB and link DC, we only need to select any one of them to analyze the reciprocal screws exerted on link BC. For example, we select crank AB as the investigation object shown in Figure 4.2.

The Plücker coordinates of revolute pair A can be obtained in the absolute coordinate system shown in Figure 4.1 and the local coordinate system shown in Figure 4.2 which is just superposed with the absolute one:

$$\$_1 = [1 \quad 0 \quad 0 \quad 0 \quad 0 \quad 0]^T, \quad \$_2 = [1 \quad 0 \quad 0 \quad l_1 \sin\alpha_1 \quad -l_1\cos\alpha_1 \quad 0]^T.$$

Therefore, the kinematic screws of crank AB can be expressed as:

$$\$_{AB} = [\$_1 \quad \$_2]. \tag{4.13}$$

So, the reciprocal screws of kinematic chain AB can be gained with algebra method:

$$\$^r_{AB} = \left[\$^r_1 \quad \$^r_2 \quad \$^r_3 \quad \$^r_4\right] \tag{4.14}$$

where

$\$^r_1 = [1 \quad 0 \quad 0 \quad 0 \quad 0 \quad 0]^T$, which represents a pure force along the x-axis;

$\$^r_2 = [0 \quad 0 \quad 0 \quad 0 \quad 1 \quad 0]^T$, which represents a moment of couple around the y-axis;

$\$^r_3 = [0 \quad 0 \quad 0 \quad 0 \quad 0 \quad 1]^T$, which represents a moment of couple around the z-axis;

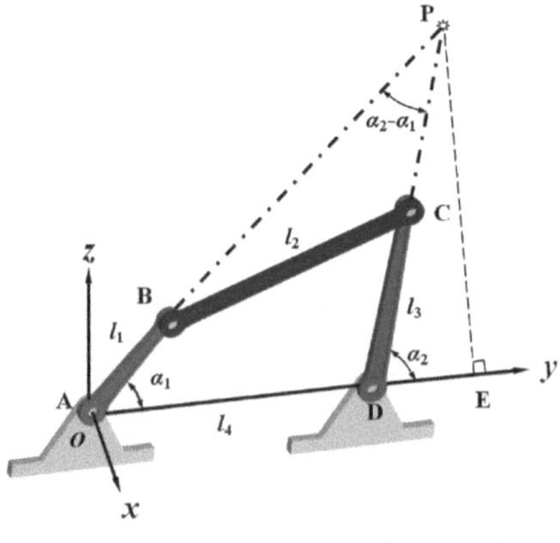

FIGURE 4.1 A planar four-bar mechanism and the absolute coordinate system.

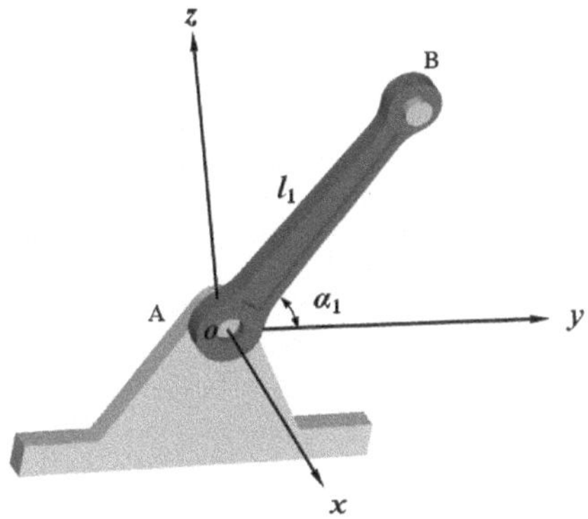

FIGURE 4.2 Local coordinate system for crank AB.

and $\$_4^r = [0 \ \ \cos\alpha_1 \ \ \sin\alpha_1 \ \ 0 \ \ 0 \ \ 0]^T$, which represents a pure force along the centerline of crank *AB*.

Similarly, we can solve the reciprocal screws of crank *DC* to be:

$$\$_{DC}^r = \left[\$_5^r \ \ \$_6^r \ \ \$_7^r \ \ \$_8^r\right] \tag{4.15}$$

where

$$\$_5^r = [1 \ \ 0 \ \ 0 \ \ 0 \ \ 0 \ \ 0]^T, \$_6^r = [0 \ \ 0 \ \ 0 \ \ 0 \ \ 1 \ \ 0]^T,$$
$$\$_7^r = [0 \ \ 0 \ \ 0 \ \ 0 \ \ 0 \ \ 1]^T, \$_8^r = [0 \ \ \cos\alpha_2 \ \ \sin\alpha_2 \ \ l_4\sin\alpha_2 \ \ 0 \ \ 0]^T.$$

Therefore, the constraints exerted on link *BC* can be denoted by:

$$\$_{BC}^r = \left[\$_1^r \ \ \$_2^r \ \ \$_3^r \ \ \$_4^r \ \ \$_8^r\right]. \tag{4.16}$$

As a result, the free motion link *BC* gained can be found according to the reciprocal screw theory:

$$\$_{BC} = [\sin(\alpha_2 - \alpha_1) \ \ 0 \ \ 0 \ \ 0 \ \ l_4\sin\alpha_1\sin\alpha_2 \ \ -l_4\cos\alpha_1\sin\alpha_2]^T. \tag{4.17}$$

Equation (4.17) indicates a twist motion. When $\alpha_2 \neq \alpha_1$, equation (4.17) can be simplified as:

$$\$_{BC} = \left[1 \ \ 0 \ \ 0 \ \ 0 \ \ \frac{l_4\sin\alpha_1\sin\alpha_2}{\sin(\alpha_2-\alpha_1)} \ \ -\frac{l_4\cos\alpha_1\sin\alpha_2}{\sin(\alpha_2-\alpha_1)}\right]^T. \tag{4.18}$$

With equation (2.17), we find that the projection point of the origin on the screw axis should be:

$$\boldsymbol{r}_P = l_4\left[0 \ \ \frac{\cos\alpha_1\sin\alpha_2}{\sin(\alpha_2-\alpha_1)} \ \ \frac{\sin\alpha_1\sin\alpha_2}{\sin(\alpha_2-\alpha_1)}\right]^T. \tag{4.19}$$

Therefore, link BC (the end effector) will instantaneously make a rotation around the axis passing through point $\left(0 \quad \frac{l_4 \cos\alpha_1 \sin\alpha_2}{\sin(\alpha_2-\alpha_1)} \quad \frac{l_4 \sin\alpha_1 \sin\alpha_2}{\sin(\alpha_2-\alpha_1)}\right)$ with the direction of $(1 \quad 0 \quad 0)^T$.

In the triangle ADP shown in Figure 4.1, we can get the following equation with the law of sines:

$$\frac{l_4}{\sin(\alpha_2-\alpha_1)} = \frac{PD}{\sin\alpha_1}. \tag{4.20}$$

Therefore,

$$PD = \frac{l_4 \sin\alpha_1}{\sin(\alpha_2-\alpha_1)}. \tag{4.21}$$

In the right triangle DEP, there is:

$$PE = PD\sin\alpha_2 = \frac{l_4 \sin\alpha_1 \sin\alpha_2}{\sin(\alpha_2-\alpha_1)}. \tag{4.22}$$

Similarly, we obtain the length of OE in right triangle OEP:

$$OE = PE\cot\alpha_1 = \frac{l_4 \cos\alpha_1 \sin\alpha_2}{\sin(\alpha_2-\alpha_1)}. \tag{4.23}$$

From equations (4.22) and (4.23), we know that the coordinates of point P shown in Figure 4.1 are $\left(0 \quad \frac{l_4 \cos\alpha_1 \sin\alpha_2}{\sin(\alpha_2-\alpha_1)} \quad \frac{l_4 \sin\alpha_1 \sin\alpha_2}{\sin(\alpha_2-\alpha_1)}\right)$, which is expressed by equation (4.19). Therefore, P is the right projection point of origin on this instantaneous screw axis denoted by equation (4.18).

When $\alpha_2 = \alpha_1$, equation (4.17) can be simplified as:

$$\$_{BC} = [0 \quad 0 \quad 0 \quad 0 \quad \sin\alpha_1 \quad -\cos\alpha_1]^T. \tag{4.24}$$

Therefore, link BC will instantaneously make a translational movement along the direction $(0 \quad \sin\alpha_1 \quad -\cos\alpha_1)^T$, which is just perpendicular to the centerline of the crank. As a matter of fact, when $\alpha_2 = \alpha_1$, the mechanism becomes a parallel four-bar linkage as shown in Figure 4.3, link BC will always make translational movement that is perpendicular to the centerline of the crank AB.

4.3.2 A Planar Parallel Mechanism with an Output Member

A planar nine-bar mechanism is shown in Figure 4.4 [2]. When link EF is specified as the output member, this planar mechanism can be thought of as a parallel mechanism that is made up of two kinematic chains, $ABCDE$ and $IHJF$. The origin of the coordinate system is superimposed with pair A, the x-axis is along the horizontal direction and the y-axis is along the vertical direction shown in Figure 4.4.

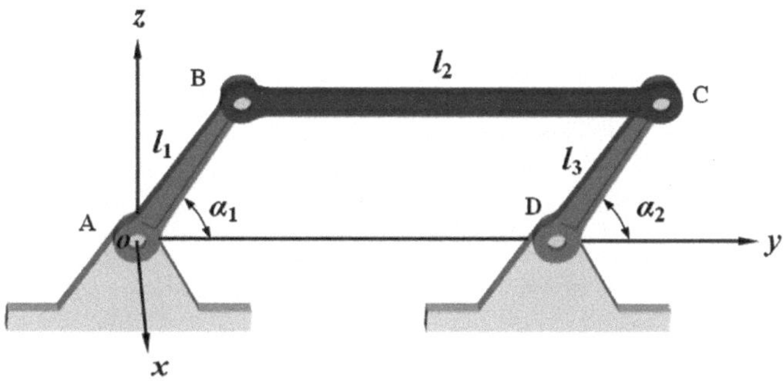

FIGURE 4.3 Planar parallel four-bar linkage.

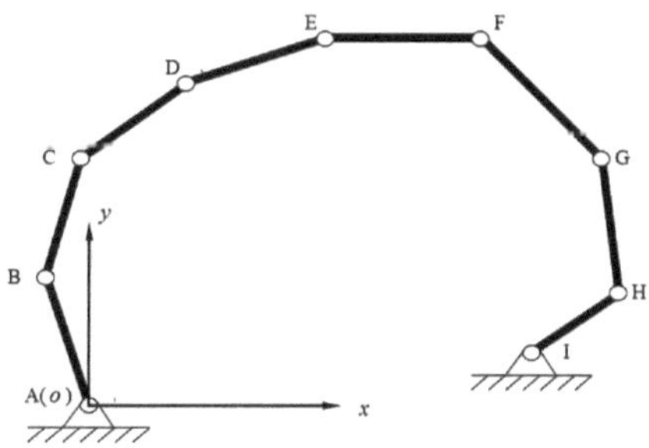

FIGURE 4.4 Coordinate system of a 9-bar planar mechanism.

The twist matrix of the kinematic chain *ABCDE* are:

$$\$_{ABCDE} = [\$_A \quad \$_B \quad \$_C \quad \$_D \quad \$_E] \tag{4.25}$$

where

$$\$_A = [0 \quad 0 \quad 1 \quad 0 \quad 0 \quad 0]^T, \quad \$_B = [0 \quad 0 \quad 1 \quad y_B \quad -x_B \quad 0]^T,$$
$$\$_C = [0 \quad 0 \quad 1 \quad y_C \quad -x_C \quad 0]^T, \quad \$_D = [0 \quad 0 \quad 1 \quad y_D \quad -x_D \quad 0]^T,$$
$$\$_E = [0 \quad 0 \quad 1 \quad y_E \quad -x_E \quad 0]^T.$$

The rank of twist matrix of the chain can be gained with the algebra method:

$$\text{Rank}(\$_{ABCDE}) = 3. \tag{4.26}$$

The rank deficiency of the matrix is:

$$n_{ABCDE} - \text{Rank}(\$_{ABCDE}) = 5 - 3 = 2. \tag{4.27}$$

The reciprocal screws of the kinematic chain can be obtained with equation (2.98) and expressed in the form of the matrix below:

$$\mathbf{\$}^r_{ABCDE} = \begin{bmatrix} \$^r_1 & \$^r_2 & \$^r_3 \end{bmatrix} \tag{4.28}$$

where

$$\$^r_1 = [0 \ \ 0 \ \ 1 \ \ 0 \ \ 0 \ \ 0]^T, \$^r_2 = [0 \ \ 0 \ \ 0 \ \ 1 \ \ 0 \ \ 0]^T,$$
$$\$^r_3 = [0 \ \ 0 \ \ 0 \ \ 0 \ \ 1 \ \ 0]^T.$$

$\mathbf{\$}^r_{ABCDE}$ are just the planar constraints.

The twist matrix of the kinematic chain *IHGF* is:

$$\mathbf{\$}_{IHGF} = [\$_I \ \ \$_H \ \ \$_G \ \ \$_F] \tag{4.29}$$

where

$$\$_I = [0 \ \ 0 \ \ 1 \ \ y_I \ \ -x_I \ \ 0]^T, \ \ \$_H = [0 \ \ 0 \ \ 1 \ \ y_H \ \ -x_H \ \ 0]^T,$$
$$\$_G = [0 \ \ 0 \ \ 1 \ \ y_G \ \ -x_G \ \ 0]^T, \ \ \$_F = [0 \ \ 0 \ \ 1 \ \ y_F \ \ -x_F \ \ 0]^T.$$

$$\text{Rank}(\mathbf{\$}_{ABCDE}) = 3 \tag{4.30}$$

The rank deficiency of the matrix is:

$$n_{IHGF} - \text{Rank}(\mathbf{\$}_{IHGF}) = 4 - 3 = 1. \tag{4.31}$$

The reciprocal screws of the kinematic chain can be obtained and expressed in the form of screw matrix:

$$\mathbf{\$}^r_{IHGF} = \begin{bmatrix} \$^r_1 & \$^r_2 & \$^r_3 \end{bmatrix} \tag{4.32}$$

where

$$\$^r_1 = [0 \ \ 0 \ \ 1 \ \ 0 \ \ 0 \ \ 0]^T, \ \ \$^r_2 = [0 \ \ 0 \ \ 0 \ \ 1 \ \ 0 \ \ 0]^T,$$
$$\$^r_3 = [0 \ \ 0 \ \ 0 \ \ 0 \ \ 1 \ \ 0]^T.$$

$\mathbf{\$}^r_{IHGF}$ also denotes the same planar constraints.

The constraint spaces that the reciprocal screws of the output member can be spanned are

$$\mathbf{S}_C = \left\{ \boldsymbol{T}_z \ \ \boldsymbol{R}_x \ \ \boldsymbol{R}_y \right\}. \tag{4.33}$$

Therefore, the DOF of the output member *EF* is:

$$F = 6 - \text{Rank}\left(\mathbf{S}_C\right) = 6 - 3 = 3. \tag{4.34}$$

The procedure of analyzing the configuration DOF of the mechanism with an output member will be discussed in Chapter 9.

4.3.3 A 3-UPU Spatial Parallel Manipulator

In what follows, we will study the DOF of the end effector as Figure 4.5 shows. Because all the kinematic chains connecting the end effector with the base are isomorphic, what we should do is to select one of them to study the reciprocal screws of every kinematic chain, say, we select limb $B_1P_1M_1$ as the chain to analyze, the sketch of which is shown in Figure 4.6.

The universal pair B_1 can be decomposed as two orthogonal revolute pairs. The individual Plücker coordinates are:

$$\$_1=[1\quad 0\quad 0\quad 0\quad 0\quad 0]^T,\quad \$_2=[0\quad 0\quad 1\quad 0\quad 0\quad 0]^T.$$

Presume that the angle from the limb to the y-axis is β_1, the Plücker coordinates of the slider P_1 are:

$$\$_3=[0\quad 0\quad 0\quad 0\quad \cos\beta_1\quad \sin\beta_1]^T.$$

Assume $B_1M_1 = l_1$, the coordinates of the universal pairs M_1 are $(0\quad l_1\cos\beta_1\quad l_1\sin\beta_1)$.

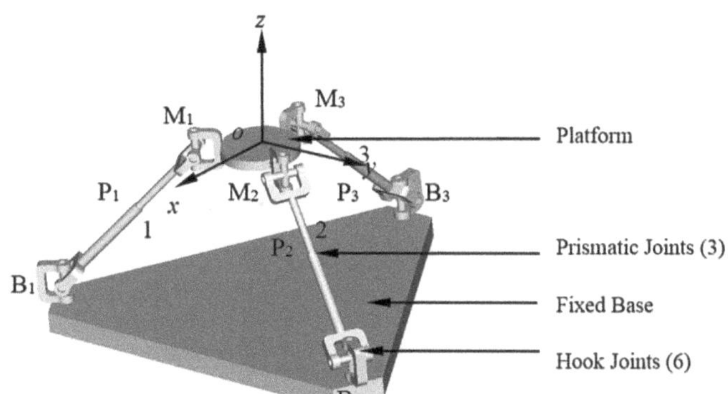

FIGURE 4.5 One spatial parallel mechanism made up of 3-UPU.

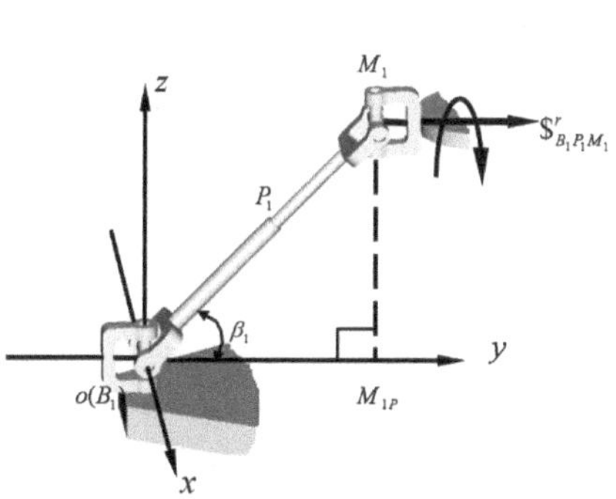

FIGURE 4.6 Reciprocal screw system of the kinematic chain UPU.

Similarly, M_1 can also be decomposed as two orthogonal revolute pairs, the Plücker coordinates are:

$$\$_4 = [1 \ \ 0 \ \ 0 \ \ 0 \ \ l_1 \sin\beta_1 \ \ -l_1 \cos\beta_1]^T, \quad \$_5 = [0 \ \ 0 \ \ 1 \ \ l_1 \cos\beta_1 \ \ 0 \ \ 0]^T.$$

Therefore, the twist matrix of kinematic chain $B_1P_1M_1$ can be denoted as:

$$\$_{B_1P_1M_1} = [\$_1 \ \ \$_2 \ \ \$_3 \ \ \$_4 \ \ \$_5]. \tag{4.35}$$

So the reciprocal twist matrix of kinematic chain $B_1P_1M_1$ are:

$$\$^r_{B_1P_1M_1} = [0 \ \ 0 \ \ 0 \ \ 0 \ \ 1 \ \ 0]^T. \tag{4.36}$$

According to the physical meaning of the reciprocal screws, $\$^r_{B_1P_1M_1}$ is just a pure couple that is perpendicular to the universal pair plane.

The movement constrained by $\$^r_{B_1P_1M_1}$ is:

$$\$^m_{B_1P_1M_1} = [0 \ \ 1 \ \ 0 \ \ 0 \ \ 0 \ \ 0]^T. \tag{4.37}$$

The same is true for the other two limbs of the end effector. So, the reciprocal screws applied to the end effector are three pure moments of couples, which are all perpendicular to the direction of the *Z*-axis, shown in Figure 4.7.

Suppose that the angles from $\$^r_{B_iP_iM_i}, i = 1,2,3$ to the *x*-axis of the coordinate system, *OXYZ*, of the end effector shown in Figure 4.7 are denoted with α_i, there will be:

$$\$^C_P = \text{Span}\left\{\left[\$^r_{B_1P_1M_1} \ \ \$^r_{B_2P_2M_2} \ \ \$^r_{B_3P_3M_3}\right]\right\} \tag{4.38}$$

where

$$\$^r_{B_1P_1M_1} = [\cos\alpha_1 \ \ \sin\alpha_1 \ \ 0 \ \ 0 \ \ 0 \ \ 0]^T,$$
$$\$^r_{B_2P_2M_2} = [\cos\alpha_2 \ \ \sin\alpha_2 \ \ 0 \ \ 0 \ \ 0 \ \ 0]^T,$$
$$\$^r_{B_3P_3M_3} = [\cos\alpha_3 \ \ \sin\alpha_3 \ \ 0 \ \ 0 \ \ 0 \ \ 0]^T.$$
$$S_C = \text{Span}\left\{\left[\$^m_{B_1P_1M_1} \ \ \$^m_{B_2P_2M_2} \ \ \$^m_{B_3P_3M_3}\right]\right\} \tag{4.39}$$

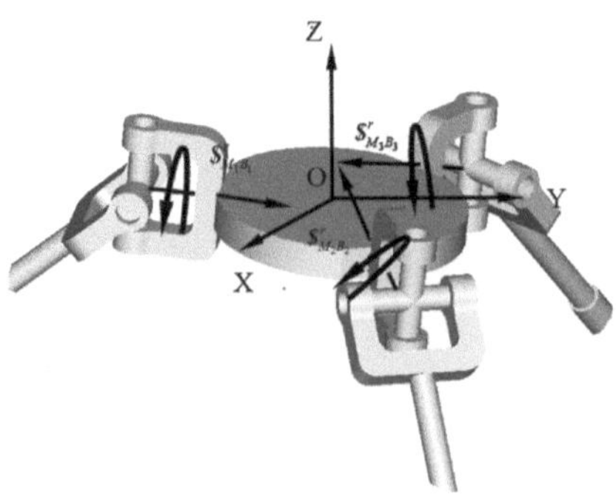

FIGURE 4.7 Reciprocal screws applied to the end effector.

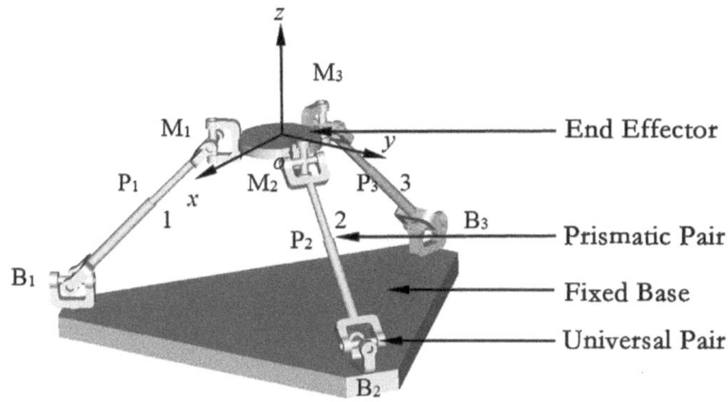

FIGURE 4.8 Another spatial parallel manipulator made up of 3-UPU.

where

$$\$^{m}_{B_1P_1M_1} = [0 \;\; 0 \;\; 0 \;\; \cos\alpha_1 \;\; \sin\alpha_1 \;\; 0]^T,$$
$$\$^{m}_{B_2P_2M_2} = [0 \;\; 0 \;\; 0 \;\; \cos\alpha_2 \;\; \sin\alpha_2 \;\; 0]^T,$$
$$\$^{m}_{B_3P_3M_3} = [0 \;\; 0 \;\; 0 \;\; \cos\alpha_3 \;\; \sin\alpha_3 \;\; 0]^T.$$

We can find that $\$^{m}_{B_iP_iM_i}, i = 1,2,3$ and $\$^{r}_{B_iP_iM_i}, i = 1,2,3$ are transformed mutually.

In the case of Figure 4.5, obviously there is $\dim S_C = \dim \$^C_P = 2$, therefore $\mathbf{S}_C$ can only be spanned fully in two-dimension subspaces $\mathbf{S}_C = \{\boldsymbol{R}_x \;\; \boldsymbol{R}_y\}$. That is,

$$r = \dim \mathbf{S}_C = \dim \$^C_P = 2. \tag{4.40}$$

The free mobility spaces of the end effector are:

$$\mathbf{S}_F = \{\boldsymbol{T}_x \;\; \boldsymbol{T}_y \;\; \boldsymbol{T}_z \;\; \boldsymbol{R}_z\}. \tag{4.41}$$

Therefore, the dimension of the mobility spaces of the end effector can be obtained by investigating the constraint spaces, $\dim \mathbf{S}_C$.

The DOF of the end effector is: $F = 6 - r = 6 - 2 = 4$ (3 Translational DOFs + 1 Rotational DOF).

Below, we will discuss the DOF of the end effector shown in Figure 4.8. Because the kinematic chains of the end effector are also isomorphic, we can also only analyze the kinematic chain of $B_1P_1M_1$. The local coordinate system is created as Figure 4.9 shows.

Universal pair B_1 can be decomposed as two orthogonal revolute pairs, the Plücker coordinates are:

$$\$_1 = [1 \;\; 0 \;\; 0 \;\; 0 \;\; 0 \;\; 0]^T, \quad \$_2 = [0 \;\; 1 \;\; 0 \;\; 0 \;\; 0 \;\; 0]^T.$$

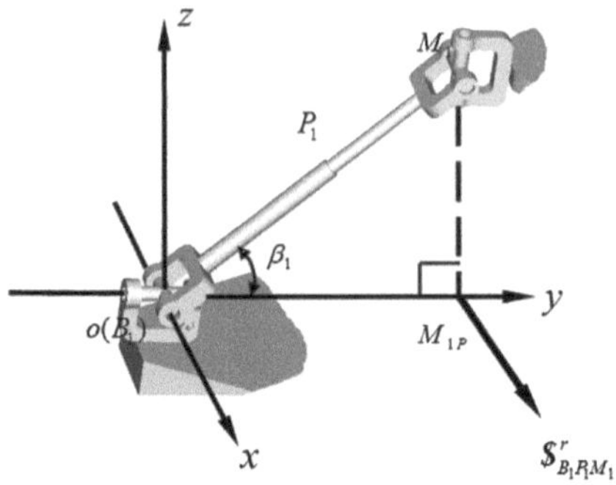

FIGURE 4.9 Reciprocal screw system of another kind of UPU kinematic chain.

Similarly, if we presume that the angle from the limb to the y-axis is denoted by β_1, the Plücker coordinates of pair P_1 are:

$$\$_3 = [0 \quad 0 \quad 0 \quad 0 \quad \cos\beta_1 \quad \sin\beta_1]^T.$$

We again assume that $B_1M_1 = l_1$, the coordinates of universal pair M_1 are $(0 \quad l_1\cos\beta_1 \quad l_1\sin\beta_1)$; M_1 can also be decomposed as two revolute pairs, the Plücker coordinates of which are:

$$\$_4 = [1 \quad 0 \quad 0 \quad 0 \quad l_1\sin\beta_1 \quad -l_1\cos\beta_1]^T, \quad \$_5 = [0 \quad 0\ 1 \quad l_1\cos\beta_1 \quad 0 \quad 0]^T.$$

Therefore, the kinematic screws of $B_1P_1M_1$ can be denoted as:

$$\$_{B_1P_1M_1} = [\$_1 \quad \$_2 \quad \$_3 \quad \$_4 \quad \$_5]. \tag{4.42}$$

So the reciprocal screws of $B_1P_1M_1$ are:

$$\$^r_{B_1P_1M_1} = [1 \quad 0 \quad 0 \quad 0 \quad 0 \quad -l_1\cos\beta_1]^T. \tag{4.43}$$

According to screw theory, $\$^r_{B_1P_1M_1}$ is a force screw passing through the point $(c \quad l_1\cos\beta_1 \quad 0)$ and being parallel to the x-axis shown in Figure 4.9, where c can be any value, so we can assume $c = 0$ for simplicity.

As for the rest of the kinematic chains, we can also draw similar conclusions. Therefore, the constraints applied to the end effector are three coplanar forces (in the plane $B_1B_2B_3$), which are all perpendicular to the z-axis shown in Figure 4.10.

We can extend the rigid end effector to the reciprocal screws' plane $B_1B_2B_3$ and presume that the projections of the three vertexes of the end effector in plane $B_1B_2B_3$ are the triangular $M_{1p}M_{2p}M_{3p}$, as Figure 4.11 shows.

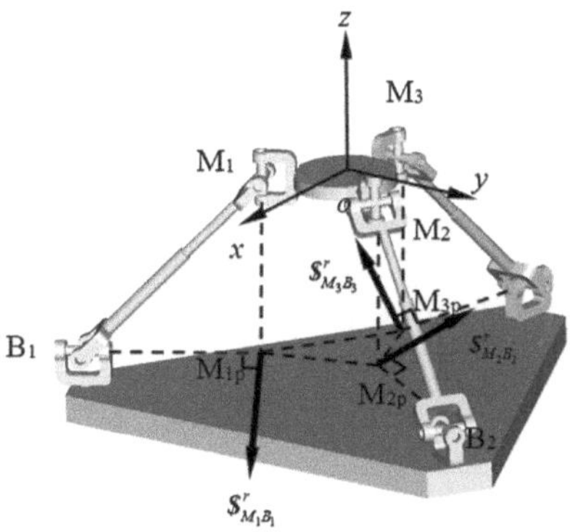

FIGURE 4.10 Reciprocal screws applied to the parallel manipulator.

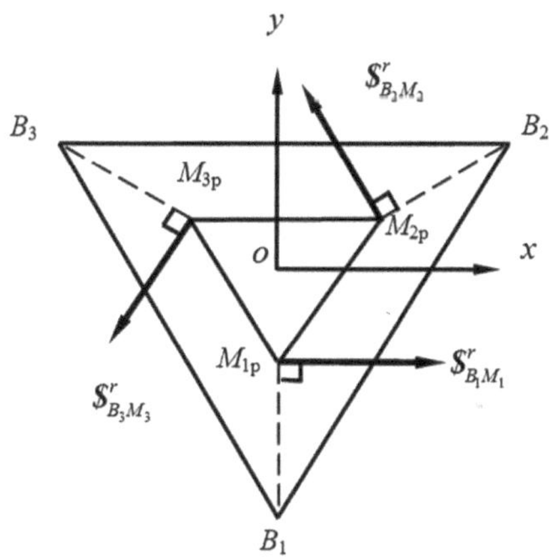

FIGURE 4.11 Coplanar reciprocal screws applied to the platform.

If we presume $\boldsymbol{r}_{M_{ip}} = (x_i \quad y_i \quad 0), i = 1,2,3$, then

$$\boldsymbol{F}_i = \frac{k_i}{\left\|\boldsymbol{r}_{M_{ip}}\right\|}(-y_i \quad x_i \quad 0), i = 1,2,3 \tag{4.44}$$

where k_i denotes the tensity of the ith kinematic chain and $\left\|\boldsymbol{r}_{M_{ip}}\right\| = \sqrt{x_i^2 + y_i^2}$.

Assume that the reciprocal screws of the ith kinematic chain are denoted as:

$$\$^r_{B_i P_i M_i} = [\boldsymbol{F}_i \quad \boldsymbol{r}_{M_i} \times \boldsymbol{F}_i]^T. \tag{4.45}$$

That is,

$$\$^r_{B_i P_i M_i} = k_i \left[-\frac{y_i}{\left\|\boldsymbol{r}_{M_{ip}}\right\|} \quad \frac{x_i}{\left\|\boldsymbol{r}_{M_{ip}}\right\|} \quad 0 \quad 0 \quad 0 \quad \left\|\boldsymbol{r}_{M_{ip}}\right\| \right]^T. \tag{4.46}$$

Because $\$_P^C = \text{Span}\left\{\left[\$^r_{B_1P_1M_1} \quad \$^r_{B_2P_2M_2} \quad \$^r_{B_3P_3M_3}\right]\right\}$, then

$$\$_P^C = \begin{bmatrix} 1 & 0 & 0 \\ 0 & 1 & 0 \\ 0 & 0 & 0 \\ 0 & 0 & 0 \\ 0 & 0 & 1 \\ 0 & 0 & 0 \end{bmatrix}. \tag{4.47}$$

Obviously, there is

$$\dim \$_P^C = 3. \tag{4.48}$$

When we presume that $\$_1^m = [0 \ 0 \ 0 \ 1 \ 0 \ 0]^T$, $\$_2^m = [0 \ 0 \ 0 \ 0 \ 1 \ 0]^T$, and $\$_6^m = [0 \ 0 \ 1 \ 0 \ 0 \ 0]^T$, where $\$_i^m, i = 1,2,6$ is the transformed screws corresponding to $\$_i^C, i = 1,2,6$, the constrained spaces, $\mathbf{S}_C$, spanned by the reciprocal screws can be obtained:

$$\mathbf{S}_C = \left\{\boldsymbol{T}_x \quad \boldsymbol{T}_y \quad \boldsymbol{R}_z\right\} \tag{4.49}$$

where $\boldsymbol{T}_x = \$_1^m, \boldsymbol{T}_y = \$_2^m, \boldsymbol{R}_z = \$_6^m$.

$$r = \dim \mathbf{S}_C = \dim \$_P^C = 3 \tag{4.50}$$

Naturally, $\mathbf{S}_C$ can prevent the end effector from making translational movements along the x-axis, the y-axis, and the rotational movement around the z-axis.

Therefore, the free mobility spaces of the end effector are:

$$\mathbf{S}_F = \left\{\boldsymbol{T}_z \quad \boldsymbol{R}_x \quad \boldsymbol{R}_y\right\}. \tag{4.51}$$

The number of DOFs of the end effector is:

$$F = 6 - r = 6 - 3 = 3 \tag{4.52}$$

(1 translational DOF + 2 rotational DOFs).

According to the above analysis, the numbers of pairs and links of the mechanisms in Figures 4.5 and 4.8 are equal, individually; the results of the DOFs, however, are quite different only because of the different connecting relationships of the pairs. With the method proposed in this paper, we cannot only obtain the result of the DOFs, but also find the types and directions of the movements that the end effector could execute.

4.3.4 A 4-PUU Spatial Parallel Mechanism with a Manipulator

The spatial parallel mechanism, shown in Figure 4.12, is made up of 4-PUU (one prismatic pair and two universal pairs) kinematic chains [2]. To study the DOF of the mechanism with the manipulator, we can first decompose the spatial parallel mechanism to be four kinematic chains connecting the manipulator with the base, and then study the reciprocal screws of each kinematic chain.

Because all the kinematic chains connecting the manipulator with the base are identical, we can simplify the process of analysis. For example, we can analyze chain P_1B_1 and create the local Cartesian coordinates shown in Figure 4.13; B_{1P} is the projection of B_1 in the *xoy* plane, and OB_{1P} is the projective line of OB_1 in the *xoy* plane, β_1 is the angle from line OB_{1P} to line OB_1,α_1 is the angle from the *x*-axis to line OB_{1P}.

The Plücker coordinates of the prismatic pair P_1 are:

$$\$_1 = [0 \quad 0 \quad 0 \quad 1 \quad 0 \quad 0]^T. \tag{4.53}$$

The universal pair P_1 can be decomposed as two orthogonal revolute pairs. The individual Plücker coordinates are:

$$\$_2 = [\sin\alpha_1 \quad -\cos\alpha_1 \quad 0 \quad 0 \quad 0 \quad 0]^T, \quad \$_3 = [0 \quad 0 \quad 0 \quad 1 \quad 0 \quad 0]^T.$$

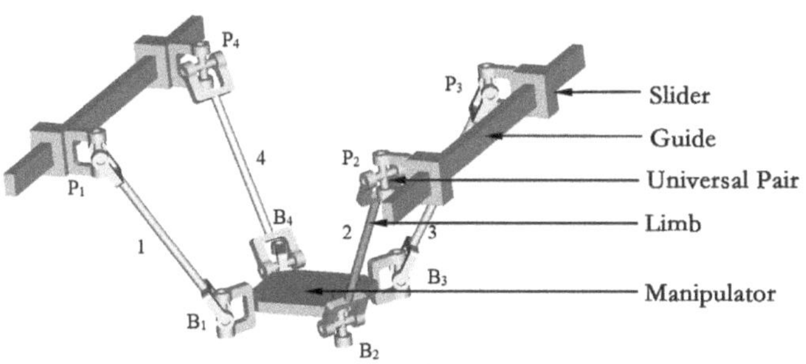

FIGURE 4.12 A spatial parallel mechanism made up of 4-PUU.

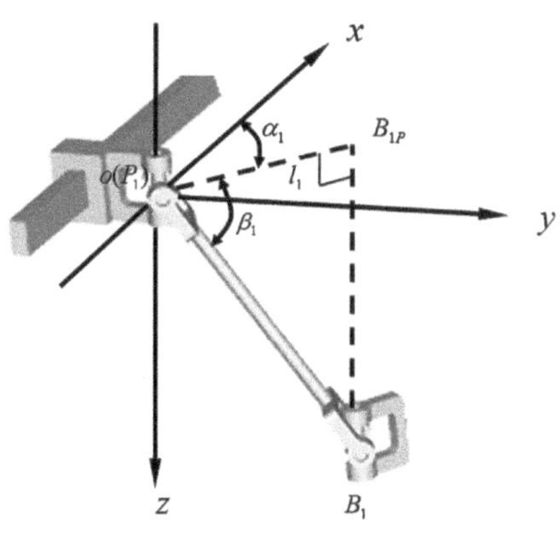

FIGURE 4.13 PUU kinematic chain.

And the same are true for the universal pair B_1. The coordinates of B_1 are:

$$B_1(l\cos\beta_1\cos\alpha_1,\ \ l\cos\beta_1\sin\alpha_1,\ \ l\sin\beta_1).$$

The Plücker coordinates can be denoted as:

$$\begin{cases} \$_4 = [\sin\alpha_1 \quad -\cos\alpha_1 \quad 0 \quad l\sin\beta_1\cos\alpha_1 \quad l\sin\beta_1\sin\alpha_1 \quad -l\cos\beta_1]^T \\ \$_5 = [0 \quad 0 \quad 1 \quad l\cos\beta_1\sin\alpha_1 \quad -l\cos\beta_1\cos\alpha_1 \quad 0]^T \end{cases} \tag{4.54}$$

Therefore, the kinematic screws of the kinematic chain P_1B_1 can be expressed as:

$$\$_{P_1B_1} = [\$_1 \quad \$_2 \quad \$_3 \quad \$_4 \quad \$_5]. \tag{4.55}$$

The reciprocal twist matrix of kinematic chain P_1B_1 can be determined:

$$\$^r_{P_1B_1} = [0 \quad 0 \quad 0 \quad \cos\alpha_1 \quad \sin\alpha_1 \quad 0]^T. \tag{4.56}$$

According to the physical meaning of the reciprocal screws of twists, $\$^r_{P_1B_1}$ denotes a pure moment of couple that is perpendicular to the plane of universal pair.

The movement constrained by $\$^r_{P_1B_1}$ is $[\cos\alpha_1 \quad \sin\alpha_1 \quad 0 \quad 0 \quad 0 \quad 0]^T$. This is a pure rotation about the direction $(\cos\alpha_1 \quad \sin\alpha_1 \quad 0)^T$, which is perpendicular to the planes of the universal pairs P_1 and B_1 shown in Figure 4.14. That is, the manipulator shall not execute any rotation around the direction $(\cos\alpha_1 \quad \sin\alpha_1 \quad 0)^T$. The same are true for the other three kinematic chains of the manipulator. Therefore, the reciprocal screws applied to the manipulator are four pure moments of couples, which are all perpendicular to the direction of the z-axis, shown in Figure 4.15.

The movements constrained by the reciprocal screws applied to the manipulator are:

$$\$^C_{Platform} = \left[\$^C_{P_1B_1} \quad \$^C_{P_2B_2} \quad \$^C_{P_3B_3} \quad \$^C_{P_4B_4}\right] \tag{4.57}$$

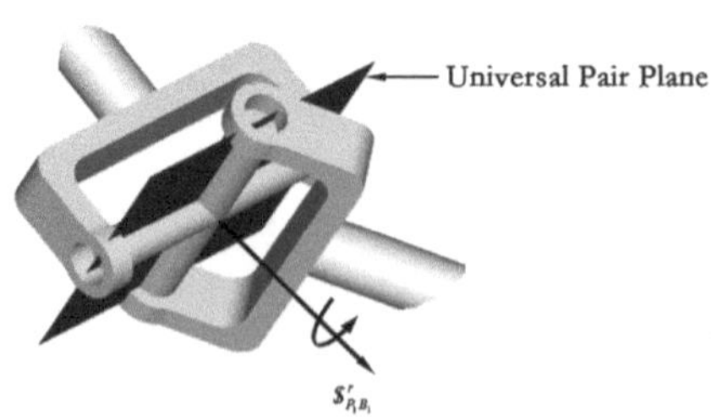

FIGURE 4.14 Reciprocal screw of limb P_1B_1 applied to the mechanism.

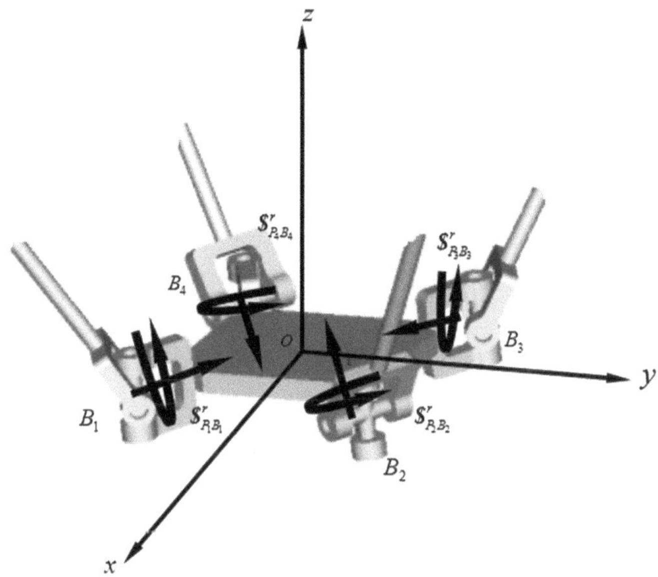

FIGURE 4.15 Reciprocal screws exerted on the manipulator $B_1B_2B_3B_4$.

where

$$\$^C_{P_1B_1} = [\cos\alpha_1 \ \sin\alpha_1 \ 0\ 0\ 0\ 0]^T, \$^C_{P_2B_2} = [\cos\alpha_2 \ \sin\alpha_2 \ 0\ 0\ 0\ 0]^T;$$
$$\$^C_{P_3B_3} = [\cos\alpha_3 \ \sin\alpha_3 \ 0\ 0\ 0\ 0]^T, \$^C_{P_4B_4} = [\cos\alpha_4 \ \sin\alpha_4 \ 0\ 0\ 0\ 0]^T;$$

And $\alpha_i\,(i = 1,2,3,4)$ is the angle from the x-axis of the absolute coordinate system to the projective line of the ith kinematic chain in the plane xoy.

Obviously, $1 \le \dim \$^C_{Platform} \le 2$, when $\dim \$^C_{Platform} = 1$, there must be:

$$\alpha_1 = \alpha_4 = \frac{\pi}{2}, \quad \alpha_2 = \alpha_3 = \frac{3\pi}{2}. \tag{4.58}$$

This case shows kinds of singularities which can be avoided by adjusting the structure parameters.

In general, $\dim \$^C_{Platform} = 2$ and $\$^C_{Platform}$ can be fully spanned in a subspaces $\mathbf{S}_C$:

$$\mathbf{S}_C = \{\boldsymbol{R}_x \ \ \boldsymbol{R}_y\}. \tag{4.59}$$

Therefore, there is

$$F = 6 - \dim \$^C_{Platform} = 6 - 2 = 4. \tag{4.60}$$

If we select the four sliders shown in Figure 4.14 as the actuators, the F can be analyzed as follows. When the actuators are given, the four sliders can be seen as fixed pairs. Therefore, the twist matrix of kinematic chain P_1B_1 can be rewritten as:

$$\$^{drive}_{P_1B_1} = [\$_2 \quad \$_3 \quad \$_4 \quad \$_5]. \tag{4.61}$$

So the reciprocal screws of kinematic chain P_1B_1 can be determined:

$$\$^{r1}_{P_1B_1} = [0 \quad 0 \quad 0 \quad \cos\alpha_1 \quad \sin\alpha_1 \quad 0]^T \tag{4.62}$$

$$\$^{r2}_{P_1B_1} = [\cos\beta_1\cos\alpha_1 \quad \cos\beta_1\sin\alpha_1 \quad -\sin\beta_1 \quad 0 \quad 0 \quad 0]^T \tag{4.63}$$

According to the physical meaning of the reciprocal screws of kinematic screws, $\$^{r1}_{P_1B_1}$ denotes a pure moment of couple that is perpendicular to the universal pair plane, and $\$^{r2}_{P_1B_1}$ is a pure force along the axis of kinematic chain P_1B_1 shown in Figure 4.14.

The same are true for the other three kinematic chains of the manipulator. Therefore, the reciprocal screws acting on the manipulator are four pure forces along the individual axes of the kinematic chains, and four pure moments of couples, which are all perpendicular to the direction of the z-axis in the absolute coordinate system shown in Figure 4.15.

From the above analysis, we know that the four pure moments of couples can prevent the manipulator from rolling around the x-axis and the y-axis in the absolute coordinate system, and can be spanned to a subspace $\mathbf{S}_C$. Therefore, we can only analyze whether the four pure forces $\$^{drive}_{Platform}$ (reciprocal screws) applied to the manipulator should be spanned fully to the annihilated complementary spaces of $\mathbf{S}_C$, denoted by $\mathbf{S}_C^{\perp}$. If so, the manipulator must be stable.

$$\$^{drive}_{Platform} = \left[\$^{r2}_{P_1B_1} \quad \$^{r2}_{P_2B_2} \quad \$^{r2}_{P_3B_3} \quad \$^{r2}_{P_4B_4}\right] \tag{4.64}$$

where

$$\begin{aligned}
\$^{r2}_{P_1B_1} &= [\cos\beta_1\cos\alpha_1 \quad \cos\beta_1\sin\alpha_1 \quad -\sin\beta_1 \quad 0 \quad 0 \quad 0]^T, \\
\$^{r2}_{P_2B_2} &= [\cos\beta_2\cos\alpha_2 \quad \cos\beta_2\sin\alpha_2 \quad -\sin\beta_2 \quad 0 \quad 0 \quad 0]^T, \\
\$^{r2}_{P_3B_3} &= [\cos\beta_3\cos\alpha_3 \quad \cos\beta_3\sin\alpha_3 \quad -\sin\beta_3 \quad 0 \quad 0 \quad 0]^T, \\
\$^{r2}_{P_4B_4} &= [\cos\beta_4\cos\alpha_4 \quad \cos\beta_4\sin\alpha_4 \quad -\sin\beta_4 \quad 0 \quad 0 \quad 0]^T.
\end{aligned}$$

Therefore, if $\$^{drive}_{Platform}$ can be spanned fully to $\mathbf{S}_C^{\perp}$, the manipulator must be stable, the reciprocal screws $\$^{r2}_{P_1B_1}$,$\$^{r2}_{P_2B_2}$,$\$^{r2}_{P_3B_3}$, and $\$^{r2}_{P_4B_4}$ should prevent the manipulator from executing translational movements along three orthogonal axes and one rotational movement around the z-axis in the absolute coordinate system under any external forces along the three orthogonal axes and the torques around the z-axis. Assume $k_i\,(i = 1{,}2{,}3{,}4)$ denote the intensity of the four reciprocal screws, F_x, F_y and F_z to be the three components of external forces applied to the manipulator and M_z to be the components of external torque around the z-axis applied to the manipulator. Assuming $B_1B_4 = 2a, B_1B_2 = 2b$, there are:

$$\begin{cases} \sum F_x = 0; & (k_1\cos\theta_1 + k_2\cos\theta_2 + k_3\cos\theta_3 + k_4\cos\theta_4)\cos\beta + F_x = 0 \\ \sum F_y = 0; & (k_1\sin\theta_1 + k_2\sin\theta_2 + k_3\sin\theta_3 + k_4\sin\theta_4)\cos\beta + F_y = 0 \\ \sum F_z = 0; & (k_1 + k_2 + k_3 + k_4)\sin\beta + F_z = 0 \\ \sum M_z = 0; & (-a\sin\theta_1 + b\cos\theta_1)k_1 + (a\sin\theta_2 + b\cos\theta_2)k_2 \\ & +(a\sin\theta_3 - b\cos\theta_3)k_3 + (-a\sin\theta_4 - b\cos\theta_4)k_4 + T_z = 0 \end{cases} \tag{4.65}$$

where $\theta_i\,(i = 1,2,3,4)$ denotes the angles from the projection lines of the four kinematic chains to the x-axis in the absolute coordinate system. The above equations can be transformed into:

$$\mathbf{A}\boldsymbol{F}_i = \boldsymbol{F}_f \tag{4.66}$$

where

$$\mathbf{A} = \begin{bmatrix} \cos\theta_1 & \cos\theta_2 & \cos\theta_3 & \cos\theta_4 \\ \sin\theta_1 & \sin\theta_2 & \sin\theta_3 & \sin\theta_4 \\ 1 & 1 & 1 & 1 \\ a\sin\theta_1 + b\cos\theta_1 & a\sin\theta_2 - b\cos\theta_2 & -a\sin\theta_3 - b\cos\theta_3 & -a\sin\theta_4 + b\cos\theta_4 \end{bmatrix},$$

$$\boldsymbol{F}_i = [\,k_1 \quad k_2 \quad k_3 \quad k_4\,]^T, \quad \boldsymbol{F}_f = [\,-F_x/\cos\beta \quad -F_y/\cos\beta \quad -F_z/\sin\beta \quad -T_z\,]^T.$$

The rank of matrix **A** is the dimension that the reciprocal screws $\$^{drive}_{Platform}$ can be spanned. Obviously, if the equation below holds, the manipulator should be uncontrollable:

$$|\mathbf{A}| = 0. \tag{4.67}$$

That is,

$$\begin{aligned} &a[(\sin\theta_1 - \sin\theta_2)\sin(\theta_4 - \theta_3) + (\sin\theta_4 - \sin\theta_3)\sin(\theta_2 - \theta_1)] \\ &+b[(\cos\theta_4 - \cos\theta_1)\sin(\theta_2 - \theta_3) + \cos(\theta_3 - \theta_2) + \sin(\theta_4 - \theta_1)] = 0 \end{aligned} \tag{4.68}$$

When the above equation holds, $\dim \$^{drive}_{Platform} < 4$, which means that the four actuators could not control the position and orientation of the end effector. Chapters 9 and 10 will further discuss this problem. When $\alpha_i \neq \frac{\pi}{2}, \alpha_i \neq \frac{3\pi}{2}$ and $|\mathbf{A}| \neq 0$, $\dim \$^{drive}_{Platform} = 4$, the four actuators can stably control the manipulator.

Similarly, we can analyze the DOF of the mechanism with the manipulator shown in Figure 4.16. Some detailed procedures are omitted.

Suppose that $\$^{C}_{Platform}$ denotes the reciprocal screws applied to the manipulator, in ordinary cases, there is also:

$$\dim \$^{C}_{Platform} = 2. \tag{4.69}$$

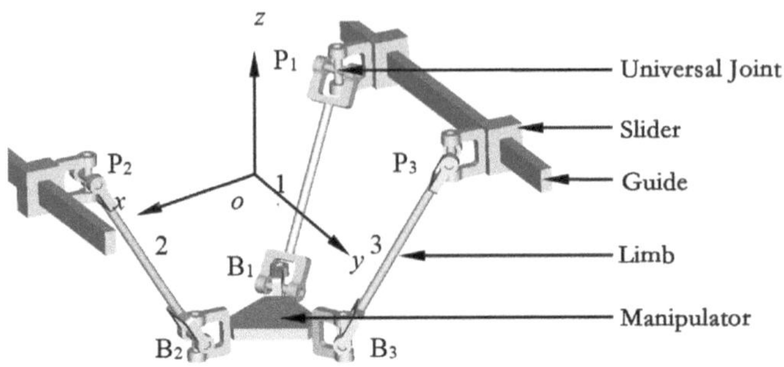

FIGURE 4.16 A spatial parallel mechanism made up of 3-PUU.

Therefore, $\dim \$^{C}_{Platform} = 2$ in the ordinary cases, and then

$$F = 6 - \dim \$^{C}_{Platform} = 6 - 2 = 4. \tag{4.70}$$

With the above procedure, we can also obtain the DOF of the end effector of other mechanism.

4.3.5 A Schoenflies-Type Parallel Manipulator

A spatial parallel manipulator, shown in Figure 4.17, is made up of 4-UPU (one prismatic pair and two universal pairs) kinematic chains [5]. According to the method presented above, the local coordinate system $o_c x_c y_c z_c$ and the absolute coordinate system $oxyz$ are created as Figure 4.17 shows. The z_c-axis is perpendicular to the plane of the manipulator $M_1M_2M_3M_4$, the origin is superposed with the geometric center of the $M_1M_2M_3M_4$, x_c, and y_c axes parallel to the two orthogonal sides of the manipulator, and the z-axis perpendicular to the base plane $B_1B_2B_3B_4$, the origin is superposed with

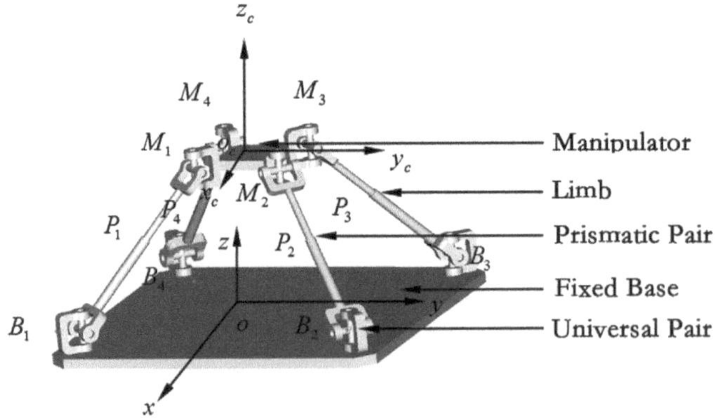

FIGURE 4.17 A kind of spatial parallel manipulator made up of 4-UPU kinematic chain.

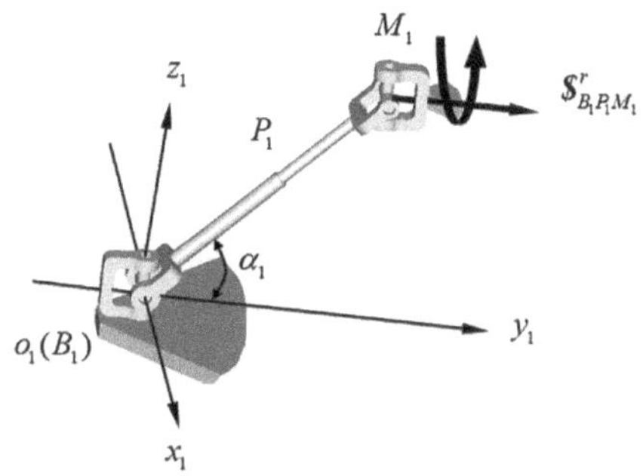

FIGURE 4.18 The coordinate system for kinematic chain $B_1P_1M_1$.

the geometric center of $B_1B_2B_3B_4$, the x- and y-axes are parallel to the two orthogonal sides of the base.

First, we will analyze the DOF of the manipulator as Figure 4.17 shows [6]. Let us analyze the reciprocal screws of all of the kinematic chains. Because the kinematic chains joined to the manipulator with the base are all identical, we only need to select any one of them to study the reciprocal screws of the kinematic chains; for example, we select kinematic chain $B_1P_1M_1$ shown in Figure 4.18 as the object to study. In the local coordinate system, the origin is superposed with universal pair B_1, the x-axis and z-axis pass through the two orthogonal axes of the universal pair, individually. Assuming the angle from the base plane to the limb $B_1P_1M_1$ is α_1, the length of the limb is l_1, we can analyze the reciprocal screws of the kinematic chain below.

The universal pair B_1 can be decomposed as two orthogonal revolute pairs. The individual Plücker coordinates are:

$$\$_1 = [1 \quad 0 \quad 0 \quad 0 \quad 0 \quad 0]^T, \quad \$_2 = [0 \quad 0 \quad 1 \quad 0 \quad 0 \quad 0]^T.$$

The Plücker coordinates of slider P_1 are:

$$\$_1 = [0 \quad 0 \quad 0 \quad 0 \quad \cos\alpha_1 \quad \sin\alpha_1]^T.$$

The same is true of universal pair M_1. The coordinates of M_1 are:

$$M_1(0 \quad l_1\sin\alpha_1 \quad l_1\cos\alpha_1).$$

The Plücker coordinates can be denoted as:

$$\$_4 = [1 \quad 0 \quad 0 \quad 0 \quad l_1\cos\alpha_1 \quad -l_1\sin\alpha_1]^T, \$_5 = [0 \quad 0 \quad 1 \quad l_1\sin\alpha_1 \quad 0 \quad 0]^T.$$

Therefore, the kinematic screws of limb $B_1P_1M_1$ can be expressed as:

$$\$_{B_1P_1M_1} = [\$_1 \quad \$_2 \quad \$_3 \quad \$_4 \quad \$_5]. \tag{4.71}$$

The reciprocal screws of branch $B_1P_1M_1$ can be deduced:

$$\$^r_{B_1P_1M_1} = [0 \quad 0 \quad 0 \quad 0 \quad 1 \quad 0]^T. \tag{4.72}$$

According to the physical meaning of the reciprocal screws of kinematic screws, $\$^r_{B_1P_1M_1}$ denotes a pure moment of couple around the y_1-axis. That is, $\$^r_{B_1P_1M_1}$ is perpendicular to the plane determined by the two orthogonal axes of universal pair B_1 or M_1. The same is true for the other three kinematic chains of the manipulator. So, the reciprocal screws applied to the manipulator are four pure moments of couples, which are all perpendicular to the z_c-axis. If we assume the angles of the reciprocal screws of the $i^{th}(i = 1,2,3,4)$ kinematic chain to the x_c-axis are denoted as β_i, the reciprocal screws applied to the end effector in the $o_cx_cy_cz_c$ coordinate system can be denoted as:

$$\$^r = \begin{bmatrix} \$^r_1 & \$^r_2 & \$^r_3 & \$^r_4 \end{bmatrix} \tag{4.73}$$

where

$$\$^r_1 = [0 \quad 0 \quad 0 \quad \cos\beta_1 \quad \sin\beta_1 \quad 0]^T, \quad \$^r_2 = [0 \quad 0 \quad 0 \quad \cos\beta_2 \quad \sin\beta_2 \quad 0]^T,$$
$$\$^r_3 = [0 \quad 0 \quad 0 \quad \cos\beta_3 \quad \sin\beta_3 \quad 0]^T, \quad \$^r_4 = [0 \quad 0 \quad 0 \quad \cos\beta_4 \quad \sin\beta_4 \quad 0]^T.$$

The dimension of these reciprocal screws can be found through singular value decomposition (SVD). According to algebra, the SVD of an $m \times n$ matrix $\mathbf{C}$ has the form:

$$\mathbf{C} = \mathbf{U}\,\mathbf{\Sigma}\mathbf{V}^T \tag{4.74}$$

where $\mathbf{U}$ represents an $m \times m$ orthogonal matrix, $\mathbf{V}$ denotes an $n \times n$ orthogonal matrix, $\mathbf{\Sigma}$ indicates an $m \times n$ diagonal matrix, with

$$\sigma_{ij} = \begin{cases} 0 & i \neq j \\ \sigma_i \geq 0 & i = j. \end{cases} \tag{4.75}$$

The diagonal entries σ_i are called the singular values of $\mathbf{C}$ and are usually ordered so that $\sigma_i \geq \sigma_{i+1}, i = 1,2,\ldots,n-1$. σ_i^2 are the eigenvalues of $\mathbf{CC}^T$; as a result, σ_i can be gained by solving

$$\left|\mathbf{CC}^T - \sigma^2\mathbf{I}\right| = 0 \tag{4.76}$$

where $\mathbf{I}$ is an $n \times n$ identity matrix.

The columns u_i of $\mathbf{U}$ and v_i of $\mathbf{V}$ are the corresponding left and right singular vectors. The dimension that the lines of $\mathbf{C}$ can span is equal to the number of $\sigma_i > 0$.

According to equation (4.76), there will be:

$$\left|\$^r(\$^r)^T - \lambda\mathbf{I}\right| = 0. \tag{4.77}$$

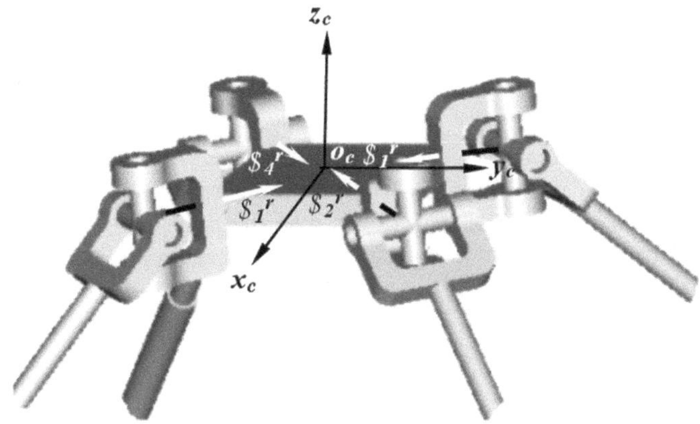

FIGURE 4.19 Reciprocal screws applied to end effector of the manipulator.

Equation (4.77) can be expanded:

$$\frac{1}{2}\lambda^2\begin{bmatrix}2\lambda^2-8\lambda+6-\cos 2\left(\beta_4-\beta_1\right)-\cos 2\left(\beta_2-\beta_4\right)-\cos 2\left(\beta_3-\beta_4\right)\\ -\cos 2\left(\beta_3-\beta_2\right)-\cos 2\left(\beta_1-\beta_2\right)-\cos 2\left(\beta_1-\beta_3\right)\end{bmatrix}=0. \tag{4.78}$$

Obviously, $\lambda_3=\lambda_4=0$ are two solutions of equation (4.78), λ_1 and λ_2 are the two roots of quadratic equation:

$$\lambda^2-4\lambda+\frac{p}{2}=0 \tag{4.79}$$

where

$$p=6-\cos 2\left(\beta_4-\beta_1\right)-\cos 2\left(\beta_2-\beta_4\right)-\cos 2\left(\beta_3-\beta_4\right)$$
$$-\cos 2\left(\beta_3-\beta_2\right)-\cos 2\left(\beta_1-\beta_2\right)-\cos 2\left(\beta_1-\beta_3\right).$$

From Figure 4.19, we can find that the range of $\beta_i\,(i=1{,}2{,}3{,}4)$ will be:

$$\frac{\pi}{2}<\beta_1<\pi,\pi<\beta_2<\frac{3\pi}{2},\frac{3\pi}{2}<\beta_3<0,0<\beta_4<\frac{\pi}{2}.$$

Therefore, $p\neq 0$, and equation (4.79) should have two positive roots, $\lambda_1\geq\lambda_2$. As a result, $\sigma_1\geq\sigma_2>0$ and one basis for $\$^r$ can be selected as:

$$\boldsymbol{C}=\begin{bmatrix}0&0&0&1&0&0\\0&0&0&0&1&0\end{bmatrix} \tag{4.80}$$

where $\sigma_1=\sqrt{\lambda_1}$ and $\sigma_2=\sqrt{\lambda_2}$.

Equation (4.80) means that the reciprocal screws will constrain two rotations around the x-axis and the y-axis. So

$$r=\dim\text{Span}\left\{\$^r_{B_1P_1M_1}\ \$^r_{B_2P_2M_2}\ \$^r_{B_3P_3M_3}\ \$^r_{B_4P_4M_4}\right\}=\dim C=2. \tag{4.81}$$

Therefore,

$$F = 6 - r = 6 - 2 = 4 \tag{4.82}$$

And the manipulator shown in Figure 4.17 is a kind of Schoenflies-type parallel manipulator and has four DOFs, including three orthogonal translational movements and one rotational moment around the z_c-axis.

Now we can select $(x_c \quad y_c \quad z_c)$ and the rotational angular around z-axis, β, as the least number of pose parameters.

If we presume the length of M_1M_2 is 2α and the length of M_1M_4 is $2b$, the local coordinates of the four vertexes of the manipulator can be obtained:

$$\begin{cases} \boldsymbol{r}_{M_1}^L = [b \quad -a \quad 0]^T \\ \boldsymbol{r}_{M_2}^L = [b \quad a \quad 0]^T \\ \boldsymbol{r}_{M_3}^L = [-b \quad a \quad 0]^T \\ \boldsymbol{r}_{M_4}^L = [-b \quad -a \quad 0]^T \end{cases} \tag{4.83}$$

The transformation matrix can be expressed as:

$$A = \begin{bmatrix} \cos\beta & \sin\beta & 0 \\ -\sin\beta & \cos\beta & 0 \\ 0 & 0 & 1 \end{bmatrix}. \tag{4.84}$$

If we presume that the geometric center point on the manipulator connected by the n limbs is the origin of the local coordinate system $o_cx_cy_cz_c$, the end vertexes of each kinematic chain joined with the manipulator are denoted by $M_i, (i = 1,2,\ldots,n)$, the end vertexes of each kinematic chain joined with the fixed base are denoted by $B_i, (i = 1,2,\ldots,n)$. Assume that the pose angles are ψ, θ, φ, where ψ denotes precession angle, θ indicates the nutation angle and φ represents the spinning angle. However, we need not to select all of the above variables in reality. On the contrary, it is sufficient to select a set of variables corresponding to the DOFs that the manipulator can move freely. In fact, the coordinates of $M_i, (i = 1,2,\ldots,n)$ in the local coordinate system $o_cx_cy_cz_c$, denoted by $\boldsymbol{r}_{M_i}^L$, are often very simple.

In the absolute coordinate system $oxyz$, the vectors' coordinates of the end vertexes of each kinematic chain connected with the fixed base are $\boldsymbol{r}_{B_1} \quad \boldsymbol{r}_{B_2} \quad \cdots \quad \boldsymbol{r}_{B_n}$, the coordinates of the geometric center of the manipulator, C, are (x_c, y_c, z_c) and the vectors of the vertexes M_i and point C are denoted by $\boldsymbol{r}_{M_i}$ and $\boldsymbol{r}_C$. Therefore,

$$\boldsymbol{r}_{M_i} = \boldsymbol{r}_C + \mathbf{A}\boldsymbol{r}_{M_i}^L \tag{4.85}$$

where **A** represents the transformation matrix. According to the geometry characteristics, we can obtain a set of constraint equations:

$$\begin{cases} \left|\boldsymbol{r}_{B_1} - \boldsymbol{r}_{M_1}\right| = l_1 \\ \left|\boldsymbol{r}_{B_2} - \boldsymbol{r}_{M_2}\right| = l_2 \\ \vdots \\ \left|\boldsymbol{r}_{B_F} - \boldsymbol{r}_{M_F}\right| = l_F \end{cases} \tag{4.86}$$

where $l_i, (i = 1,2,\ldots,F)$ denotes the length of each kinematic chain. We call equation set (4.86) pose equations.

The pose relationship of the manipulator with the fixed base is given with equations (4.86) and (4.85). Assume function:

$$G(\text{var}_1, \text{var}_2, \ldots, \text{var}_F) = \begin{bmatrix} \left|\boldsymbol{r}_{B_1} - \boldsymbol{r}_{M_1}\right|^2 - l_1^2 \\ \left|\boldsymbol{r}_{B_2} - \boldsymbol{r}_{M_2}\right|^2 - l_2^2 \\ \vdots \\ \left|\boldsymbol{r}_{B_F} - \boldsymbol{r}_{M_F}\right|^2 - l_F^2 \end{bmatrix} = 0. \tag{4.87}$$

We can gain the Jacobian matrix:

$$\mathbf{J}_{F\times F} = \frac{\partial G(\text{var}_1, \text{var}_2, \ldots, \text{var}_F)}{\partial(\text{var}_1, \text{var}_2, \ldots, \text{var}_F)} \tag{4.88}$$

where $\text{var}_i, i = 1,2,\ldots,F$ represents the independent variables. So, the singularity criterion is:

$$|J_{F\times F}| = 0. \tag{4.89}$$

In reality, the DOFs of the manipulator are not more than six; therefore, the order of Jacobian matrix $\mathbf{J}_{F\times F}$ is not more than six. So, equation (4.89) is an efficient method to solve the singularity problems of the manipulator, especially for those manipulators with limited DOFs.

With equation (4.85), we can obtain:

$$\begin{cases} \boldsymbol{r}_{M_1} = [x_c + b\cos\beta - a\sin\beta \quad y_c - a\cos\beta - b\sin\beta \quad z_c]^T \\ \boldsymbol{r}_{M_2} = [x_c + b\cos\beta + a\sin\beta \quad y_c + a\cos\beta - b\sin\beta \quad z_c]^T \\ \boldsymbol{r}_{M_3} = [x_c - b\cos\beta + a\sin\beta \quad y_c + a\cos\beta + b\sin\beta \quad z_c]^T \\ \boldsymbol{r}_{M_4} = [x_c - b\cos\beta - a\sin\beta \quad y_c - a\cos\beta + b\sin\beta \quad z_c]^T. \end{cases} \tag{4.90}$$

If we assume the length of B_1B_2 is $2c$ and the length of B_1B_4 is $2d$, the absolute coordinates of the four vertexes of the base can be obtained:

$$\begin{cases} \boldsymbol{r}_{B_1} = [d \quad -c \quad 0]^T \\ \boldsymbol{r}_{B_2} = [d \quad c \quad 0]^T \\ \boldsymbol{r}_{B_3} = [-d \quad c \quad 0]^T \\ \boldsymbol{r}_{B_4} = [-d \quad -c \quad 0]^T. \end{cases} \tag{4.91}$$

With equation (4.86), we can obtain:

$$\begin{cases} |\boldsymbol{r}_{B_1} - \boldsymbol{r}_{M_1}| = \sqrt{[d-(x_c + b\cos\beta - a\sin\beta)]^2 + (c + y_c - a\cos\beta - b\sin\beta)^2 + z_c^2} \\ |\mathbf{r}_{B_2} - \boldsymbol{r}_{M_2}| = \sqrt{[d-(x_c + b\cos\beta + a\sin\beta)]^2 + [c-(y_c + a\cos\beta - b\sin\beta)]^2 + z_c^2} \\ |\boldsymbol{r}_{B_3} - \boldsymbol{r}_{M_3}| = \sqrt{(d + x_c - b\cos\beta + a\sin\beta)^2 + [c-(y_c + a\cos\beta + b\sin\beta)]^2 + z_c^2} \\ |\boldsymbol{r}_{B_4} - \boldsymbol{r}_{M_4}| = \sqrt{(d + x_c - b\cos\beta - a\sin\beta)^2 + (c + y_c - a\cos\beta + b\sin\beta)^2 + z_c^2}. \end{cases} \tag{4.92}$$

That is,

$$\begin{cases} (x_c + b\cos\beta - a\sin\beta - d)^2 + (y_c - a\cos\beta - b\sin\beta + c)^2 + z_c^2 = l_1^2 \\ (x_c + b\cos\beta + a\sin\beta - d)^2 + (y_c + a\cos\beta - b\sin\beta - c)^2 + z_c^2 = l_2^2 \\ (x_c - b\cos\beta + a\sin\beta + d)^2 + (y_c + a\cos\beta + b\sin\beta - c)^2 + z_c^2 = l_3^2 \\ (x_c - b\cos\beta - a\sin\beta + d)^2 + (y_c - a\cos\beta + b\sin\beta + c)^2 + z_c^2 = l_4^2 \end{cases} \tag{4.93}$$

where $l_i\,(i = 1,2,3,4)$ represents the length of the ith kinematic chain.

Obtaining the pose equations (4.93), we can now study the singularity problems. From the above analysis, we can find that the parameters $(x_c \quad y_c \quad z_c)$ and β are four independent variables. Therefore, we can assume a function $G(\text{var}_1, \text{var}_2, \ldots, \text{var}_F) = 0$:

$$G(x_c, y_c, z_c, \beta) = \begin{bmatrix} (x_c + b\cos\beta - a\sin\beta - d)^2 + (y_c - a\cos\beta - b\sin\beta + c)^2 + z_c^2 - l_1^2 \\ (x_c + b\cos\beta + a\sin\beta - d)^2 + (y_c + a\cos\beta - b\sin\beta - c)^2 + z_c^2 - l_2^2 \\ (x_c - b\cos\beta + a\sin\beta + d)^2 + (y_c + a\cos\beta + b\sin\beta - c)^2 + z_c^2 - l_3^2 \\ (x_c - b\cos\beta - a\sin\beta + d)^2 + (y_c - a\cos\beta + b\sin\beta + c)^2 + z_c^2 - l_4^2 \end{bmatrix}. \tag{4.94}$$

According to equation (4.88), there are:

$$\mathbf{J}_{4\times4} = \frac{\partial G(x_c, y_c, z_c, \beta)}{\partial(x_c, y_c, z_c, \beta)} = 2\begin{bmatrix} x_c + b\cos\beta - a\sin\beta - d & y_c - a\cos\beta - b\sin\beta + c & z_c & (ac + bd + ay_c - bx_c)\sin\beta + (ad - bc - ax_c - by_c)\cos\beta \\ x_c + b\cos\beta + a\sin\beta - d & y_c + a\cos\beta - b\sin\beta - c & z_c & (ac + bd - bx_c - ay_c)\sin\beta + (bc - ad + ax_c - by_c)\cos\beta \\ x_c - b\cos\beta + a\sin\beta + d & y_c + a\cos\beta + b\sin\beta - c & z_c & (ac + bd + bx_c - ay_c)\sin\beta + (ad - bc + ax_c + by_c)\cos\beta \\ x_c - b\cos\beta - a\sin\beta + d & y_c - a\cos\beta + b\sin\beta + c & z_c & (ac + bd + ay_c + bx_c)\sin\beta + (bc - ad + by_c - ax_c)\cos\beta \end{bmatrix}. \tag{4.95}$$

So, the singularity criteria will be:

$$|\mathbf{J}_{4\times4}| = 0. \tag{4.96}$$

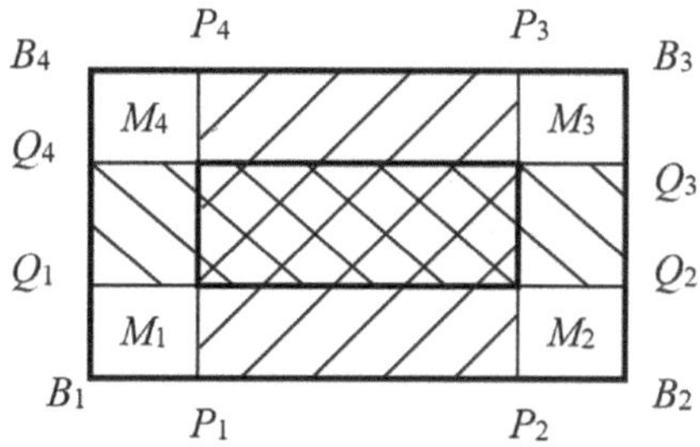

FIGURE 4.20 Relationship of the projected areas of the manipulator and the base.

Therefore, we can obtain:

$$\begin{vmatrix} x_C + b\cos\beta - a\sin\beta - d & y_C - a\cos\beta - b\sin\beta + c & z_C & (ac + bd + ay_C - bx_C)\sin\beta + (ad - bc - ax_C - by_C)\cos\beta \\ x_C + b\cos\beta + a\sin\beta - d & y_C + a\cos\beta - b\sin\beta - c & z_C & (ac + bd - bx_C - ay_C)\sin\beta + (bc - ad + ax_C - by_C)\cos\beta \\ x_C - b\cos\beta + a\sin\beta + d & y_C + a\cos\beta + b\sin\beta - c & z_C & (ac + bd + bx_C - ay_C)\sin\beta + (ad - bc + ax_C + by_C)\cos\beta \\ x_C - b\cos\beta - a\sin\beta + d & y_C - a\cos\beta + b\sin\beta + c & z_C & (ac + bd + ay_C + bx_C)\sin\beta + (bc - ad + by_C - ax_C)\cos\beta \end{vmatrix} = 0.$$

That is,

$$8z_c\left(bc - ad\right)\left[\left(ab + cd\right)\cos\beta - \left(ad + bc\right)\cos^2\beta\right] = 0. \tag{4.97}$$

The solutions of (4.97) are:

$$\begin{cases} z_C = 0 \\ \frac{a}{b} = \frac{c}{d} \\ \beta = \pm\frac{\pi}{2} \\ \beta = \arccos\left(\frac{ab+cd}{ad+bc}\right), \quad (ab + cd \le ad + bc). \end{cases} \tag{4.98}$$

If the areas of $M_1M_2M_3M_4, B_1B_2B_3B_4, P_1P_2P_3P_4, Q_1Q_2Q_3Q_4$ and $Q_1M_1P_1P_2M_2Q_2Q_3M_3P_3P_4M_4Q_4$ in Figure 4.20 are denoted by S_M, S_B, S_P, S_O, and S_H individually, there are $S_B > S_H$ and $S_H = S_P + S_O - S_M$. Therefore, $S_M + S_B > S_P + S_O$. In the case of Figure 4.20, $4(ab + cd) = S_M + S_B, 4(ad + bc) = S_P + S_Q$, and $4(ab + cd)$ denote the sum of the areas of the manipulator plane $M_1M_2M_3M_4$ and the base plane $B_1B_2B_3B_4$; $4(ad + bc)$ denote the sum of the hatched areas $P_1P_2P_3P_4$ and $Q_1Q_2Q_3Q_4$.

$S_M + S_B > S_P + S_Q$, that is,

$$ab + cd > ad + bc. \tag{4.99}$$

Therefore, the fourth solution in equation (4.98) does not hold, and as a result, the singularity criteria of the manipulator are reduced to:

$$\begin{cases} z_C = 0 \\ \frac{a}{b} = \frac{c}{d} \\ \beta = \pm\frac{\pi}{2}. \end{cases} \tag{4.100}$$

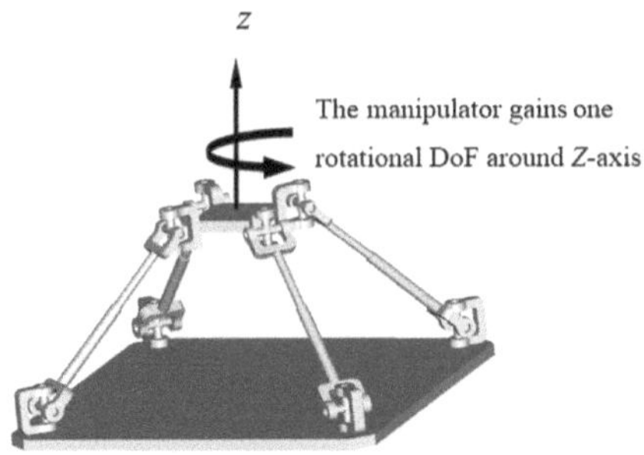

FIGURE 4.21 Similar singularity of the spatial parallel manipulator.

Next, we will discuss the three singularity cases individually:

(1) $z_C = 0$

When $z_C = 0$, the manipulator plane will be superimposed with that of the base, which is called *boundary point singularity* for the case when $z_C = 0$ is corresponding to the boundary vertex of the workspace, and which can be eliminated in reality;

(2) $\frac{a}{b} = \frac{c}{d}$

When $\frac{a}{b} = \frac{c}{d}$, the shape of the manipulator plane will be similar to that of the base, which is the so called *similar singularity*. In this case, the rotational DOF around the z-axis of the manipulator is uncontrollable, which is shown in Figure 4.21;

(3) $\beta = \pm\frac{\pi}{2}$

When $\beta = \pm\frac{\pi}{2}$, the rotational DOF around the z-axis of the manipulator will also be uncontrollable, which is shown in Figure 4.22. This kind singularity is so called $\frac{\pi}{2}$*-singularity*.In applications, the*similar singularity* can be eliminated by adjusting the structure parameters of the manipulator

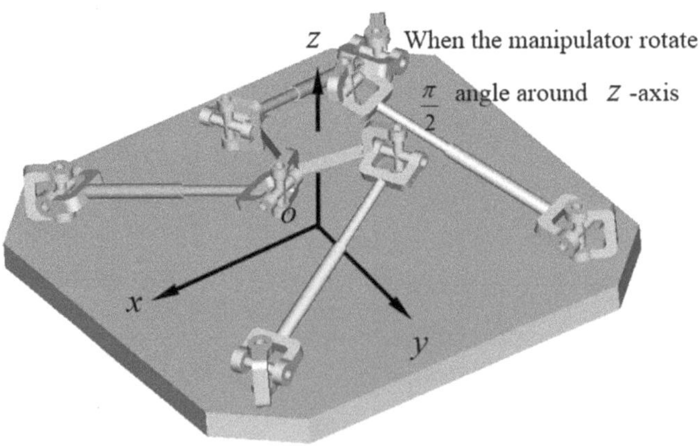

FIGURE 4.22 $\frac{\pi}{2}$-Singularity of the spatial parallel manipulator.

to safeguard that $\frac{a}{b} \neq \frac{c}{d}$, but the $\frac{\pi}{2}$-*singularity* case cannot be avoided unless $\beta \neq \pm\frac{\pi}{2}$.

4.3.6 A Spatial Parallel Five-Link Mechanism

A spatial parallel five-link mechanism shown in Figure 4.23 consists of five identical SS (Sphere pair) kinematic chains. The parameters are given in Table 4.1. With the above steps, the reciprocal screws of each link can be solved. As a result, the terminal constraint exerted on the end effector in the absolute coordinate system can be easily obtained. They are a set of pure forces shown in Figure 4.23.

Because the reciprocal screw is along the direction of the link, the unit vector of each link, for example, the unit vector of B_1B_0, can be denoted as:

$$s_2 = \frac{\boldsymbol{r}_{B_1} - \boldsymbol{r}_{B_0}}{\left\|\boldsymbol{r}_{B_1} - \boldsymbol{r}_{B_0}\right\|}$$
$$= (-0.214115091028876 \quad -0.160586318271657 \quad 0.963517909629941).$$

Therefore,

$$s_{20} = \boldsymbol{r}_{B_1} \times s_2$$
$$= (77.0814327703953 \quad -43.3583059333473 \quad 9.9028229600855).$$

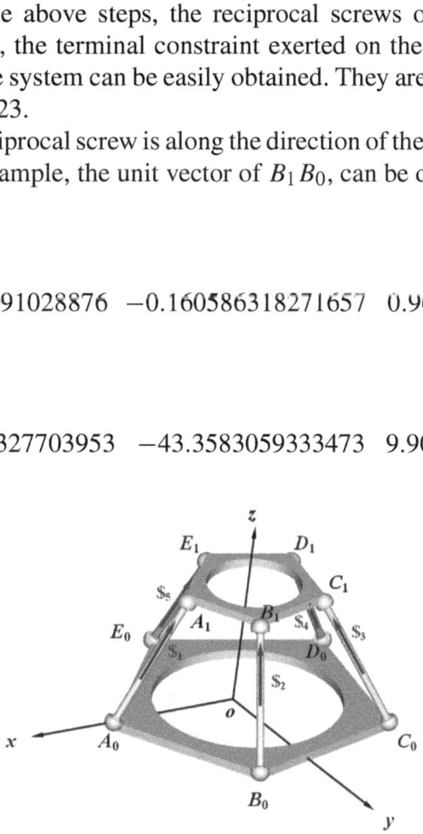

FIGURE 4.23 Spatial parallel multi-link mechanism.

TABLE 4.1 Structure Parameters of the Mechanism for Rear Suspension

Sphere Pair	x	y	z	Sphere Pair	x	y	z
A_0	120	0	0	A_1	45	20	165
B_0	45	80	0	B_1	5	50	180
C_0	−65	70	0	C_1	−30	45	170
D_0	−75	−60	0	D_1	−20	−60	160
E_0	60	−70	0	E_1	−10	40	165

According to equation (2.98), the force screw at point B_1 exerted on the end effector $A_1B_1C_1D_1E_1$ can be denoted in the absolute coordinate system as:

$$\$_2^r = \begin{bmatrix} -0.214115091028876 \\ -0.160586318271657 \\ 0.963517909629941 \\ 77.0814327703953 \\ -43.3583059333473 \\ 9.9028229600855 \end{bmatrix}. \tag{4.101}$$

In a similar manner, the other force screws exerted on the end effector $A_1B_1C_1D_1E_1$ can be obtained in the absolute coordinate system. Therefore, the reciprocal screws can be expressed in matrix form as:

$$\$^r = \begin{bmatrix} -0.411306372830311 & -0.214115091028876 & 0.19959308197927 & 0.325079768511048 & -0.332861924055832 \\ 0.109681699421416 & -0.160586318271657 & -0.14256648712805 & 0 & 0.523068737802021 \\ 0.904874020226685 & 0.963517909629941 & 0.969452112470742 & 0.945686599304867 & 0.784603106703032 \\ 0 & 77.0814327703953 & 67.8616478729519 & -56.741195958292 & -54.9222174692122 \\ -108.584882427202 & -43.3583059333473 & 63.0143873105982 & 70.926494947865 & -47.0761864021819 \\ 13.16180393057 & 9.9028229600855 & -4.70469407522566 & 19.5047861106629 & 8.08378958421306 \end{bmatrix}. \tag{4.102}$$

According to reciprocal screw theory, the instant free motion allowed can be deduced:

$$\$_m = \begin{bmatrix} 1 \\ -1.179760 \\ 2.018676 \\ 386.283267 \\ 252.347574 \\ -25.938049 \end{bmatrix}. \tag{4.103}$$

Therefore, the instant pitch of the screw is:

$$h = \frac{\mathbf{s}\cdot(\mathbf{s}_0 + h\mathbf{s})}{\|\mathbf{s}\|^2} = 5.599784. \tag{4.104}$$

The instant screw axis is:

$$[\,0.3924 \quad -0.4639 \quad 0.7938 \quad 151.8952 \quad 99.1353 \quad -10.0898\,]^T. \tag{4.105}$$

The projection point of the coordinate origin on this instant screw axis is:

$$(\,-74.039852 \quad 124.591460 \quad 109.491532\,). \tag{4.106}$$

That is, the end effector instantaneously makes a twist movement, $\$_m$, with the screw axis of (4.106), pitch of (4.104), and passing through point

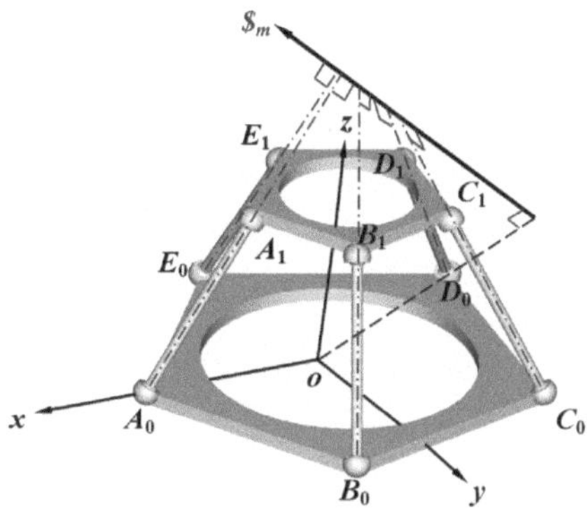

FIGURE 4.24 The instant virtual axis of a spatial parallel multi-link mechanism.

of (4.106), which is shown in Figure 4.24. The instantaneous screw axis, $\$_m$, is just corresponding to the instantaneous center of planar case which is governed by the Aronhold–Kennedy theorem.

4.4 THE EQUIVALENT SUBSTITUTIONS FOR HYBRID KINEMATIC CHAINS

Theorem 4.1. *The resultant twist(s) of a parallel kinematic chain can be equivalently substituted by the twist(s) of a series kinematic chain, which just expresses the DOF of the end effector of the parallel kinematic chains.*

This theorem can be immediately proved with the reciprocal screw theory. We present some examples below.

4.4.1 The Hybrid Kinematic Chain of a Bionic Mechanism

A spatial hybrid structure for a bionic mechanism is shown in Figure 4.25. It has four hybrid legs shown in Figure 4.26. Next, we will establish a Cartesian coordinate system to analyze the hybrid kinematic chain.

The hybrid kinematic chain consists of a prismatic pair E_1 and a parallel structure $A_1B_1C_1D_1$ and a spherical pair F_1 in series. The parallel structure $A_1B_1C_1D_1$ has two identical kinematic chains made up of one universal pair and one revolute pair. Obviously, this parallel structure will always remain coplanar.

Next, let us consider the motions and constraints transformed from C_1D_1 to A_1B_1 by this parallel structure $A_1B_1C_1D_1$. The local Cartesian coordinate system $o_1x_1y_1z_1$ for the kinematic chain D_1A_1 is shown in Figure 4.27.

$$\$_{\mathrm{D}_1} = [1 \;\; 0 \;\; 0 \;\; 0 \;\; 0 \;\; 0]^T, \;\; \$_{\mathrm{A}_1} = [1 \;\; 0 \;\; 0 \;\; 0 \;\; l_{D_1A_1} \;\; 0]^T.$$

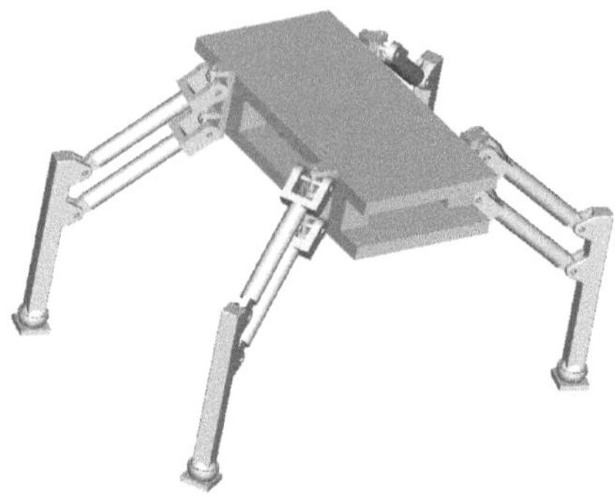

FIGURE 4.25 A bionic mechanism consisting of four hybrid legs.

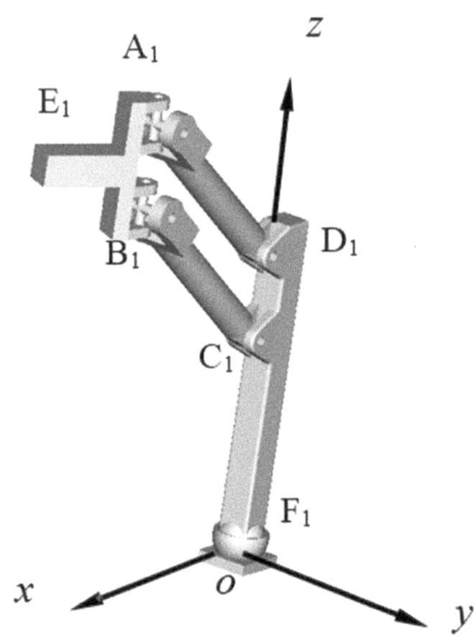

FIGURE 4.26 A hybrid kinematic chain of the bionic mechanism.

The terminal constraint matrix of the kinematic chain consists of the reciprocal screws of the kinematic chain:

$$\$_{D_1A_1} = [\$_1^r \ \ \$_2^r \ \ \$_3^r \ \ \$_4^r] \tag{4.107}$$

where

$$\$_1^r = [1 \ \ 0 \ \ 0 \ \ 0 \ \ 0 \ \ 0]^T, \ \ \$_2^r = [0 \ \ 0 \ \ 1 \ \ 0 \ \ 0 \ \ 0]^T,$$
$$\$_3^r = [0 \ \ 0 \ \ 0 \ \ 0 \ \ 1 \ \ 0]^T, \ \ \$_4^r = [0 \ \ 0 \ \ 0 \ \ 0 \ \ 0 \ \ 1]^T.$$

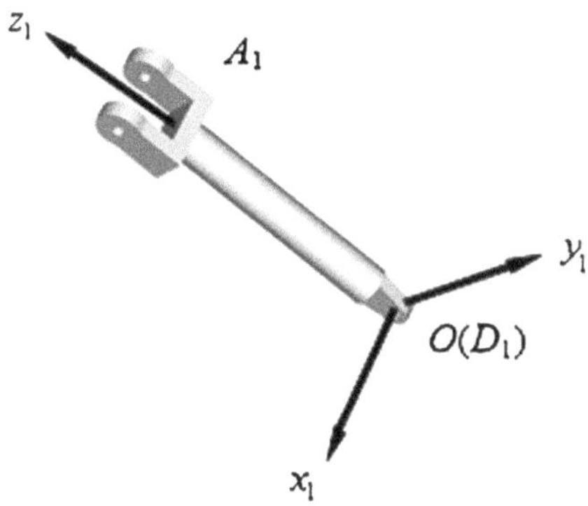

FIGURE 4.27 Sub-kinematic chain D_1A_1.

$\$^r_1$, $\$^r_3$ and $\$^r_4$ form the planar constraints ($y_1o_1z_1$), and $\$^r_2$ restricts a translation along the axis of link D_1A_1.

Similarly, within the local Cartesian coordinate system shown in Figure 4.26, we can also find that the terminal constraints of the sub-kinematic chain C_1B_1 provide the same planar constraints ($y_1o_1z_1$) and a force along the axis of link C_1B_1.

Therefore, the terminal constraints applied to link A_1B_1 from sub-kinematic chains D_1A_1 and C_1B_1 can be spanned into a five-dimension subspace, a basis of which can be denoted in the Cartesian coordinate system shown in Figure 4.26:

$$\$^C_P = \begin{bmatrix} 1 & 0 & 0 & 0 & 0 \\ 0 & 0 & 0 & 0 & 0 \\ 0 & 1 & 0 & 0 & 0 \\ 0 & 0 & 1 & 0 & 0 \\ 0 & 0 & 0 & 1 & 0 \\ 0 & 0 & 0 & 0 & 1 \end{bmatrix}. \tag{4.108}$$

Therefore, the free mobility space of link A_1B_1 is:

$$\mathbf{S}_F = \left\{\boldsymbol{T}_{y_1}\right\} \tag{4.109}$$

where $\boldsymbol{T}_{y_1} = [0 \quad 0 \quad 0 \quad 0 \quad 1 \quad 0]^T$.

With coordinate transformation, we can obtain the form of $\mathbf{S}_F$ in the Cartesian coordinate system *oxyz*:

$$\mathbf{S}_F = \{\boldsymbol{T}_{xz}\} \tag{4.110}$$

where $\boldsymbol{T}_{xz} = [0 \quad 0 \quad 0 \quad z_{D_1} - z_{A_1} \quad 0 \quad x_{A_1} - x_{D_1}\,]^T$.

Therefore, $\boldsymbol{T}_{xz}$ is the motion transformed from links C_1D_1 to A_1B_1 by the parallel structure $A_1B_1C_1D_1$. That is, the twist of the parallel kinematic chains $A_1B_1C_1D_1$ can be equivalently substituted by $\boldsymbol{T}_{xz}$.

Now, let's come back to the twists of the hybrid kinematic chain shown in Figure 4.26. The twists of spherical pair F_1 are:

$$\$_1 = [1 \ \ 0 \ \ 0 \ \ 0 \ \ 0 \ \ 0]^T, \ \ \$_2 = [0 \ \ 1 \ \ 0 \ \ 0 \ \ 0 \ \ 0]^T,$$
$$\$_3 = [0 \ \ 0 \ \ 1 \ \ 0 \ \ 0 \ \ 0]^T.$$

The twists of link A_1B_1 are made up of $\boldsymbol{T}_{xz}$ and a rotation about the z-axis:

$$\$_5 = [0 \ \ 0 \ \ 1 \ \ 0 \ \ -x_{A_1} \ \ 0]^T.$$

If we presume that the direction ratios of the prismatic pair E_1 are $[x_{E_1} \ \ y_{E_1} \ \ z_{E_1}]^T$, the Plücker coordinates can be expressed as:

$$\$_6 = [0 \ \ 0 \ \ 0 \ \ x_{E_1} \ \ y_{E_1} \ \ z_{E_1}]^T.$$

Therefore, the equivalent twist matrix of the original hybrid kinematic chain $\$_e$ is:

$$\$_e = [\$_1 \ \ \$_2 \ \ \$_3 \ \ T_{xz} \ \ \$_5 \ \ \$_6]. \tag{4.111}$$

Immediately, we know that the terminal constraint of this hybrid kinematic chain is zero, and the body of the bionic mechanism has full degrees of freedom (6 DOFs).

4.4.2 Hybrid Kinematic Chain of Spatial Mechanism with Four Branches

The mechanism shown in Figure 4.28 is made up of four hybrid kinematic chains, which consist of the same kinematic chain of one slider and two parallel kinematic chains in series.

In order to investigate the characteristics of the hybrid branch, we set up a local coordinate system, as Figure 4.29 shows. The origin of the coordinates is superposed with one of the spherical pairs, say pair A_1, the axis of the kinematic chain is superposed with the y-axis. So, the kinematic screw matrix of the two pairs of kinematic chain A_1B_1 can be gained:

$$\$_{ss} = [\$_1 \ \ \$_2 \ \ \$_3 \ \ \$_4 \ \ \$_5 \ \ \$_6] \tag{4.112}$$

where

$$\$_1 = [1 \ \ 0 \ \ 0 \ \ 0 \ \ 0 \ \ 0]^T, \ \ \$_2 = [0 \ \ 1 \ \ 0 \ \ 0 \ \ 0 \ \ 0]^T,$$
$$\$_3 = [0 \ \ 0 \ \ 1 \ \ 0 \ \ 0 \ \ 0]^T, \ \ \$_4 = [1 \ \ 0 \ \ 0 \ \ 0 \ \ 0 \ \ -l]^T,$$
$$\$_5 = [0 \ \ 1 \ \ 0 \ \ 0 \ \ 0 \ \ 0]^T, \ \ \$_6 = [0 \ \ 0 \ \ 1 \ \ l \ \ 0 \ \ 0]^T.$$

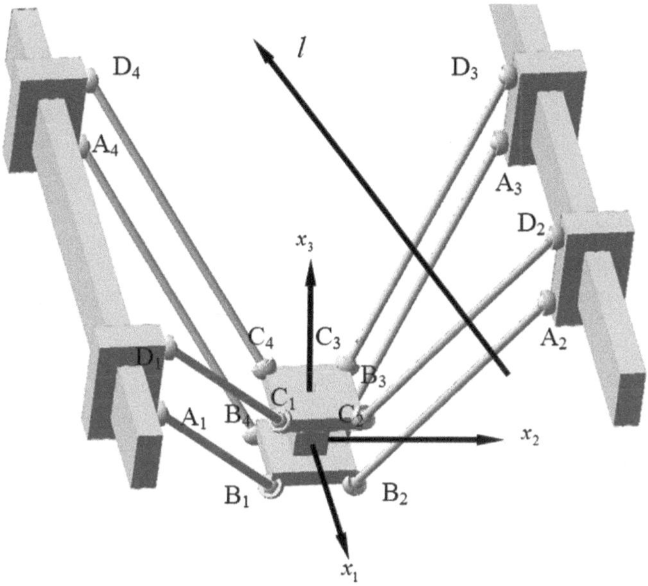

FIGURE 4.28 A kind of spatial hybrid mechanism with four branches.

FIGURE 4.29 One kinematic chain of the parallel members in the spatial hybrid mechanism.

According to the reciprocal screw theory, the twists are reciprocal to the wrenches of the kinematic chain. That is, the reciprocal product of the twists and the terminal constraints of the chain are zero. Therefore, the following equation holds:

$$\$_{SS}^T \boldsymbol{\Delta} \$_{SS}^r = 0. \tag{4.113}$$

The rank of the twists, $\$_{SS}$, of the kinematic chain is:

$$\text{Rank}(\$_{SS}) = 5. \tag{4.114}$$

The reciprocal screw system of $\$_{SS}$ only consists of one force screw:

$$\$_{SS}^r = [0 \quad 1 \quad 0 \quad 0 \quad 0 \quad 0]^T \tag{4.115}$$

which is a pure force whose direction is along the y-axis (the axis of the two spherical pairs), and which constrains the translational movements along the positive and negative directions of y-axis.

In the case that the two kinematic chains are in the same plane, their terminal constraints will restrict the following two movements:

$$\$_1 = [1 \ 0 \ 0 \ 0 \ 0 \ 0]^T, \ \$_2 = [0 \ 0 \ 0 \ 0 \ 1 \ 0]^T.$$

According to the screw theory, the equivalent kinematic screws of the parallel kinematic chains can be deduced if they always remain coplanar:

$$\$_{ep} = [\$_{ep1} \ \ \$_{ep2} \ \ \$_{ep3} \ \ \$_{ep4}] \tag{4.116}$$

where

$$\$_{\text{ep1}} = [0 \ 1 \ 0 \ 0 \ 0 \ 0]^T, \ \$_{\text{ep2}} = [0 \ 0 \ 1 \ 0 \ 0 \ 0]^T,$$
$$\$_{\text{ep3}} = [0 \ 0 \ 0 \ 1 \ 0 \ 0]^T, \ \$_{\text{ep4}} = [0 \ 0 \ 0 \ 0 \ 0 \ 1]^T.$$

That is, the kinematic screw system $\$_{ep}$ can execute the same function of the initial parallel members if the two parallel kinematic chains always remain coplanar.

In the following section, we will prove that the two parallel kinematic chains of each branch in the mechanism shown in Figure 4.28 can always remain coplanar no matter what kind of movements are executed.

Assume that all of the links in each hybrid branch of the rigid end effector remain coplanar at first, and after certain movements, their initial relative spatial relationship is changed. That is, the coplanar relationship of the members of the branch does not hold any longer. According to the theorem of Euler for the rigid body revolving around a given point [7], we know that the latest position of the rigid end effector can be equivalently gained by a finite rotation around a certain pivot from the initial position.

Assume that line l shown in Figure 4.28 can act as the very pivot of the rigid end effector. The finite angle $\theta(\theta \neq 0)$ that the rigid end effector revolved about pivot l might cause each hybrid branch to alter the coplanar relationship.

Because of the symmetry of the branches, it is enough for us to investigate the relationship of the spatial position of one hybrid branch, for example, the branch $A_1B_1C_1D_1$, which should also be applied to any other branches.

Generally, the distances from the eight vertexes of the rigid end effector to a prescribed line are not always equivalent except for the three pivots (shown in Figure 4.28) x_1,x_2, and x_3, which pass through the geometry centers of the three adjacent perpendicular planes and are perpendicular to them, individually. Therefore, if the rigid end effector makes a rotation about the given line, the displacements of the eight vertexes are not always equivalent, which is impossible for a rigid body except for the three pivots x_1,x_2, and x_3. In the following, we will discuss the three cases individually.

4.4.2.1 A Finite Angle the Rigid End Effector Revolved About the x_1-Axis

Figure 4.30 shows the projection of the rigid end effector $B_1B_2B_3B_4 - C_1C_2C_3C_4$ along the x_1-axis. The dashed line circle denotes the possible trajectory of the four vertexes B_1, B_2, C_2 and C_1 of the projection rectangle $B_1B_2C_2C_1$. When the rigid end effector spins any finite angle of $\theta\,(\theta \neq 0)$ about the x_1-axis, the length of rigid link A_1B_1 and C_1D_1 will be changed into A_1B_{11} and C_1D_{11}, respectively, which are impossible for the rigid bodies. Therefore, any finite rotation about the x_1-axis should not exist.

4.4.2.2 A Finite Angle the Rigid End Effector Revolved About the x_2-Axis

Figure 4.31 shows the projection of the rigid end effector $B_1B_2B_3B_4 - C_1C_2C_3C_4$ along the x_2-axis. The dashed line circle o also denotes the possible trajectory of the four vertexes B_1, B_4, C_4, and C_1 of the projection rectangle $B_1B_4C_4C_1$. The dashed line circle o_1 denotes the transection of the projection plane and the spherical trajectory of vertex B_1 revolved about the given point

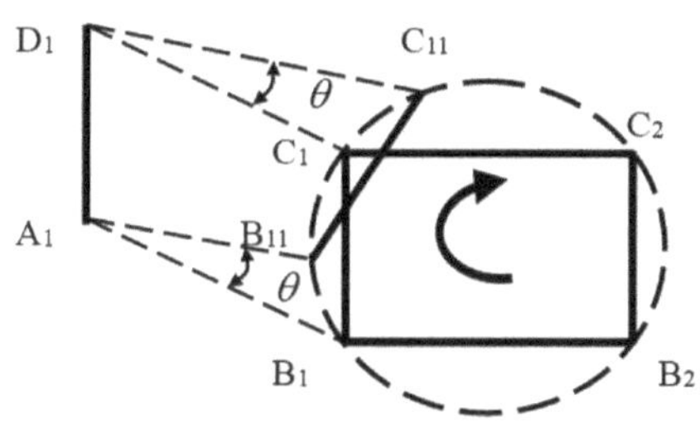

FIGURE 4.30 The projection of rigid end effector along the x_1-axis.

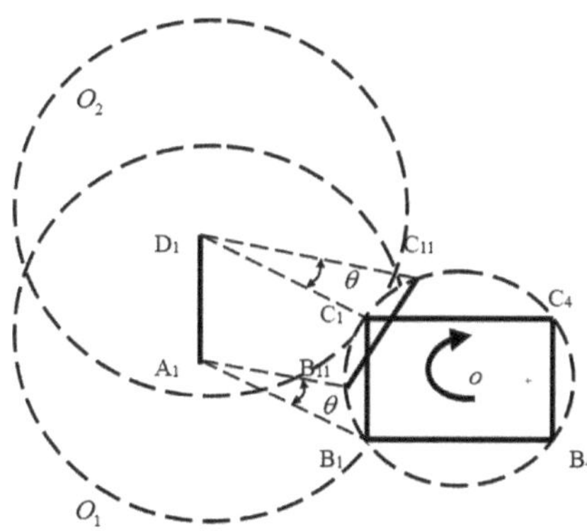

FIGURE 4.31 Projection of the rigid end effector along the x_2-axis.

A_1. Similarly, the dashed line circle o_2 denotes the transection of the projection plane and the spherical trajectory of vertex C_1 revolved about the given point D_1. When the rigid end effector spins an angle of $\theta(\theta \neq 0)$ about the x_2-axis, the vertex B_1 and C_1 will rotate to the points B_{11} and C_{11}, individually. However, the trajectory of point B_1 in plane $B_1B_4C_4C_1$ should only be dashed line circle o_1 and the trajectory of point C_1 in plane $B_1B_4C_4C_1$ should only be the dashed line circle o_2. Therefore, any finite spin about the x_2-axis should not exist, either.

4.4.2.3 A Finite Angle the Rigid End Effector Revolved About the x_3-Axis

This kind of spin is feasible and the coplanar relationship of each branch should always be held. In fact, it is just one rotational DOF of the hybrid end effector.In conclusion, the members of each branch of the hybrid end effector will always remain coplanar, no matter what kind of movement executed.So the kinematic screws of the hybrid can be denoted as:

$$\$_{h1} = [\$_{slider}\ \ \$_{ep1}\ \ \$_{ep2}\ \ \$_{ep3}\ \ \$_{ep4}]. \tag{4.117}$$

With formula (4.117) and Figure 4.30, we can obtain the screws of $\$_{h1}$ as follows:

$$\begin{aligned}
&\$_{slider} = [0\ \ 0\ \ 0\ \ a\ \ b\ \ c]^T,\ \ \$_{\text{ep1}} = [0\ \ 1\ \ 0\ \ 0\ \ 0\ \ 0]^T,\\
&\$_{\text{ep2}} = [0\ \ 0\ \ 1\ \ 0\ \ 0\ \ 0]^T,\ \ \$_{\text{ep3}} = [0\ \ 0\ \ 0\ \ 1\ \ 0\ \ 0]^T,\\
&\$_{\text{ep4}} = [0\ \ 0\ \ 0\ \ 0\ \ 0\ \ 1]^T
\end{aligned}$$

where $a,b,c, \neq 0$.

Therefore, the reciprocal screws of the branch can be deduced:

$$\$^r_{h1} = [0\ \ 0\ \ 0\ \ 1\ \ 0\ \ 0]^T \tag{4.118}$$

because $\$^r_{h1}$ is a pure couple of movement around the x-axis in Figure 4.32. That is, $\$^r_{h1}$ is a pure couple of movement, which is perpendicular to the plane decided by the parallel kinematic chains of the hybrid branch $A_1B_1C_1D_1$.

Similarly, we can obtain the reciprocal screws of the other three hybrid branches of the end effector. The reciprocal screws exerted on the end effector are shown in Figure 4.33. The four reciprocal screws are all pure couple constraints and all perpendicular to the z-axis, which can only constrain the rotations about the x-axis and y-axis at most. Therefore, the DOF of the end effector is $F = 6 - 2 = 4$ at least. In most cases, because the four pure couples perpendicular to the z-axis should constrain the rotations about the x-axis and the y-axis, the DOF of the hybrid system is $F = 6 - 2 = 4$.

However, because the actuators of the four sliders are not always independent, the DOFs of the hybrid system may be larger than four. That is, we also need to analyze the cases when the actuators of the four sliders are

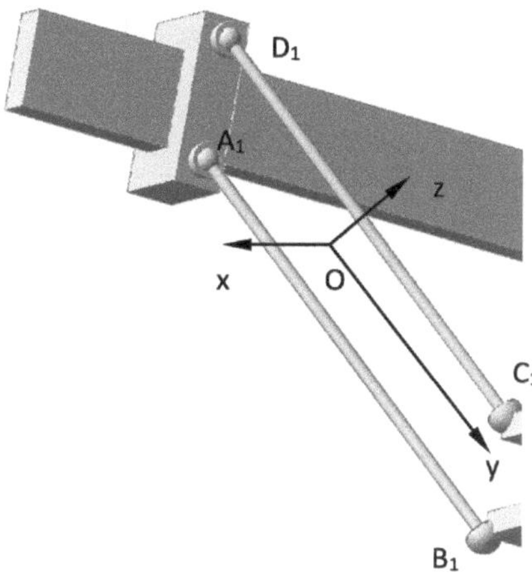

FIGURE 4.32 One branch of the spatial hybrid mechanism.

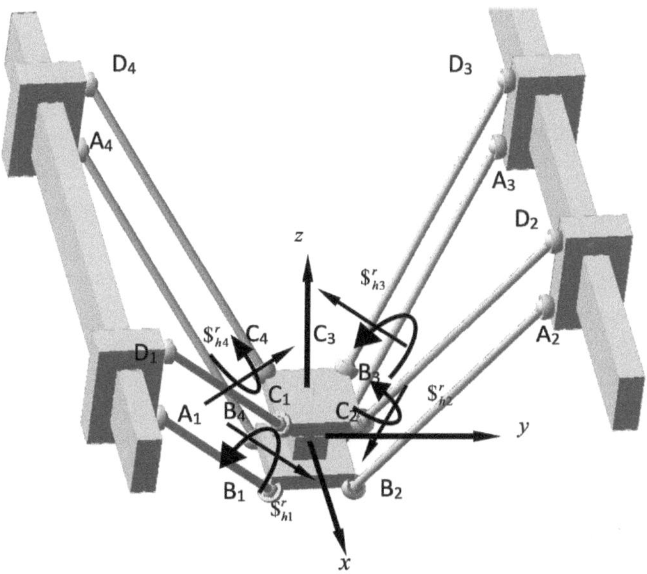

FIGURE 4.33 Reciprocal screws applied to the end effector.

prescribed. The twist system of the hybrid branch 1 shown in Figure 4.31 can be expressed in the twist matrix:

$$\$_{h1actu} = \begin{bmatrix} \$_{ep1} & \$_{ep2} & \$_{ep3} & \$_{ep4} \end{bmatrix}. \tag{4.119}$$

The reciprocal screw matrix can be gained as follows:

$$\$^r_{h1actu} = \begin{bmatrix} \$^{r1}_{h1} & \$^{r2}_{h1} \end{bmatrix} \tag{4.120}$$

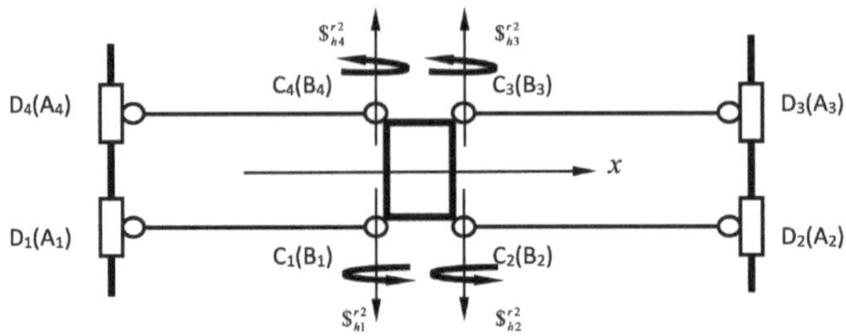

FIGURE 4.34 The end effector of the hybrid mechanism when singularity occurs.

where $\$^{r1}_{h1} = [0 \ \ 1 \ \ 0 \ \ 0 \ \ 0 \ \ 0]^T, \$^{r2}_{h1} = [0 \ \ 0 \ \ 0 \ \ 1 \ \ 0 \ \ 0]^T, \$^{r1}_{h1}$ is a pure a force screw along the y-axis and $\$^{r2}_{h1}$ is a pure couple of moment about the x-axis, shown in Figure 4.29.

Because the singularity is defined as the cases when the DOF of the hybrid system will be changed, the singularity cases can be concluded as one of the follows:

1. $\$^{r2}_{h1}, \$^{r2}_{h2}$ are in a line and $\$^{r2}_{h3}, \$^{r2}_{h4}$ are in a line;
2. $\$^{r2}_{h1}, \$^{r2}_{h4}$ are in a line and $\$^{r2}_{h2}, \$^{r2}_{h3}$ are in a line;
3. $\$^{r2}_{h1}, \$^{r1}_{h2}, \$^{r1}_{h3}$ and $\$^{r1}_{h4}$ intersect with the same line x_3 in Figure 4.27.

Considering the geometry characteristics and the above three cases, we can find that only cases 2 and 3 are feasible. Case 2 is shown in Figure 4.34; case 3 is often called *similarity singularity* [8], in which the instantaneous singularity occurs because the four actuators are not linearly independent.

Considering the geometry characteristics of the end effector, we can investigate the newly increased DOF for the end effector in case 2. Supposing the projection of the hybrid mechanism along the x-direction in Figure 4.34 is shown in Figure 4.35, the rotation can be investigated as follows.

Figure 4.35 shows the projection of rigid end effector $B_1B_2B_3B_4 - C_1C_2C_3C_4$ along the x-direction. The dashed line circle O denotes the trajectory of the four vertexes B_1, B_4, C_4, and C_1 of the projection rectangle $B_1B_4C_4C_1$. The dashed line circle O_1 denotes the transection of the projection plane and the spherical trajectory of vertex B_1 revolved around the given point A_1. Similarly, the dashed line circle O_2 denotes the transection of the projection plane and the spherical trajectory of vertex C_1 revolved around the given point D_1. When the rigid end effector spins any angle of $\theta (\theta \neq 0)$ around the x-direction, the vertex B_1 and C_1 will rotate to the point B_{11} and C_{11} individually. However, the possible trajectory of point B_1 in plane $B_1B_4C_4C_1$ should only be the dashed line circle O_1 and the possible trajectory of point C_1 in plane $B_1B_4C_4C_1$ should only be the dashed line circle O_2. Therefore, this kind of finite spin around x-direction should not exist. And the singularity case 2 of the hybrid mechanism shown in Figure 4.34 does not hold in reality. So, we can draw a conclusion that

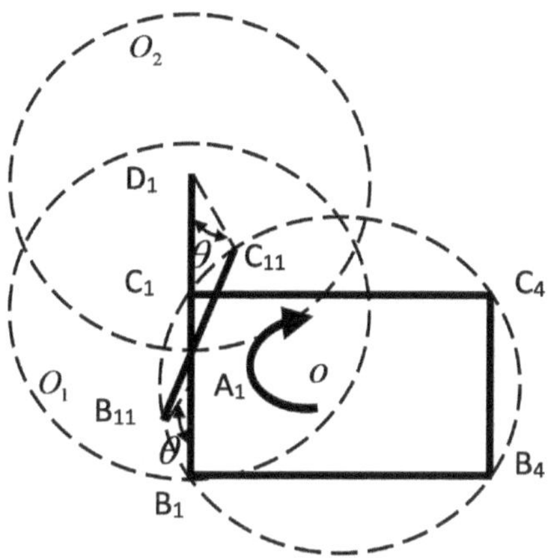

FIGURE 4.35 The end effector of the hybrid mechanism when singularity occurs.

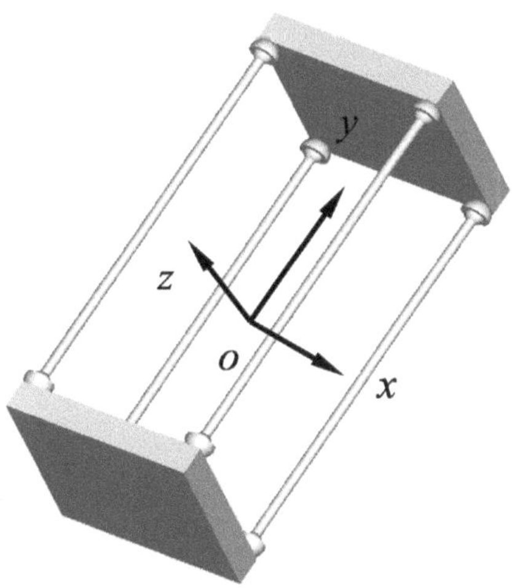

FIGURE 4.36 A parallel kinematic chain.

the singularity of the mechanism shown in Figure 4.28 will appear only when case 3 occurs in its workspace.

Similarly, the equivalent twists of the parallel kinematic chain shown in Figure 4.36 have the following form when the end effector is parallel to the base:

$$\$_e = [\$_{e1} \quad \$_{e2} \quad \$_{e3}] \tag{4.121}$$

where

$$\$_{e1} = [0 \quad 0 \quad 0 \quad 1 \quad 0 \quad 0]^T, \$_{e2} = [0 \quad 0 \quad 0 \quad 0 \quad 0 \quad 1]^T,$$
$$\$_{e3} = [0 \quad 1 \quad 0 \quad 0 \quad h \quad 0]^T \quad \text{and} \quad h \neq 0.$$

4.4.3 Hybrid Mechanism with Three Translational DOFs and One Rotational DOF

As Figure 4.37 shows, the hybrid mechanism has three translational DOFs and one rotational DOF. The end effector is connected to the fixed guides by three identical hybrid kinematic chains. In each hybrid kinematic chain, a prismatic pair is serially connected with a parallel kinematic chain made up of two SS chains.

With a similar process to that in Section 4.4.2, we can prove that the hybrid kinematic chains of the end effector will always remain coplanar. Next, we will focus on the equivalent substitutions of such a hybrid kinematic chain, shown in Figure 4.37.

According to the analysis in Section 4.4.2, each limb of the parallel kinematic chain shown in Figure 4.32 can be equivalently substituted with a *SU* kinematic chain. The reciprocal screws of these two limbs, $\$^r_{AB}$ and $\$^r_{DC}$, will not be linear dependent. Therefore, the equivalent kinematic chain made up of the parallel limbs only consists of four linear independent twists whose reciprocal screws restrict the following motions:

$$\$_1 = [1 \quad 0 \quad 0 \quad 0 \quad 0 \quad 0]^T, \quad \$_2 = [0 \quad 0 \quad 0 \quad 0 \quad 1 \quad 0]^T.$$

According to the reciprocal screw theory, the equivalent twists of the parallel limbs are also formulated by equation (2.98).

In the coordinate system shown in Figure 4.38, if we presume the moving direction ratios of the slider P_1 are $[x_P \quad y_P \quad z_P]^T$, the Plücker coordinates of the prismatic pair can be denoted as:

$$\$_P = [0 \quad 0 \quad 0 \quad x_P \quad y_P \quad z_P]^T.$$

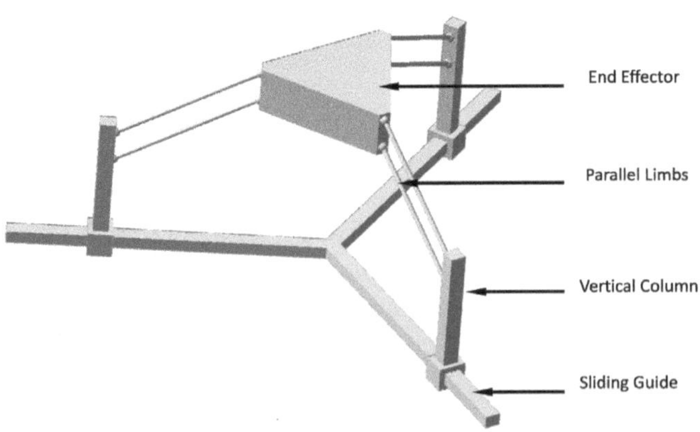

FIGURE 4.37 A hybrid mechanism with three translational DOFs and one rotational DOF.

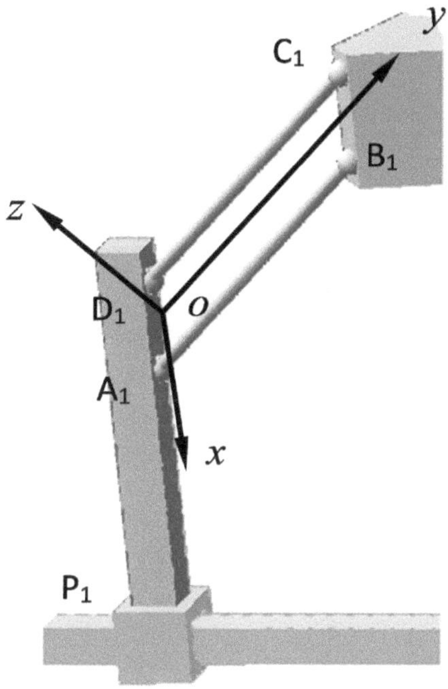

FIGURE 4.38 One hybrid kinematic chain of the mechanism.

The equivalent twists of the hybrid kinematic chain can be expressed in the twist matrix form:

$$\$_{h1} = [\$_P \quad \$_{e1} \quad \$_{e2} \quad \$_{e3} \quad \$_{e4}]. \tag{4.122}$$

The reciprocal screw of the twists (4.7) is:

$$\$^r_{h1} = [0 \quad 0 \quad 0 \quad 1 \quad 0 \quad 0]^T. \tag{4.123}$$

The physical meanings of $\$^r_{h1}$ is that it restricts the end effector of the hybrid kinematic chain from rotating about the x-axis of Figure 4.39. That is, the end effector could not rotate about the norm of the plane $A_1B_1C_1D_1$.

Similarly, we can find the terminal constraints of the other two hybrid kinematic chains. Therefore, the constraint space spanned by the terminal constraints of the three hybrid kinematic chains can be obtained:

$$\$^C_P = \text{Span}\left\{\left[\$^r_{B_1P_1M_1} \quad \$^r_{B_2P_2M_2} \quad \$^r_{B_3P_3M_3}\right]\right\}. \tag{4.124}$$

That is

$$\$^C_P = \begin{bmatrix} 0 & 0 & 0 & 1 & 0 & 0 \\ 0 & 0 & 0 & 0 & 1 & 0 \end{bmatrix}^T. \tag{4.125}$$

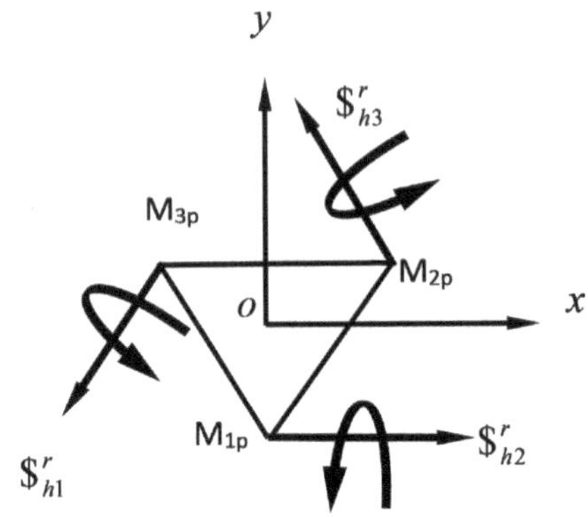

FIGURE 4.39 Wrenches acted on the end effector from the hybrid kinematic chains.

Obviously, we have

$$\dim \$_P^C = 2. \tag{4.126}$$

Assume

$$\$_1^m = [1 \;\; 0 \;\; 0 \;\; 0 \;\; 0 \;\; 0]^T, \;\; \$_2^m = [0 \;\; 1 \;\; 0 \;\; 0 \;\; 0 \;\; 0]^T.$$

We can obtain the subspace, $\mathbf{S}_C$, restricted by the terminal constraints:

$$\mathbf{S}_C = \{\boldsymbol{R}_x \;\; \boldsymbol{R}_y\} \tag{4.127}$$

where $\boldsymbol{R}_x = \$_1^m, \boldsymbol{R}_y = \$_2^m$.

Therefore,

$$r = \dim \mathbf{S}_C = \dim \$_P^C = 2. \tag{4.128}$$

And $\mathbf{S}_C$ could restrict the end effector from rotating about the x- and y-axes. Consequently, the free mobility space of the end effector is obtained:

$$\mathbf{S}_F = \{\boldsymbol{T}_x \;\; \boldsymbol{T}_y \;\; \boldsymbol{T}_z \;\; \boldsymbol{R}_z\}. \tag{4.129}$$

Therefore, the DOF of the end effector is:

$$F = 6 - r = 6 - 2 = 4(3 \text{ Translational DOFs} + 1 \text{ Rotational DOF}).$$

Based on the analysis above, the number of the DOFs of the end effector shown in Figure 4.34 is four. If we select the three sliders as the actuators, the end effector is uncontrollable. From the viewpoint of the mobility space and the DOF of the end effector, the mechanism shown in Figure 4.37 is equivalent to the one shown in Figure 4.40. However, the kinematic chains of the mechanism in Figure 4.40 are serial *PUU* chains which are much simpler than the hybrid ones in Figure 4.37.

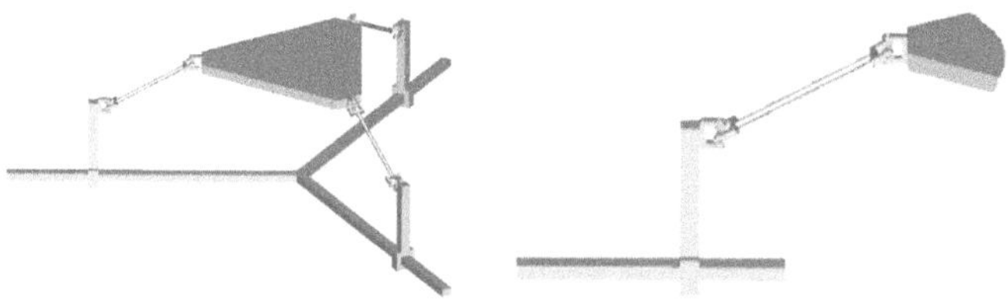

FIGURE 4.40 An equivalent spatial parallel mechanism to the hybrid one shown in Figure 4.34 and its one kinematic chain.

4.5 CONCLUSIONS

Through singular value decomposition of reciprocal screws, this chapter addresses the mobility space, which defines the type, number, and direction characteristics of the independent motions that the manipulator should execute and then proposes an effective and efficient method to determine the instant virtual axis of spatial parallel mechanism. Then the constraint screws exerted on the end effector are obtained in matrix form. We firstly investigate the reciprocal screws applied to the end effector and obtain the spaces spanned by the reciprocal screws of the kinematic chains via SVD, and then the free mobility spaces of the end effector and its DOF. With this method, we can both obtain the numeral properties of the DOF of the end effectors of spatial parallel manipulators and find their type and direction properties. The application example demonstrates that this methodology can be utilized to exploit the all-sided mobility characteristics of parallel manipulators, including the singularity and DOF.

REFERENCES

[1] J.-S. Zhao, K. Zhou, et al., A new method to study the degree of freedom of spatial parallel mechanisms, The International Journal of Advanced Manufacturing Technology 23 (2004) 288–294.

[2] J.S. Zhao, K. Zhou, Z.J. Feng, A theory of degrees of freedom for mechanisms, Mechanism and Machine Theory 39 (2004) 621–643.

[3] J.S. Zhao, Z.J. Feng, et al., Reanalysis on the degree of freedom of the platforms in spatial parallel mechanisms with constraints spaces, The International Journal of Advanced Manufacturing Technology 28 (2006) 190–196.

[4] J.S. Zhao, Z.J. Feng, et al., The free mobility of a parallel manipulator, Robotica 24 (2006) 635–641.

[5] J.S. Zhao, Y.Z. Fu, et al., Mobility properties of a schoenflies type parallel Manipulator, Robotics and Computer-Integrated Manufacturing 22 (2006) 275–284.

[6] J.S. Zhao, K. Zhou, et al., A new method to study the degree of freedom of spatial parallel mechanisms, The International Journal of Advanced Manufacturing Technology 23 (2004) 288–294.

[7] S. Jia, Dynamics of Rigid Body, Advanced Education Press, China, 1987.

[8] Y. Fang, L.W. Tsai, Structure synthesis of a class of 4-DOF and 5-DOF parallel manipulators with identical kinematic chain structures, The International Journal of Robotics Research 21 (2002) 799–810.

Chapter 5

Workspace of the End Effector of a Robot Mechanism

In engineering applications, we often attach a great importance to the study of the workspace of a manipulator, especially spatial parallel manipulators. The workspace of a manipulator is the reachable domain of its end effector. The workspace of any manipulator is bounded in the 3D space [1]. The workspace boundary is a curve on a plane or a surface in space that defines the extent of reach of the end effector [2]. The workspace of a manipulator has been defined in literature as the totality of positions that a particular identified point on the end effector of the manipulator can reach. In this chapter, we define the reachable workspace as:

Definition 5.1. The reachable workspace of a mechanism is the totality of positions that a particular identified point of the end effector can reach, which is referred to as the workspace.

In general, the particular identical point of the end effector is often specified to the geometric center of the end effector. Many scholars have studied the workspace of the spatial manipulators [1–3]. Dibakar and Mruthyunjaya [4] presented a general methodology for the synthesis of the external boundary of the workspaces of a planar manipulator with arbitrary topology. Both the desired workspace and the manipulator workspaces are identified by their boundaries and are treated as simple closed polygons. Gao and Liu [5] classified three-degree-of-freedom (DOF) planar parallel manipulators, presented various of shapes of the workspaces and the relationships between the shapes of the workspaces, and the link lengths of each of the classifications. Cervantes-Sánchez and Rendón-Sánchez [6] presented a method for generating the reachable workspace of a class of planar manipulators with two DOFs. Cervantes-Sánchez *et al.* [7] presented an approach for numerically obtaining the workspace and the singularity curves of a planar 5R-type manipulator. To determine the workspace of a manipulator, many scholars have proposed algebraic methods [8,9] and geometric methods [3,10]. Wang *et al.* [11] presented an algorithm, called the boundary search method, to determine the workspace of a parallel machine tool. They also analyzed the boundary of

Advanced Theory of Constraint and Motion Analysis for Robot Mechanisms.
http://dx.doi.org/10.1016/B978-0-12-420162-0.00005-9

the workspace in order to expand its operating scope. Bonev and Ryu [12] presented a discretization method for computing the orientation workspace of 6-DOF parallel manipulators.

The above literatures mainly focused on searching the workspace of a manipulator, but seldom discussed the relationship between the shapes of the workspace and the symmetric geometry structure of the manipulator. In fact, there is an inevitable relationship between the reachable workspace and the symmetric geometry structure of the manipulator. Therefore, this chapter will first discuss the coordinate transformation method to deal with the workspace of the manipulator. The basis of this method is to choose the least variables according to the mobility analysis of a manipulator, and then obtain the workspace parametric expressions of the reference points of the end effector through coordinate transformation. Then the relationship of the workspace and symmetric geometry of the manipulator will be investigated. The mathematical derivation proves that the workspace of a manipulator with symmetric structure must be symmetry, too. And the symmetry of the structure and workspace is a kind of homomorphic mapping. The quantitative description of the symmetric relationship provides an important theoretical basis for the analysis and creative design of mechanisms. On this basis, this chapter sets the Schoenfiles parallel mechanism as an example to study the dexterous workspace of manipulators.

5.1 WORKSPACE BASED ON MOBILITY ANALYSIS

Based on the analysis of the type, orientation, and number of the DOFs of a parallel manipulator, we can choose the least variables, and then the constraint of the joints and links can be expressed in the unified parametric equations by coordinate transformation. This method can be applied as a general solution of the workspace of parallel manipulators.

The workspace of the manipulator can be expressed by a group of parametric equations, so the workspace could be uniquely defined by these equations. Assuming that the geometric center point on the platform connected with n branches is C, the local coordinate system, $Ox_1y_1z_1$, is shown in Figure 5.1, whose origin is superposed with point C. The end vertexes of each branch connected with the platform are denoted by $B_i (i = 1,2,\ldots,n)$, the end vertexes of each branch connected with the fix base or sliders are denoted by P_i $(i = 1,2,\ldots,n)$. Assume that the attitude angles are ψ,θ,ϕ, where ψ denotes precession angle, θ denotes nutation angle, and ϕ denotes spinning angle.

In the absolute coordinate system, the vector coordinates of the end vertexes of each branch connected with the fixed base or sliders are $\boldsymbol{r}_1,\boldsymbol{r}_2,\ldots,\boldsymbol{r}_n$, the coordinates of C are (x_C,y_C,z_C) and the vectors of the vertexes B_i and point C are denoted by $\boldsymbol{r}_{Bi}$ and $\boldsymbol{r}_C$, respectively. In the local coordinate system, the vectors of the vertexes B_i is denoted by $\boldsymbol{r}_{Bi}^L$.

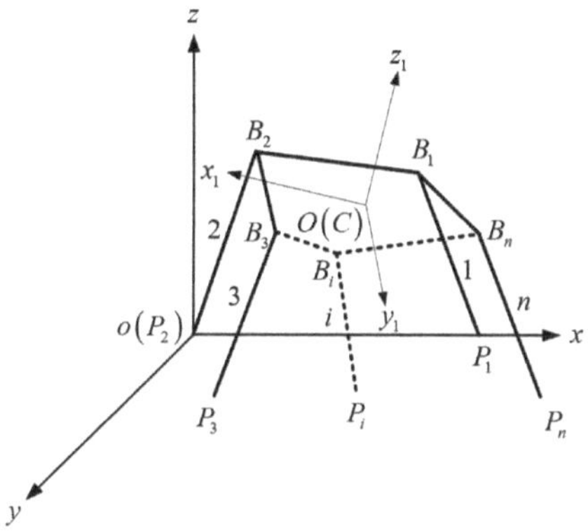

FIGURE 5.1 Local coordinate system of a platform.

With the pose angles, we can easily get the transformation matrix **R**:

$$\mathbf{R}=\begin{bmatrix} c\psi c\phi - s\psi c\theta s\phi & -c\psi s\phi - s\psi c\theta c\phi & s\psi s\theta \\ s\psi c\phi + c\psi c\theta s\phi & -s\psi s\phi + c\psi c\theta c\phi & -c\psi s\theta \\ s\theta s\phi & s\theta c\phi & c\theta \end{bmatrix} \tag{5.1}$$

where c and s are used as abbreviations for cosine and sine, respectively. So, we obtain

$$\boldsymbol{r}_{Bi}=\boldsymbol{r}_C+\boldsymbol{r}_{CBi}=\boldsymbol{r}_C+\mathbf{R}\boldsymbol{r}_{Bi}^L \tag{5.2}$$

where $i=1,2,\ldots,n,\boldsymbol{r}_{CBi}$ denotes the vector of CB_i in the absolute coordinates.

Because $\boldsymbol{r}_{Bi}^L$ has very simple expressions in the local coordinate system, we can therefore obtain $\boldsymbol{r}_{Bi}$ in the absolute coordinate system.

For a spatial parallel manipulator, the workspace could be described with a group of rigid constraint equations. In general, the particular identical point of the end effector, C, satisfies

$$\mathbf{W}_R=\left\{(x,y,z)\,|\mathbf{g}\left(x,y,z,\theta_x,\theta_y,\theta_z\right)=0\right\} \tag{5.3}$$

where $\mathbf{W}_R$ represents the reachable workspace.

Solving the equation (5.3), we obtain the following parametric solutions:

$$\begin{cases} x_C=f_1(\mathrm{var}_1\ \mathrm{var}_2\ \cdots\ \mathrm{var}_m) \\ y_C=f_2(\mathrm{var}_1\ \mathrm{var}_2\ \cdots\ \mathrm{var}_m) \\ z_C=f_3(\mathrm{var}_1\ \mathrm{var}_2\ \cdots\ \mathrm{var}_m) \\ \psi=f_4(\mathrm{var}_1\ \mathrm{var}_2\ \cdots\ \mathrm{var}_m) \\ \theta=f_5(\mathrm{var}_1\ \mathrm{var}_2\ \cdots\ \mathrm{var}_m) \\ \varphi=f_6(\mathrm{var}_1\ \mathrm{var}_2\ \cdots\ \mathrm{var}_m) \end{cases} \tag{5.4}$$

where var_i $\left(i=1,2,\ldots,m\right)$ denotes the independent input variable.

In applications, equation set (5.4) may only include two to six equations. With equation set (5.4), the workspace of the geometric center of the platform can be obtained.

Definition 5.2. The dexterous workspace of a mechanism is defined as the region which can be reached by the reference point with any orientation within a specified range.

If the dexterous workspace of a mechanism is defined as the domain where the reference point C of the end effector can reach with any orientation in a prescribed pose range, the dexterous workspace of the point can be expressed as:

$$\mathbf{W}_D = \{(x,y,z)\,|\mathbf{g}(x,y,z,\theta_x,\theta_y,\theta_z) = 0, \theta_x \in [\theta_{x1},\theta_{x2}], \theta_y \in [\theta_{y1},\theta_{y2}], \theta_z \in [\theta_{z1},\theta_{z2}]\} \tag{5.5}$$

where $\mathbf{W}_D$ represents the desired workspace, θ_{x_1} denotes the lower limit of θ_x, and θ_{x_2} denotes the upper limit of θ_x, and the meanings of other symbols can be similarly deduced.

Generally, $m = F$ holds, where F indicates the DOFs of the end effector. Thus, equations (5.4) and (5.5) include two to six constraint equations. According to equation set (5.4), we obtain the reachable workspace of the manipulator. Considering the range of var_i, we can assume that $\text{var}_i \in [a_i,b_i]\ (i = 1,2,\ldots,m)$. As a result, equation set (5.4) can be transferred to

$$\begin{cases} a_1 \le \mathbf{g}_1(x,y,z,\theta_x,\theta_y,\theta_z) \le b_1 \\ a_2 \le \mathbf{g}_2(x,y,z,\theta_x,\theta_y,\theta_z) \le b_2 \\ \vdots \\ a_M \le \mathbf{g}_M(x,y,z,\theta_x,\theta_y,\theta_z) \le b_M. \end{cases} \tag{5.6}$$

For the manipulators using angular input as a driver, the constraint equations of the reference point of the end effector often have the following general form:

$$\mathbf{g}(\theta_1,\theta_2,\ldots,\theta_m) = \begin{bmatrix} [u_{11}(x) + a_{11}\cos\theta_1 + b_{11}\sin\theta_1]^2 + [v_{11}(y) + c_{11}\cos\theta_1 + d_{11}\sin\theta_1]^2 \\ +[w_{11}(z) + e_{11}\cos\theta_1 + f_{11}\sin\theta_1]^2 - l_1^2 \\ [u_{21}(x) + a_{21}\cos\theta_2 + b_{21}\sin\theta_2]^2 + [v_{21}(y) + c_{21}\cos\theta_2 + d_{21}\sin\theta_2]^2 \\ +[w_{21}(z) + e_{21}\cos\theta_2 + f_{21}\sin\theta_2]^2 - l_2^2 \\ \vdots \\ [u_{m1}(x) + a_{m1}\cos\theta_m + b_{m1}\sin\theta_m]^2 + [v_{m1}(y) + c_{m1}\cos\theta_m + d_{m1}\sin\theta_m]^2 \\ +[w_{m1}(z) + e_{m1}\cos\theta_m + f_{m1}\sin\theta_m]^2 - l_m^2 \end{bmatrix} \tag{5.7}$$

where $u_{i1}(\cdot), v_{i1}(\cdot)$ and $w_{i1}(\cdot)$ represent the functions of x, y, and z, respectively; $i = 1,2,\ldots,m$, m denotes the number of constraint equations; $a_{i1}, b_{i1}, \ldots, f_{i1}$ represent the constant coefficients, θ_i represents the ith independent angular input.

Equation (5.7) can be equivalently transformed into a function of m angular inputs:

$$\boldsymbol{g}\left(\theta_1,\theta_2,\ldots,\theta_m\right)=\begin{bmatrix} p_{11}-p_{12}\cos\theta_1-p_{13}\sin\theta_1 \\ p_{21}-p_{22}\cos\theta_2-p_{23}\sin\theta_2 \\ \vdots \\ p_{m1}-p_{m2}\cos\theta_3-p_{m3}\sin\theta_m \end{bmatrix} \tag{5.8}$$

where $p_{ij}(i=1,2,\ldots,m;\ j=1,2,3)$ are the functions of position and pose parameters, $x,y,z,\theta_x,\theta_y,\theta_z$.

For a set of angular inputs $\left(\theta_1,\theta_2,\ldots,\theta_m\right)$, there are at least one set of $x,y,z,\theta_x,\theta_y,\theta_z$ such that $\boldsymbol{g}\left(\theta_1,\theta_2,\ldots,\theta_m\right)=0$ holds; and *vice versa*, for any point within the reachable workspace of the end effector, there must be a set of real parameters $\left(\theta_1,\theta_2,\ldots,\theta_m\right)$ such that the position and pose parameters $x,y,z,\theta_x,\theta_y,\theta_z$ of the point satisfy the equation $\boldsymbol{g}\left(\theta_1,\theta_2,\ldots,\theta_m\right)=0$. However, the additional constraints of $\cos\theta_i\in\left[-1,1\right]$ and $\sin\theta_i\in\left[-1,1\right]$ are very difficult to be directly used to search the reachable workspace. The fact that $\tan\alpha\left(-\frac{\pi}{2}<\alpha<\frac{\pi}{2}\right)$ belongs to $\left(-\infty,\infty\right)$ allows us to easily crack this difficult problem. So, we can try to obtain the criteria for the boundary surfaces of the reachable and dexterous workspaces through trigonometric transformations, such as $\cos\theta_i=\frac{1-\tan^2\frac{\theta_i}{2}}{1+\tan^2\frac{\theta_i}{2}}$ and $\sin\theta_i=\frac{2\tan\frac{\theta_i}{2}}{1+\tan^2\frac{\theta_i}{2}}$.

Suppose that $t_i=\tan\frac{\theta_i}{2}(i=1,2,\ldots,m)$, we can immediately transform equation (5.8) into the following form:

$$\boldsymbol{g}\left(\theta_1,\theta_2,\ldots,\theta_m\right)=\begin{bmatrix} \left(p_{11}+p_{12}\right)t_1^2-2p_{13}t_1+p_{11}-p_{12} \\ \left(p_{21}+p_{22}\right)t_2^2-2p_{23}t_2+p_{21}-p_{22} \\ \vdots \\ \left(p_{m1}+p_{m2}\right)t_m^2-2p_{m3}t_3+p_{m1}-p_{m2} \end{bmatrix}. \tag{5.9}$$

The criteria for the equation $\boldsymbol{g}\left(\theta_1,\theta_2,\ldots,\theta_m\right)=0$ to have real solutions are:

$$\begin{cases} \Delta_1=\mathbf{4}\left(p_{12}^2+p_{13}^2-p_{11}^2\right)\geq 0 \\ \Delta_2=\mathbf{4}\left(p_{22}^2+p_{23}^2-p_{21}^2\right)\geq 0 \\ \vdots \\ \Delta_m=\mathbf{4}\left(p_{m2}^2+p_{m3}^2-p_{m1}^2\right)\geq 0 \end{cases} \tag{5.10}$$

where p_{i1},p_{i2},p_{i3} are functions of the mere position and pose parameters $x,y,z,\theta_x,\theta_y,\theta_z$.

With the criteria of formula (5.10), we can obtain the reachable and dexterous workspaces of any parallel mechanisms whose inputs are rotations. In the following sections, we will illustrate the validity and efficiency of the above algorithm for the mechanisms under rotational actuations.

5.1.1 Workspace of a Kind of 3-UPU Parallel Manipulator

A novel spatial parallel manipulator, shown in Figure 5.2, is made up of 3-UPU (one prismatic joint and two universal joints) kinematic chains. First, we will create an absolute coordinate system $oxyz$ as Figure 5.2 shows. The plane xoy is perpendicular to the three vertical guides and the x-axis is parallel to the midline of triangle $B_1B_2B_3$. Assuming that the radius of the circumcircle of triangle $B_1B_2B_3$ is R, the coordinates of the three sliders are as follows:

$$P_1\left(R\ 0\ z_1\right),\ P_2\left(-\frac{1}{2}R\ -\frac{\sqrt{3}}{2}R\ z_2\right),\ P_3\left(-\frac{1}{2}R\ \frac{\sqrt{3}}{2}R\ z_3\right).$$

If we presume that the geometric center of manipulator $M_1M_2M_3$ is C, the local coordinate system $o_cx_cy_cz_c$ can be shown in Figure 5.2, whose origin is superposed with point C. The vertexes of the manipulator are denoted by M_i, $(i = 1,2,3)$, and the sliders are denoted by P_i, $(i = 1,2,3)$.

In the absolute coordinate system, the vector coordinates of the sliders are $\boldsymbol{r}_1$,$\boldsymbol{r}_2$, and $\boldsymbol{r}_3$, and the vector coordinates of C are (x, y, z).

In what follows, we will first analyze the DOFs of the manipulator as Figure 5.2 shows. Because all the kinematic chains connecting the manipulator with the base are identical, we only need to select any one of them to study the reciprocal screws of the kinematic chains; for example, we select kinematic chain P_2M_2 as the analyzing object shown in Figure 5.3.

The Plücker coordinates of slider P_2 are:

$$\$_1 = [0\,0\,0\,0\,0\,1]^T.$$

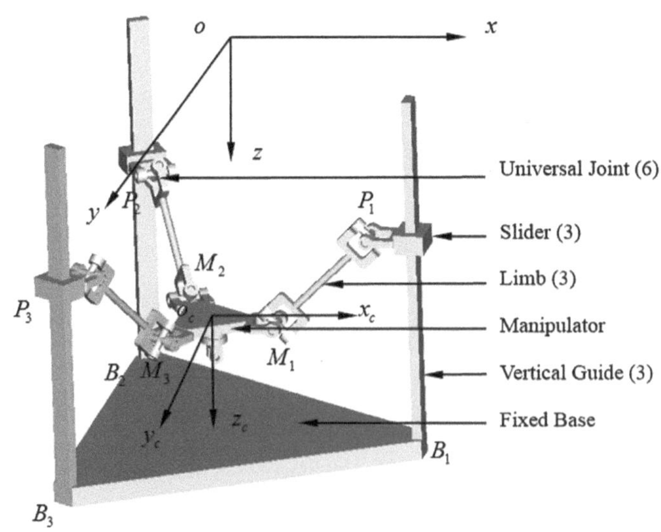

FIGURE 5.2 A spatial parallel manipulator with 3-PUU.

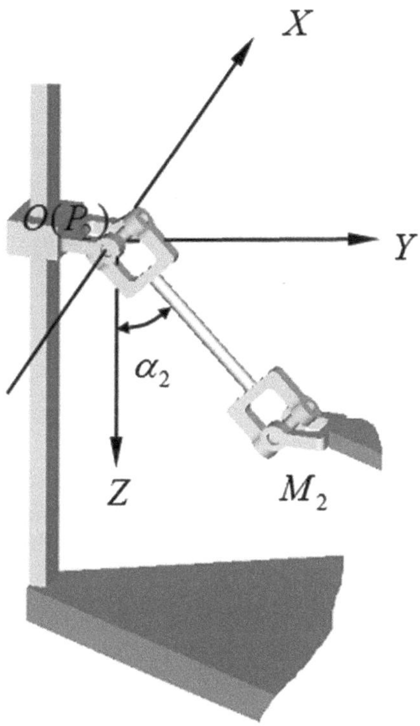

FIGURE 5.3 Coordinate system for the kinematic chain.

The universal joint P_2 can be decomposed as two orthogonal revolute pairs. The individual Plücker coordinates are:

$$\$_2 = [\,0 \; -\cos\alpha_2 \; \sin\alpha_2 \; 0 \; 0 \; 1\,]^T,\; \$_3 = [\,1 \; 0 \; 0 \; 0 \; 0 \; 0\,]^T.$$

The same is true for the universal joint M_2. The coordinates of M_2 are $M_2(0 \; l\sin\alpha_2 \; l\cos\alpha_2)$. The Plücker coordinates can be denoted as:

$$\$_4 = \begin{bmatrix} s_4^1 \\ s_4^0 \end{bmatrix},\; \$_5 = \begin{bmatrix} s_5^1 \\ s_5^0 \end{bmatrix}$$

where $s_4^1 = [\,0 \;\; -\cos\alpha_2 \;\; \sin\alpha_2\,]^T, s_5^1 = [\,1 \;\; 0 \;\; 0\,]^T, s_4^0 = r_{M_2} \times s_4^1 = [\,l \;\; 0 \;\; 0\,]^T, s_5^0 = r_{M_2} \times s_5^1 = [\,0 \;\; l\cos\alpha_2 \;\; -l\sin\alpha_2\,]^T$.

Therefore, the kinematic screws of the kinematic chain P_2M_2 can be expressed as:

$$\$_{P_2M_2} = [\,\$_1 \;\; \$_2 \;\; \$_3 \;\; \$_4 \;\; \$_5\,]^T.$$

The reciprocal screw of kinematic chain P_2M_2 can be gained:

$$\$^r_{P_2M_2} = k[\,0\ 0\ 0\ 0\ \sin\alpha_2\ \cos\alpha_2\,]^T.$$

According to the physical meaning of the reciprocal screws of kinematic screws, $\$^r_{P_2M_2}$, denotes a pure moment of couple that is perpendicular to the universal joint plane.

The movement is constrained by $\$^r_{P_2M_2}$ is $\$_{P_2M_2} = [0\ \sin\alpha_2\ \cos\alpha_2\ 0\ 0\ 0]^T$, this is a pure rotation around the axis $[\,0\ \ \sin\alpha_2\ \ \cos\alpha_2\,]^T$, which is perpendicular to the planes of the universal joints P_2 and M_2. That is, the manipulator does not execute any rotation around the axis $[\,0\ \sin\alpha_2\ \cos\alpha_2\,]^T$.

The same is true for the other two kinematic chains of the manipulator. So, the reciprocal screws applied to the manipulator are three pure moments of couples shown in Figure 5.4.

The three pure moments of couples will prevent the manipulator from rolling around the x-, y-, and z-axes, and the three pure moments of couples applied to the manipulator will always be linearly independent if the structure parameters are appropriately designed and can be spanned fully in a three-dimensional linear subspace in the normal spaces. Therefore, the number of DOFs of the manipulator is:

$$F = 6 - d = 6 - 3 = 3.$$

According to the reciprocal screw theory and the above analysis, the three orthogonal rotational motions are forbidden by the three reciprocal screws and the manipulator can only execute three independent translations along the x-, y-, and z-axes.

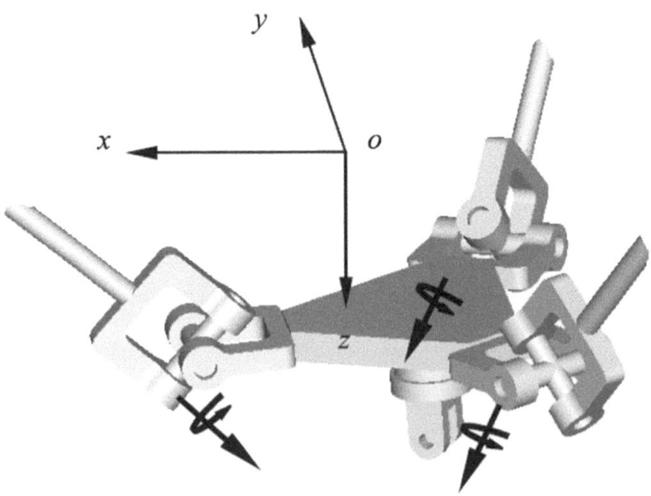

FIGURE 5.4 Reciprocal screws applied to the manipulator.

Therefore, the manipulator $M_1M_2M_3$ will always be parallel to the base $B_1B_2B_3$. Therefore, the transformation matrix $\mathbf{R} = \mathbf{I}$, where $\mathbf{I}$ denotes the identity matrix:

$$\mathbf{I} = \begin{bmatrix} 1 & 0 & 0 \\ 0 & 1 & 0 \\ 0 & 0 & 1 \end{bmatrix}.$$

Assuming that the radius of the circumcircle of triangle $M_1M_2M_3$ is $r(r < R)$, the coordinates of the three sliders in the local coordinate system are as follows:

$$M_1\left(r,0,0\right),\ M_2\left(-\frac{1}{2}r,-\frac{\sqrt{3}}{2}r,0\right),\ M_3\left(-\frac{1}{2}r,\frac{\sqrt{3}}{2}r,0\right).$$

Considering the geometry characteristics, we can obtain the following equation:

$$\boldsymbol{r}_{P_iM_i} = \boldsymbol{r}_{o_c} + \mathbf{R}\boldsymbol{r}_{o_cP_i} - \boldsymbol{r}_{P_i} = \boldsymbol{r}_{o_c} + \boldsymbol{r}_{o_cP_i} - \boldsymbol{r}_{P_i}$$

where $\left|\boldsymbol{r}_{P_iM_i}\right| = l\ \left(i = 1,2,3\right)$, and

$$\boldsymbol{r}_{P_1} = \begin{bmatrix} R & 0 & z_1 \end{bmatrix}^T, \boldsymbol{r}_{P_1} = \begin{bmatrix} -\frac{1}{2}R & -\frac{\sqrt{3}}{2}R & z_2 \end{bmatrix}^T,$$
$$\boldsymbol{r}_{P_3} = \begin{bmatrix} -\frac{1}{2}R & \frac{\sqrt{3}}{2}R & z_3 \end{bmatrix}^T.$$

So, if the absolute coordinates of point o_c are (x, y, z), then the vector of point o_c can be denoted by:

$$\boldsymbol{r}_{o_c} = [x \quad y \quad z]^T$$

And

$$\boldsymbol{r}_{o_cP_1} = \begin{bmatrix} r & 0 & 0 \end{bmatrix}^T, \boldsymbol{r}_{o_cP_2} = \begin{bmatrix} -\frac{1}{2}r & -\frac{\sqrt{3}}{2}r & 0 \end{bmatrix}^T,$$
$$\boldsymbol{r}_{o_cP_3} = \begin{bmatrix} -\frac{1}{2}r & \frac{\sqrt{3}}{2}r & 0 \end{bmatrix}^T.$$

Therefore,

$$\begin{cases} \left(x + r - R\right)^2 + y^2 + \left(z - z_1\right)^2 = l^2 \\ \left(x - \frac{1}{2}r + \frac{1}{2}R\right)^2 + \left(y - \frac{\sqrt{3}}{2}r + \frac{\sqrt{3}}{2}R\right)^2 + \left(z - z_2\right)^2 = l^2 \\ \left(x - \frac{1}{2}r + \frac{1}{2}R\right)^2 + \left(y + \frac{\sqrt{3}}{2}r - \frac{\sqrt{3}}{2}R\right)^2 + \left(z - z_3\right)^2 = l^2. \end{cases}$$

From the geometry relationship of the mechanism shown in Figure 5.2, we can find that $z - z_i > 0$ always holds for $i = 1,2,3$. Therefore,

$$\begin{cases} z_1 = z - \sqrt{l^2 - (x + r - R)^2 - y^2} \\ z_2 = z - \sqrt{l^2 - \left(x - \frac{1}{2}r + \frac{1}{2}R\right)^2 - \left(y - \frac{\sqrt{3}}{2}r + \frac{\sqrt{3}}{2}R\right)^2} \\ z_3 = z - \sqrt{l^2 - \left(x - \frac{1}{2}r + \frac{1}{2}R\right)^2 - \left(y + \frac{\sqrt{3}}{2}r - \frac{\sqrt{3}}{2}R\right)^2}. \end{cases}$$

Considering $0 \le z_i \le D_i$ $(i = 1,2,3)$, $D_i \in [0, Z]$, there will be:

$$\begin{cases} (x + r - R)^2 + y^2 + z^2 \ge l^2 \\ \left(x - \frac{1}{2}r + \frac{1}{2}R\right)^2 + \left(y - \frac{\sqrt{3}}{2}r + \frac{\sqrt{3}}{2}R\right)^2 + z^2 \ge l^2 \\ \left(x - \frac{1}{2}r + \frac{1}{2}R\right)^2 + \left(y + \frac{\sqrt{3}}{2}r - \frac{\sqrt{3}}{2}R\right)^2 + z^2 \ge l^2 \\ \sqrt{l^2 - (x + r - R)^2 - y^2} \ge z - D_1 \\ \sqrt{l^2 - \left(x - \frac{1}{2}r + \frac{1}{2}R\right)^2 - \left(y - \frac{\sqrt{3}}{2}r + \frac{\sqrt{3}}{2}R\right)^2} \ge z - D_2 \\ \sqrt{l^2 - \left(x - \frac{1}{2}r + \frac{1}{2}R\right)^2 - \left(y + \frac{\sqrt{3}}{2}r - \frac{\sqrt{3}}{2}R\right)^2} \ge z - D_3. \end{cases}$$

Therefore, we can discuss the workspace of the manipulator in the following two cases:

1. When $z < D_i$ $(i = 1,2,3)$, there will be

$$\begin{cases} (x + r - R)^2 + y^2 + z^2 \ge l^2 \\ \left(x - \frac{1}{2}r + \frac{1}{2}R\right)^2 + \left(y - \frac{\sqrt{3}}{2}r + \frac{\sqrt{3}}{2}R\right)^2 + z^2 \ge l^2 \\ \left(x - \frac{1}{2}r + \frac{1}{2}R\right)^2 + \left(y + \frac{\sqrt{3}}{2}r - \frac{\sqrt{3}}{2}R\right)^2 + z^2 \ge l^2 \\ (x + r - R)^2 + y^2 \le l^2 \\ \left(x - \frac{1}{2}r + \frac{1}{2}R\right)^2 + \left(y - \frac{\sqrt{3}}{2}r + \frac{\sqrt{3}}{2}R\right)^2 \le l^2 \\ \left(x - \frac{1}{2}r + \frac{1}{2}R\right)^2 + \left(y + \frac{\sqrt{3}}{2}r - \frac{\sqrt{3}}{2}R\right)^2 \le l^2. \end{cases} \tag{5.11}$$

2. When $z \geq D_i$ $(i = 1,2,3)$, there will be

$$\begin{cases} (x+r-R)^2 + y^2 + z^2 \geq l^2 \\ \left(x - \frac{1}{2}r + \frac{1}{2}R\right)^2 + \left(y - \frac{\sqrt{3}}{2}r + \frac{\sqrt{3}}{2}R\right)^2 + z^2 \geq l^2 \\ \left(x - \frac{1}{2}r + \frac{1}{2}R\right)^2 + \left(y + \frac{\sqrt{3}}{2}r - \frac{\sqrt{3}}{2}R\right)^2 + z^2 \geq l^2 \\ (x+r-R)^2 + y^2 + (z-D_1)^2 \leq l^2 \\ \left(x - \frac{1}{2}r + \frac{1}{2}R\right)^2 + \left(y - \frac{\sqrt{3}}{2}r + \frac{\sqrt{3}}{2}R\right)^2 + (z-D_2)^2 \leq l^2 \\ \left(x - \frac{1}{2}r + \frac{1}{2}R\right)^2 + \left(y + \frac{\sqrt{3}}{2}r - \frac{\sqrt{3}}{2}R\right)^2 + (z-D_3)^2 \leq l^2. \end{cases} \tag{5.12}$$

Assuming $r = 200$ mm, $R = 1200$ mm, and $l = 2000$ mm, we can analyze the workspace of the manipulator as follows.

As Figure 5.5a (left) shows, the workspace in the range of $S_1(z < D_i, i = 1,2,3)$ presents the solutions of equation (5.11); the workspace in the range of $S_2(z \geq D_i, i = 1,2,3)$ presents the solution of equation (5.12). The reachable workspace of the manipulator is obtained as Figure 5.5a (right) shows. In fact, the workspace of the manipulator can also be looked at as a volume swept by the tri-spherical surfaces shown in Figure 5.5b (left) along the z-axis.

When the reachable workspace is projected to the plane xoy, we can find that the hatched area, shown in Figure 5.6, is enveloped by three intersected circles, whose centers are $(R-r, 0)$, $\left(-\frac{1}{2}(R-r), -\frac{\sqrt{3}}{2}(R-r)\right)$, and $\left(-\frac{1}{2}(R-r), \frac{\sqrt{3}}{2}(R-r)\right)$, respectively. The hatched area can be expressed as:

$$\begin{cases} (x+r-R)^2 + y^2 \leq l^2 \\ \left(x - \frac{1}{2}r + \frac{1}{2}R\right)^2 + \left(y - \frac{\sqrt{3}}{2}r + \frac{\sqrt{3}}{2}R\right)^2 \leq l^2 \\ \left(x - \frac{1}{2}r + \frac{1}{2}R\right)^2 + \left(y + \frac{\sqrt{3}}{2}r - \frac{\sqrt{3}}{2}R\right)^2 \leq l^2. \end{cases} \tag{5.13}$$

5.1.2 Workspace of a Planar 3-RRR Parallel Manipulator

A planar parallel mechanism, as shown in Figure 5.7, is characterized by the fact that the end effector $C_1C_2C_3$ is an equilateral triangle whose three vertexes are connected with the vertexes of the fixed equilateral triangle $A_1A_2A_3$ through three identical RRR kinematic chains, $A_iB_iC_i$, $i = 1,2,3$. Suppose that the reference point of the end effector $C_1C_2C_3$ is represented by its geometry center, C. We can now investigate the reachable workspace and dexterous workspace of the mechanism.

Suppose that the coordinates of the reference point of the end effector, denoted by C, in the fixed coordinate system $oxyz$ are $\boldsymbol{r} = [x\ y\ 0]^T$, the angular displacement about the z-axis is θ, then the absolute coordinates of the three

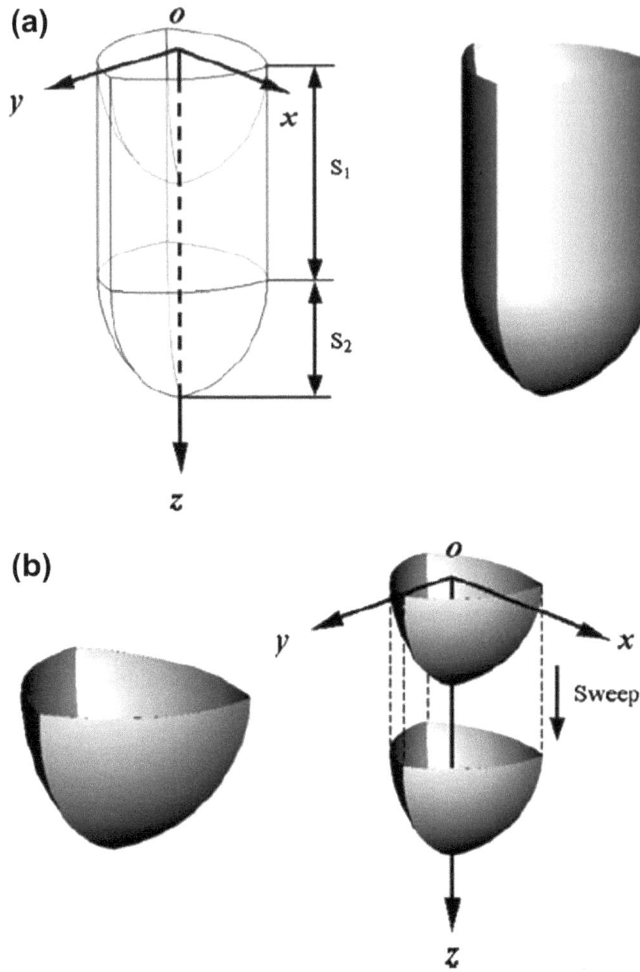

FIGURE 5.5 Workspace of the manipulator.

vertexes of the triangle end effector $C_1C_2C_3$ can be obtained, individually:

$$\boldsymbol{r}_{C_i} = \boldsymbol{r} + \mathbf{R}\boldsymbol{r}^L_{CC_i}$$

where $i = 1,2,3$, $\mathbf{R} = \begin{bmatrix} \cos\theta & -\sin\theta & 0 \\ \sin\theta & \cos\theta & 0 \\ 0 & 0 & 1 \end{bmatrix}$, $\boldsymbol{r}^L_{CC_1} = \begin{bmatrix} -\frac{\sqrt{3}}{2}r \\ -\frac{1}{2}r \\ 0 \end{bmatrix}$, $\boldsymbol{r}^L_{CC_2} = \begin{bmatrix} \frac{\sqrt{3}}{2}r \\ -\frac{1}{2}r \\ 0 \end{bmatrix}$, and $\boldsymbol{r}^L_{CC_3} = \begin{bmatrix} 0 \\ r \\ 0 \end{bmatrix}$, and r denotes the circumcircle radius of the triangle $C_1C_2C_3$, and $r^L_{CC_i}(i = 1,2,3)$ denotes the vector coordinates of $\overrightarrow{CC_i}$.

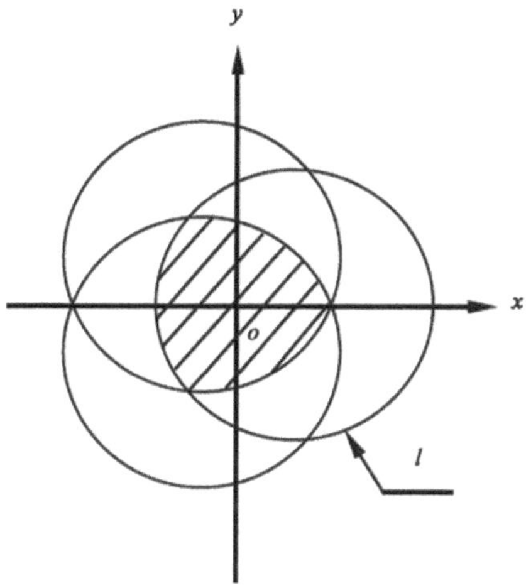

FIGURE 5.6 Projection of the workspace shown in Figure 5.5 in the plane *xoy*.

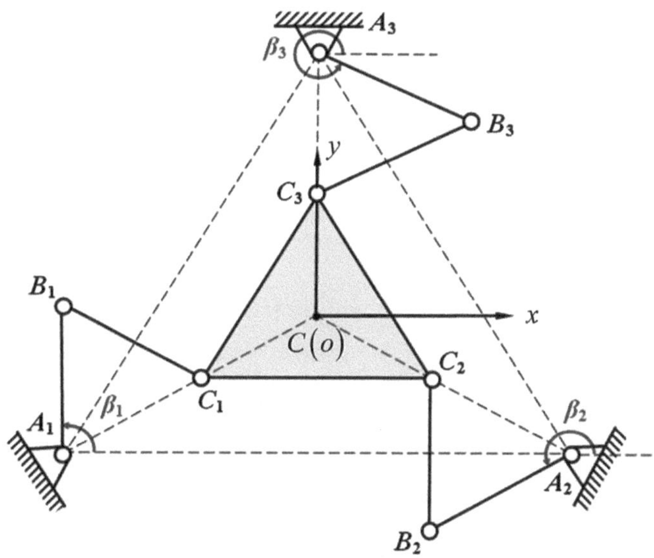

FIGURE 5.7 A planar 3-RRR parallel mechanism.

The coordinates of points B_1, B_2, and B_3 can be expressed with the three angular displacements, β_1, β_2, and β_3:

$$\boldsymbol{r}_{B_i} = \boldsymbol{r}_{A_i} + l_1 \begin{bmatrix} \cos\beta_i \\ \sin\beta_i \\ 0 \end{bmatrix}$$

where $i = 1,2,3$. So, the constraint equations are:

$$\mathbf{g}\left(x,y,\theta\right) = \begin{bmatrix} \left(x - \frac{\sqrt{3}}{2}r\cos\theta + \frac{1}{2}r\sin\theta - x_{A_1} - l_1\cos\theta_1\right)^2 \\ \quad + \left(y - \frac{\sqrt{3}}{2}r\sin\theta - \frac{1}{2}r\cos\theta - y_{A_1} - l_1\sin\theta_1\right)^2 - l_2^2 \\ \left(x + \frac{\sqrt{3}}{2}r\cos\theta + \frac{1}{2}r\sin\theta - x_{A_2} - l_1\cos\theta_2\right)^2 \\ \quad + \left(y + \frac{\sqrt{3}}{2}r\sin\theta - \frac{1}{2}r\cos\theta - y_{A_2} - l_1\sin\theta_2\right)^2 - l_2^2 \\ \left(x - r\sin\theta - x_{A_3} - l_1\cos\theta_3\right)^2 \\ \quad + \left(y + r\cos\theta - y_{A_3} - l_1\sin\theta_3\right)^2 - l_2^2 \end{bmatrix}. \tag{5.14}$$

Therefore, from equation (5.3), the reachable workspace of the specified point C is expressed as:

$$\mathbf{W}_R = \left\{\left(x,y\right) \middle| \mathbf{g}\left(x,y,\theta\right) = 0\right\}. \tag{5.15}$$

According to equation (5.5), the dexterous workspace of point C can be represented by:

$$\mathbf{W}_D = \left\{\left(x,y\right) \middle| \mathbf{g}\left(x,y,\theta\right) = 0, \theta \in \left[\theta_{z_1},\theta_{z_2}\right]\right\}. \tag{5.16}$$

Following, we will investigate the searching problems of the reachable workspace and the dexterous workspace. Firstly, the reachable workspace should be obtained. The primary steps to fulfill such a task might be:

Step 1: Establish parametric equations of polar coordinates of the point within the workspace:

$$\begin{cases} x = \rho\cos\left(i\Delta\gamma\right) \\ y = \rho\sin\left(i\Delta\gamma\right) \\ \theta = 0 \end{cases}$$

where ρ is the searching polar radius with a polar angle $i\Delta\gamma$ from the x-axis, and $0 \le i \le \frac{2\pi}{\Delta\gamma}$, and $\Delta\gamma$ is the angular interval and i is a natural number.

Step 2: Suppose the workspace is divided evenly with $k-1$ radial lines, where k is the largest natural number that is less than or equal to $\frac{2\pi}{\Delta\gamma}$ and therefore, $0 \le i \le k$.

Step 3: Search the boundary point in the ith radial line. Let $\rho = \rho\left(i\Delta\gamma\right)$ increase from zero with a proper interval $\Delta\rho$, namely $\rho\left(i\Delta\gamma\right) = j\Delta\rho, j = 0,1,2,\ldots$.

Merlet [13,14], Bonev and Ryu [3], and Dash *et al.* [15] proposed such a numerical algorithm for generating the reachable workspace of parallel manipulators. However, the stop criterion for such an increase is not obvious for the mechanism with rotational actuations. Therefore, we will discuss this problem in the following section.

The reachable workspace of the mechanism is represented by equation (5.15). Supposing the position and pose parameters (x,y,θ) of the end effector to be parametric coefficients, we can immediately transform the relationship constraint equation (5.14) into a function of θ_1,θ_2, and θ_3 in accordance with equation (5.8):

$$\mathbf{g}(\theta_1,\theta_2,\theta_3) = \begin{bmatrix} p_{11} - p_{12}\cos\theta_1 - p_{13}\sin\theta_1 \\ p_{21} - p_{22}\cos\theta_2 - p_{23}\sin\theta_2 \\ p_{31} - p_{32}\cos\theta_3 - p_{33}\sin\theta_3 \end{bmatrix}$$

where

$$\begin{aligned}
p_{11} &= \left(x - \frac{\sqrt{3}}{2}r\cos\theta + \frac{1}{2}r\sin\theta - x_{A_1}\right)^2 \\
&\quad + \left(y - \frac{\sqrt{3}}{2}r\sin\theta - \frac{1}{2}r\cos\theta - y_{A_1}\right)^2 + l_1^2 - l_2^2, \\
p_{12} &= 2l_1\left(x - \frac{\sqrt{3}}{2}r\cos\theta + \frac{1}{2}r\sin\theta - x_{A_1}\right), \\
p_{13} &= 2l_1\left(y - \frac{\sqrt{3}}{2}r\sin\theta - \frac{1}{2}r\cos\theta - y_{A_1}\right); \\
p_{21} &= \left(x + \frac{\sqrt{3}}{2}r\cos\theta + \frac{1}{2}r\sin\theta - x_{A_2}\right)^2 \\
&\quad + \left(y + \frac{\sqrt{3}}{2}r\sin\theta - \frac{1}{2}r\cos\theta - y_{A_2}\right)^2 + l_1^2 - l_2^2, \\
p_{22} &= 2l_1\left(x + \frac{\sqrt{3}}{2}r\cos\theta + \frac{1}{2}r\sin\theta - x_{A_2}\right), \\
p_{23} &= 2l_1\left(y + \frac{\sqrt{3}}{2}r\sin\theta - \frac{1}{2}r\cos\theta - y_{A_2}\right); \\
p_{31} &= \left(x - r\sin\theta - x_{A_3}\right)^2 + \left(y + r\cos\theta - y_{A_3}\right)^2 + l_1^2 - l_2^2, \\
p_{32} &= 2l_1\left(x - r\sin\theta - x_{A_3}\right), p_{33} = 2l_1\left(y + r\cos\theta - y_{A_3}\right).
\end{aligned}$$

Obviously, within the reachable workspace, the three angular inputs θ_1,θ_2, and θ_3 should be real values. Equivalently, equation (5.9) should have real solutions for θ_1,θ_2, and θ_3 when we let $\mathbf{g}(\theta_1,\theta_2,\theta_3) = 0$. Assuming $t_1 = \tan\frac{\theta_1}{2}, t_2 = \tan\frac{\theta_2}{2}$ and $t_3 = \tan\frac{\theta_3}{2}$, we immediately have:

$$\begin{cases} (p_{11} + p_{12})\, t_1^2 - 2p_{13}t_1 + p_{11} - p_{12} = 0 \\ (p_{21} + p_{22})\, t_2^2 - 2p_{23}t_2 + p_{21} - p_{22} = 0 \\ (p_{31} + p_{32})\, t_3^2 - 2p_{33}t_3 + p_{31} - p_{32} = 0. \end{cases} \tag{5.17}$$

Consequently, the boundaries of the reachable workspace are now transformed into the existing criteria for real solutions of the three quadric equations of (5.14). With equation (5.10), the criteria for the boundaries of the reachable workspace of the mechanism are now expressed as:

$$\begin{cases} \Delta_1 = \mathbf{4}\left(p_{12}^2 + p_{13}^2 - p_{11}^2\right) \geq 0 \\ \Delta_2 = \mathbf{4}\left(p_{22}^2 + p_{23}^2 - p_{21}^2\right) \geq 0 \\ \Delta_3 = \mathbf{4}\left(p_{32}^2 + p_{33}^2 - p_{31}^2\right) \geq 0. \end{cases} \tag{5.18}$$

Therefore, the stop criterion for increasing the polar radius in the radial direction can be guaranteed with inequality function set (5.18) while the free variables are merely restricted in (x, y, θ), the three position and pose parameters. Select the largest value of $\rho\,(i\Delta\gamma)$ as the radius of boundary point in the ith radial direction under the constraint of inequality function set (5.18). Consequently, the closed boundary curve of the reachable workspace will be obtained as i increases from zero to k.

Based on the above process, we can also use the following steps to obtain the dexterous workspace of the reference point of the end effector.

Step 4: Within step 3, let the orientation angle θ increase evenly from the lower limit to the upper limit at the point $(\rho, i\Delta\gamma)$. The dexterous workspace boundary point in the radial direction occurs when at least one inequality equation of (5.18) does not hold for a certain orientation within $[\theta_1, \theta_2]$ as $\rho\,(i\Delta\gamma)$ increases in the radial direction. Similarly, the closed boundary curve of the dexterous workspace is obtained as i increases from zero to k.

Here we performed some simulations on the mechanism shown in Figure 5.7 with different parameters shown in Table 5.1. The workspaces and the corresponding dexterous workspace defined by the prescribed range of pose angles are shown in Figure 5.8. Simulations indicate that the algorithm proposed in this chapter is valid and effective for searching the reachable and dexterous workspaces.

5.1.3 The Workspace of a 3-RUU Parallel Manipulator

In this section, we investigate the reachable workspace of a spatial 3-RUU parallel mechanism shown in Figure 5.9. For the sake of convenience, we still

TABLE 5.1 Structure Parameters for the Planar 3-RRR Parallel Mechanism Shown in Figure 5.7

	Item				
Sequence	r (mm)	R (mm)	l_1 (mm)	l_2 (mm)	Range of Pose Angles
1	100	800	300	600	$[-\frac{\pi}{6},\frac{\pi}{6}]\ [-\frac{\pi}{3},\frac{\pi}{3}]\ [-\frac{2\pi}{3},\frac{2\pi}{3}]$
2	100	800	500	500	$[-\frac{\pi}{6},\frac{\pi}{6}]\ [-\frac{\pi}{3},\frac{\pi}{3}]\ [-\frac{2\pi}{3},\frac{2\pi}{3}]$
3	100	800	800	600	$[-\frac{\pi}{6},\frac{\pi}{6}]\ [-\frac{\pi}{3},\frac{\pi}{3}]\ [-\frac{2\pi}{3},\frac{2\pi}{3}]$

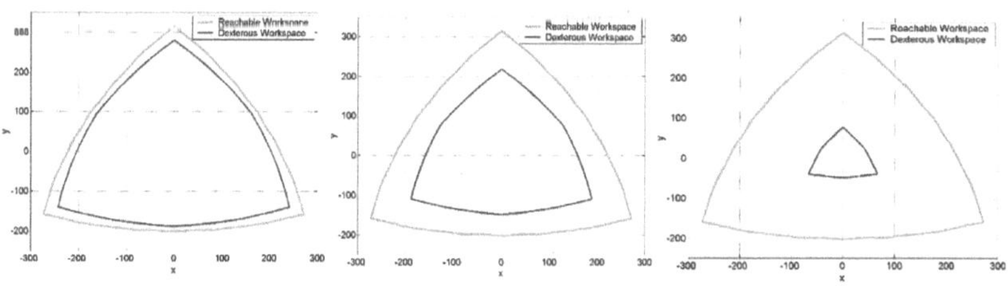

(a) Reachable and dexterous workspaces of the mechanism with the 1st set of parameters

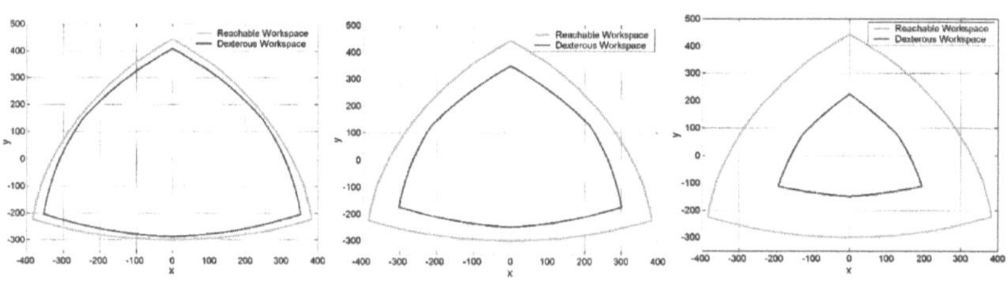

(b) Reachable and dexterous workspaces of the mechanism with the 2nd set of parameters

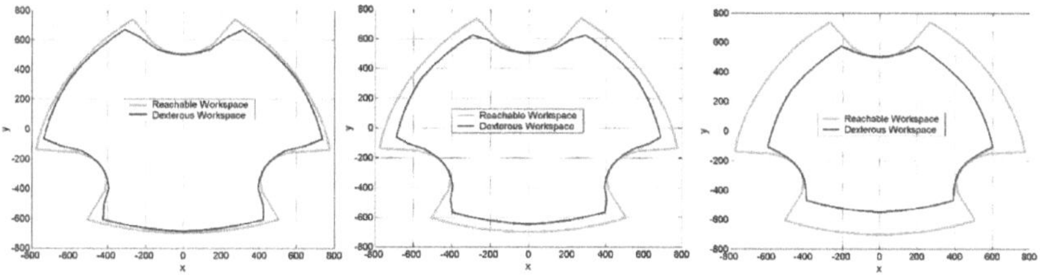

(c) Reachable and dexterous workspaces of the mechanism with the 3rd set of parameters

FIGURE 5.8 Different workspaces of a planar 3-RRR parallel mechanism.

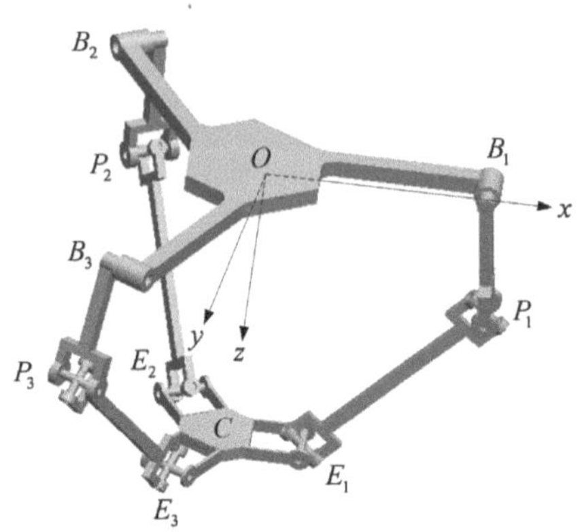

FIGURE 5.9 A 3-RUU parallel mechanism.

assume that the reference point of the end effector located in the geometry center of the end effector is denoted by C. Also presume that the origin of the fixed coordinate system, denoted by O, is superimposed with the center of the fixed base $B_1B_2B_3$, the x-axis is along OB_1, the z-axis is perpendicular to the fixed base and directing the end effector, and the y-axis is determined with the right-hand rule.

Suppose that the circumcircle radii of the equilateral triangles $B_1B_2B_3$ and $E_1E_2E_3$ are denoted by R and r, individually. Also, suppose the lengths of limbs B_iP_i and P_iE_i to be represented by l_1 and l_2, respectively. The absolute coordinates of the three vertexes of the fixed base can be directly represented with:

$$B_1\left(R \;\; 0 \;\; 0\right),\; B_2\left(-\tfrac{1}{2}R \;\; -\tfrac{\sqrt{3}}{2}R \;\; 0\right) \text{ and } B_2\left(-\tfrac{1}{2}R \;\; \tfrac{\sqrt{3}}{2}R \;\; 0\right).$$

It is not difficult to find that the end effector $E_1E_2E_3$ has three translations. Therefore, the absolute coordinates of the three vertexes of the end effector can also be obtained if we assume the coordinates of the reference point C to be $(x \; y \; z)$:

$$E_1\left(x+r \;\; y \;\; z\right),\; E_2\left(x-\tfrac{1}{2}r \;\; y-\tfrac{\sqrt{3}}{2}r \;\; z\right) \text{ and}$$
$$E_3\left(x-\tfrac{1}{2}r \;\; y+\tfrac{\sqrt{3}}{2}r \;\; z\right).$$

The coordinates of $P_i(i = 1,2,3)$ are also obtained if the three angular inputs at B_i is indicated by θ_i:

$$P_1(R+l_1\cos\theta_1 \;\; 0 \;\; l_1\sin\theta_1),$$

$$P_2\left(-\tfrac{1}{2}\left(R+l_1\cos\theta_2\right) \;\; -\tfrac{\sqrt{3}}{2}\left(R+l_1\cos\theta_2\right) \;\; l_1\sin\theta_2\right),$$
$$P_3\left(-\tfrac{1}{2}\left(R+l_1\cos\theta_3\right) \;\; \tfrac{\sqrt{3}}{2}\left(R+l_1\cos\theta_3\right) \;\; l_1\sin\theta_3\right).$$

According to the fact that the length of P_iE_i ($i = 1,2,3$) is l_2, we immediately gain a set of relationship constraint equations with the parameters x, y, z being independent variables:

$$\mathbf{g}(x,y,z) = \begin{bmatrix} (x + r - R - l_1 \cos\theta_1)^2 + y^2 + (z - l_1 \sin\theta_1)^2 - l_2^2 \\ \left[x + \frac{1}{2}(R - r) + \frac{1}{2} l_1 \cos\theta_2\right]^2 + \left[y + \frac{\sqrt{3}}{2}(R - r) + \frac{\sqrt{3}}{2} l_1 \cos\theta_2\right]^2 \\ + (z - l_1 \sin\theta_2)^2 - l_2^2 \\ \left[x + \frac{1}{2}(R - r) + \frac{1}{2} l_1 \cos\theta_3\right]^2 + \left[y - \frac{\sqrt{3}}{2}(R - r) - \frac{\sqrt{3}}{2} l_1 \cos\theta_3\right]^2 \\ + (z - l_1 \sin\theta_3)^2 - l_2^2 \end{bmatrix}.$$

For any point $(x \; y \; z)$ within the workspace, the inverse displacement solutions of the angular inputs, θ_1,θ_2 and θ_3, in equation $g(x,y,z) = 0$ should be real. As a result, we can obtain the constraints of the reachable workspace of the end effector with equation (5.7).

$$\mathbf{g}(\theta_1,\theta_2,\theta_3) = \begin{bmatrix} \left[(x + r - R)^2 + y^2 + z^2 + l_1^2 - l_2^2\right] - 2l_1 (x + r - R) \cos\theta_1 - 2zl_1 \sin\theta_1 \\ \left\{\left[x + \frac{1}{2}(R - r)\right]^2 + \left[y + \frac{\sqrt{3}}{2}(R - r)\right]^2 + z^2 + l_1^2 - l_2^2\right\} \\ + l_1 \left\{\left[x + \frac{1}{2}(R - r)\right] + \sqrt{3}\left[y + \frac{\sqrt{3}}{2}(R - r)\right]\right\} \cos\theta_2 - 2zl_1 \sin\theta_2 \\ \left\{\left[x + \frac{1}{2}(R - r)\right]^2 + \left[y - \frac{\sqrt{3}}{2}(R - r)\right]^2 + z^2 + l_1^2 - l_2^2\right\} \\ + l_1 \left\{\left[x + \frac{1}{2}(R - r)\right] - \sqrt{3}\left[y - \frac{\sqrt{3}}{2}(R - r)\right]\right\} \cos\theta_3 - 2zl_1 \sin\theta_3 \end{bmatrix} \tag{5.19}$$

Also suppose

$$p_{11} = (x + r - R)^2 + y^2 + z^2 + l_1^2 - l_2^2,$$
$$p_{12} = 2l_1 (x + r - R), p_3 = 2zl_1;$$
$$p_{21} = \left[x + \frac{1}{2}(R - r)\right]^2 + \left[y + \frac{\sqrt{3}}{2}(R - r)\right]^2 + z^2 + l_1^2 - l_2^2,$$
$$p_{22} = l_1 \left\{\left[x + \frac{1}{2}(R - r)\right] + \sqrt{3}\left[y + \frac{\sqrt{3}}{2}(R - r)\right]\right\};$$
$$p_{31} = \left[x + \frac{1}{2}(R - r)\right]^2 + \left[y - \frac{\sqrt{3}}{2}(R - r)\right]^2 + z^2 + l_1^2 - l_2^2,$$
$$p_{32} = l_1 \left\{\left[x + \frac{1}{2}(R - r)\right] - \sqrt{3}\left[y - \frac{\sqrt{3}}{2}(R - r)\right]\right\}.$$

Rearranging equation (5.19) yields:

$$\boldsymbol{g}(\theta_1,\theta_2,\theta_3) = \begin{bmatrix} p_{11} - p_{12}\cos\theta_1 - p_3\sin\theta_1 \\ p_{21} + p_{22}\cos\theta_2 - p_3\sin\theta_2 \\ p_{31} + p_{32}\cos\theta_3 - p_3\sin\theta_3 \end{bmatrix}.$$

Assuming $t_1 = \tan\frac{\theta_1}{2}, t_2 = \tan\frac{\theta_2}{2}$ and $t_3 = \tan\frac{\theta_3}{2}$, and letting $\boldsymbol{g}(\theta_1,\theta_2,\theta_3) = 0$, we immediately have:

$$\begin{cases} (p_{11} + p_{12})\, t_1^2 - 2p_3t_1 + p_{11} - p_{12} = 0 \\ (p_{21} - p_{22})\, t_2^2 - 2p_3t_2 + p_{21} + p_{22} = 0 \\ (p_{31} - p_{32})\, t_3^2 - 2p_3t_3 + p_{31} + p_{32} = 0. \end{cases} \tag{5.20}$$

Consequently, the boundaries of the reachable workspace are now transformed into the criteria for the existence of real solutions of the three quadric equations of equation (5.20). According to equation (5.10), the boundaries of the reachable workspace of the mechanism are restricted by:

$$\begin{cases} \Delta_1 = \mathbf{4}\left(p_{12}^2 + p_3^2 - p_{11}^2\right) \geq 0 \\ \Delta_2 = \mathbf{4}\left(p_{22}^2 + p_3^2 - p_{21}^2\right) \geq 0 \\ \Delta_3 = \mathbf{4}\left(p_{32}^2 + p_3^2 - p_{31}^2\right) \geq 0. \end{cases} \tag{5.21}$$

As a result, the searching process for the reachable workspace of the mechanism shown in Figure 5.9 is immediately obtained:

Step 1: Solve the highest and lowest position of the reference point, C. The coordinates of the two points are presumed as $C_H(0\;0\;z_H)$, $C_L(0\;0\;z_L)$.

Step 2: Slice the Cartesian space evenly along the z-axis from point $C_H(0\;0\;z_H)$ to $C_L(0\;0\;z_L)$. Suppose the interval of two consecutive adjacent slices to be denoted by Δz. The reachable workspace is therefore sliced into $m = \left[\frac{z_L - z_H}{\Delta z}\right]$ planes, where $[\cdot]$ indicates the largest number that is less than or equal to "$\cdot$".

Step 3: Search the boundary curves within the ith $(1 \leq i\, m)$ plane with the criteria indicated by equation (5.21). The searching process is similar to the steps proposed in Section 5.1.2.

We also made some simulations on the mechanism shown in Figure 5.9 here with different parameters shown in Table 5.2. The reachable workspace shapes are shown in Figure 5.10. However, because the mechanism shown in Figure 5.9 only has three translational DOFs, there is no need to discuss the dexterous workspace.

TABLE 5.2 Structure Parameters for the Spatial 3-RUU Parallel Mechanism Shown in Figure 5.9

	Item			
Sequence	**r(mm)**	**r(mm)**	**l_1(mm)**	**l_2(mm)**
1	300	500	500	1000
2	300	1000	500	500
3	400	1000	400	800
4	400	800	400	800

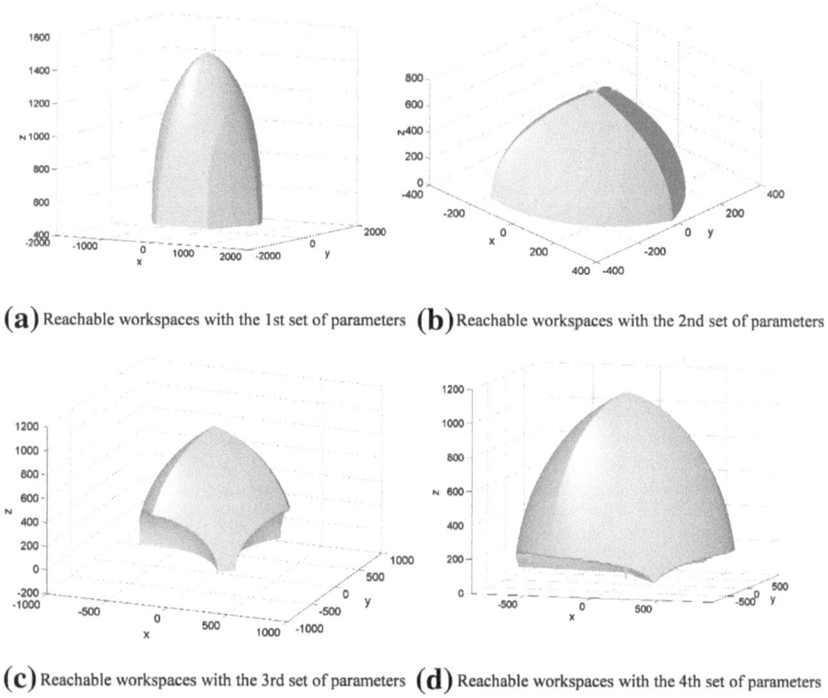

(a) Reachable workspaces with the 1st set of parameters (b) Reachable workspaces with the 2nd set of parameters

(c) Reachable workspaces with the 3rd set of parameters (d) Reachable workspaces with the 4th set of parameters

FIGURE 5.10 Different workspaces of the spatial 3-RUU parallel mechanism.

5.2 SYMMETRICAL CHARACTERISTICS OF THE WORKSPACE FOR SPATIAL PARALLEL MECHANISMS WITH SYMMETRIC STRUCTURE

In applications, most spatial parallel manipulators have symmetric identical kinematic chains. These chains are symmetric about a certain plane, about

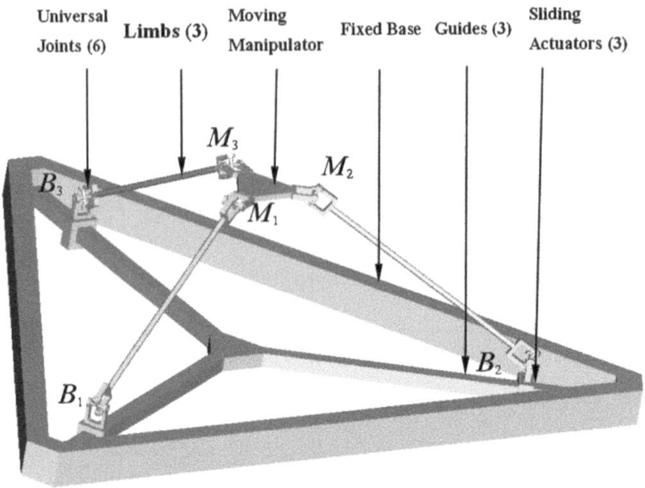

FIGURE 5.11 A spatial parallel manipulator made up of 3-PUU kinematic chains.

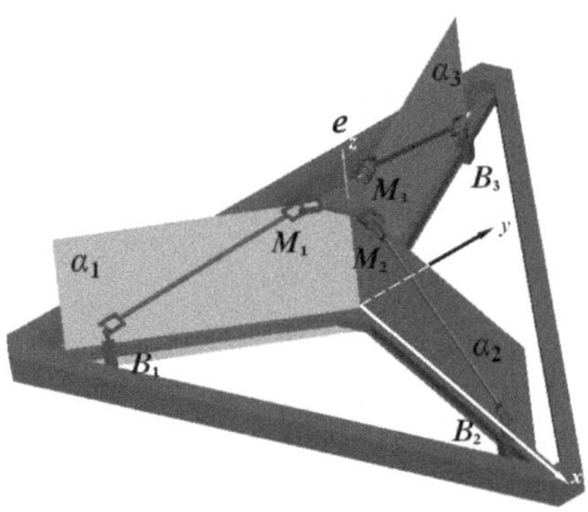

FIGURE 5.12 Structural symmetry of the spatial parallel manipulator.

one axis or about one specified point. For example, the manipulator shown in Figure 5.11 has three identical kinematic chains, each one of which consists of two universal joints and one prismatic joint in series. The structural symmetric characteristics can be found in Figure 5.12, where α_1,α_2, and α_3 are the three symmetric planes and e is a symmetric axis that is intersected by these three symmetric planes. As to the point of symmetry, we can find that the middle point of AD is the symmetry center of the planar inverse parallel 4-bar manipulator shown in Figure 5.13.

As mentioned above, in applications, most spatial parallel mechanisms have a symmetric structure. The symmetries can be depicted by a subgroup of a certain symmetric group S_n [16], no matter what kinds of symmetries the

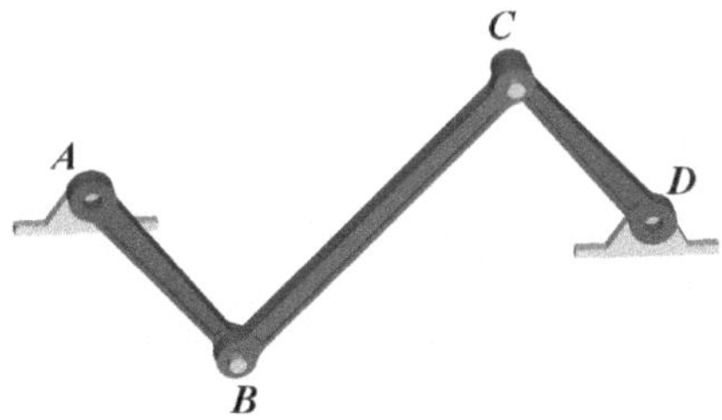

FIGURE 5.13 A planar crossing parallel 4-bar manipulator.

mechanism structure has—about a certain plane, about one axis or about one specified point. In order to exploit the relationship between the symmetries of the workspace and the structural characteristics of the kinematic chains, we proposed a symmetrical workspace theorem for the spatial parallel mechanism [17]:

Theorem 5.1. *If the symmetry group of a workspace is denoted by G_W, and the symmetry group of the kinematic chain structure of the end effector in a particular configuration is denoted by G_M, then G_M must be a subgroup of G_W; namely, the following relationship will always hold:*

$$G_M \subseteq G_W \tag{5.22}$$

Prior to proving the validation of the theorem, we also make the following assumptions [17]:

(1) The absolute coordinates of the geometric center of the mechanism are denoted by $\boldsymbol{r} = [x\ y\ z]^T$;

(2) The total number of symmetric kinematic chains (need not require a collinear for each chain) of the spatial parallel mechanism is indicated by $n (n \geq 2)$;

(3) The mechanism's $ith (i = 1,2,\ldots,n)$ vertex linked by the ith kinematic chain is denoted by M_i, and its local coordinates are denoted by $\left(x^L_{M_i}\ y^L_{M_i}\ z^L_{M_i}\right)$, where the superscript L represents "local;"

(4) The base's $ith (i = 1,2,\ldots,n)$ vertex linked by the ith kinematic chain is denoted by B_i, its absolute coordinates are denoted by $\left(x^A_{B_i}\ y^A_{B_i}\ z^A_{B_i}\right)$, where the superscript A indicates "absolute."

As a result, the absolute coordinates of the ith vertex, M_i, can be obtained:

$$\boldsymbol{r}^A_{M_i} = \boldsymbol{r} + \mathbf{R}\boldsymbol{r}^L_{M_i} \tag{5.23}$$

where **R** *is a transformation matrix from the local coordinate system to the absolute one, $r^A_{M_i} = \left[x^A_{M_i}\ y^A_{M_i}\ z^A_{M_i}\right]^T$ represents the absolute coordinates of*

M_i,$\boldsymbol{r} = [x\ y\ z]^T$ represents the absolute coordinates of the geometry center of the mechanism, and $\boldsymbol{r}_{M_i}^L = \left[x_{M_i}^L\ y_{M_i}^L\ z_{M_i}^L\right]^T$ are the local coordinates of M_i.

Proof. Mathematically, the workspace of the end effector of a mechanism, Ω, is a mapping of the threedimensional Euclidean spaces $\mathbb{R}^3$:

$$\boldsymbol{g} : \mathbb{R}^3 \rightarrow \Omega$$

The workspace Ω can therefore be denoted as:

$$\mathbf{W}_R = \{\boldsymbol{r}\,|\boldsymbol{g}(\boldsymbol{r}) = \mathbf{0}\} \tag{5.24}$$

where $\boldsymbol{g}(\cdot)$ represents a set of constraint equations of the workspace boundaries:

$$\boldsymbol{g}(r) = \begin{bmatrix} \|\boldsymbol{r}_{M_1}^A - \boldsymbol{r}_{B_1}^A\| - l_1 \\ \|\boldsymbol{r}_{M_2}^A - \boldsymbol{r}_{B_2}^A\| - l_2 \\ \vdots \\ \|\boldsymbol{r}_{M_m}^A - \boldsymbol{r}_{B_m}^A\| - l_m \end{bmatrix} \tag{5.25}$$

where m denotes the number of constraint equations.

According to equation (5.23), equation (5.25) can be expanded:

$$\boldsymbol{g}(r) = \begin{bmatrix} \|\boldsymbol{r} + \mathbf{R}\boldsymbol{r}_{M_1}^L - \boldsymbol{r}_{B_1}^A\| - l_1 \\ \|\boldsymbol{r} + \mathbf{R}\boldsymbol{r}_{M_2}^L - \boldsymbol{r}_{B_2}^A\| - l_2 \\ \vdots \\ \|\boldsymbol{r} + \mathbf{R}\boldsymbol{r}_{M_m}^L - \boldsymbol{r}_{B_m}^A\| - l_m \end{bmatrix} \tag{5.26}$$

Suppose the symmetry operations in a particular configuration form a group $G_M G_M$. Obviously, G_M must be a subgroup of a symmetry group $S(V)$, which is a distance-preserving group and consists of the motions ϕ_i for which $\phi_i(v_j) = v_j$.

Considering the actions of symmetry group G_M on the workspace Ω are defined by equation (5.24), for any symmetry operation $\phi_i \in G, i = 1,2,\ldots,n$, we have

$$
\begin{aligned}
g\left(\phi_i\left(\boldsymbol{r}\right)\right) &= \begin{bmatrix} \left\|\phi_i\left(\boldsymbol{r}\right) + \mathbf{R}\boldsymbol{r}_{M_1}^L - \boldsymbol{r}_{B_1}^A\right\| - l_1 \\ \left\|\phi_i\left(\boldsymbol{r}\right) + \mathbf{R}\boldsymbol{r}_{M_2}^L - \boldsymbol{r}_{B_2}^A\right\| - l_2 \\ \vdots \\ \left\|\phi_i\left(\boldsymbol{r}\right) + \mathbf{R}\boldsymbol{r}_{M_m}^L - \boldsymbol{r}_{B_m}^A\right\| - l_m \end{bmatrix} \\
&= \begin{bmatrix} \left\|\phi_i\left(\boldsymbol{r}\right) + \mathbf{R}\phi_i\phi_i^{-1}\left(\boldsymbol{r}_{M_1}^L\right) - \phi_i\phi_i^{-1}\left(\boldsymbol{r}_{B_1}^A\right)\right\| - \phi_i\phi_i^{-1}\left(l_1\right) \\ \left\|\phi_i\left(\boldsymbol{r}\right) + \mathbf{R}\phi_i\phi_i^{-1}\left(\boldsymbol{r}_{M_2}^L\right) - \phi_i\phi_i^{-1}\left(\boldsymbol{r}_{B_2}^A\right)\right\| - \phi_i\phi_i^{-1}\left(l_2\right) \\ \vdots \\ \left\|\phi_i\left(\boldsymbol{r}\right) + \mathbf{R}\phi_i\phi_i^{-1}\left(\boldsymbol{r}_{M_m}^L\right) - \phi_i\phi_i^{-1}\left(\boldsymbol{r}_{B_m}^A\right)\right\| - \phi_i\phi_i^{-1}\left(l_m\right) \end{bmatrix}.
\end{aligned}
\tag{5.27}
$$

Because $\phi_i^{-1} \in G_M$ and the binary operation is closed on G_M, and ϕ_i^{-1} denotes a different permutation, $\sigma(1)\sigma(2)\cdots\sigma(m)$, of the number series $(1,2,\ldots,m)$. In equation (5.27), the actions of ϕ_i and ϕ_i^{-1} on vectors $\boldsymbol{r}$, $r_{M_j}^L$ and $r_{B_j}^A$, and on scalar l_j denote the corresponding permutations (or symmetry operations), where $j = 1,2,\ldots,m$. As a result, equation (5.27) can be transformed into:

$$
\begin{aligned}
g\left(\phi_i\left(\boldsymbol{r}\right)\right) &= \phi_i \begin{bmatrix} \left\|\boldsymbol{r} + \mathbf{R}\phi_i^{-1}\left(\boldsymbol{r}_{M_1}^L\right) - \phi_i^{-1}\left(\boldsymbol{r}_{B_1}^A\right)\right\| - \phi_i^{-1}\left(l_1\right) \\ \left\|\boldsymbol{r} + \mathbf{R}\phi_i^{-1}\left(\boldsymbol{r}_{M_2}^L\right) - \phi_i^{-1}\left(\boldsymbol{r}_{B_2}^A\right)\right\| - \phi_i^{-1}\left(l_2\right) \\ \vdots \\ \left\|\boldsymbol{r} + \mathbf{R}\phi_i^{-1}\left(\boldsymbol{r}_{M_m}^L\right) - \phi_i^{-1}\left(\boldsymbol{r}_{B_m}^A\right)\right\| - \phi_i^{-1}\left(l_m\right) \end{bmatrix} \\
&= \phi_i \begin{bmatrix} \left\|\boldsymbol{r} + \mathbf{R}\boldsymbol{r}_{M_{\sigma(1)}}^L - \boldsymbol{r}_{B_{\sigma(1)}}^A\right\| - l_{\sigma(1)} \\ \left\|\boldsymbol{r} + \mathbf{R}\boldsymbol{r}_{M_{\sigma(2)}}^L - \boldsymbol{r}_{B_{\sigma(2)}}^A\right\| - l_{\sigma(2)} \\ \vdots \\ \left\|\boldsymbol{r} + \mathbf{R}\boldsymbol{r}_{M_{\sigma(m)}}^L - \boldsymbol{r}_{B_{\sigma(m)}}^A\right\| - l_{\sigma(m)} \end{bmatrix}.
\end{aligned}
\tag{5.28}
$$

From equation (5.28), we can find that the constraints $\begin{bmatrix} \left\|\boldsymbol{r} + \mathbf{R}\boldsymbol{r}_{M_{\sigma(1)}}^L - \boldsymbol{r}_{B_{\sigma(1)}}^A\right\| - l_{\sigma(1)} \\ \left\|\boldsymbol{r} + \mathbf{R}\boldsymbol{r}_{M_{\sigma(2)}}^L - \boldsymbol{r}_{B_{\sigma(2)}}^A\right\| - l_{\sigma(2)} \\ \vdots \\ \left\|\boldsymbol{r} + \mathbf{R}\boldsymbol{r}_{M_{\sigma(m)}}^L - \boldsymbol{r}_{B_{\sigma(m)}}^A\right\| - l_{\sigma(m)} \end{bmatrix}$ are equivalent to equation (5.26) for the differences only result from the new permutations of the sequence numbers of the constraint equations. As a result, equation (5.28) can be simplified as:

$$g\left(\phi_i\left(r\right)\right)=\phi_i\left(g\left(r\right)\right). \tag{5.29}$$

If the kinematic chain structure of the end effector is kept unaltered by certain symmetry operation $\phi_i, i = 1,2,\ldots,n$, then the workspace $\Omega = \{r \,|F(r) = 0\}$ is also unaltered under the symmetry operation ϕ_i, and is therefore unaltered under the symmetry group G_M.

Because G_W characterizes all the symmetries of the workspace, there must be $G_M \subseteq G_W$. As a matter of fact, the relationship depicted by equation (5.22) can also be generalized to the dexterous workspace. That is, $G_M \subseteq G_D$ if G_D denote the dexterous workspace of the end effector, which can be similarly proved.

Theorem 5.2. *If the symmetry group of the dexterous workspace is denoted by G_D, and the symmetry group of the kinematic chain structure of the end effector in a particular configuration is denoted by G_M; then G_M must be a subgroup of G_D, namely, the following relationship will always hold:*

$$G_M \subseteq G_D. \tag{5.30}$$

It is not difficult to prove Theorem 5.2 which is omitted here.

Theorem 5.3. *If the order of the symmetry group of the mechanism structure is n, then the searching angle range can be reduced to $\left[0,\frac{2\pi}{n}\right)$. That is, the searching range can be reduced to $\frac{1}{n}$ of the initial one.*

Proof. As the symmetry group is a finite permutation group, any elements in the symmetry group can be expressed as the power exponent of the generating elements. When the order of the group is n, for any elements, $\varphi_i \in G$, the $\varphi_i^n = e$ holds. e is the unit element of the group. Therefore, Theorem 5.3 is proved.

5.3 APPLICATIONS OF THE SYMMETRICAL WORKSPACE THEOREM

This section will investigate the relationship between the symmetry reachable workspace and the symmetry structure characteristic of some typical parallel manipulators.

5.3.1 Workspace of a Gough-Stewart Parallel Mechanism

A general Gough–Stewart parallel mechanism is shown in Figure 5.14. Assume the origin of the absolute coordinate system *oxyz* shown in Figure 5.15 is superimposed with the geometry center of the fixed base plane $B_1B_2B_3B_4B_5B_6$,

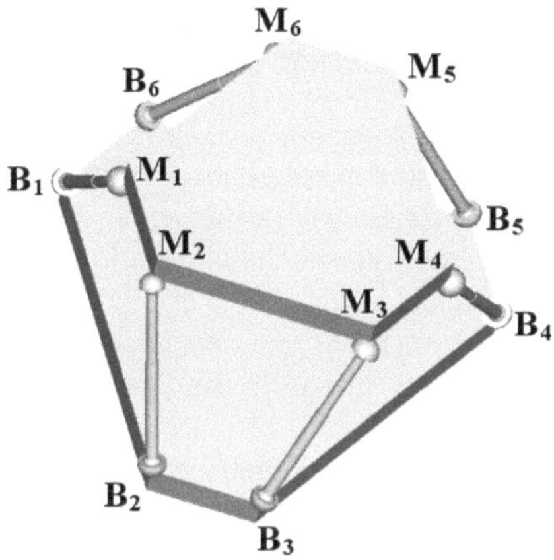

FIGURE 5.14 Gough–Stewart parallel mechanism.

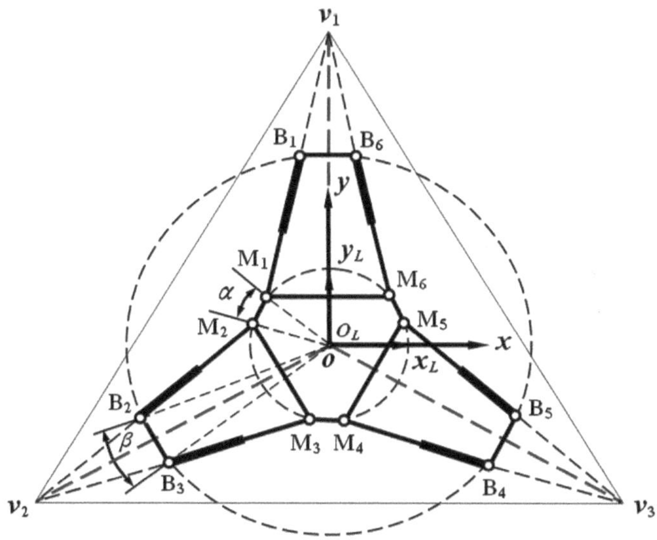

FIGURE 5.15 Top view of Gough–Stewart mechanism.

the x-axis is parallel to B_1B_6, the y-axis is perpendicular to B_1B_6, and the z-axis is perpendicular to the base plane $B_1B_2B_3B_4B_5B_6$. In a similar manner, we also presume that the origin of the local coordinate system $o_Lx_Ly_Lz_L$ shown in Figure 5.15 is superimposed with the geometry center of the mobile end effector $M_1M_2M_3M_4M_5M_6$ the x_L-axis is parallel to M_1M_6, the y_L-axis is perpendicular to M_1M_6, and the z_L-axis is perpendicular to the mobile plane $M_1M_2M_3M_4M_5M_6$.

In the absolute coordinate system, the vertexes of the fixed base are denoted by $B_i, i = 1,2,\ldots,6$, and the geometry center is o. The vectors of these vertexes are represented by $\boldsymbol{r}_{B_i}$, respectively. In the local coordinate system, the coordinates of vertexes M_i of the mobile mechanism are represented by $\boldsymbol{r}_{M_i}^L$, the geometry center of the mobile end effector is denoted by C, and the absolute coordinates are $(x\ y\ z)$.

The structure of the spatial parallel mechanism shown in Figure 5.14 has very interesting characteristics: the hexagons $M_1M_2M_3M_4M_5M_6$ and $B_1B_2B_3B_4B_5B_6$ are similar polygons; the kinematic chains of the end effector $M_1M_2M_3M_4M_5M_6$ are identical *SPS* chains and have the same actuations; the intersections of M_1B_1 and M_6B_6, M_2B_2, and M_3B_3, M_4B_4, and M_5B_5 in the top view shown in Figure 5.15 just form the three vertexes of an equilateral triangle $v_1v_2v_3$.

It is not difficult to find that the symmetry operations of the Gough–Stewart mechanism shown in Figure 5.15 forms a symmetry group S_3, the elements of which are the permutations of a set $V = \{v_1, v_2, v_3\}$. The exact number of the permutations is $3! = 6$. Therefore, $S_3 = \{\phi_1, \phi_2, \phi_3, \phi_4, \phi_5, \phi_6\}$, where $\phi_1 = (1)$, $\phi_2 = (123)$, $\phi_3 = (132)$, $\phi_4 = (23)$, $\phi_5 = (13)$, and $\phi_6 = (12)$. It is not difficult to prove that S_3 is just a dihedral group D_6, which describes all the symmetries of the mechanism structure shown in Figure 5.15. The first three elements of S_3 represent the three rotations 0, $\frac{2\pi}{3}$, and $\frac{4\pi}{3}$ about the z-axis; and the last three denote three symmetries about ov_1, ov_2, and ov_3, respectively.

Now, let's consider the actions of the symmetric group S_3 on the workspace of the end effector.

The workspace of the mechanism can be denoted as:

$$\Omega = \{\boldsymbol{r}\,|\,\boldsymbol{g}(\boldsymbol{r}) = 0\} \tag{5.31}$$

where $\boldsymbol{g}(\cdot)$ are a set of constraint equations of the workspace boundary surfaces:

$$\boldsymbol{g}(\boldsymbol{r}) = \begin{bmatrix} \|\boldsymbol{r}_{M_1}^A - \boldsymbol{r}_{B_1}^A\| - l_1 \\ \|\boldsymbol{r}_{M_2}^A - \boldsymbol{r}_{B_2}^A\| - l_2 \\ \vdots \\ \|\boldsymbol{r}_{M_6}^A - \boldsymbol{r}_{B_6}^A\| - l_6 \end{bmatrix}. \tag{5.32}$$

According to equation (5.23), equation (5.32) can be expanded:

$$\boldsymbol{g}(\boldsymbol{r}) = \begin{bmatrix} \|\boldsymbol{r} + \mathbf{R}\boldsymbol{r}_{M_1}^L - \boldsymbol{r}_{B_1}^A\| - l_1 \\ \|\boldsymbol{r} + \mathbf{R}\boldsymbol{r}_{M_2}^L - \boldsymbol{r}_{B_2}^A\| - l_2 \\ \vdots \\ \|\boldsymbol{r} + \mathbf{R}\boldsymbol{r}_{M_6}^L - \boldsymbol{r}_{B_6}^A\| - l_6 \end{bmatrix}. \tag{5.33}$$

In what follows, we will demonstrate that the workspace of the end effector contains the symmetries of the mechanism structure. That is, the actions of the dihedral group of the mechanism on the workspace expressed in equation (5.31) will make the workspace superimpose with itself.

(1) The action of ϕ_1 on Ω

Obviously, $\phi_1(\Omega) = \Omega$ holds.

(2) The action of ϕ_2 on Ω

Because

$$
g(\phi_2(\boldsymbol{r})) = \begin{bmatrix}
\|\phi_2(\boldsymbol{r}) + \mathbf{R}\boldsymbol{r}_{M_1}^L - \boldsymbol{r}_{B_1}^A\| - l_1 \\
\|\phi_2(\boldsymbol{r}) + \mathbf{R}\boldsymbol{r}_{M_2}^L - \boldsymbol{r}_{B_2}^A\| - l_2 \\
\|\phi_2(\boldsymbol{r}) + \mathbf{R}\boldsymbol{r}_{M_3}^L - \boldsymbol{r}_{B_3}^A\| - l_3 \\
\|\phi_2(\boldsymbol{r}) + \mathbf{R}\boldsymbol{r}_{M_4}^L - \boldsymbol{r}_{B_4}^A\| - l_4 \\
\|\phi_2(\boldsymbol{r}) + \mathbf{R}\boldsymbol{r}_{M_5}^L - \boldsymbol{r}_{B_5}^A\| - l_5 \\
\|\phi_2(\boldsymbol{r}) + \mathbf{R}\boldsymbol{r}_{M_6}^L - \boldsymbol{r}_{B_6}^A\| - l_6
\end{bmatrix}
$$

$$
= \begin{bmatrix}
\|\phi_2(\boldsymbol{r}) + \mathbf{R}\phi_2\phi_2^{-1}(\boldsymbol{r}_{M_1}^L) - \phi_2\phi_2^{-1}(\boldsymbol{r}_{B_1}^A)\| - \phi_2\phi_2^{-1}(l_1) \\
\|\phi_2(\boldsymbol{r}) + \mathbf{R}\phi_2\phi_2^{-1}(\boldsymbol{r}_{M_2}^L) - \phi_2\phi_2^{-1}(\boldsymbol{r}_{B_2}^A)\| - \phi_2\phi_2^{-1}(l_2) \\
\|\phi_2(\boldsymbol{r}) + \mathbf{R}\phi_2\phi_2^{-1}(\boldsymbol{r}_{M_3}^L) - \phi_2\phi_2^{-1}(\boldsymbol{r}_{B_3}^A)\| - \phi_2\phi_2^{-1}(l_3) \\
\|\phi_2(\boldsymbol{r}) + \mathbf{R}\phi_2\phi_2^{-1}(\boldsymbol{r}_{M_4}^L) - \phi_2\phi_2^{-1}(\boldsymbol{r}_{B_4}^A)\| - \phi_2\phi_2^{-1}(l_4) \\
\|\phi_2(\boldsymbol{r}) + \mathbf{R}\phi_2\phi_2^{-1}(\boldsymbol{r}_{M_5}^L) - \phi_2\phi_2^{-1}(\boldsymbol{r}_{B_5}^A)\| - \phi_2\phi_2^{-1}(l_5) \\
\|\phi_2(\boldsymbol{r}) + \mathbf{R}\phi_2\phi_2^{-1}(\boldsymbol{r}_{M_6}^L) - \phi_2\phi_2^{-1}(\boldsymbol{r}_{B_6}^A)\| - \phi_2\phi_2^{-1}(l_6)
\end{bmatrix} \tag{5.34}
$$

while $\phi_2 = (123)$ and $\phi_2^{-1} = (132) = \phi_3$, so equation (5.34) equals:

$$
g(\phi_2(\boldsymbol{r})) = \phi_2 \begin{bmatrix}
\|\boldsymbol{r} + \mathbf{R}\phi_3(\boldsymbol{r}_{M_1}^L) - \phi_3(\boldsymbol{r}_{B_1}^A)\| - \phi_3(l_1) \\
\|\boldsymbol{r} + \mathbf{R}\phi_3(\boldsymbol{r}_{M_2}^L) - \phi_3(\boldsymbol{r}_{B_2}^A)\| - \phi_3(l_2) \\
\|\boldsymbol{r} + \mathbf{R}\phi_3(\boldsymbol{r}_{M_3}^L) - \phi_3(\boldsymbol{r}_{B_3}^A)\| - \phi_3(l_3) \\
\|\boldsymbol{r} + \mathbf{R}\phi_3(\boldsymbol{r}_{M_4}^L) - \phi_3(\boldsymbol{r}_{B_4}^A)\| - \phi_3(l_4) \\
\|\boldsymbol{r} + \mathbf{R}\phi_3(\boldsymbol{r}_{M_5}^L) - \phi_3(\boldsymbol{r}_{B_5}^A)\| - \phi_3(l_5) \\
\|\boldsymbol{r} + \mathbf{R}\phi_3(\boldsymbol{r}_{M_6}^L) - \phi_3(\boldsymbol{r}_{B_6}^A)\| - \phi_3(l_6)
\end{bmatrix}
$$

$$= \phi_2 \begin{bmatrix} \left\| \boldsymbol{r} + \mathbf{R}\boldsymbol{r}_{M_5}^{L} - \boldsymbol{r}_{B_5}^{A} \right\| - l_5 \\ \left\| \boldsymbol{r} + \mathbf{R}\boldsymbol{r}_{M_6}^{L} - \boldsymbol{r}_{B_6}^{A} \right\| - l_6 \\ \left\| \boldsymbol{r} + \mathbf{R}\boldsymbol{r}_{M_1}^{L} - \boldsymbol{r}_{B_1}^{A} \right\| - l_1 \\ \left\| \boldsymbol{r} + \mathbf{R}\boldsymbol{r}_{M_2}^{L} - \boldsymbol{r}_{B_2}^{A} \right\| - l_2 \\ \left\| \boldsymbol{r} + \mathbf{R}\boldsymbol{r}_{M_3}^{L} - \boldsymbol{r}_{B_3}^{A} \right\| - l_3 \\ \left\| \boldsymbol{r} + \mathbf{R}\boldsymbol{r}_{M_4}^{L} - \boldsymbol{r}_{B_4}^{A} \right\| - l_4 \end{bmatrix} = \phi_2 \left(g \left(r \right) \right). \quad (5.35)$$

Therefore, the action of ϕ_2 on Ω makes Ω rotate $\frac{2\pi}{3}$ about the z-axis and superimpose with itself. The action of ϕ_3 on Ω can be similarly found to be a $\frac{4\pi}{3}$-rotation about the z-axis and superimpose with itself.

(3) The action of ϕ_4 on Ω

Because

$$g\left(\phi_4\left(\boldsymbol{r}\right)\right) = \begin{bmatrix} \left\| \phi_4\left(\boldsymbol{r}\right) + \mathbf{R}\boldsymbol{r}_{M_1}^{L} - \boldsymbol{r}_{B_1}^{A} \right\| - l_1 \\ \left\| \phi_4\left(\boldsymbol{r}\right) + \mathbf{R}\boldsymbol{r}_{M_2}^{L} - \boldsymbol{r}_{B_2}^{A} \right\| - l_2 \\ \left\| \phi_4\left(\boldsymbol{r}\right) + \mathbf{R}\boldsymbol{r}_{M_3}^{L} - \boldsymbol{r}_{B_3}^{A} \right\| - l_3 \\ \left\| \phi_4\left(\boldsymbol{r}\right) + \mathbf{R}\boldsymbol{r}_{M_4}^{L} - \boldsymbol{r}_{B_4}^{A} \right\| - l_4 \\ \left\| \phi_4\left(\boldsymbol{r}\right) + \mathbf{R}\boldsymbol{r}_{M_5}^{L} - \boldsymbol{r}_{B_5}^{A} \right\| - l_5 \\ \left\| \phi_4\left(\boldsymbol{r}\right) + \mathbf{R}\boldsymbol{r}_{M_6}^{L} - \boldsymbol{r}_{B_6}^{A} \right\| - l_6 \end{bmatrix} \quad (5.36)$$

while ϕ_4 is a symmetry operation of the mechanism and $\phi_4^{-1} = (23) = \phi_4$, equation (5.36) can be further transformed into

$$g\left(\phi_4\left(\boldsymbol{r}\right)\right) = \begin{bmatrix} \left\| \phi_4\left(\boldsymbol{r}\right) + \mathbf{R}\phi_4\phi_4^{-1}\left(\boldsymbol{r}_{M_1}^{L}\right) - \phi_4\phi_4^{-1}\left(\boldsymbol{r}_{B_1}^{A}\right) \right\| - \phi_4\phi_4^{-1}\left(l_1\right) \\ \left\| \phi_4\left(\boldsymbol{r}\right) + \mathbf{R}\phi_4\phi_4^{-1}\left(\boldsymbol{r}_{M_2}^{L}\right) - \phi_4\phi_4^{-1}\left(\boldsymbol{r}_{B_2}^{A}\right) \right\| - \phi_4\phi_4^{-1}\left(l_2\right) \\ \left\| \phi_4\left(\boldsymbol{r}\right) + \mathbf{R}\phi_4\phi_4^{-1}\left(\boldsymbol{r}_{M_3}^{L}\right) - \phi_4\phi_4^{-1}\left(\boldsymbol{r}_{B_3}^{A}\right) \right\| - \phi_4\phi_4^{-1}\left(l_3\right) \\ \left\| \phi_4\left(\boldsymbol{r}\right) + \mathbf{R}\phi_4\phi_4^{-1}\left(\boldsymbol{r}_{M_4}^{L}\right) - \phi_4\phi_4^{-1}\left(\boldsymbol{r}_{B_4}^{A}\right) \right\| - \phi_4\phi_4^{-1}\left(l_4\right) \\ \left\| \phi_4\left(\boldsymbol{r}\right) + \mathbf{R}\phi_4\phi_4^{-1}\left(\boldsymbol{r}_{M_5}^{L}\right) - \phi_4\phi_4^{-1}\left(\boldsymbol{r}_{B_5}^{A}\right) \right\| - \phi_4\phi_4^{-1}\left(l_5\right) \\ \left\| \phi_4\left(\boldsymbol{r}\right) + \mathbf{R}\phi_4\phi_4^{-1}\left(\boldsymbol{r}_{M_6}^{L}\right) - \phi_4\phi_4^{-1}\left(\boldsymbol{r}_{B_6}^{A}\right) \right\| - \phi_4\phi_4^{-1}\left(l_6\right) \end{bmatrix}$$

$$= \phi_4 \begin{bmatrix} \left\| \boldsymbol{r} + \mathbf{R}\phi_4\left(\boldsymbol{r}_{M_1}^{L}\right) - \phi_4\left(\boldsymbol{r}_{B_1}^{A}\right) \right\| - \phi_4\left(l_1\right) \\ \left\| \boldsymbol{r} + \mathbf{R}\phi_4\left(\boldsymbol{r}_{M_5}^{L}\right) - \phi_4\left(\boldsymbol{r}_{B_5}^{A}\right) \right\| - \phi_4\left(l_5\right) \\ \left\| \boldsymbol{r} + \mathbf{R}\phi_4\left(\boldsymbol{r}_{M_4}^{L}\right) - \phi_4\left(\boldsymbol{r}_{B_4}^{A}\right) \right\| - \phi_4\left(l_4\right) \\ \left\| \boldsymbol{r} + \mathbf{R}\phi_4\left(\boldsymbol{r}_{M_3}^{L}\right) - \phi_4\left(\boldsymbol{r}_{B_3}^{A}\right) \right\| - \phi_4\left(l_3\right) \\ \left\| \boldsymbol{r} + \mathbf{R}\phi_4\left(\boldsymbol{r}_{M_2}^{L}\right) - \phi_4\left(\boldsymbol{r}_{B_2}^{A}\right) \right\| - \phi_4\left(l_2\right) \\ \left\| \boldsymbol{r} + \mathbf{R}\phi_4\left(\boldsymbol{r}_{M_6}^{L}\right) - \phi_4\left(\boldsymbol{r}_{B_6}^{A}\right) \right\| - \phi_4\left(l_6\right) \end{bmatrix}. \quad (5.37)$$

Therefore, equation (5.37) is equivalent to

$$g\left(\phi_4\left(r\right)\right)=\phi_4\begin{bmatrix}\left\|r+\mathbf{R}r_{M_6}^L-r_{B_6}^A\right\|-l_6\\ \left\|r+\mathbf{R}r_{M_5}^L-r_{B_5}^A\right\|-l_5\\ \left\|r+\mathbf{R}r_{M_4}^L-r_{B_4}^A\right\|-l_4\\ \left\|r+\mathbf{R}r_{M_3}^L-r_{B_3}^A\right\|-l_3\\ \left\|r+\mathbf{R}r_{M_2}^L-r_{B_2}^A\right\|-l_2\\ \left\|r+\mathbf{R}r_{M_1}^L-r_{B_1}^A\right\|-l_1\end{bmatrix}=\phi_4\left(g\left(r\right)\right). \tag{5.38}$$

Consequently, the action of ϕ_4 on Ω makes Ω reflect about the *yoz*-plane. The actions of ϕ_5 and ϕ_6 on Ω can be similarly found to be reflections about the planes of v_2oz and v_3oz, individually.

Therefore, the workspace of the Gough–Stewart mechanism has all the symmetries of the mechanism. The workspace can be investigated as follows.

Supposing that the coordinate transformations, **R**, from the mobile end effector to the fixed base are 231-type, then

$$\mathbf{R}=\begin{bmatrix}\cos\theta & 0 & \sin\theta\\ 0 & 1 & 0\\ -\sin\theta & 0 & \cos\theta\end{bmatrix}\begin{bmatrix}\cos\psi & -\sin\psi & 0\\ \sin\psi & \cos\psi & 0\\ 0 & 0 & 1\end{bmatrix}\begin{bmatrix}1 & 0 & 0\\ 0 & \cos\phi & -\sin\phi\\ 0 & \sin\phi & \cos\phi\end{bmatrix}$$

$$=\begin{bmatrix}\cos\psi\cos\theta & -\sin\psi\cos\theta\cos\phi+\sin\theta\sin\phi & \sin\psi\cos\theta\sin\phi+\sin\theta\cos\phi\\ \sin\psi & \cos\psi\cos\phi & -\cos\psi\sin\phi\\ -\cos\psi\sin\theta & \sin\psi\sin\theta\cos\phi+\cos\theta\sin\phi & -\sin\psi\sin\theta\sin\phi+\cos\theta\cos\phi\end{bmatrix}. \tag{5.39}$$

Substituting equation (5.39) into equation (5.23) results in

$$r_{M_i}^A=r+\mathbf{R}r_{M_i}^L \tag{5.40}$$

where $i=1,2,\ldots,6$.

Assuming $r_{M_i}^L=\begin{bmatrix}x_{M_i}\\ y_{M_i}\\ z_{M_i}\end{bmatrix}, i=1,2,\ldots,6$ and $r=\begin{bmatrix}x\\ y\\ z\end{bmatrix}$, equation (5.40) can be expanded to yield:

$$r_{M_i}^A=r+\mathbf{R}r_{M_i}^L$$

$$=\begin{bmatrix}x+x_{M_i}^L\cos\psi\cos\theta+y_{M_i}^L(-\sin\psi\cos\theta\cos\phi+\sin\theta\sin\phi)\\ \quad +z_{M_i}^L(\sin\psi\cos\theta\sin\phi+\sin\theta\cos\phi)\\ y+x_{M_i}^L\sin\psi+y_{M_i}^L\cos\psi\cos\phi-z_{M_i}^L\cos\psi\sin\phi\\ z-x_{M_i}^L\cos\psi\sin\theta+y_{M_i}^L(\sin\psi\sin\theta\cos\phi+\cos\theta\sin\phi)\\ \quad +z_{M_i}^L(-\sin\psi\sin\theta\sin\phi+\cos\theta\cos\phi)\end{bmatrix}. \tag{5.41}$$

Substituting equation (5.41) into equation (5.25) yields a set of constraint equations of the workspace of the end effector. To obtain the workspace boundary surfaces, we can resort to programming using a computer. Therefore, suppose

$$\begin{bmatrix} f_1 \\ f_2 \\ f_3 \\ f_4 \\ f_5 \\ f_6 \end{bmatrix} = \begin{bmatrix} f_1(x,y,z) \\ f_2(x,y,z) \\ f_3(x,y,z) \\ f_4(x,y,z) \\ f_5(x,y,z) \\ f_6(x,y,z) \end{bmatrix} \tag{5.42}$$

where

$$\begin{aligned} f_1(x,y,z) = {} & [x + x_{M_1}^L \cos\psi\cos\theta + y_{M_1}^L(-\sin\psi\cos\theta\cos\phi + \sin\theta\sin\phi) \\ & + z_{M_1}^L(\sin\psi\cos\theta\sin\phi + \sin\theta\cos\phi) - x_{B_1}]^2 \\ & + [y + x_{M_1}^L \sin\psi + y_{M_1}^L \cos\psi\cos\phi - z_{M_1}^L \cos\psi\sin\phi - y_{B_1}]^2 \\ & + [z - x_{M_1}^L \cos\psi\sin\theta + y_{M_1}^L(\sin\psi\sin\theta\cos\phi + \cos\theta\sin\phi) \\ & + z_{M_1}^L(-\sin\psi\sin\theta\sin\phi + \cos\theta\cos\phi) - z_{B_1}]^2 \end{aligned}$$

$$\begin{aligned} f_2(x,y,z) = {} & [x + x_{M_2}^L \cos\psi\cos\theta + y_{M_2}^L(-\sin\psi\cos\theta\cos\phi + \sin\theta\sin\phi) \\ & + z_{M_2}^L(\sin\psi\cos\theta\sin\phi + \sin\theta\cos\phi) - x_{B_2}]^2 \\ & + [y + x_{M_2}^L \sin\psi + y_{M_2}^L \cos\psi\cos\phi - z_{M_2}^L \cos\psi\sin\phi - y_{B_2}]^2 \\ & + [z - x_{M_2}^L \cos\psi\sin\theta + y_{M_2}^L(\sin\psi\sin\theta\cos\phi + \cos\theta\sin\phi) \\ & + z_{M_2}^L(-\sin\psi\sin\theta\sin\phi + \cos\theta\cos\phi) - z_{B_2}]^2 \end{aligned}$$

$$\begin{aligned} f_3(x,y,z) = {} & [x + x_{M_3}^L \cos\psi\cos\theta + y_{M_3}^L(-\sin\psi\cos\theta\cos\phi + \sin\theta\sin\phi) \\ & + z_{M_3}^L(\sin\psi\cos\theta\sin\phi + \sin\theta\cos\phi) - x_{B_3}]^2 \\ & + [y + x_{M_3}^L \sin\psi + y_{M_3}^L \cos\psi\cos\phi - z_{M_3}^L \cos\psi\sin\phi - y_{B_3}]^2 \\ & + [z - x_{M_3}^L \cos\psi\sin\theta + y_{M_3}^L(\sin\psi\sin\theta\cos\phi + \cos\theta\sin\phi) \\ & + z_{M_3}^L(-\sin\psi\sin\theta\sin\phi + \cos\theta\cos\phi) - z_{B_3}]^2 \end{aligned}$$

$$\begin{aligned} f_4(x,y,z) = {} & [x + x_{M_4}^L \cos\psi\cos\theta + y_{M_4}^L(-\sin\psi\cos\theta\cos\phi + \sin\theta\sin\phi) \\ & + z_{M_4}^L(\sin\psi\cos\theta\sin\phi + \sin\theta\cos\phi) - x_{B_4}]^2 \\ & + [y + x_{M_4}^L \sin\psi + y_{M_4}^L \cos\psi\cos\phi - z_{M_4}^L \cos\psi\sin\phi - y_{B_4}]^2 \\ & + [z - x_{M_4}^L \cos\psi\sin\theta + y_{M_4}^L(\sin\psi\sin\theta\cos\phi + \cos\theta\sin\phi) \\ & + z_{M_4}^L(-\sin\psi\sin\theta\sin\phi + \cos\theta\cos\phi) - z_{B_4}]^2 \end{aligned}$$

$$
\begin{aligned}
f_5(x,y,z) = {} & [x + x^L_{M_5}\cos\psi\cos\theta + y^L_{M_5}(-\sin\psi\cos\theta\cos\phi + \sin\theta\sin\phi) \\
& + z^L_{M_5}(\sin\psi\cos\theta\sin\phi + \sin\theta\cos\phi) - x_{B_5}]^2 \\
& + [y^L_{M_5} + x^L_{M_5}\sin\psi + y^L_{M_5}\cos\psi\cos\phi - z^L_{M_5}\cos\psi\sin\phi - y_{B_5}]^2 \\
& + [z - x^L_{M_5}\cos\psi\sin\theta + y^L_{M_5}(\sin\psi\sin\theta\cos\phi + \cos\theta\sin\phi) \\
& + z^L_{M_5}(-\sin\psi\sin\theta\sin\phi + \cos\theta\cos\phi) - z_{B_5}]^2
\end{aligned}
$$

$$
\begin{aligned}
f_6(x,y,z) = {} & [x + x^L_{M_6}\cos\psi\cos\theta + y^L_{M_6}(-\sin\psi\cos\theta\cos\phi + \sin\theta\sin\phi) \\
& + z^L_{M_6}(\sin\psi\cos\theta\sin\phi + \sin\theta\cos\phi) - x_{B_6}]^2 \\
& + [y + x^L_{M_6}\sin\psi + y^L_{M_6}\cos\psi\cos\phi - z^L_{M_6}\cos\psi\sin\phi - y_{B_6}]^2 \\
& + [z - x^L_{M_6}\cos\psi\sin\theta + y^L_{M_6}(\sin\psi\sin\theta\cos\phi + \cos\theta\sin\phi) \\
& + z^L_{M_6}(-\sin\psi\sin\theta\sin\phi + \cos\theta\cos\phi) - z_{B_6}]^2.
\end{aligned}
$$

Assume that the coordinates of $M_i, i = 1,2,\ldots,6$ in the local coordinate system $o_L x_L y_L z_L$ are:

$$
\begin{aligned}
& M_1\left(r\cos\left(\tfrac{5\pi}{6} - \tfrac{\alpha}{2}\right)\ r\sin\left(\tfrac{5\pi}{6} - \tfrac{\alpha}{2}\right)\ 0\right), \\
& M_2\left(r\cos\left(\tfrac{5\pi}{6} + \tfrac{\alpha}{2}\right)\ r\sin\left(\tfrac{5\pi}{6} + \tfrac{\alpha}{2}\right)\ 0\right), \\
& M_3\left(r\cos\left(\tfrac{3\pi}{2} - \tfrac{\alpha}{2}\right)\ r\sin\left(\tfrac{3\pi}{2} - \tfrac{\alpha}{2}\right)\ 0\right), \\
& M_4\left(r\cos\left(\tfrac{3\pi}{2} + \tfrac{\alpha}{2}\right)\ r\sin\left(\tfrac{3\pi}{2} + \tfrac{\alpha}{2}\right)\ 0\right), \\
& M_5\left(r\cos\left(\tfrac{\pi}{6} - \tfrac{\alpha}{2}\right)\ r\sin\left(\tfrac{\pi}{6} - \tfrac{\alpha}{2}\right)\ 0\right), \\
& M_6\left(r\cos\left(\tfrac{\pi}{6} + \tfrac{\alpha}{2}\right)\ r\sin\left(\tfrac{\pi}{6} + \tfrac{\alpha}{2}\right)\ 0\right).
\end{aligned}
$$

The coordinates of $B_i, i = 1,2,\ldots,6$ in the absolute coordinate system $oxyz$ are:

$$
\begin{aligned}
& B_1\left(R\cos\left(\tfrac{\pi}{2} + \tfrac{\beta}{2}\right)\ R\sin\left(\tfrac{\pi}{2} + \tfrac{\beta}{2}\right)\ 0\right), \\
& B_2\left(R\cos\left(\tfrac{7\pi}{6} - \tfrac{\beta}{2}\right)\ R\sin\left(\tfrac{7\pi}{6} - \tfrac{\beta}{2}\right)\ 0\right), \\
& B_3\left(R\cos\left(\tfrac{7\pi}{6} + \tfrac{\beta}{2}\right)\ R\sin\left(\tfrac{7\pi}{6} + \tfrac{\beta}{2}\right)\ 0\right), \\
& B_4\left(R\cos\left(-\tfrac{\pi}{6} - \tfrac{\beta}{2}\right)\ R\sin\left(-\tfrac{\pi}{6} - \tfrac{\beta}{2}\right)\ 0\right), \\
& B_5\left(R\cos\left(-\tfrac{\pi}{6} + \tfrac{\beta}{2}\right)\ R\sin\left(-\tfrac{\pi}{6} + \tfrac{\beta}{2}\right)\ 0\right), \\
& B_6\left(R\cos\left(\tfrac{\pi}{2} - \tfrac{\beta}{2}\right)\ R\sin\left(\tfrac{\pi}{2} - \tfrac{\beta}{2}\right)\ 0\right).
\end{aligned}
$$

Suppose the lengths of $B_1M_1, B_2M_2, B_3M_3, B_4M_4, B_5M_5, B_6M_6$ are variables subjected to $l_i \in [l_{\min}, l_{\max}], i = 1,2,\ldots,6$. Given the values of the parameters: $\alpha, \beta, r, R, l_{min}, l_{max}$, we can obtain the boundary surfaces of the workspace of the end effector with the searching procedure which can be found in [11–14]:

1. Solve the lowest position $(x_{Low}\ y_{Low}\ z_{Low})$ of the identified point of the end effector when each of its legs has shrunk to the least length; and solve the highest position $(x_{Hig}\ y_{Hig}\ z_{Hig})$ of the identified point of the end effector when each of its legs has stretched to the largest length.
2. Establish parametric equations of cylindrical coordinates of the point within the workspace:
$$\begin{cases} x = \rho \cos(i\Delta\gamma) \\ y = \rho \sin(i\Delta\gamma) \\ z = z_{Low} + j\Delta z \end{cases} \tag{5.43}$$
where ρ is the searching polar radius at plane $z = z_{Low} + j\Delta z$ with a polar angle $i\Delta\gamma$, and $0 \le i < \frac{2\pi}{\Delta\gamma}$, and $0 < j < \frac{z_{Hig}-z_{Low}}{\Delta z}$, and $\Delta\gamma$ is the angular interval and Δz is the distance interval along the z-axis, and i and j are all natural numbers.
3. Suppose the workspace is sliced with an interval Δz along the z-axis into $k-1$ planes, which are all parallel to the xoy-plane, and where k is the largest natural number that is less than or equal to $\frac{z_{Hig}-z_{Low}}{\Delta z}$. And therefore, $0 < j \le k$.
4. Search the boundary curve at the jth plane. Let $\rho = \rho(j\Delta z, i\Delta\gamma)$ increase from zero with an expected interval $\Delta\rho$, namely $\rho(j\Delta z, i\Delta\gamma) = q_j\Delta\rho, q_j = 1,2,\ldots$. The stop criterion for such increasing is that when equation (5.43) is substituted into equation (5.42), there is at least one $f_p, p = 1,2,\ldots,6$, such that $f_p > l_{\max}^2$, or $f_p < l_{\min}^2$ for a set of pose angles θ, ψ, ϕ. Select the largest parameter $\rho(j\Delta z, i\Delta\gamma)$ to plot the boundary curve in the jth plane with equation (5.43) as i increase from zero to $\frac{2\pi}{\Delta\gamma}$.
5. When the curves in all of the slices are obtained, the boundary surfaces of the reachable workspace of the end effector are gained.

Different workspace shapes of the Gough–Stewart type mechanism are shown in Figures 5.16, 5.17 and 5.18. Although the workspace shapes are different with different structure parameters, the symmetry relationships between the workspace and the mechanism structure are still governed by equation (5.22). Therefore, the searching processes can be reduced to search the range of a generating group of the symmetric group of the mechanism. For a symmetric mechanism, if the order of the symmetry group of the mechanism structure is n, then the searching range of angle about the z-axis can be reduced to $\left[0, \frac{2\pi}{n}\right)$, for example, the searching range of angle about the z-axis can be reduced to $\left[0, \frac{\pi}{3}\right)$. And as a result, the searching task and time are greatly reduced.

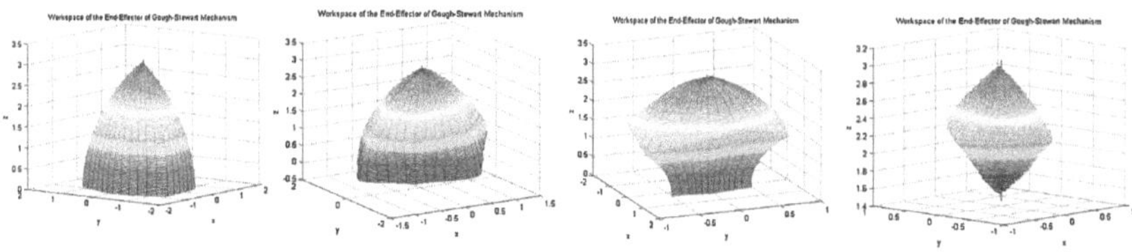

FIGURE 5.16 Different workspaces of the end effector of a Gough–Stewart mechanism.

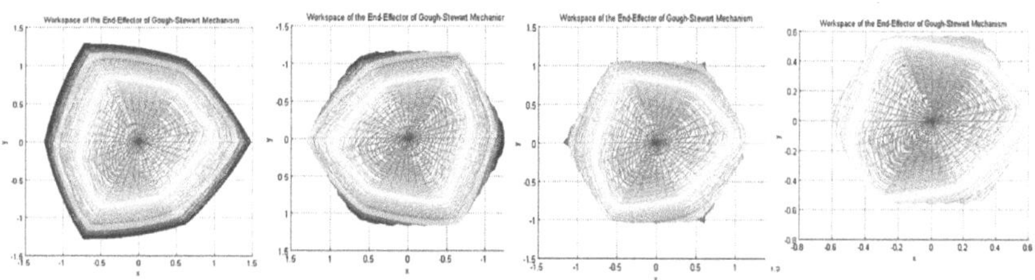

FIGURE 5.17 Top views of the workspaces of the end effector corresponding to Figure 5.16.

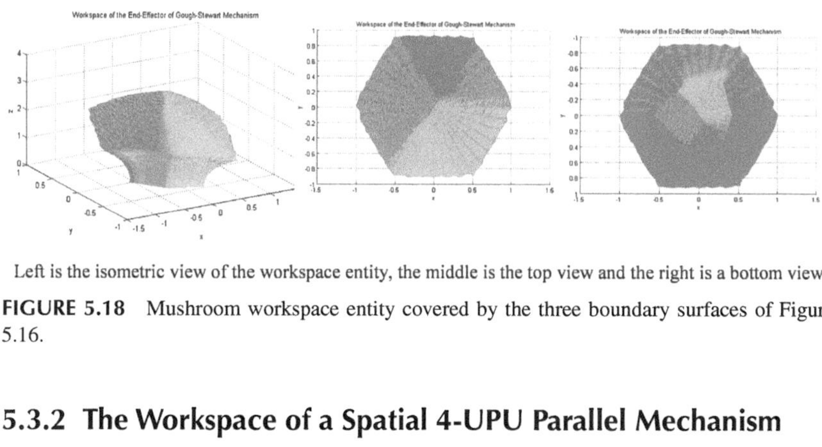

Left is the isometric view of the workspace entity, the middle is the top view and the right is a bottom view.

FIGURE 5.18 Mushroom workspace entity covered by the three boundary surfaces of Figure 5.16.

5.3.2 The Workspace of a Spatial 4-UPU Parallel Mechanism

A Schoenflies-type parallel mechanism is shown in Figure 5.19. The mechanism proposed herein is actuated by four identical UPU (universal joint+ prismatic joint + universal joint) kinematic chains [18].

The top view of the Schoenflies-type spatial parallel mechanism is shown in Figure 5.20. Therefore, we can consider a symmetry group S_4, the elements of which are the permutations of a set $V = \{B_1, B_2, B_3, B_4\}$. The exact number of the permutations is $4! = 24$. However, not all of the 24 permutations can form real symmetries of the mechanism shown in Figure 5.20. As a matter of fact, the possible symmetries of the mechanism shown in Figure 5.20 are eight at most. In order to avoid the similar singularity, we know that $\frac{\|M_1M_2\|}{\|M_1M_4\|} \neq \frac{\|B_1B_2\|}{\|B_1B_4\|}$

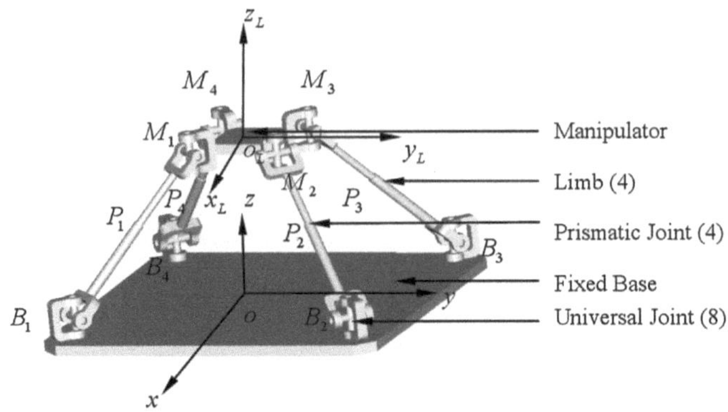

FIGURE 5.19 A kind of Schoenflies-type spatial parallel mechanism.

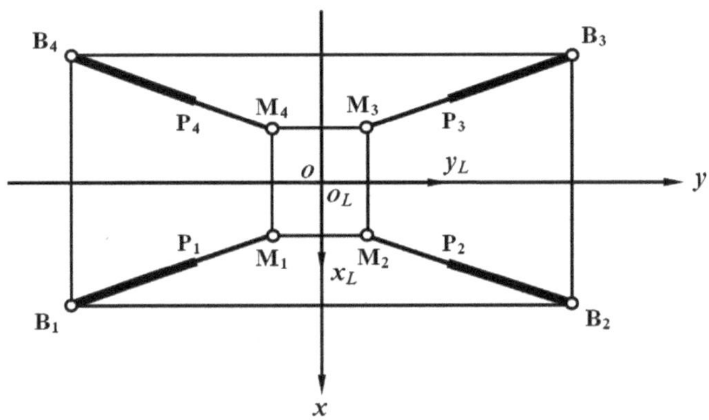

FIGURE 5.20 Top view of the Schoenflies-type spatial parallel mechanism.

[18]. Therefore, the symmetry group of the kinematic chains of the end effector is $G = \{\phi_1, \phi_2, \phi_3, \phi_4\}$, where $\phi_1 = (1)$, $\phi_2 = (13)(24)$, $\phi_3 = (12)(34)$, $\phi_4 = (14)(23)$, and G is a subgroup of dihedral group D_8. G describes all the symmetries of the mechanism structure shown in Figure 5.20.

The first two elements of G express the two rotations of angles 0 and π about the z-axis; and the last two denote two symmetries about the xoz- and yoz-planes, respectively.

Now, let us consider the actions of group G on the workspace of the end effector. The workspace of the mechanism can be represented by

$$g(\boldsymbol{r}) = \begin{bmatrix} \|\boldsymbol{r}_{M_1}^A - \boldsymbol{r}_{B_1}^A\| - l_1 \\ \|\boldsymbol{r}_{M_2}^A - \boldsymbol{r}_{B_2}^A\| - l_2 \\ \|\boldsymbol{r}_{M_3}^A - \boldsymbol{r}_{B_3}^A\| - l_3 \\ \|\boldsymbol{r}_{M_4}^A - \boldsymbol{r}_{B_4}^A\| - l_4 \end{bmatrix}. \tag{5.44}$$

According to equation (5.23), equation (5.44) can be expanded to:

$$\boldsymbol{g}(\boldsymbol{r}) = \begin{bmatrix} \left\|\boldsymbol{r} + \mathbf{R}\boldsymbol{r}_{M_1}^L - r_{B_1}^A\right\| - l_1 \\ \left\|\boldsymbol{r} + \mathbf{R}\boldsymbol{r}_{M_2}^L - r_{B_2}^A\right\| - l_2 \\ \left\|\boldsymbol{r} + \mathbf{R}\boldsymbol{r}_{M_3}^L - r_{B_3}^A\right\| - l_3 \\ \left\|\boldsymbol{r} + \mathbf{R}\boldsymbol{r}_{M_4}^L - r_{B_4}^A\right\| - l_4 \end{bmatrix} \tag{5.45}$$

(1) The action of ϕ_1 on Ω

Obviously, $\phi_1(\Omega) = \Omega$ holds.

(2) The action of ϕ_2 on Ω

Because

$$\boldsymbol{g}(\phi_2(\boldsymbol{r})) = \begin{bmatrix} \left\|\phi_2(\boldsymbol{r}) + \mathbf{R}\boldsymbol{r}_{M_1}^L - \boldsymbol{r}_{B_1}^A\right\| - l_1 \\ \left\|\phi_2(\boldsymbol{r}) + \mathbf{R}\boldsymbol{r}_{M_2}^L - \boldsymbol{r}_{B_2}^A\right\| - l_2 \\ \left\|\phi_2(\boldsymbol{r}) + \mathbf{R}\boldsymbol{r}_{M_3}^L - \boldsymbol{r}_{B_3}^A\right\| - l_3 \\ \left\|\phi_2(\boldsymbol{r}) + \mathbf{R}\boldsymbol{r}_{M_4}^L - \boldsymbol{r}_{B_4}^A\right\| - l_4 \end{bmatrix} \tag{5.46}$$

while ϕ_2 is a symmetry operation of the mechanism and $\phi_2^{-1} = (13)(24) = \phi_2$, equation (5.46) is further equivalent to

$$\boldsymbol{g}(\phi_2(\boldsymbol{r})) = \begin{bmatrix} \left\|\phi_2(\boldsymbol{r}) + \mathbf{R}\phi_2\phi_2\left(\boldsymbol{r}_{M_1}^L\right) - \phi_2\phi_2\left(\boldsymbol{r}_{B_1}^A\right)\right\| - \phi_2\phi_2(l_1) \\ \left\|\phi_2(\boldsymbol{r}) + \mathbf{R}\phi_2\phi_2\left(\boldsymbol{r}_{M_2}^L\right) - \phi_2\phi_2\left(\boldsymbol{r}_{B_2}^A\right)\right\| - \phi_2\phi_2(l_2) \\ \left\|\phi_2(\boldsymbol{r}) + \mathbf{R}\phi_2\phi_2\left(\boldsymbol{r}_{M_3}^L\right) - \phi_2\phi_2\left(\boldsymbol{r}_{B_3}^A\right)\right\| - \phi_2\phi_2(l_3) \\ \left\|\phi_2(\boldsymbol{r}) + \mathbf{R}\phi_2\phi_2\left(\boldsymbol{r}_{M_4}^L\right) - \phi_2\phi_2\left(\boldsymbol{r}_{B_4}^A\right)\right\| - \phi_2\phi_2(l_4) \end{bmatrix}$$

$$= \phi_2 \begin{bmatrix} \left\|r + \mathbf{R}\boldsymbol{r}_{M_3}^L - \boldsymbol{r}_{B_3}^A\right\| - l_3 \\ \left\|r + \mathbf{R}\boldsymbol{r}_{M_4}^L - \boldsymbol{r}_{B_4}^A\right\| - l_4 \\ \left\|r + \mathbf{R}\boldsymbol{r}_{M_1}^L - \boldsymbol{r}_{B_1}^A\right\| - l_1 \\ \left\|r + \mathbf{R}\boldsymbol{r}_{M_2}^L - \boldsymbol{r}_{B_2}^A\right\| - l_2 \end{bmatrix} = \phi_2(\boldsymbol{g}(r)). \tag{5.47}$$

Therefore, the action of ϕ_2 on Ω makes Ω rotate π about the z-axis and superimpose with itself.

(3) The action of ϕ_3 on Ω

Because

$$\boldsymbol{g}(\phi_3(\boldsymbol{r})) = \begin{bmatrix} \left\|\phi_3(\boldsymbol{r}) + \mathbf{R}\boldsymbol{r}_{M_1}^L - \boldsymbol{r}_{B_1}^A\right\| - l_1 \\ \left\|\phi_3(\boldsymbol{r}) + \mathbf{R}\boldsymbol{r}_{M_2}^L - \boldsymbol{r}_{B_2}^A\right\| - l_2 \\ \left\|\phi_3(\boldsymbol{r}) + \mathbf{R}\boldsymbol{r}_{M_3}^L - \boldsymbol{r}_{B_3}^A\right\| - l_3 \\ \left\|\phi_3(\boldsymbol{r}) + \mathbf{R}\boldsymbol{r}_{M_4}^L - \boldsymbol{r}_{B_4}^A\right\| - l_4 \end{bmatrix} \tag{5.48}$$

while ϕ_3 is a symmetry operation of the mechanism and $\phi_3^{-1} = (12)(34) = \phi_3$, equation (5.44) is further equivalent to

$$g\left(\phi_3(\boldsymbol{r})\right) = \begin{bmatrix} \left\|\phi_3(\boldsymbol{r}) + \mathbf{R}\phi_3\phi_3\left(\boldsymbol{r}_{M_1}^L\right) - \phi_3\phi_3\left(\boldsymbol{r}_{B_1}^A\right)\right\| - \phi_3\phi_3(l_1) \\ \left\|\phi_3(\boldsymbol{r}) + \mathbf{R}\phi_3\phi_3\left(\boldsymbol{r}_{M_2}^L\right) - \phi_3\phi_3\left(\boldsymbol{r}_{B_2}^A\right)\right\| - \phi_3\phi_3(l_2) \\ \left\|\phi_3(\boldsymbol{r}) + \mathbf{R}\phi_3\phi_3\left(\boldsymbol{r}_{M_3}^L\right) - \phi_3\phi_3\left(\boldsymbol{r}_{B_3}^A\right)\right\| - \phi_3\phi_3(l_3) \\ \left\|\phi_3(\boldsymbol{r}) + \mathbf{R}\phi_3\phi_3\left(\boldsymbol{r}_{M_4}^L\right) - \phi_3\phi_3\left(\boldsymbol{r}_{B_4}^A\right)\right\| - \phi_3\phi_3(l_4) \end{bmatrix}$$

$$= \phi_3 \begin{bmatrix} \left\|\boldsymbol{r} + \mathbf{R}\boldsymbol{r}_{M_2}^L - \boldsymbol{r}_{B_2}^A\right\| - l_2 \\ \left\|\boldsymbol{r} + \mathbf{R}\boldsymbol{r}_{M_1}^L - \boldsymbol{r}_{B_1}^A\right\| - l_1 \\ \left\|\boldsymbol{r} + \mathbf{R}\boldsymbol{r}_{M_4}^L - \boldsymbol{r}_{B_4}^A\right\| - l_4 \\ \left\|\boldsymbol{r} + \mathbf{R}\boldsymbol{r}_{M_3}^L - \boldsymbol{r}_{B_3}^A\right\| - l_3 \end{bmatrix} = \phi_3\left(g(r)\right). \tag{5.49}$$

Therefore, the action of ϕ_3 on Ω makes it reflect about the *xoz*-plane. The action of ϕ_4 on Ω can be similarly found to be a reflection about the plane of *yoz*.

Therefore, the workspace of the Schoenflies-type parallel mechanism shown in Figure 5.21 contains all the symmetries of those of the mechanism. The workspace of the end effector can be analyzed below.

The end effector has four DOFs (one rotational DOF and three orthogonal translational DOFs), which can be analyzed according to [19]. Correspondingly, we can only select four independent variables, three of which are the absolute coordinates of the geometry center of the end effector, denoted by (x, y, z); and the other is the rotational angle about the z-axis, denoted by α. As a result, the transformation matrix from the local coordinate system to the absolute one can be expressed as:

$$\mathbf{R} = \begin{bmatrix} \cos\alpha & -\sin\alpha & 0 \\ \sin\alpha & \cos\alpha & 0 \\ 0 & 0 & 1 \end{bmatrix}.$$

Suppose that the coordinates of the ith $(i = 1,2,3,4)$ vertex of the end effector in the local coordinate system are denoted by $M_i\left(x_{M_i}^L, y_{M_i}^L, z_{M_i}^L\right)$, the

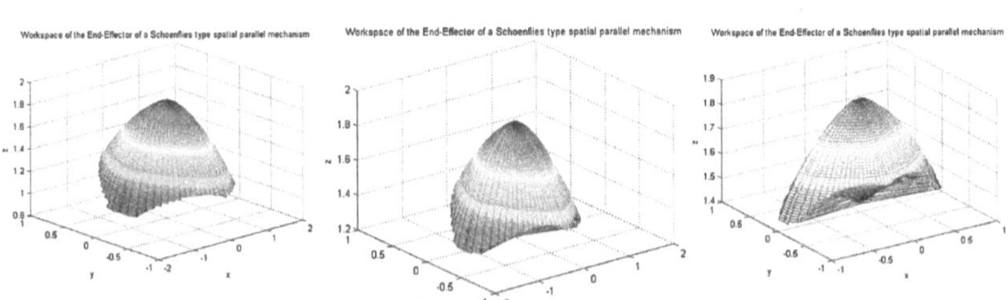

FIGURE 5.21 Different workspaces of a Schoenflies-type parallel mechanism.

absolute coordinates of $M_i \left(x^A_{M_i}, y^A_{M_i}, z^A_{M_i}\right)$ can be obtained:

$$\begin{bmatrix} x^A_{M_i} \\ y^A_{M_i} \\ z^A_{M_i} \end{bmatrix} = \begin{bmatrix} x \\ y \\ z \end{bmatrix} + \mathbf{R} \begin{bmatrix} x^L_{M_i} \\ y^L_{M_i} \\ z^L_{M_i} \end{bmatrix} = \begin{bmatrix} x \\ y \\ z \end{bmatrix} + \begin{bmatrix} \cos\alpha & -\sin\alpha & 0 \\ \sin\alpha & \cos\alpha & 0 \\ 0 & 0 & 1 \end{bmatrix} \begin{bmatrix} x^L_{Pi} \\ y^L_{Pi} \\ z^L_{Pi} \end{bmatrix}$$

$$= \begin{bmatrix} x + x^L_{M_i}\cos\alpha - y^L_{M_i}\sin\alpha \\ y + x^L_{M_i}\sin\alpha + y^L_{M_i}\cos\alpha \\ z + z^L_{M_i} \end{bmatrix}. \tag{5.50}$$

The actuation condition of the ith kinematic chain will be:

$$\|M_i B_i\|^2 - L_i^2 = 0. \tag{5.51}$$

Expanding the above equation yields:

$$\left(x + x^L_{M_i}\cos\alpha - y^L_{M_i}\sin\alpha - x^A_{B_i}\right)^2 + \left(y + x^L_{M_i}\sin\alpha + y^L_{M_i}\cos\alpha - y^A_{B_i}\right)^2 + \left(z + z^L_{M_i} - z^A_{B_i}\right)^2 - L_i^2 = 0.$$

According to equation (5.26), we have:

$$\mathbf{g}(x,y,z,\alpha) = \begin{bmatrix} \left(x + x^L_{M_1}\cos\alpha - y^L_{M_1}\sin\alpha - x^A_{B_1}\right)^2 + \left(y + x^L_{M_1}\sin\alpha + y^L_{M_1}\cos\alpha - y^A_{B_1}\right)^2 \\ +\left(z + z^L_{M_1} - z^A_{B_1}\right)^2 - L_1^2 \\ \left(x + x^L_{M_2}\cos\alpha - y^L_{M_2}\sin\alpha - x^A_{B_2}\right)^2 + \left(y + x^L_{M_2}\sin\alpha + y^L_{M_2}\cos\alpha - y^A_{B_2}\right)^2 \\ +\left(z + z^L_{M_2} - z^A_{B_2}\right)^2 - L_2^2 \\ \left(x + x^L_{M_3}\cos\alpha - y^L_{M_3}\sin\alpha - x^A_{B_3}\right)^2 + \left(y + x^L_{M_3}\sin\alpha + y^L_{M_3}\cos\alpha - y^A_{B_3}\right)^2 \\ +\left(z + z^L_{M_3} - z^A_{B_3}\right)^2 - L_3^2 \\ \left(x + x^L_{M_4}\cos\alpha - y^L_{M_4}\sin\alpha - x^A_{B_4}\right)^2 + \left(y + x^L_{M_4}\sin\alpha + y^L_{M_4}\cos\alpha - y^A_{B_4}\right)^2 \\ +\left(z + z^L_{M_4} - z^A_{B_4}\right)^2 - L_4^2 \end{bmatrix}.$$

Given the values of the parameters $\left(x^L_{M_i}\ y^L_{M_i}\ z^L_{M_i}\right)$, $\left(x^A_{B_i}\ y^A_{B_i}\ z^A_{B_i}\right)$ and the actuations $l_i \in [l_{\min}, l_{\max}]$, we can obtain the workspace of the end effector with a similar procedure to that given in Section 5.3.1. However, the searching range of the angle α can be reduced to $\left[0, \frac{\pi}{2}\right)$ for the mechanism structure has a symmetry plane xoz. Accordingly, the workspace shapes are shown in Figures 5.21, 5.22, and 5.23.

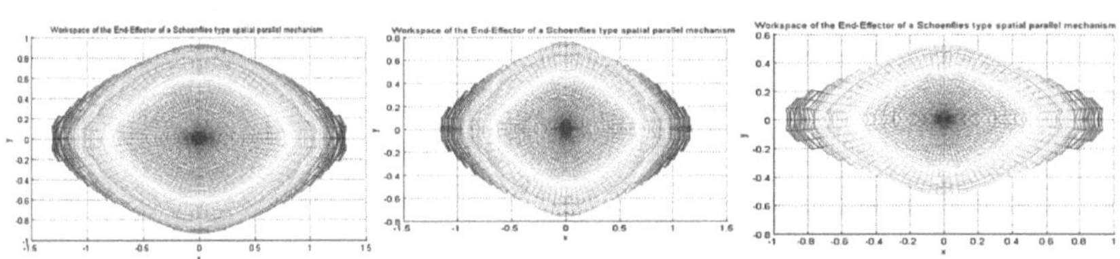

FIGURE 5.22 Top views of the workspaces corresponding to Figure 5.21.

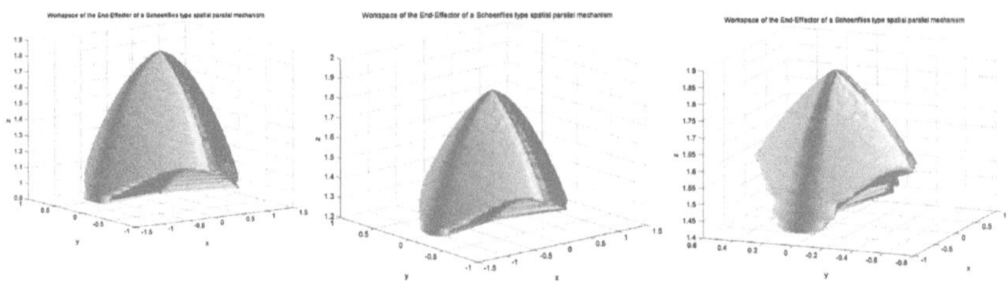

FIGURE 5.23 Workspace entities covered by three boundary surfaces of Figure 5.21.

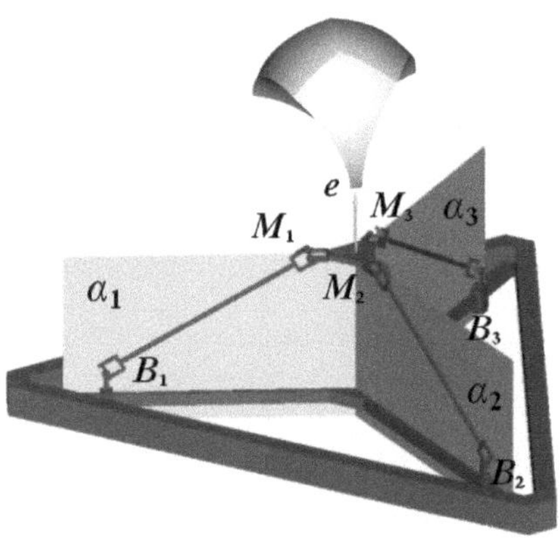

FIGURE 5.24 Symmetric workspace corresponding to the structure of the 3-PUU manipulator.

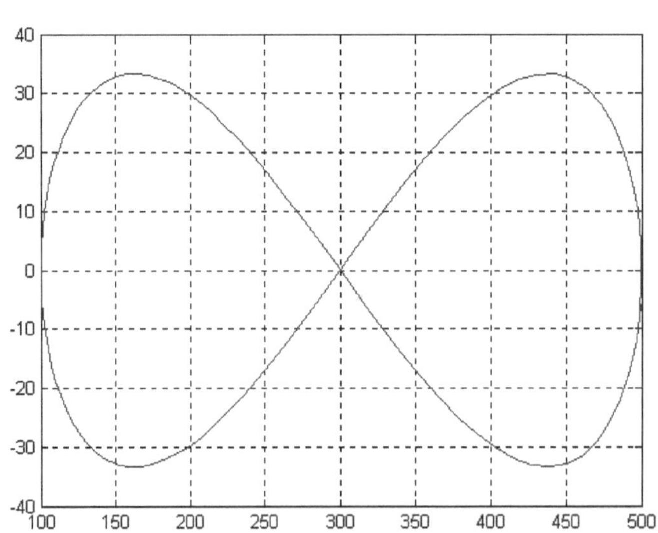

FIGURE 5.25 Workspace of the inverse parallel 4-bar manipulator.

Similarly, the reachable workspace of the manipulator shown in Figure 5.12 is illustrated in Figure 5.24.

The reachable workspace of the 4-bar manipulator shown in Figure 5.13 is a planar closed curve as Figure 5.25 shows, which is symmetric about the middle point of the fixed base *AD*.

5.4 CONCLUSIONS

This chapter addresses the determinations of the reachable and dexterous workspaces of spatial parallel mechanisms. The relationship constraint equations established from geometry and mobility characteristics of a mechanism are transformed into the constraints for the boundary curves or boundary surfaces of the reachable and dexterous workspaces of a reference point of the end effector. Then an analogous symmetric theorem of workspace for spatial parallel manipulators with symmetric identical kinematic chains is proposed. The symmetry group of the kinematic chain structure must be a subgroup of that of the workspace of the end effector. Through mathematic proof, it reveals the analogous relationship between the workspace shape and the structure of the manipulator. Examples demonstrate that this theorem can be utilized to estimate the geometry characteristics of the workspace and to guide the conceptual design of spatial parallel manipulators, especially for those with full DOFs. The theorem offers a concrete method for the workspace analysis of complicated spatial parallel manipulators. For a symmetric mechanism, the searching range can be reduced to $\frac{1}{n}$ of the initial one if the order of the symmetry group of the mechanism structure is n.

REFERENCES

[1] S. Dibakar, T.S. Mruthyunjaya, A centro-based characterization of singularities in the workspace of planar closed-loop manipulators, Mechanism and Machine Theory 33 (1998) 1091–1104.

[2] S. Dibakar, T.S. Mruthyunjaya, A computational geometry approach for determination of boundary of workspaces of planar manipulators with arbitrary topology, Mechanism and Machine Theory 34 (1999) 149–169.

[3] I.A. Bonev, J. Ryu, A geometrical method for computing the constant-orientation workspace of 6-PRRS parallel manipulators, Mechanism and Machine Theory 36 (2001) 1–13.

[4] S. Dibakar, T.S. Mruthyunjaya, Synthesis of workspaces of planar manipulators with arbitrary topology using shape representation ansimulated annealing, Mechanism and Machine Theory 34 (1999) 391–420.

[5] F. Gao, X.J. Liu, X. Chen, The relationships between the shapes of the workspaces and the link lengths of 3-DOF symmetrical planar parallel manipulators, Mechanism and Machine Theory 36 (2001) 205–220.

[6] J.J. Cervantes-Sánchez, J.G. Rendón-Sánchez, A simplified approach for obtaining the workspace of a class of 2-dof planar parallel manipulators, Mechanism and Machine Theory 34 (1999) 1057–1073.

[7] J.J. Cervantes-Sánchez, J.C. Hernández-Rodríguez, J.G. Rendón-Sánchez, On the workspace, assembly configurations and singularity curves of the RRRRR-type planar manipulator, Mechanism and Machine Theory 35 (2000) 1117–1139.

[8] M. Ceccarelli, A formulation for the workspace boundary of general n-revolute manipulators, Mechanism and Machine Theory 31 (1996) 637–646.
[9] M. Ceccarelli, C. Lanni, A multi-objective optimum design of general 3R manipulators for prescribed workspace limits, Mechanism and Machine Theory 39 (2004) 119–132.
[10] C.M. Gosselin, M. Guillot, The synthesis of manipulators with prescribed workspace, Journal of Mechanical Design 113 (1991) 451–455.
[11] Z. Wang, Z. Wang, W. Liu, Y. Lei, A study on workspace, boundary workspace analysis and work piece positioning for parallel machine tools, Mechanism and Machine Theory 36 (2001) 605–622.
[12] I.A. Bonev, J. Ryu, A new approach to orientation workspace analysis of 6-DOF parallel manipulators, Mechanism and Machine Theory 36 (2001) 15–28.
[13] J.P. Merlet, Determination of 6D workspaces of gough-type parallel manipulator and comparison between different geometries, International Journal of Robotics Research 18 (1999) 902–916.
[14] J.P. Merlet, Designing a parallel manipulator for a specific workspace, International Journal of Robotics Research 16 (1997) 545–556.
[15] A.K. Dash, I.M. Chen, et al., Workspace generation and planning singularity-free path for parallel manipulators, Mechanism and Machine Theory 40 (2005) 776–805.
[16] J.J. Rotman, Advanced Modern Algebra, Pearson Education Inc. Publishing as Prentice Hall, 2002.
[17] J.S. Zhao, M. Chen, et al., Workspace of mechanisms with symmetry identical kinematic chains, Mechanism and Machine Theory 41 (2006) 632–645.
[18] J.S. Zhao, Y.Z. Fu, et al., Mobility properties of a schoenflies type parallel mechanism, Robotics and Computer-Integrated Manufacturing 22 (2006) 124–133.
[19] J.S. Zhao, K. Zhou, et al., A new method to study the degree of freedom of spatial parallel mechanisms, The International Journal of Advanced Manufacturing Technology 23 (2004) 288–294.

Chapter 6

Singularity Analysis of the End Effector of a Mechanism within Its Workspace

With the extensive research of complex spatial manipulators, especially the parallel mechanism, we come to realize that the parallel mechanism has the advantages of higher structure stiffness, smaller movement quality, faster respose, and so on, but it also has an important disadvantage: kinematic singularity properties within its workspace, which seriously restrict its development. The kinematic singularity occurs where the degree of freedom (DOF) of the end effector changes, including the changes of number, orientation and type. In general, singularities of serial manipulators are characterized by the loss or gain of one or more DOFs and it has been shown that a serial manipulator will be uncontrollable at kinematic singularities where the velocity Jacobian loses rank [1]. Therefore, the kinematic singularity analysis has become a very important issue in the mechanisms filed, and many scholars have been studying the kinematic singularity using several different methods.

Burdick [2] introduced the useful motion of generic manipulator singularities and enumerated the generic set. Abdul-Majid Wazwaz [3] investigated the singularly perturbation problem by using the second-order differential equation with two singular boundary points. But this methodology strongly requires the power of the linear superposition for linear ODE (ordinary differential equation). Bhattacharya and Ghosh [4] proposed and compared two on-line singularity avoidance schemes. Mattson and Pennock [5] formulated the forward and inverse velocity problems using two Jacobian matrices associated with the independent inputs and the payload outputs, discussed the singular and configuration of the system, and pointed out that the singular configurations would occur when the system lost an instantaneous DOF. Karger [6] discussed the singular positions and self-motions of a special class of planar parallel manipulators.

To obtain the singular points, Cervantes-Sánchez et al. [7] have used the properties of the Jacobian matrix. The singularities [8] can be defined by algebraic approaches, by using the symbolic expression of the Jacobian matrix and its determinant, and are interpreted geometrically. Reciprocal screws can be

Advanced Theory of Constraint and Motion Analysis for Robot Mechanisms.
http://dx.doi.org/10.1016/B978-0-12-420162-0.00006-0

found for each of the n velocity-degeneracy conditions. Mattson and Pennock [5] formulated the forward and inverse velocity problems by using two Jacobian matrices associated with the independent inputs and the payload outputs, discussed the singular configuration of the system, and pointed out that the singular configurations would occur when the system lost an instantaneous DOF. However, it is also a kind of singularity when the parallel manipulator gains an instantaneous DOF.

In addition to the above methods, the screw theory is a very useful method to deal with the kinematic singularity. Phillips and Hunt [9] and Hunt [10] made important contemporary contributions, discoveries, and applications of theory of screws to the kinematics of mechanisms. Nokleby and Podhorodeski [11] proposed a method for determining the velocity-degenerate configurations of joint-redundant manipulators. They use the reciprocal screws to determine all degenerate (singular) configurations resulting in a single DOF motion loss. In their method, six joint screws must be chosen to form one six-joint sub-group Jacobian $\mathbf{J}_{\text{sub}}$. Setting the determinant of $\mathbf{J}_{\text{sub}}$ to zero allows all conditions that cause the six-joint sub-group to become velocity degenerate to be identified. Many scholars [8,11–13] have addressed the singularities of manipulators with screw theory. The physical meaning of the reciprocal screws of kinematic screws [14,15] is also available to analyze the singularities. The so-called physical meaning of the reciprocal screws of kinematic screws can be defined as the terminal constraints of each kinematic chain.

This chapter proposes a method to study the kinematic singularities of parallel manipulators according to the screw theory, which need not analyze the complex Jacobian matrices and discuss the type of singularities. This method could be widely used to solve the singularity problems of spatial parallel manipulators based on the analysis of their kinematic chains.

6.1 STATIC EQUILIBRIUM EQUATIONS OF THE PARALLEL MANIPULATORS

According to the theory of reciprocal screws, the reciprocal screws of the kinematic screws denote constraint forces, including couples and forces. As an equalized configuration, the platform can keep balance if applying the reciprocal screws no matter what kinds of external forces imposed on the platform unless the singular position occurs.

As a result, the singular position and orientation can be found by introducing a set of static equilibrium equations and probing the identical criteria of the equations, which will naturally convert the singularity problem into solving the rank of the coefficient matrix **A**.

Assuming that the mechanism consists of n branches and the DOF of the mechanism is F, the reciprocal screw set of the ith $(i = 1,2,\ldots,n)$ branch consists of m_i screws after exerting F actuators, which are denoted by $\$_{ij}$ $(i = 1,2,\ldots,n;\ j = 1,2,\ldots,m_i)$. With the static equilibrium equations, we can get the following equation set:

$$\begin{cases} \sum \boldsymbol{R}_x = 0; \quad \sum_{i=1}^{n}\sum_{j=1}^{m_i} \boldsymbol{\$}_{ij}^{Fx} + N_x = 0 \\ \sum \boldsymbol{R}_y = 0; \quad \sum_{i=1}^{n}\sum_{j=1}^{m_i} \boldsymbol{\$}_{ij}^{Fy} + N_y = 0 \\ \sum \boldsymbol{R}_z = 0; \quad \sum_{i=1}^{n}\sum_{j=1}^{m_i} \boldsymbol{\$}_{ij}^{Fz} + N_z = 0 \\ \sum \boldsymbol{M}_x = 0; \quad \sum_{i=1}^{n}\sum_{j=1}^{m_i} \boldsymbol{\$}_{ij}^{Mx} + T_x = 0 \\ \sum \boldsymbol{M}_y = 0; \quad \sum_{i=1}^{n}\sum_{j=1}^{m_i} \boldsymbol{\$}_{ij}^{My} + T_y = 0 \\ \sum \boldsymbol{M}_z = 0; \quad \sum_{i=1}^{n}\sum_{j=1}^{m_i} \boldsymbol{\$}_{ij}^{Mz} + T_z = 0 \end{cases} \tag{6.1}$$

where $\sum \boldsymbol{R}_x$ denotes the resultant of forces of the system along the x-axis. $\boldsymbol{\$}_{ij}^{Fx}$ denotes the component reciprocal screw of the jth reciprocal screw in i^{th} branch mechanism along the x-axis. N_x denotes the external forces along the x-axis. $\boldsymbol{\$}_{ij}^{Mx}$ denotes the component reciprocal screw moment of the jth reciprocal screw in i^{th} branch mechanism around x-axis. T_x denotes the external torque around the x-axis. The meanings of the other symbols are similar to the above.

Equation (6.1) can be converted into the following form:

$$\mathbf{A}\boldsymbol{R}_{inv} = \boldsymbol{N} \tag{6.2}$$

where $\boldsymbol{R}_{inv} = \begin{bmatrix} \boldsymbol{\$}_{1j}^{r} & \cdots & \boldsymbol{\$}_{1m_1}^{r} & \cdots & \boldsymbol{\$}_{n1}^{r} & \cdots & \boldsymbol{\$}_{nm_n}^{r} \end{bmatrix}^T$ denotes the matrix of reciprocal screws of the system. $\boldsymbol{N} = \begin{bmatrix} F_x & F_y & F_z & T_x & T_y & T_z \end{bmatrix}^T$ denotes the external forces or the deforms of the external forces.

If the whole mechanism is controllable and stable under any kinds of the external forces, the rank of the coefficient matrix **A** should be equal to the dimension of **N**. That is,

$$\text{Rank}(\mathbf{A}) = \dim(\boldsymbol{N}). \tag{6.3}$$

Considering the arbitrary of $\boldsymbol{N}$, the dimension of $\boldsymbol{N}$ should equal the number of the DOF F.

$$\dim(\boldsymbol{N}) = F \tag{6.4}$$

For a general spatial mechanism, N need not be a six-line vector. The number of the lines of vector N can be one through six based on the analysis of the reciprocal screws of all the branches. If equation (6.3) holds, the whole mechanism will not be singular, or else, the singularity must be occurred. So, the singularity problem is converted into searching the criteria that equation (6.3) does not hold. Therefore, we can adopt all kinds of linear algebra methods to search the criteria that equation (6.3) does not hold.

For most spatial parallel mechanisms have the same branches, the solving of the reciprocal screws can be abbreviated to solve only one branch. The whole process of this methodology, in reality, is very simple. Moreover, the physical meaning of this method is very intuitive.

6.2 SYMMETRY OF SINGULARITY DISTRIBUTION WITHIN THE REACHABLE WORKSPACE

In Chapter 5, we studied the relationship between the reachable workspace and the symmetric structure of the mechanisms; this section will further study the symmetry similarities between the singularities and workspace of a mechanism and the mechanism structure. Firstly, a definition of symmetry group of a geometry element should be introduced.

Theorem 6.1. *Suppose that the symmetry group of the singularities of the reference point of an end effector is represented by G_S and the symmetry group of the end effector and its kinematic chain structure is represented by G_M, the relationship $G_M \subseteq G_S$ also holds. Further, if the symmetry group of the reachable workspace of the reference point of the end effector is represented by G_W, the relationship $G_W \subseteq G_S$ must hold too.*

Prior to proving the above theorem, the natural coordinate method should be briefly introduced to facilitate the proof. Assume

$$\boldsymbol{x} = [x_1 \quad y_1 \quad z_1 \quad \cdots \quad x_n \quad y_n \quad z_n \quad \boldsymbol{u}_{1x} \quad \boldsymbol{u}_{1y} \quad \boldsymbol{u}_{1z} \quad \cdots \quad \boldsymbol{u}_{mx} \quad \boldsymbol{u}_{my} \quad \boldsymbol{u}_{mz}]^T \tag{6.5}$$

where x_i, y_i, z_i $(i = 1,2,\ldots,n)$ denote the Cartesian coordinates of the selected points, $\boldsymbol{u}_{jx}, \boldsymbol{u}_{jy}, \boldsymbol{u}_{jz} (j = 1,2,\ldots,m)$ denote the row Cartesian coordinates of the selected directions, n denotes the number of the selected points, and m denotes the number of the selected directions.

The relationship constraint equations can be denoted as

$$\boldsymbol{P}(\boldsymbol{x}) = 0. \tag{6.6}$$

The Jacobian matrix of equation (6.6) is immediately gained

$$\boldsymbol{P}'(\boldsymbol{x}) = \frac{\partial \boldsymbol{P}(\boldsymbol{x})}{\partial \boldsymbol{x}}. \tag{6.7}$$

For a determinate mechanical system, the number of independent constraint equations and the number of independent variables are always identical. Therefore, the Jacobian matrix will always be square and, as a result, the singularity criterion of the mechanism is

$$\left|\boldsymbol{P}'(\boldsymbol{x})\right| = 0 \tag{6.8}$$

where $|\cdot|$ represents the determinant of a matrix "$\cdot$".

To simplify the proof, the following lemma is proposed:

The Cartesian coordinates, velocities, and accelerations of any other point in the end effector plane can be expressed as the linear combinations of those of the prescribed non-collinear three points in the plane.

Assume the absolute coordinate of the reference point C is $\boldsymbol{x}_C = [x \;\; y \;\; z]^T$. The coordinate of point C can be expressed as the linear combinations of $\boldsymbol{x}$.

$$\boldsymbol{x}_C = h(\boldsymbol{x})$$

where $h(\cdot)$ denotes the linear combinations which is dependent on the coordinates of the point and $\boldsymbol{x}$. Therefore, the velocity and acceleration of point C can be expressed as

$$\begin{cases} \dot{\boldsymbol{x}}_C = h\left(\dot{\boldsymbol{x}}\right) \\ \ddot{\boldsymbol{x}}_C = h\left(\ddot{\boldsymbol{x}}\right) \end{cases}$$

where $\dot{\boldsymbol{x}}_C$ and $\ddot{\boldsymbol{x}}_C$ represent the velocity and acceleration, respectively.

Proof. Mathematically, the workspace of the reference point of the end effector of a mechanism, denoted by $\mathbf{W}_R$, is a mapping from the natural coordinate spaces $\mathbf{R}^{3(n+m)}$ into the Cartesian coordinate spaces $\mathbf{R}^3$.

$$F : \mathbb{R}^{3(n+m)} \rightarrow \mathbf{W}_R$$

The workspace $\mathbf{W}_R$ can therefore be represented by

$$\mathbf{W}_R = \{\boldsymbol{r} = h(\boldsymbol{x}) | \boldsymbol{P}(\boldsymbol{x}) = 0\} \tag{6.9}$$

While the singularities within $\mathbf{W}_R$ can be represented by

$$\mathbf{S} = \left\{ \boldsymbol{r} = h(\boldsymbol{x}) \left| \begin{array}{l} \boldsymbol{P}(\boldsymbol{x}) = 0 \\ \left|\boldsymbol{P}'(\boldsymbol{x})\right| = 0 \end{array} \right. \right\} \tag{6.10}$$

where $\mathbf{S}$ indicates the set of singularities within the workspace $\mathbf{W}_R$,$\boldsymbol{P}'(\boldsymbol{x})$ is presented by equation (6.7), and $\boldsymbol{P}(\cdot)$ is the relationship constraint equations of the mechanism, which can be alternatively denoted by

$$\boldsymbol{P}(\boldsymbol{x}) = \begin{bmatrix} \boldsymbol{G}(\boldsymbol{x}) \\ \boldsymbol{R}(\boldsymbol{x}) \\ \boldsymbol{A}(\boldsymbol{x}) \end{bmatrix}$$

where $\boldsymbol{G}(\boldsymbol{x})$,$\boldsymbol{R}(\boldsymbol{x})$, and $\boldsymbol{A}(\boldsymbol{x})$ represent the geometry constraint equations, rigid body constraint equations, and the actuation constraint equations, respectively.

Suppose that the symmetry operations in a particular configuration form a group G_M. Next, the actions of the symmetry group G_M on the workspace $\mathbf{W}_R$ defined by equation (6.9) will be investigated.

Because the structural symmetries of the geometry constraint equations $\boldsymbol{G}(\boldsymbol{x})$, the rigid body constraint equations $\boldsymbol{R}(\boldsymbol{x})$, and the actuation constraint equations $\boldsymbol{A}(\boldsymbol{x})$ are all commanded by the group G_M, for any element $g_M \in G_M$, the action of g_M on $\boldsymbol{P}(\boldsymbol{x})$ only results in a new permutation of its constraint equations; the action of g_M on $\boldsymbol{x}$ also only leads to a new permutation of its elements. Therefore, the action of g_M on $h(\boldsymbol{x})$ will make $h(\boldsymbol{x})$ unaltered, too.

Consequently, the sole difference between $g_M(\boldsymbol{P}(\boldsymbol{x}))$ and $\boldsymbol{P}(\boldsymbol{x})$ is the sequence of their constraint equations, as a result, $g_M(\boldsymbol{P}(\boldsymbol{x})) = \boldsymbol{P}(\boldsymbol{x}) = 0$; the differences between $g_M(\boldsymbol{P}'(\boldsymbol{x}))$ and $\boldsymbol{P}'(\boldsymbol{x})$ are that the rows have a different permutation, and therefore, there is an identical determinate equation

$$\left|\boldsymbol{P}'(g_M(\boldsymbol{x}))\right| = (-1)^{\tau(F(g_M))}\left|\boldsymbol{P}'(\boldsymbol{x})\right| = 0$$

where $\tau\left(\boldsymbol{P}\left(g_M\right)\right)$ represents the total permutations of $\boldsymbol{P}'\left(g_M(\boldsymbol{x})\right)$ with respect to $\boldsymbol{P}'(\boldsymbol{x})$. So, one immediately obtains

$$g_M\left(\mathbf{W}_R\right) = \left\{\boldsymbol{r} = g_M\left(h(\boldsymbol{x})\right)\left|g_M\left(\boldsymbol{P}(\boldsymbol{x})\right) = 0\right.\right\} = \left\{\boldsymbol{r} = h(\boldsymbol{x})\left|\boldsymbol{P}(\boldsymbol{x}) = 0\right.\right\} = \mathbf{W}_R$$

and

$$\begin{aligned} g_M(\boldsymbol{S}) &= \left\{\boldsymbol{r} = g_M\left(h(\boldsymbol{x})\right)\left|\begin{array}{l} g_M\left(\boldsymbol{P}(\boldsymbol{x})\right) = 0 \\ \left|g_M\left(\boldsymbol{P}'(\boldsymbol{x})\right)\right| = 0 \end{array}\right.\right\} \\ &= \left\{\boldsymbol{r} = h(\boldsymbol{x})\left|\begin{array}{l} \boldsymbol{P}(\boldsymbol{x}) = 0 \\ \left|\boldsymbol{P}'(\boldsymbol{x})\right| = 0 \end{array}\right.\right\} = \boldsymbol{S} \end{aligned}$$

where $\boldsymbol{P}'\left(g_M(\boldsymbol{x})\right) = \begin{bmatrix} \boldsymbol{G}'\left(g_M(\boldsymbol{x})\right) \\ \boldsymbol{R}'\left(g_M(\boldsymbol{x})\right) \\ \boldsymbol{A}'\left(g_M(\boldsymbol{x})\right) \end{bmatrix}$.

Therefore, $g_M \in G_W$ and $g_M \in G_S$. So, one obtains

$$G_M \subseteq G_W \text{ and } G_M \subseteq G_S.$$

Theorem 6.2. *Suppose that the symmetry group of the singularities of the reference point of an end effector is represented by G_S and the symmetry group of the end effector and its kinematic chain structure is represented by G_M, the symmetry group of the reachable workspace of the reference point of the end effector is represented by G_W, the relationship $G_M \subseteq G_W \subseteq G_S$ must hold.*

Proof. For $\forall g_W \in G_W$, there exists

$$g_W\left(\mathbf{W}_R\right) = \left\{\boldsymbol{r} = g_W\left(h(\boldsymbol{x})\right)\left|g_W\left(\boldsymbol{P}(\boldsymbol{x})\right) = 0\right.\right\} = \left\{\boldsymbol{r} = h(\boldsymbol{x})\left|\boldsymbol{P}(\boldsymbol{x}) = 0\right.\right\} = \mathbf{W}_R \tag{6.11}$$

and

$$\left|\boldsymbol{P}'\left(g_W(\boldsymbol{x})\right)\right| = (-1)^{\tau\left(F\left(g_W\right)\right)}\left|\boldsymbol{P}'(\boldsymbol{x})\right| = 0. \tag{6.12}$$

Equations (6.11) and (6.12) indicate that $g_W \in G_S$. As a result, one obtains

$$G_W \subseteq G_S.$$

Therefore,

$$G_M \subseteq G_W \subseteq G_S.$$

By the way, it is not difficulty to prove that if the order of the symmetry group of the mechanism structure is n, then the searching range of polar angle can be reduced from 2π to $\frac{2\pi}{n}$.

6.3 APPLICATIONS AND DISCUSSION

6.3.1 The Singularity Distribution of a Planar 3-RRR Parallel Mechanism within Its Reachable Workspace

A planar parallel mechanism shown in Figure 6.1 is characterized with that its end effector, $C_1C_2C_3$, is an equilateral triangle whose three vertexes are connected with the vertexes of the fixed equilateral triangle, $A_1A_2A_3$, through three identical RRR kinematic chains, $A_iB_iC_i (i = 1,2,3)$, individually. Suppose that the end effector, $C_1C_2C_3$, is represented by its geometry center, C. The singularity distribution within its workspace will be investigated in what follows.

Obviously, the symmetries of the planar mechanism shown in Figure 6.1 can be interpreted by a symmetry group S_3, the elements of which are the

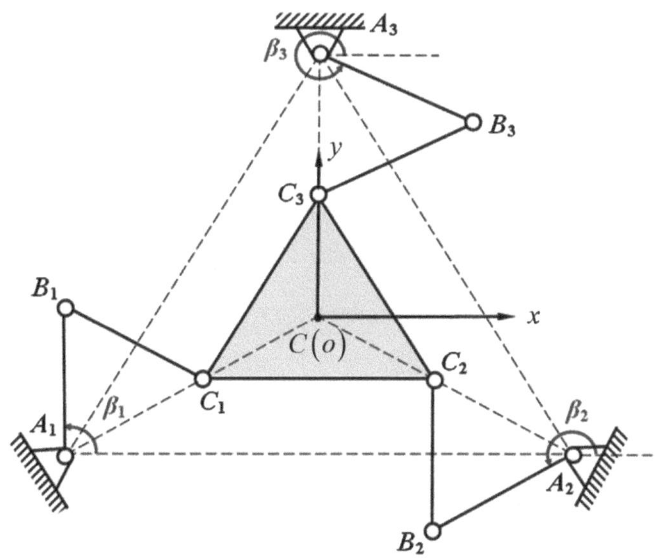

FIGURE 6.1 A planar 3-RRR parallel mechanism.

permutations of a set $V = \{A_1, A_2, A_3\}$. Hence, $S_3 = \{\phi_1, \phi_2, \phi_3, \phi_4, \phi_5, \phi_6\}$, where $\phi_1 = (1), \phi_2 = (123), \phi_3 = (132), \phi_4 = (23), \phi_5 = (13), \phi_6 = (12)$. S_3 describes all symmetries of the mechanism structure shown in Figure 6.1. From the viewpoint of geometry symmetry, the first three elements of S_3 represent the three rotations $0, \frac{2\pi}{3}$, and $\frac{4\pi}{3}$ about the z-axis which is perpendicular to the $A_1A_2A_3$-plane; and the last three denote three symmetries about CA_1, CA_2, and CA_3, respectively. Therefore, with the theorem proposed in this paper, we can reduce the searching range of the polar angle to $\frac{1}{6}$ of that of the initial. However, the whole region is particularly searched to verify the validation of the theorem.

Suppose the coordinates of the reference point of the end effector, C, in the fixed coordinate system are $\boldsymbol{r} = [x \quad y \quad 0]^T$, the rotational angle about the z-axis with respect to the x-axis is denoted by θ, and the radius of circumcircle of the triangle, $C_1C_2C_3$, is R_1; then, the absolute coordinates of the three vertexes of the triangle end effector, $C_1C_2C_3$, can be obtained, individually

$$\boldsymbol{r}_{C_i} = \boldsymbol{r} + \mathbf{R}\boldsymbol{r}_{CC_i} \tag{6.13}$$

where $i = 1,2,3$, $\mathbf{R} = \begin{bmatrix} \cos\theta & -\sin\theta & 0 \\ \sin\theta & \cos\theta & 0 \\ 0 & 0 & 1 \end{bmatrix}$, $\boldsymbol{r}_{CC_1} = \begin{bmatrix} -\frac{\sqrt{3}}{2}R_1 \\ -\frac{1}{2}R_1 \\ 0 \end{bmatrix}$, $\boldsymbol{r}_{CC_2} = \begin{bmatrix} \frac{\sqrt{3}}{2}R_1 \\ -\frac{1}{2}R_1 \\ 0 \end{bmatrix}$.

Suppose that the lengths of A_iB_i and B_iC_i are l_1 and l_2, respectively; the radius of circumcircle of the triangle $A_1A_2A_3$ is R_2; and the free rotational angle range about the z-axis is $\theta_z = [\theta_L, \theta_U]$. Also presume that $\boldsymbol{r}_{A_1} = \left[-\frac{\sqrt{3}}{2}R_2 \quad -\frac{1}{2}R_2 \quad 0\right]^T$, $\boldsymbol{r}_{A_2} = \left[\frac{\sqrt{3}}{2}R_2 \quad -\frac{1}{2}R_2 \quad 0\right]^T$, and $\boldsymbol{r}_{A_3} = [0 \quad R_2 \quad 0]^T$. The coordinates of points B_1, B_2, and B_3, can be expressed with the three input angular displacements, θ_1, θ_2 and θ_3.

$$\boldsymbol{r}_{B_i} = \boldsymbol{r}_{A_i} + l_1 \begin{bmatrix} \cos\theta_i \\ \sin\theta_i \\ 0 \end{bmatrix} \tag{6.14}$$

where $i = 1,2,3$.

The constraint equations for link $B_iC_i (i = 1,2,3)$ can be expressed by

$$\left\|\boldsymbol{r}_{C_i} - \boldsymbol{r}_{B_i}\right\|^2 - l_2^2 = 0. \tag{6.15}$$

Substituting equations (6.13) and (6.14) into equation (6.15), one obtains

$$\mathbf{g}(x,y,\theta)=\begin{bmatrix}\left(x-\frac{\sqrt{3}}{2}R_1\cos\theta+\frac{1}{2}R_1\sin\theta+\frac{\sqrt{3}}{2}R_2-l_1\cos\theta_1\right)^2\\ +\left(y-\frac{\sqrt{3}}{2}R_1\sin\theta-\frac{1}{2}R_1\cos\theta+\frac{1}{2}R_2-l_1\sin\theta_1\right)^2-l_2^2\\ \left(x+\frac{\sqrt{3}}{2}R_1\cos\theta+\frac{1}{2}R_1\sin\theta-\frac{\sqrt{3}}{2}R_2-l_1\cos\theta_2\right)^2\\ +\left(y+\frac{\sqrt{3}}{2}R_1\sin\theta-\frac{1}{2}R_1\cos\theta+\frac{1}{2}R_2-l_1\sin\theta_2\right)^2-l_2^2\\ \left(x-R_1\sin\theta-l_1\cos\theta_3\right)^2+\left(y+R_1\cos\theta-R_2-l_1\sin\theta_3\right)^2-l_2^2\end{bmatrix}. \tag{6.16}$$

According to equation (6.9), the reachable workspace of the specified point C is

$$\mathbf{W}_R=\left\{(x,y)\,|\mathbf{g}(x,y,\theta)=0\right\}.$$

The Jacobian matrix of the constraint equation (6.16) can be expressed as

$$\frac{\partial g(x,y,\theta)}{\partial(x,y,\theta)}=2\begin{bmatrix}x-\frac{\sqrt{3}}{2}R_1\cos\theta+\frac{1}{2}R_1\sin\theta+\frac{\sqrt{3}}{2}R_2-l_1\cos\theta_1 & y-\frac{\sqrt{3}}{2}R_1\sin\theta-\frac{1}{2}R_1\cos\theta+\frac{1}{2}R_2-l_1\sin\theta_1 & t_1\\ x+\frac{\sqrt{3}}{2}R_1\cos\theta+\frac{1}{2}R_1\sin\theta-\frac{\sqrt{3}}{2}R_2-l_1\cos\theta_2 & y+\frac{\sqrt{3}}{2}R_1\sin\theta-\frac{1}{2}R_1\cos\theta+\frac{1}{2}R_2-l_1\sin\theta_2 & t_2\\ x-R_1\sin\theta-l_1\cos\theta_3 & y+R_1\cos\theta-R_2-l_1\sin\theta_3 & t_3\end{bmatrix}$$

where

$$\begin{aligned}t_1&=R_1x\sin\left(\theta+\frac{\pi}{6}\right)+R_1y\sin\left(\theta-\frac{\pi}{3}\right)\\&\quad+l_1R_1\sin\left(\theta_1-\theta-\frac{\pi}{6}\right)+R_1R_2\sin\theta,\\ t_2&=-R_1x\sin\left(\theta-\frac{\pi}{6}\right)+R_1y\sin\left(\theta+\frac{\pi}{3}\right)\\&\quad+l_1R_1\sin\left(\theta-\theta_2-\frac{\pi}{6}\right)+R_1R_2\sin\theta,\\ t_3&=-R_1x\cos\theta-R_1y\sin\theta+l_1R_1\cos\left(\theta-\theta_3\right)+R_1R_2\sin\theta.\end{aligned}$$

According to equation (6.10), the singularities within the workspace can be obtained

$$\mathbf{W}=\left\{(x,y)\left|\begin{matrix}\boldsymbol{P}(x,y,\theta)=0\\ \left|\frac{\partial\boldsymbol{P}(x,y,\theta)}{\partial(x,y,\theta)}\right|=0\end{matrix}\right.\right\}. \tag{6.17}$$

Next, the searching problem of singularities within the reachable workspace is investigated. Firstly, the reachable workspace should be obtained. The basic steps to fulfill such a task might be found in [16,17], etc.

Step 1: Suppose that [·] represents the largest natural number that is less than or equal to "·". Establish parametric equations of polar coordinates for a point within the workspace

$$\begin{cases} x = \rho \cos\left(i\Delta\alpha\right) \\ y = \rho \sin\left(i\Delta\alpha\right) \\ \theta = j\Delta\theta \end{cases} \tag{6.18}$$

where ρ is the searching polar radius with a polar angle $i\Delta\alpha$ from the x-axis, and $0 \leq i \leq \left[\frac{2\pi}{\Delta\alpha}\right]$, $\Delta\alpha$ is the angular interval for the polar angle, i is a natural number; $0 \leq j \leq \left[\frac{\theta_{\max}-\theta_{\min}}{\Delta\theta}\right]$, and $\Delta\theta$ is the angular interval from the minimum pose angle, $\theta_{\min}$, to the maximum pose angle, $\theta_{\max}$, and j is also a natural number.

Step 2: Suppose that at the pose of $\theta = \theta_{\min} + j\Delta\theta$, the *xoy*-plane is sliced evenly with k radial lines from the origin of the coordinate system, where $k = \left[\frac{2\pi}{\Delta\alpha}\right]$. And therefore, $0 \leq i \leq k$.

Step 3: To search the boundary points for the desired workspace and the reachable workspace in the *ith* radial line, we can let $\rho = \rho\left(i\Delta\alpha\right)$ increase from zero with a proper interval $\Delta\rho$, namely $\rho\left(i\Delta\alpha\right) = m\Delta\rho, m = 0,1,2,\ldots$

Step 4: For each point $\left(\rho\left(i\Delta\alpha\right), \Delta\alpha\right)$ in the radial line, let j increase evenly from zero to $\left[\frac{\theta_{\max}-\theta_{\min}}{\Delta\theta}\right]$, and correspondingly, the pose angle, θ, will also increase evenly from $\theta_{\min}$ to $\theta_{\max}$. If there is at least one θ_j such that when equation (6.18) is substituted into equation (6.16), there is at least one $\theta_i\,(i = 1,2,3)$ that has no real solution; the increase in the ith radial line should be terminated. Then, select the largest parameter $\rho\left(i\Delta\alpha\right)$ as the boundary point of the desired workspace in the ith radial line. It is not difficult to find that the searched point in the ith line will be the boundary point of the reachable workspace when one lets θ change from $-\pi$ to π. Of course, we can also first solve the maximum and the minimum values of the pose angle θ within the range $\left[-\pi,\pi\right]$ at the geometry center of the mechanism structure, and then let θ change from the minimum limit to the maximum limit. It is not difficult to prove that the absolute values of the maximum and minimum pose angle θ must be identical for a mechanism with symmetry structure.

Step 5: When i increases from 0 to k, the polar angle can be correspondingly increased from 0 to 2π, and the boundary curves of the desired workspace and the reachable workspace of the end effector are gained, respectively.

According to the theorem, the searching region of the polar angle can be reduced to $\frac{1}{6}$ of that of the initial one. In order to verify the validation of the theorem proposed in this paper, the whole workspace is searched by the computer. Some numerical simulations on the mechanism shown in Figure 6.1 with different parameters shown in Table 6.1 are executed.

TABLE 6.1 Structure Parameters for the Planar 3-RRR Parallel Mechanism shown in Figure 6.1

	Item			
Sequence	R_1 (mm)	R_2 (mm)	l_1 (mm)	l_2 (mm)
1	100	700	200	600
2	100	700	200	800
3	100	800	300	600

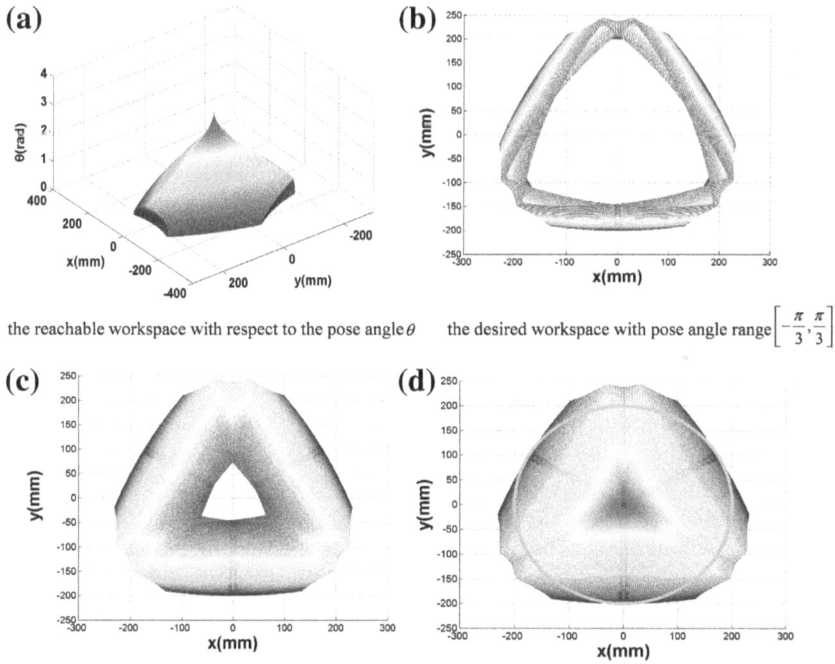

(a) the reachable workspace with respect to the pose angle θ (b) the desired workspace with pose angle range $\left[-\frac{\pi}{3}, \frac{\pi}{3}\right]$

(c) the desired workspace with pose angle range $\left[-\frac{2\pi}{3}, \frac{2\pi}{3}\right]$ (d) the continuous singular curve within the workspace

FIGURE 6.2 Desired and reachable workspaces of the mechanism with the first set of parameters.

For the first set of parameters, the reachable workspace of the center C of the end effector, $C_1C_2C_3$, is the workspace when the pose angle range is set to any real value, which is shown by Figure 6.2a. The different desired workspaces with respect to the pose angle θ are shown by the interior blank region of Figure 6.2b and the interior blank region of Figure 6.2c. Computer simulation indicates that the singularities within its reachable workspace only occur at $\theta = 0$ and the singularities form a continuous curve, which is shown Figure 6.2d.

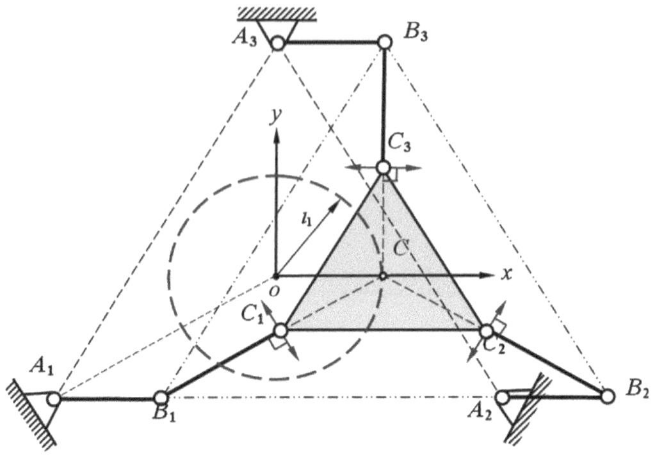

FIGURE 6.3 A singular configuration of the mechanism.

Next, we will investigate the physical meanings of this singular curve. This singular curve is denoted by the circle in Figure 6.2d, and the radius is just the length of link $A_i B_i$.

Firstly, the singularities occur when the pose angle is zero. Now, we can study the geometry characteristics of the mechanism with the first set of parameters in Table 6.1 (sequence 1). It is not difficult to find that the following relationship always holds:

$$R_2 - R_1 = l_2. \tag{6.19}$$

Under the condition that the pose angle $\theta = 0$, equation (6.19) indicates that the points B_1, B_2, and B_3 can form an equilateral triangle which is shown by the double-dashed triangle in Figure 6.3. If one does not consider the instantaneous rotation of the end effector, the double-dashed triangle, $B_1 B_2 B_3$, might be temporarily thought of as a rigid body. And as a result, the virtual four-bar linkages, $A_1 B_1 B_2 A_2$, $A_2 B_2 B_3 A_3$, and $A_1 B_1 B_3 A_3$ form three parallelogram four-bar mechanisms. From Figure 6.3, it is not difficult to find that the locus of the center, C, of the end effector, $C_1 C_2 C_3$, is of course a circle whose radius is equal to the length of l_1.

For any point on this circle locus, the end effector within the temporary rigid triangle $B_1 B_2 B_3$ (double-dashed triangle) can make an instantaneous rotation about the geometry center point, C. The instantaneous velocities of the three points, C_1,C_2, and C_3 with respect to the instantaneous velocity center, C, are shown by the three double arrows in Figure 6.3. Therefore, all the points on the dashed circle in Figure 6.3 are singularities, which demonstrate the correctness of the computer simulation.

For the second set of parameters shown in Table 6.1, the boundaries of the reachable workspace of the center, C, of the end effector, $C_1 C_2 C_3$, with respect to the pose angle, θ, is shown in Figure 6.4a. The reachable workspace,

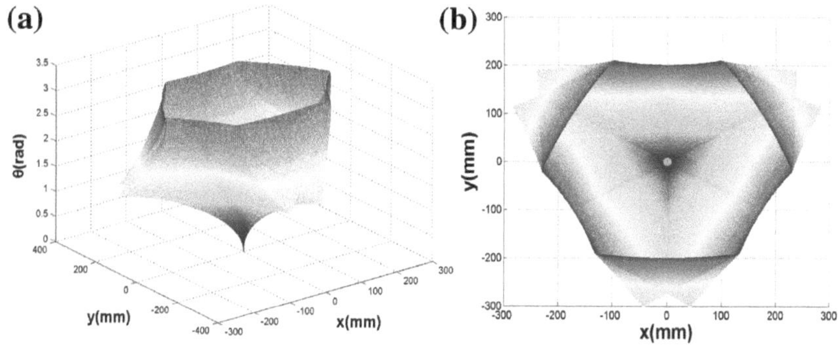

Boundary surfaces of the reachable workspace in $oxy\theta$ -space Singularity within the workspace

FIGURE 6.4 Reachable workspace and the singularity of the mechanism with the second set of parameters.

as a matter of fact, can be obtained by letting the pose angle θ be any real value between $-\pi$ and π, which is shown by the colored area of Figure 6.4b. Computer simulation indicates that singularity of the mechanism is just the origin of the xoy-system, which is on the geometry center of the desired and reachable workspaces.

Next, the physical meanings of this singular point will be also investigated. It is not difficult to find that the following relationship always holds for the second set of parameters (refer to Table 6.1, sequence 2)

$$R_2 - R_1 = l_2 - l_1. \tag{6.20}$$

Equation (6.20) indicates that the points $A_i, B_i, C_i (i = 1,2,3)$ can remain in a line, which is coincident with a midline of the equilateral triangle, $C_1C_2C_3$, that passes through the vertex, C_i. As is shown in Figure 6.5, at the configuration where $x = y = \theta = 0$, the end effector can make an instantaneous rotation about its center point, C. The instantaneous velocities of the three points, C_1, C_2, and C_3, with respect to C are shown by the three double arrows in Figure 6.5. As a matter of fact, such a configuration is a singularity of the mechanism, which is illustrated by Figure 6.4b.

For the third set of parameters (refer to Table 6.1, sequence 3), we can also make a numerical simulation. The increment in the radial direction is $\Delta\rho = 1$, angular increment for the polar angle is $\Delta\alpha = \frac{\pi}{144}$, and the angular increment for the pose angle is $\Delta\theta = \frac{\pi}{144}$. Computer simulation indicates that there is no singularity within the whole reachable workspace except when $\rho = 0, \theta = \pm \arccos \frac{13}{14}$. These two configurations are shown in Figure 6.6.

The reachable workspace in $oxy\theta$-space and in xoy-plane are shown in subfigures (a) and (b) in Figure 6.7a and b. The desired workspaces with respect to the pose angle, θ, is shown in Figure 6.7c and d. The interior blank regions

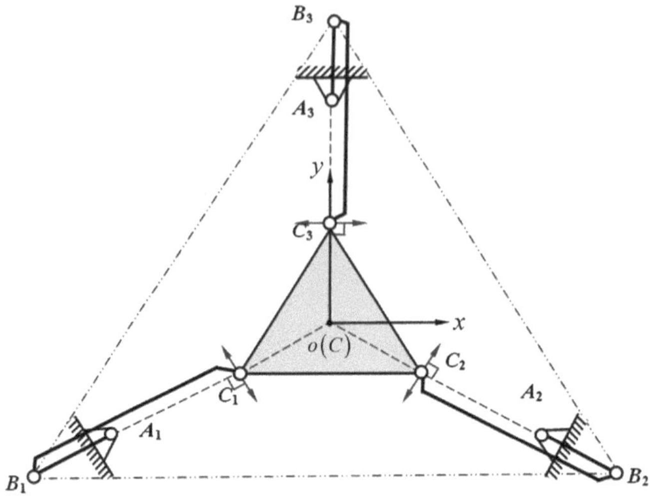

FIGURE 6.5 A singular configuration of the mechanism within the reachable workspace.

in Figure 6.7c and d represent the desired workspaces with pose angle ranges $\left[-\frac{\pi}{3}, \frac{\pi}{3}\right]$ and $\left[-\frac{2\pi}{3}, \frac{2\pi}{3}\right]$, individually.

From Figures 6.2, 6.4, and 6.7, we can draw a conclusion that the relationship of theorem 6.2 holds for all the mechanisms described by the parameters in Table 6.1; and this verifies the correctness of the theorem.

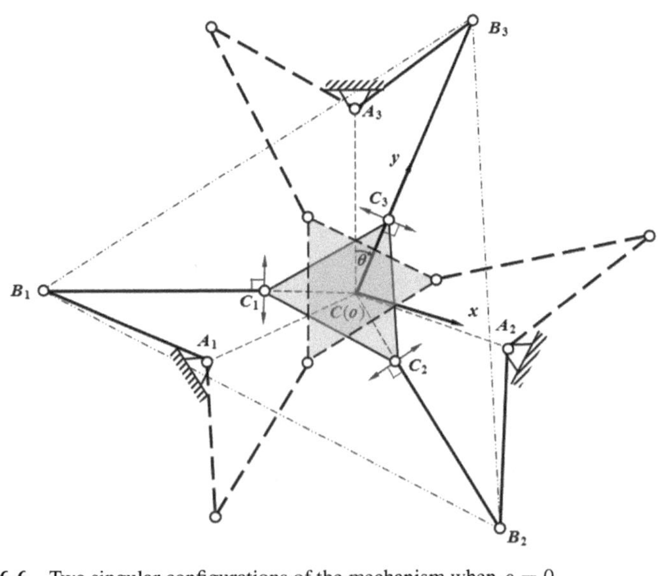

FIGURE 6.6 Two singular configurations of the mechanism when $\rho = 0$.

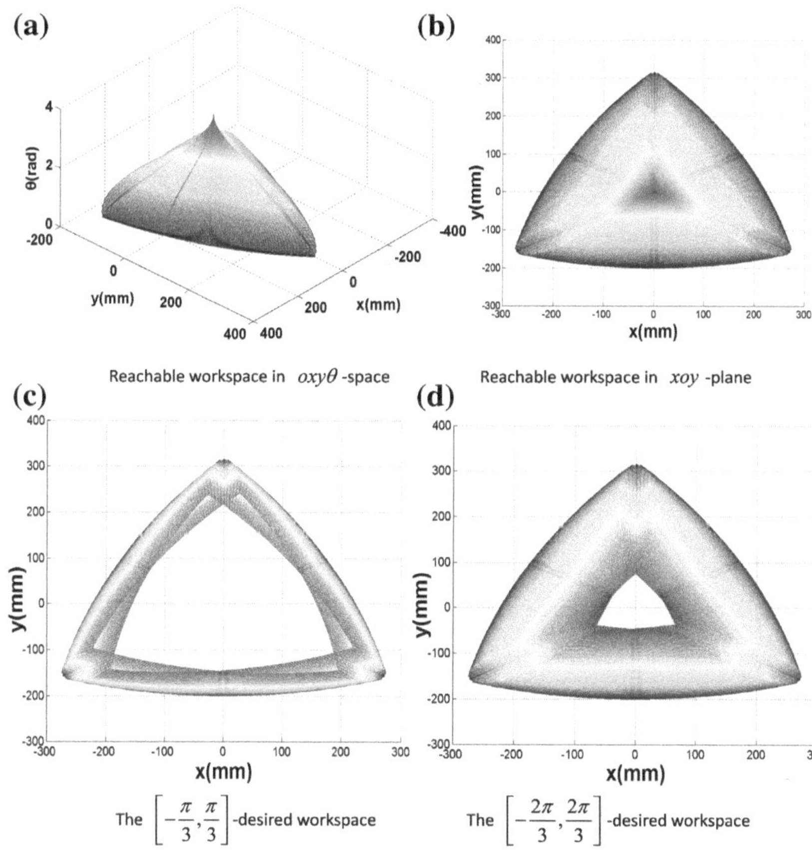

FIGURE 6.7 No singular reachable and desired workspaces.

6.3.2 Singularities Distribution of a Spatial 4-UPU Parallel Mechanism within Its Workspace

A Schoenflies-type parallel mechanism is shown in Figure 6.8 [18]. The mechanism proposed herein is actuated by four identical UPU (universal joint + prismatic joint + universal joint) kinematic chains.

The local coordinate system $o_L x_L y_L z_L$ and the absolute coordinate system $oxyz$ are created as Figure 6.8 shows. The z_L-axis is perpendicular to the plane of the end effector $M_1M_2M_3M_4$, the origin is superimposed with the geometric center of $M_1M_2M_3M_4$,x_L and y_L axes are parallel to the two orthogonal sides of the manipulator. For the absolute coordinate system, the z-axis is perpendicular to the base plane $B_1B_2B_3B_4$, and the origin is superimposed with the geometric center of $B_1B_2B_3B_4$,x-and y-axes are parallel to the two orthogonal sides of the rectangle base.

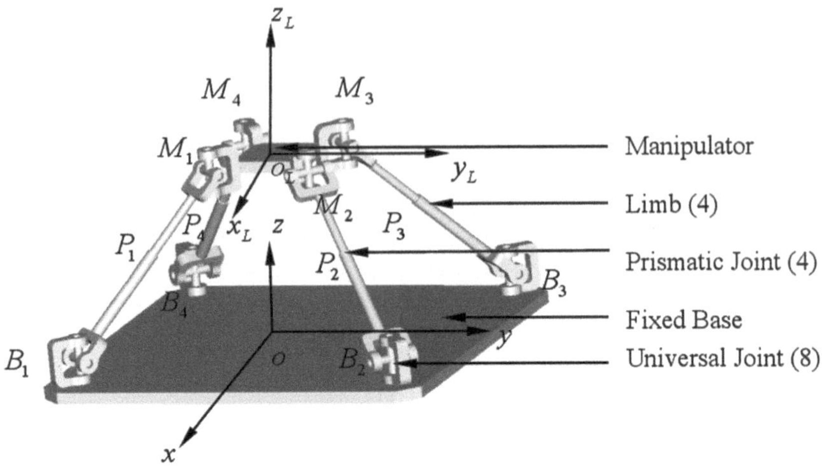

FIGURE 6.8 A kind of Schoenflies-type spatial parallel mechanism.

Firstly, the end effector has four DOFs (one rotational DOF and three orthogonal translational DOFs). Correspondingly, we can only select four independent variables, three of which are the absolute coordinates of the geometry center of the end effector, denoted by (x, y, z); and the other is the rotational angle about the z-axis, denoted by α. As a result, the transformation matrix from the local coordinate system to the absolute we can be expressed as:

$$\mathbf{R} = \begin{bmatrix} \cos\alpha & -\sin\alpha & 0 \\ \sin\alpha & \cos\alpha & 0 \\ 0 & 0 & 1 \end{bmatrix}. \tag{6.21}$$

Suppose the coordinates of the $i^{\text{th}}(i = 1,2,3,4)$ vertex of the end effector in the local coordinate system are denoted by $M_i\left(x_{M_i}^L, y_{M_i}^L, 0\right)$, the absolute coordinates of $M_i\left(x_{M_i}^A, y_{M_i}^A, z_{M_i}^A\right)$ can be obtained:

$$\begin{bmatrix} x_{M_i}^A \\ y_{M_i}^A \\ z_{M_i}^A \end{bmatrix} = \begin{bmatrix} x \\ y \\ z \end{bmatrix} + \mathbf{R}\begin{bmatrix} x_{M_i}^L \\ y_{M_i}^L \\ 0 \end{bmatrix} = \begin{bmatrix} x \\ y \\ z \end{bmatrix} + \begin{bmatrix} \cos\alpha & -\sin\alpha & 0 \\ \sin\alpha & \cos\alpha & 0 \\ 0 & 0 & 1 \end{bmatrix}\begin{bmatrix} x_{M_i}^L \\ y_{M_i}^L \\ 0 \end{bmatrix}$$
$$= \begin{bmatrix} x + x_{M_i}^L\cos\alpha - y_{M_i}^L\sin\alpha \\ y + x_{M_i}^L\sin\alpha + y_{M_i}^L\cos\alpha \\ z \end{bmatrix}. \tag{6.22}$$

The absolute coordinates of B_i can be assumed as $B_i\left(x_{B_i}^A, y_{B_i}^A, 0\right)$. The actuation condition of the ith kinematic chain will be

$$\|M_iB_i\|^2 - L_i^2 = 0. \tag{6.23}$$

Expanding the above equation yields:

$$\left(x + x_{M_i}^L \cos\alpha - y_{M_i}^L \sin\alpha - x_{B_i}^A\right)^2 + \left(y + x_{M_i}^L \sin\alpha + y_{M_i}^L \cos\alpha - y_{B_i}^A\right)^2 + z^2 - L_i^2 = 0. \tag{6.24}$$

So, the constraint equations for the workspace can be denoted by

$$\mathbf{P}(x,y,z,\alpha) = \begin{bmatrix} \left(x + x_{M_1}^L \cos\alpha - y_{M_1}^L \sin\alpha - x_{B_1}^A\right)^2 + \left(y + x_{M_1}^L \sin\alpha + y_{M_1}^L \cos\alpha - y_{B_1}^A\right)^2 + z^2 - L_1^2 \\ \left(x + x_{M_2}^L \cos\alpha - y_{M_2}^L \sin\alpha - x_{B_2}^A\right)^2 + \left(y + x_{M_2}^L \sin\alpha + y_{M_2}^L \cos\alpha - y_{B_2}^A\right)^2 + z^2 - L_2^2 \\ \left(x + x_{M_3}^L \cos\alpha - y_{M_3}^L \sin\alpha - x_{B_3}^A\right)^2 + \left(y + x_{M_3}^L \sin\alpha + y_{M_3}^L \cos\alpha - y_{B_3}^A\right)^2 + z^2 - L_3^2 \\ \left(x + x_{M_4}^L \cos\alpha - y_{M_4}^L \sin\alpha - x_{B_4}^A\right)^2 + \left(y + x_{M_4}^L \sin\alpha + y_{M_4}^L \cos\alpha - y_{B_4}^A\right)^2 + z^2 - L_4^2 \end{bmatrix}. \tag{6.25}$$

The Jacobian matrix is obtained from equation (6.25)

$$\frac{\partial \mathbf{P}(x,y,z,\alpha)}{\partial (x,y,z,\alpha)} = 2\begin{bmatrix} x + x_{M_1}^L \cos\alpha - y_{M_1}^L \sin\alpha - x_{B_1}^A & y + x_{M_1}^L \sin\alpha + y_{M_1}^L \cos\alpha - y_{B_1}^A & z & s_1 \\ x + x_{M_2}^L \cos\alpha - y_{M_2}^L \sin\alpha - x_{B_2}^A & y + x_{M_2}^L \sin\alpha + y_{M_2}^L \cos\alpha - y_{B_2}^A & z & s_2 \\ x + x_{M_3}^L \cos\alpha - y_{M_3}^L \sin\alpha - x_{B_3}^A & y + x_{M_3}^L \sin\alpha + y_{M_3}^L \cos\alpha - y_{B_3}^A & z & s_3 \\ x + x_{M_4}^L \cos\alpha - y_{M_4}^L \sin\alpha - x_{B_4}^A & y + x_{M_4}^L \sin\alpha + y_{M_4}^L \cos\alpha - y_{B_4}^A & z & s_4 \end{bmatrix} \tag{6.26}$$

where

$$s_1 = -\left(x + x_{M_1}^L \cos\alpha - y_{M_1}^L \sin\alpha - x_{B_1}^A\right)\left(x_{M_1}^L \sin\alpha + y_{M_1}^L \cos\alpha\right) + \left(y + x_{M_1}^L \sin\alpha + y_{M_1}^L \cos\alpha - y_{B_1}^A\right)\left(x_{M_1}^L \cos\alpha - y_{M_1}^L \sin\alpha\right)$$

$$s_2 = -\left(x + x_{M_2}^L \cos\alpha - y_{M_2}^L \sin\alpha - x_{B_2}^A\right)\left(x_{M_2}^L \sin\alpha + y_{M_2}^L \cos\alpha\right) + \left(y + x_{M_2}^L \sin\alpha + y_{M_2}^L \cos\alpha - y_{B_2}^A\right)\left(x_{M_2}^L \cos\alpha - y_{M_2}^L \sin\alpha\right)$$

$$s_3 = -\left(x + x_{M_3}^L \cos\alpha - y_{M_3}^L \sin\alpha - x_{B_3}^A\right)\left(x_{M_3}^L \sin\alpha + y_{M_3}^L \cos\alpha\right) + \left(y + x_{M_3}^L \sin\alpha + y_{M_3}^L \cos\alpha - y_{B_3}^A\right)\left(x_{M_3}^L \cos\alpha - y_{M_3}^L \sin\alpha\right)$$

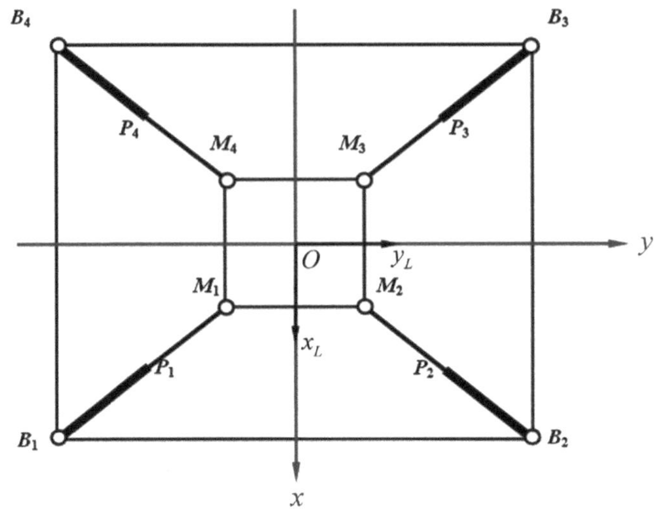

FIGURE 6.9 Top view of the Schoenflies-type spatial parallel mechanism.

$$s_4 = -\left(x + x_{M4}^L \cos\alpha - y_{M4}^L \sin\alpha - x_{B4}^A\right)\left(x_{M4}^L \sin\alpha + y_{M4}^L \cos\alpha\right) + \left(y + x_{M4}^L \sin\alpha + y_{M4}^L \cos\alpha - y_{B4}^A\right)\left(x_{M4}^L \cos\alpha - y_{M4}^L \sin\alpha\right)$$

Next, we can analyze the relationship between the distribution of singularities and the reachable workspace and the mechanism structure. The top view of the Schoenflies-type spatial parallel mechanism is shown in Figure 6.9. The elements of the symmetric group, S_4, are the permutations of a set $V = \{B_1, B_2, B_3, B_4\}$. The exact number of the permutations is $4! = 24$. However, not all of the 24 permutations can form real symmetries of the mechanism shown in Figure 6.9. The symmetry group of the kinematic chains of the end effector has now become $G = \{\phi_1, \phi_2, \ldots, \phi_8\}$, where $\phi_1 = (1), \phi_2 = (1234), \phi_3 = (2341), \phi_4 = (3412), \phi_5 = (12)(34), \phi_6 = (14)(23), \phi_7 = (13), \phi_8 = (24)$, and G is the dihedral group D_8. The first four elements of G represent the four rotations of angles $0, \frac{\pi}{2}, \frac{3\pi}{2}$ and π about the z-axis; ϕ_5 and ϕ_6 denote two symmetries about the xoz- and yoz-planes; and the ϕ_7 and ϕ_8 denote two symmetries about M_2oz and M_1oz, respectively. G describes all the symmetries of the mechanism structure shown in Figure 6.9.

However, when the end effector and the fixed base are not squares, the symmetry group of the kinematic chains of the end effector will be $G_1 = \{\phi_1, \phi_2, \phi_3, \phi_4\}$, where $\phi_1 = (1), \phi_2 = (13)(24), \phi_3 = (12)(34), \phi_4 = (14)(23)$ and, G_1 is a subgroup of the dihedral group D_8.

If one presumes that the length of M_1M_2 is $2a$ and the length of M_1M_4 is $2b$, the local coordinates of the four vertexes of the end effector can be obtained:

$$\begin{cases} r_{M_1}^L = \begin{bmatrix} b & -a & 0 \end{bmatrix}^T \\ r_{M_2}^L = \begin{bmatrix} b & a & 0 \end{bmatrix}^T \\ r_{M_3}^L = \begin{bmatrix} -b & a & 0 \end{bmatrix}^T \\ r_{M_4}^L = \begin{bmatrix} -b & -a & 0 \end{bmatrix}^T . \end{cases}$$

Similarly, if one presumes that the length of B_1B_2 is $2c$ and the length of B_1B_4 is $2d$, the absolute coordinates of the four vertexes of the fixed base can be obtained:

$$\begin{cases} r_{B_1} = \begin{bmatrix} d & -c & 0 \end{bmatrix}^T \\ r_{B_2} = \begin{bmatrix} d & c & 0 \end{bmatrix}^T \\ r_{B_3} = \begin{bmatrix} -d & c & 0 \end{bmatrix}^T \\ r_{B_4} = \begin{bmatrix} -d & -c & 0 \end{bmatrix}^T . \end{cases}$$

The Jacobian matrix (6.26) can be replaced by

$$\frac{\partial P(x,y,z,\alpha)}{\partial (x,y,z,\alpha)} = 2\begin{bmatrix} x+b\cos\alpha+a\sin\alpha-d & y+b\sin\alpha-a\cos\alpha+c & z & -(bx-ay-ac-bd)\sin\alpha+(ax+by+bc-ad)\cos\alpha \\ x+b\cos\alpha-a\sin\alpha-d & y+b\sin\alpha+a\cos\alpha-c & z & -(bx+ay-ac-bd)\sin\alpha-(ax-by+bc-ad)\cos\alpha \\ x-b\cos\alpha-a\sin\alpha+d & y-b\sin\alpha+a\cos\alpha-c & z & (bx-ay+ac+bd)\sin\alpha-(ax+by-bc+ad)\cos\alpha \\ x-b\cos\alpha+a\sin\alpha+d & y-b\sin\alpha-a\cos\alpha+c & z & (bx+ay+ac+bd)\sin\alpha+(ax-by-bc+ad)\cos\alpha \end{bmatrix}.$$

So, the singularity criteria will be

$$\left|\frac{\partial \mathbf{P}(x,y,z,\alpha)}{\partial (x,y,z,\alpha)}\right| = 0.$$

Therefore, we can obtain

$$\begin{vmatrix} x+b\cos\alpha+a\sin\alpha-d & y+b\sin\alpha-a\cos\alpha+c & z & -(bx-ay-ac-bd)\sin\alpha+(ax+by+bc-ad)\cos\alpha \\ x+b\cos\alpha-a\sin\alpha-d & y+b\sin\alpha+a\cos\alpha-c & z & -(bx+ay-ac-bd)\sin\alpha-(ax-by+bc-ad)\cos\alpha \\ x-b\cos\alpha-a\sin\alpha+d & y-b\sin\alpha+a\cos\alpha-c & z & (bx-ay+ac+bd)\sin\alpha-(ax+by-bc+ad)\cos\alpha \\ x-b\cos\alpha+a\sin\alpha+d & y-b\sin\alpha-a\cos\alpha+c & z & (bx+ay+ac+bd)\sin\alpha+(ax-by-bc+ad)\cos\alpha \end{vmatrix} = 0.$$

Expanding the above equation yields

$$z(bc-ad)\cos\alpha\left[ab+cd-(ad+bc)\cos\alpha\right] = 0. \tag{6.27}$$

The solutions of (6.27) are [18]:

(1) $z = 0$,
(2) $\frac{b}{a} = \frac{d}{c}$,
(3) $\alpha = \pm\frac{\pi}{2}$.

Below, we will discuss the three singularity cases individually:

(1) $z = 0$
When $z = 0$, the manipulator plane will be superimposed with that of the base, which is called *boundary point singularity* for the case when $z = 0$ is corresponding to the boundary vertex of the workspace and all the points whose z-coordinates are zeros within the reachable workspace are all singularities. However, it can be eliminated in reality.

(2) $\frac{a}{b} = \frac{c}{d}$
When $\frac{a}{b} = \frac{c}{d}$, the shape of the manipulator plane will be similar to that of the base, which is the so-called *similar singularity*. In this case, the rotational DOF around the z-axis of the manipulator is uncontrollable, which is shown in Figure 6.10.

(3) $\alpha = \pm\frac{\pi}{2}$
When $\alpha = \pm\frac{\pi}{2}$, the rotational DOF around z-axis of the manipulator will also be uncontrollable, which is shown in Figure 6.11. For this case, any point within the oriented workspace when $\alpha = \pm\frac{\pi}{2}$ is a singularity. In other words, the oriented workspaces when $\alpha = \pm\frac{\pi}{2}$ are continuous singular surfaces which are called $\frac{\pi}{2}$-*singular surfaces*. Substituting $\alpha = \frac{\pi}{2}$

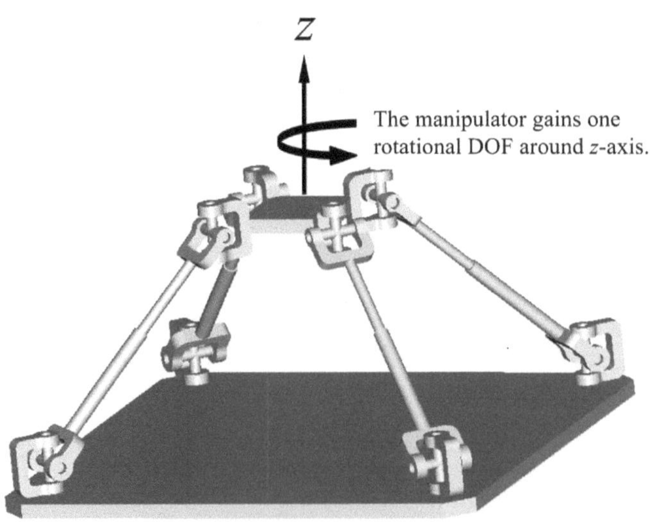

FIGURE 6.10 Similar singularity of the spatial parallel manipulator.

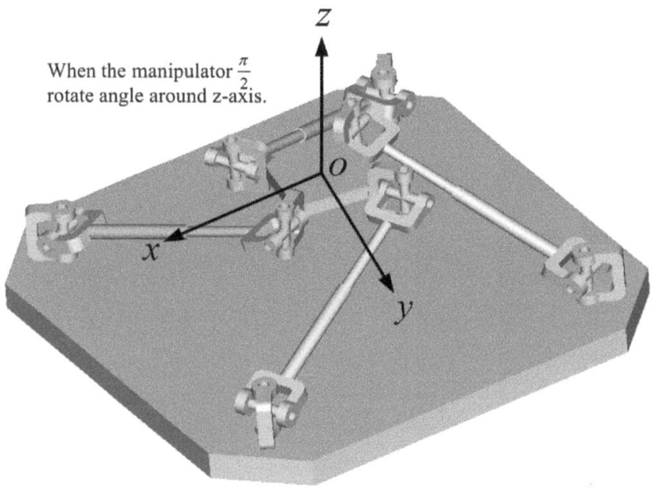

FIGURE 6.11 $\frac{\pi}{2}$-Singularity of the spatial parallel manipulator.

and $\alpha = -\frac{\pi}{2}$ into equation (6.25), individually, yields the constraint equations for the continuous singular surfaces. As an example, assume that $a = 300\sin\frac{\pi}{4}$, $b = 300\cos\frac{\pi}{4}$, $c = 1100\sin\frac{\pi}{3}$, $d = 1100\cos\frac{\pi}{3}$, and $1200 \le L_i \le 2000(i = 1,2,3,4)$. The continuous singular surfaces for the third solution are obtained by a computer simulation, and the relationship between the reachable workspace and the singular surfaces is shown in Figure 6.12. The isotropic view of the continuous singular surfaces are shown by Figure 6.12a, the top view of which is shown Figure 6.12b; the singular surfaces with respect to the reachable workspace of the end effector of the mechanism are shown by Figure 6.12c for the isotropic view and by Figure 6.12d for the top view; and the reachable workspace containing the $\frac{\pi}{2}$-continuous singular surfaces of the end effector is shown by Figure 6.12 e. It is not difficult to find that the three singular cases within the reachable workspace of the end effector shown in Figure 6.12 are all well interpreted by theorem 6.2 proposed in this chapter.

In applications, the *similar singularity* can be eliminated by adjusting the structure parameters of the manipulator to safeguard that $\frac{a}{b} \neq \frac{c}{d}$, but the $\frac{\pi}{2}$-*singularity* case cannot be avoided unless $\beta \neq \pm\frac{\pi}{2}$.

6.3.3 Singularities of a 3-UPU Parallel Manipulator

A spatial parallel manipulator shown in Figure 6.13 consists of three identical limbs, which are all made up of one prismatic joint (P) and two universal joints (U), and are all denoted by UPU kinematic chains. The manipulator has three DOFs at ordinary positions, which will be discussed below.

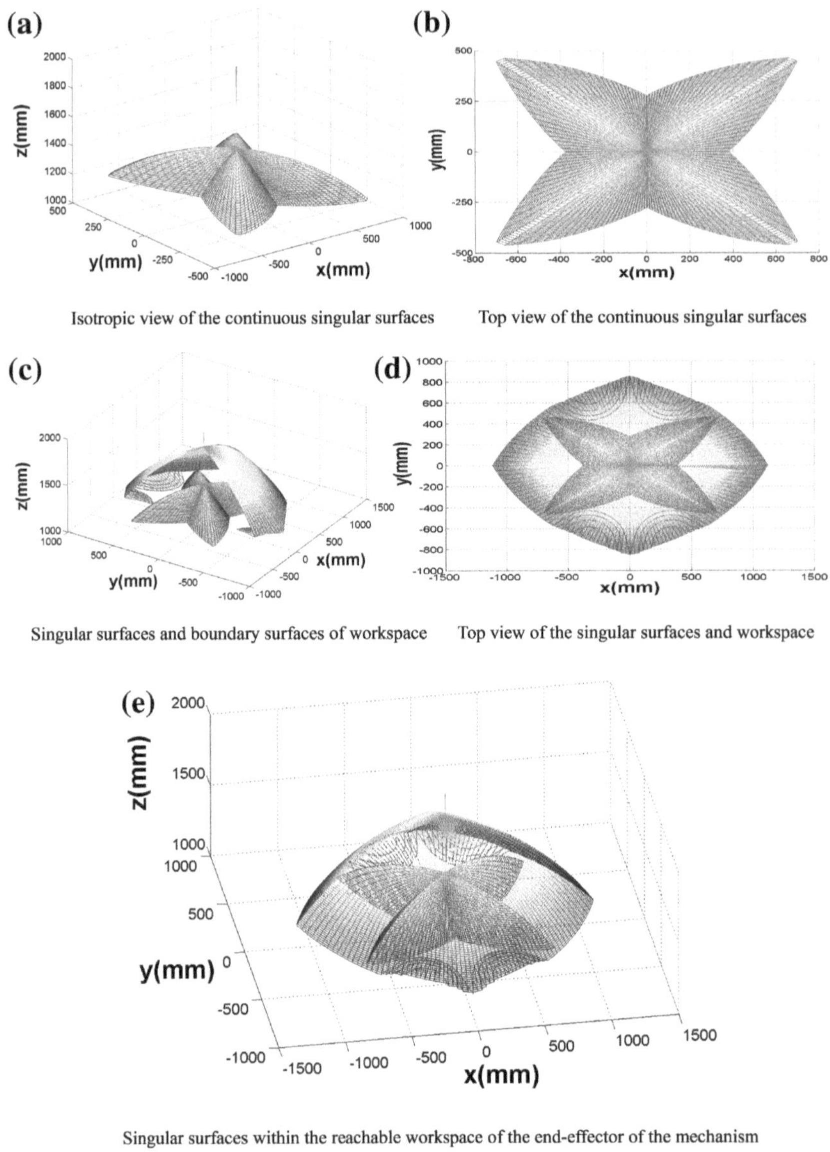

FIGURE 6.12 $\frac{\pi}{2}$-singular surfaces with respect to the reachable workspace.

In order to investigate the terminal constraints of the limbs, we should analyze the kinematic screws of every limb. Due to the symmetry of the manipulator, we can select only one of the limbs, say, limb $B_1P_1M_1$ shown in Figure 6.14, to study the kinematic characteristics. The origin of the local coordinate system $o_1x_1y_1z_1$ is superposed with the center of universal joint B_1,

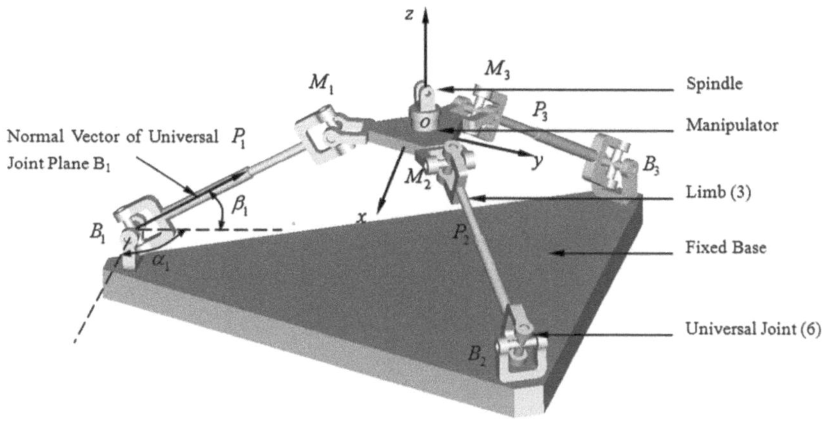

FIGURE 6.13 Architecture of a 3-PUU Parallel Manipulator.

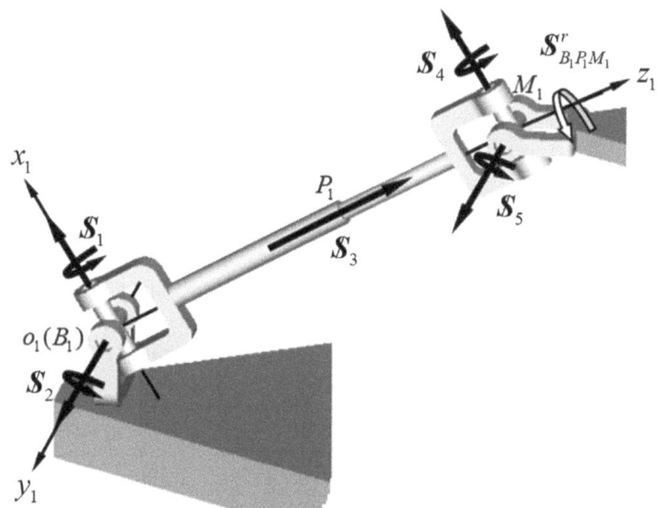

FIGURE 6.14 Screw Analysis of Limb $B_1P_1M_1$ in the Local Coordinate System.

the x_1-axis and the y_1-axis pass through the two orthogonal axes of universal joint B_1, respectively, and the z_1-axis is along the axis of the limb shown in the case of Figure 6.14. Assume the length of limb $B_1P_1M_1$ is l.

The universal joint B_1 can be decomposed as two orthogonal revolute joints around the x_1-axis and y_1-axis, respectively. The individual Plücker coordinates are:

$$\$_1 = \begin{bmatrix} 1 & 0 & 0 & 0 & 0 & 0 \end{bmatrix}^T; \quad \$_2 = \begin{bmatrix} 0 & 1 & 0 & 0 & 0 & 0 \end{bmatrix}^T.$$

The Plücker coordinates of slider P_1 are:

$$\$_3 = \begin{bmatrix} 0 & 0 & 0 & 0 & 0 & 1 \end{bmatrix}^T.$$

The coordinates of M_1 in the local coordinate system $o_1x_1y_1z_1$ are $M_1(0\ \ 0\ \ l)$.

The Plücker coordinates for the two revolute joints of M_1 can be denoted as:

$$\$_4 = \begin{bmatrix} \$_4^1 \\ \$_4^0 \end{bmatrix}, \$_5 = \begin{bmatrix} \$_5^1 \\ \$_5^0 \end{bmatrix}$$

where $\$_4^1 = [1\ \ 0\ \ 0]^T$, $\$_5^1 = [0\ \ 1\ \ 0]^T$, $\$_4^0 = r_{B_1} \times \$_4^1 = [0\ \ l\ \ 0]^T$, $\$_5^0 = r_{B_1} \times \$_5^1 = [-l\ \ 0\ \ 0]^T$.

Therefore, the kinematic screws of the limb $B_1P_1M_1$ can be expressed as:

$$\$_{B_1P_1M_1} = \begin{bmatrix} \$_1 & \$_2 & \$_3 & \$_4 & \$_5 \end{bmatrix}.$$

The reciprocal screw of the kinematic chain $B_1P_1M_1$ can be gained:

$$\$^r_{B_1P_1M_1} = \begin{bmatrix} 0 & 0 & 0 & 0 & 0 & 1 \end{bmatrix}^T.$$

According to the physical meaning of the reciprocal screws of kinematic screws, $\$^r_{B_1P_1M_1}$ denotes a pure moment of couple that is perpendicular to the universal joint plane, which is shown in Figure 6.14.

The movement constrained by $\$^r_{B_1P_1M_1}$ is a pure rotation around the z_1-axis in the local coordinate system $o_1x_1y_1z_1$. The same is true of the other two limbs of the manipulator. So, the reciprocal screws applied to the manipulator are three pure moments of couples, shown in Figure 6.15.

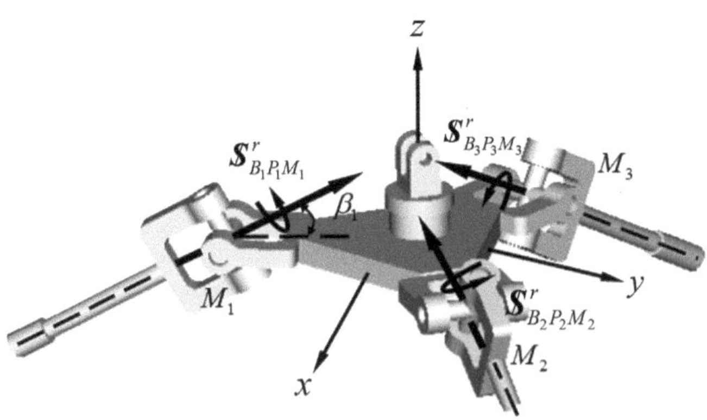

FIGURE 6.15 Terminal constraints of kinematic chains.

In the absolute coordinate system, if we presume the angle from the normal vector of the universal joint plane (see Figure 6.15) of the $i^{\text{th}}(i = 1,2,3)$ limb to the xoy plane is denoted by β_i and the angle from projection line of the normal vector to the x-axis is denoted by α_i, the reciprocal screws can be rewritten as follows:

$$\$^r_{B_i P_i M_i} = \begin{bmatrix} 0 & 0 & 0 & \cos\beta_i \cos\alpha_i & \cos\beta_i \sin\alpha_i & \sin\beta_i \end{bmatrix}^T \tag{6.28}$$

where $i = 1,2,3$.

Considering the geometric characteristics of the manipulator shown in Figure 6.13, we know $0 \leq \beta_i \leq \frac{\pi}{2}$ and can obtain:

$$\alpha_1 = \frac{\pi}{2}, \quad \alpha_2 = \frac{7\pi}{6}, \quad \alpha_3 = \frac{11\pi}{6}.$$

Therefore, equation (6.28) can be simplified:

$$\$^r = \begin{bmatrix} \$^r_{B_1 P_1 M_1} & \$^r_{B_2 P_2 M_2} & \$^r_{B_3 P_3 M_3} \end{bmatrix}. \tag{6.29}$$

where $\$^r_{B_1 P_1 M_1} = [0\ 0\ 0\ 0\ \cos\beta_1\ \sin\beta_1]^T$, $\$^r_{B_2 P_2 M_2} = [0\ 0\ 0\ -\frac{\sqrt{3}}{2}\cos\beta_2\ -\frac{1}{2}\cos\beta_2\ \sin\beta_2]^T$, $\$^r_{B_3 M_3} = [0\ 0\ 0\ \frac{\sqrt{3}}{2}\cos\beta_3\ -\frac{1}{2}\cos\beta_3\ \sin\beta_3]^T$.

At ordinary positions, the rank of $\$^r$ is:

$$\text{Rank}\left(\$^r\right) = 3.$$

Therefore, the DOF of the manipulator at ordinary positions is:

$$F = 6 - \text{Rank}(\$^r) = 6 - 3 = 3. \tag{6.30}$$

Because the singularity will occur when the DOF of the manipulator changes, the singularity criterion of the manipulator is $\text{Rank}\left(\$^r\right) < 3$. Therefore, if we presume

$$\mathbf{A} = \begin{bmatrix} 0 & \cos\beta_1 & \sin\beta_1 \\ -\frac{\sqrt{3}}{2}\cos\beta_2 & -\frac{1}{2}\cos\beta_2 & \sin\beta_2 \\ \frac{\sqrt{3}}{2}\cos\beta_3 & -\frac{1}{2}\cos\beta_3 & \sin\beta_3 \end{bmatrix}.$$

The singularity criterion of the manipulator is:

$$|\mathbf{A}| = 0.$$

That is,

$$\frac{\sqrt{3}}{2}[\cos\beta_1 \sin(\beta_2+\beta_3)+\sin\beta_1 \cos\beta_2 \cos\beta_3)]=0. \quad (6.31)$$

Equation (6.31) can be simplified as:

$$\cos\beta_1 \sin\beta_2 \cos\beta_3+\cos\beta_1 \cos\beta_2 \sin\beta_3+\sin\beta_1 \cos\beta_2 \cos\beta_3=0. \quad (6.32)$$

Since $0 \leq \beta_i \leq \frac{\pi}{2}$, $(i=1,2,3)$, the solutions of equation (6.32) can be discussed as follows.

When $\beta_i \neq \frac{\pi}{2}, (i=1,2,3)$, equation (6.32) can be transformed to:

$$\tan\beta_1+\tan\beta_2+\tan\beta_3=0 \quad (6.33)$$

Therefore, the solutions of equation (6.33) are $\beta_i=0$, $(i=1,2,3)$.

When $\beta_i=\frac{\pi}{2}$, $(i=1,2,3)$, equation (6.32) obviously holds.

From equation (6.32), we can find that the items containing β_i are circulative symmetrical. Therefore, we might as well assume $\beta_1=\frac{\pi}{2}$ to simplify the discussion.

If $\beta_1=\frac{\pi}{2}$, equation (6.32) can be simplified as:

$$\cos\beta_2 \cos\beta_3=0. \quad (6.34)$$

Obviously, the solutions of equation (6.34) are:

$$\beta_2=\frac{\pi}{2} \quad \text{or} \quad \beta_3=\frac{\pi}{2}.$$

Similarly, we can solve the solutions of equation (6.32) when $\beta_2=\frac{\pi}{2}$ or $\beta_3=\frac{\pi}{2}$. The physical meanings of the above three cases are given below:

When $\beta_i=0$, $(i=1,2,3)$, the manipulator and the three limbs are coplanar, which is shown in Figure 6.16, and is seldom used in real application. In this case, the manipulator gains one rotational DOF around z-axis, which is equivalent to a 6R-planar parallel manipulator.

When $\beta_i=\frac{\pi}{2}$, $(i=1,2,3)$, the three limbs shown in Figure 6.17 are parallel to each other and perpendicular to the manipulator synchronously. In this case, the manipulator gains two rotational DOFs around the x-axis and the y-axis, which is shown in Figure 6.17, and can be avoided by adjusting the structure parameters of the manipulator in applications.

When $\beta_i=\beta_j=\frac{\pi}{2}$, $(i,j=1,2,3, i\neq j)$ hold, the length of B_iB_j should be equal to M_iM_j, which is very particular—B_iM_i and B_jM_j shown in Figure 6.18 are parallel to each other and all perpendicular to the manipulator synchronously. In this case, the manipulator gains one rotational DOF around the x-axis as Figure 6.18 shows and can also be avoided in the concept design by adjusting the structure parameters of the manipulator.

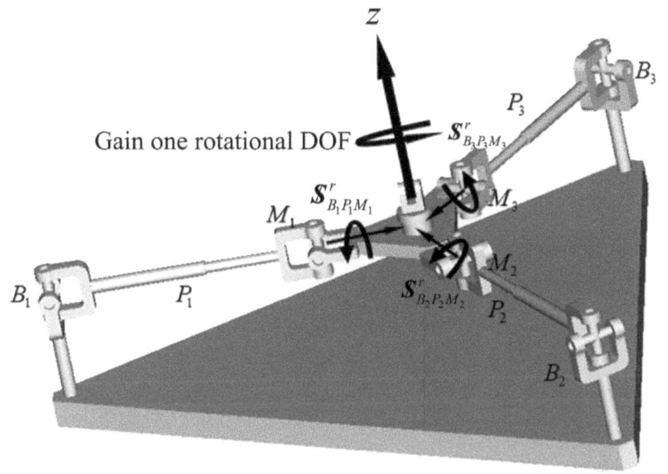

FIGURE 6.16 Singular configuration when $\beta_i = 0(i = 1,2,3)$.

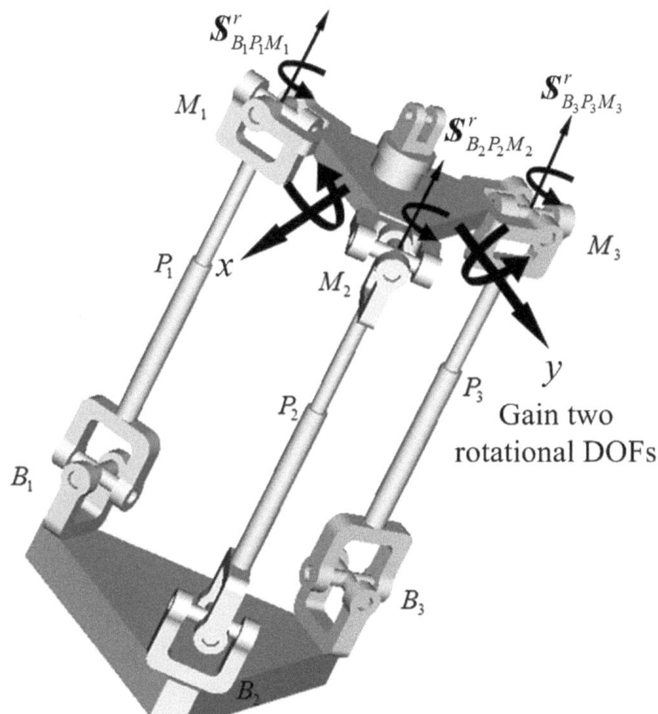

FIGURE 6.17 Singular configuration when $\beta_i = \frac{\pi}{2}(i = 1,2,3)$.

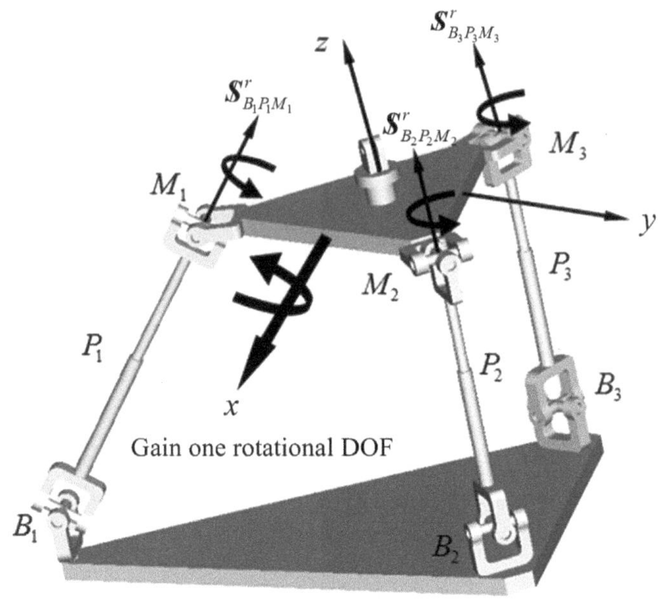

FIGURE 6.18 Singular configuration when $\beta_2 = \beta_3 = \frac{\pi}{2}$.

6.3.4 Singularities of a Hybrid Mechanism

In the following section, we will present the methodology to deal with the singularity problem of the hybrid mechanisms, shown in Figure 6.19. The procedure is relatively simple and the main steps are listed as follows [19]:

(1) The first step is to investigate the geometry character of the hybrid spatial mechanism. Then, we should select the platform and study the kinematic geometry character of the hybrid branches connected with the platform.
(2) According to the geometry characteristics and the screw theory, we make an equivalent substitution of the kinematic screws of each hybrid branch and then transform the hybrid kinematic chain into a pure serial one.
(3) According to the reciprocal screw theory, we can solve the reciprocal screws of the transformed serial kinematic chain of each branch.
(4) Next, investigate the attributes of the reciprocal screws of the branches connected with the platform, and solve the DOF of the platform at the ordinary positions.
(5) With the definition of singularity, we can obtain the criteria of the singularity of the platform according to the method proposed in this chapter.

The hybrid mechanism shown in Figure 6.19 is made up of four branches, which consists of the same kinematic chain of one slider and two parallel limbs in series.

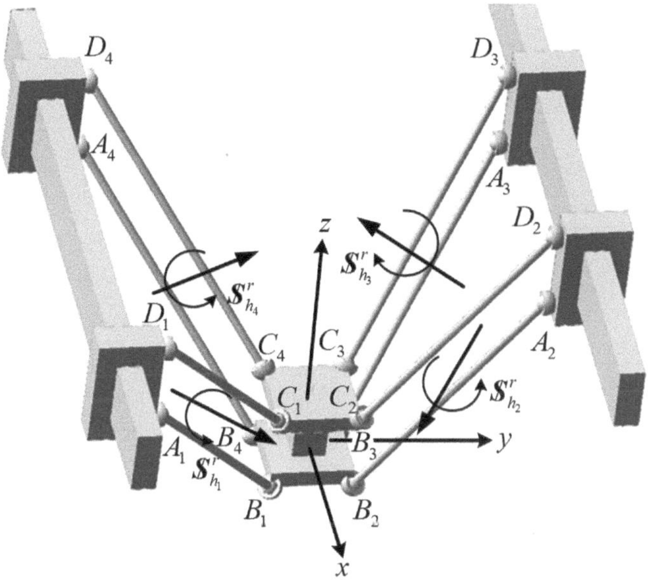

FIGURE 6.19 Architecture of the hybrid spatial mechanism.

According to Section 4.4.3, the terminal constraint of the hybrid kinematic chain, $A_i B_i$ and $D_i C_i$ ($i = 1,2,3,4$), is a constraint moment which is perpendicular to the plane $A_i B_i C_i D_i$, shown in Figure 6.19.

In addition, the terminal constraints are also perpendicular to the z-axis, which can only constrain the rotations around the x-axis and the y-axis at most. Therefore, the DOF of the platform is $F = 6 - 2 = 4$ at least [19]. That is to say, generally, the DOF of the hybrid system is $F = 6 - 2 = 4$.

However, because the actuations of the four sliders are not always independent, the DOF of the hybrid system may be larger than four. That is, we also need to analyze the singular configurations when the actuations of the four sliders are given. The kinematic screw system of the hybrid branch 1 in a local coordinate system is:

$$\$_{h_1} = \begin{bmatrix} \$_{ep1} & \$_{ep2} & \$_{ep3} & \$_{ep4} \end{bmatrix}^T$$

where $\$_{ep1} = \begin{bmatrix} 1\,0\,0\,0\,0\,0 \end{bmatrix}^T$, $\$_{ep2} = \begin{bmatrix} 0\,1\,0\,0\,0\,0 \end{bmatrix}^T$, $\$_{ep3} = \begin{bmatrix} 0\,0\,0\,0\,1\,0 \end{bmatrix}^T$, $\$_{ep1} = \begin{bmatrix} 1\,0\,0\,0\,0\,0 \end{bmatrix}^T$.

The reciprocal screw system can be gained as follows:

$$\$^r_{h_1} = \begin{bmatrix} \$^{r1}_{h_1} & \$^{r2}_{h_1} \end{bmatrix}$$

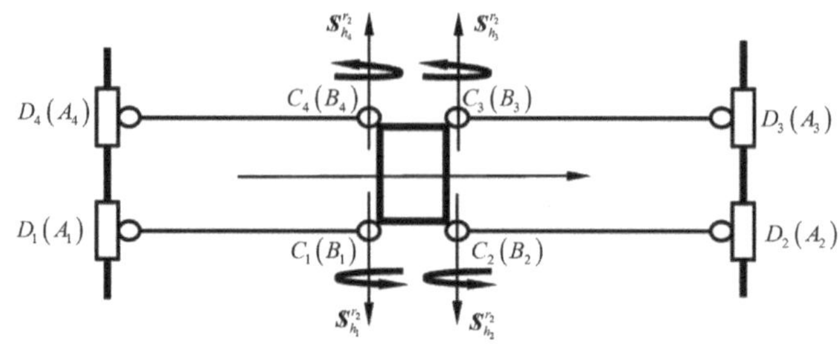

FIGURE 6.20 Singular configuration of the hybrid mechanism.

where $\$_{h_1}^{r1} = \begin{bmatrix} 1 & 0 & 0 & 0 & 0 & 0 \end{bmatrix}^T$, $\$_{h_1}^{r2} = \begin{bmatrix} 0 & 0 & 0 & 0 & 0 & 1 \end{bmatrix}^T$. $\$_{h_1}^{r1}$ is a constraint force along the x_1-axis and $\$_{h_1}^{r2}$ is a constraint moment perpendicular to $\$_{h_1}^{r1}$ shown in Figure 6.20.

Because the singularity is defined as the cases when the DOF of the hybrid system changed, the singularity cases can be concluded as one of the follows:

(1) $\$_{h_1}^{r2}, \$_{h_2}^{r2}$ are in a line and $\$_{h_3}^{r2}, \$_{h_4}^{r2}$ are in a line;
(2) $\$_{h_1}^{r2}, \$_{h_4}^{r2}$ are in a line and $\$_{h_2}^{r2}, \$_{h_3}^{r2}$ are in a line.

Considering the geometry character and the above three cases, we can find that only case 2 is feasible which is shown in Figure 6.20. In this case, one additional rotation around the x-axis occurs. Supposing the projection of the hybrid mechanism along the x-direction in Figure 6.20 is shown in Figure 6.21, the rotation can be investigated as follows.

Figure 6.21 shows the projection of rigid platform along the x-direction. The dashed-line circle o denotes the trajectory of the four vertexes B_1, B_2, C_2, and C_1 of the projection rectangle $B_1B_2C_2C_1$. The dashed line circle O_1 denotes the transection of the projection plane and the spherical trajectory of vertex B_1 revolved around the specified point A_1. Similarly, the dashed-line circle O_2 denotes the transection of the projection plane and the spherical trajectory of vertex C_1 revolved around the prescribed point D_1. When the rigid platform spins any angle of $\theta (\theta \neq 0)$ around the x-direction, the vertexes B_1 and C_1 will rotate to the point B_{11} and C_{11} individually. However, the possible trajectory of point B_1 in plane $B_1B_2C_2C_1$ should only be the dashed line circle O_1 and the possible trajectory of point C_1 in plane $B_1B_2C_2C_1$ should only be the dashed line circle O_2. Therefore, this kind of any finite spin around the x-direction should not exist. And the singularity case 2 of the hybrid mechanism does not hold in reality. So, according to the static equilibrium method proposed in this chapter, we can know that if there is no singularity occurring, the reciprocal screws, $\$_{h_1}^{r1}, \$_{h_2}^{r1}, \$_{h_3}^{r1}$ and $\$_{h_4}^{r1}$, could determine the DOF of the platform no matter whatever the external forces exerted on the platform. Assume that F_x, F_y, F_z are three component external forces applied to the platform, M_z is the component external torques around the z-axis applied to the platform and the subtended

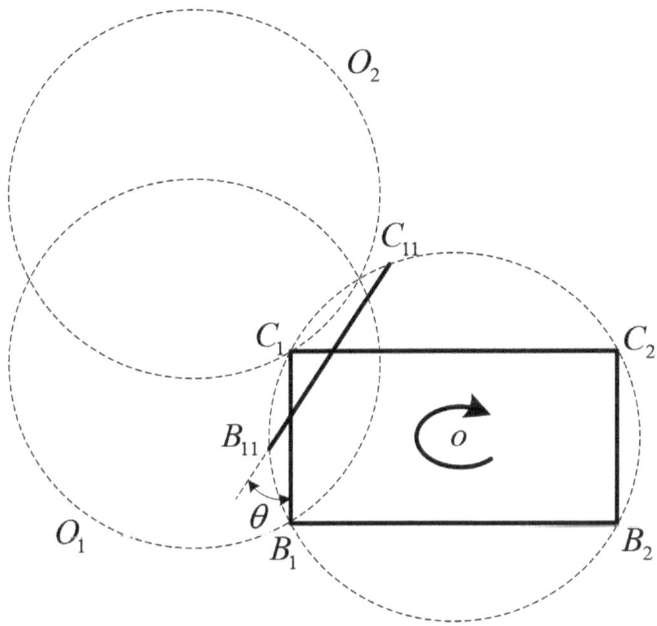

FIGURE 6.21 Projection of hybrid mechanism along the *x*-direction in Figure 6.20.

angle between the hybrid branch and the top plane of the platform is β. Suppose that $B_1B_4 = 2a, B_1B_2 = 2b$, there should be

$$\begin{cases} \Sigma \boldsymbol{F}_x = 0 \; (k_1\cos\theta_1 + k_2\cos\theta_2 + k_3\cos\theta_3 + k_4\cos\theta_4)\cos\beta + F_x = 0 \\ \Sigma \boldsymbol{F}_y = 0 \; (k_1\sin\theta_1 + k_2\sin\theta_2 + k_3\sin\theta_3 + k_4\sin\theta_4)\cos\beta + F_x = 0 \\ \Sigma \boldsymbol{F}_z = 0 \; (k_1 + k_2 + k_3 + k_4)\cos\beta + F_x = 0 \\ \Sigma \boldsymbol{M}_z = 0 \; \left[k_1(a\sin\theta_1 + b\cos\theta_1) + k_2(a\sin\theta_2 + b\cos\theta_2) \right. \\ \qquad \left. - k_3(a\sin\theta_3 + b\cos\theta_3) - k_4(a\sin\theta_4 + b\cos\theta_4)\right]\cos\beta + M_z = 0 \end{cases} \tag{6.35}$$

where $\theta_i \; (i = 1,2,3,4)$ denotes the angular coordinate of the hybrid branch to the *x*-axis in the absolute coordinate system.

Considering $0 < \beta < \frac{\pi}{2}$, the equation set (6.35) can be transformed to

$$\mathbf{A}\boldsymbol{K} = \boldsymbol{F}.$$

where

$$\mathbf{A} = \begin{bmatrix} \cos\theta_1 & \cos\theta_2 & \cos\theta_3 & \cos\theta_4 \\ \sin\theta_1 & \sin\theta_2 & \sin\theta_3 & \sin\theta_4 \\ 1 & 1 & 1 & 1 \\ a\sin\theta_1 + b\cos\theta_1 & a\sin\theta_2 - b\cos\theta_2 & -a\sin\theta_3 - b\cos\theta_3 & a\sin\theta_4 - b\cos\theta_4 \end{bmatrix},$$

$$\boldsymbol{K} = \begin{bmatrix} k_1 & k_2 & k_3 & k_4 \end{bmatrix}^T, \boldsymbol{F} = \begin{bmatrix} -\frac{F_x}{\cos\beta} & -\frac{F_y}{\cos\beta} & -\frac{F_z}{\cos\beta} & -\frac{M_z}{\cos\beta} \end{bmatrix}^T.$$

Therefore, the singularity criteria of the mechanism can be obtained as follows:

$$\det(\mathbf{A}) = 0$$

That is:

$$\begin{aligned} a&\left[(\sin\theta_1 - \sin\theta_2)\sin(\theta_4 - \theta_3) + (\sin\theta_4 - \sin\theta_3)\sin(\theta_2 - \theta_1)\right] \\ &+b\left[(\cos\theta_4 - \cos\theta_1)\sin(\theta_2 - \theta_3)\right. \\ &+\left.(\cos\theta_3 - \cos\theta_2)\sin(\theta_4 - \theta_1)\right] = 0. \end{aligned}$$

6.4 CONCLUSIONS

In this chapter, the singularity of the spatial robot mechanism in the workspace is studied and it is proved that the structural symmetry of the mechanisms themselves will inevitably lead not only to the symmetry of reachable workspace and flexible workspace of its end effector but also to that of singular workspace. A quantitative description of such a symmetric relation provides some theoretical basis for the analysis and synthesis of workspace of symmetrical mechanisms. With screw theory, a group of balance equations which can be used to analyze the singularity of mechanisms have been obtained and the conditions for singular mechanism have been derived. The order of the coefficient matrix of final equations is identical to the number of DOFs of the end effector, which constitutes great advantages for the singularity analysis of the mechanism whose end effector has less than six DOFs. The fewer the DOFs of the end effector, the lower the order of the finally converted coefficient matrix will be, and it is naturally easier to deal with them. According to equivalent replacement theorems of the kinematic chain, it can be found that it is also very easy to use this method for dealing with the singularity of compound mechanisms. Then three different parallel mechanisms and a hybrid mechanism are employed to show the applications of the above-mentioned method for the singularity of the spatial robot mechanism.

REFERENCES

[1] P. Choudhury, A. Ghosal, Singularity and controllability analysis of parallel manipulators and closed-loop mechanisms, Mechanism and Machine Theory 35 (2000) 1455–1479.

[2] J.W. Burdick, A classification of 3R regional manipulator singularities and geometries, Department of Mechanical Engineering, Mechanism and Machine Theory 30 (1995) 71–89.

[3] A.-M. Wazwaz, A comparative study on a singular perturbation problem with two singular boundary points, Applied Mathematics and Computation 99 (1999) 179–193.

[4] S. Bhattacharya, H. Hatwal, A. Ghosh, Comparison of an exact and an approximate methodology of singularity avoidance in platform type parallel manipulators, Mechanism and Machine Theory 33 (1998) 965–974.

[5] K.G. Mattson, Gordon R. Pennock, The velocity problems of two 3-R robots manipulating a four-bar linkage payload, Mechanism and Machine Theory (31) (1996) 1019–1032.

[6] A. Karger, Singularities and self-motions of equiform platforms, Mechanism and Machine Theory 36 (2001) 801–815.
[7] J.J. Cervantes-Sánchez, J.C. Hernández-Rodríguez, J.G. Rendón-Sánchez, On the workspace, assembly configurations and singularity curves of the RRRRR-type planar manipulator, Mechanism and Machine Theory 35 (2000) 1117–1139.
[8] G. Gogu, Families of 6R orthogonal robotic manipulators with only isolated and pseudo-isolated singularities, Mechanism and Machine Theory 37 (2002) 1347–1375.
[9] J.R. Phillips, K.H. Hunt, On the theorem of three axes in the spatial motion of three bodies, Australian Journal of Applied Science 15 (1964) 267–287.
[10] K.H. Hunt, Kinematic Geometry of Mechanisms, Oxford University Press, Oxford, 1978.
[11] S.B. Nokleby, R.P. Podhorodeski, Reciprocity-based resolution of velocity degeneracies (singularities) for redundant manipulators, Mechanism and Machine Theory 36 (2001) 397–409.
[12] K. Sugimoto, J. Duffy, K.H. Hunt, Special configuration of spatial mechanisms and robot arms, Mechanism and Machine Theory 17 (1982) 119–132.
[13] K. Sugimoto, J. Duffy, Determination of extreme distances of a robot hand-Part 1 A general theory, Journal of Mechanical Design 103 (1981) 631–636.
[14] J.S. Zhao, K. Zhou, et al., A new method to study the degree of freedom of spatial parallel mechanisms, The International Journal of Advanced Manufacturing Technology 23 (2004) 288–294.
[15] J.S. Zhao, K. Zhou, Z.J. Feng, A theory of degrees of freedom for mechanisms, Mechanism and Machine Theory 39 (2004) 621–643.
[16] J.P. Merlet, Determination of 6D workspaces of gough-type parallel manipulator and comparison between different geometries, International Journal of Robotics Research 18 (1999) 902–916.
[17] J.P. Merlet, Designing a parallel manipulator for a specific workspace, International Journal of Robotics Research 16 (1997) 545–556.
[18] J.S. Zhao, Y.Z. Fu, et al., Mobility properties of a schoenflies type parallel mechanism, Robotics and Computer-Integrated Manufacturing 22 (2006) 124–133.
[19] J.S. Zhao, K. Zhou, et al., The singularity study of spatial hybrid mechanism based on screw theory, The International Journal of Advanced Manufacturing Technology 25 (2005) 1053–1061.

Chapter 7

Kinematics with Four Points' Cartesian Coordinates for Spatial Parallel Manipulator

Natural coordinates were originally introduced to describe the position of multibody systems by García de Jalón *et al.* in 1981 and Serna *et al.* in 1982 for planar cases and García de Jalón *et al.* in 1986 and 1987 for spatial systems [1,2]. The outstanding characteristics are that the natural coordinates are only made up of Cartesian coordinates of points or direction cosines of unit vectors belonging to the bodies of the system. Therefore the natural coordinate method only introduces the absolute coordinates of the selected points. As a result, the velocity, acceleration, and even the jerk can be obtained by solving the linear equation system which is the derivative of the quadratic constraint equations of the manipulator. Although the number of variables required for describing the position of a body is not the minimum, the fact that constraint equations are either quadratic or linear facilitates the calculations [2,3]. Because the coordinates are fully Cartesian, the elements of the derivative matrix are linear while those of others established through matrix transformation are quadratic or even transcendental. All of these remarkable advantages in kinematic and dynamic analysis allow more and more scholars to build their kinematics and dynamics models with natural coordinates. Szkodny [4] utilized the natural coordinates of links to describe relative motion of adjacent links. These coordinates describe the manipulator in relation to a selected relative system regardless of the kinematic structure of the manipulator. Ceccarelli *et al.* [5] developed a suitable model by describing the multibody system through natural coordinates of suitable points and vectors. Szkodny [6] presented formulas connecting the driving forces of actuators with natural coordinates. Homogeneous transformations were used to describe position and orientation of the systems. These transformations afford the possibility for a joint description of the position and orientation, which is an invaluable advantage in describing kinematics and, especially dynamics, of parallel manipulators. The detailed description and application of the advantages of natural coordinates can be found in [2,7].

Advanced Theory of Constraint and Motion Analysis for Robot Mechanisms.
http://dx.doi.org/10.1016/B978-0-12-420162-0.00007-2

Compared with the widely used methods based on the transformation matrix, the prevailing virtues of the kinematics model discussed in this chapter exist in that the angular velocities and angular accelerations can be gained very easily just by solving the linear equations with the prescribed four points' Cartesian coordinates rather than those with rotation transformation matrix, in which the angular velocities and accelerations are in complicated transcendental functions. This, on the one hand, remains the merit of not using a trigonometric variable of the natural coordinate method and, on the other hand, avoids the increasing number of coordinates as the number of links increases.

7.1 KINEMATICS MODEL FOR SPATIAL PARALLEL MANIPULATORS WITH FOUR POINTS' CARTESIAN COORDINATES

The advantages of this algorithm compared with the general natural coordinate method [2] are that only four non-coplanar points' coordinates of the end effector are sufficient to build up the kinematic model of a spatial parallel manipulator. Combining the relationship constraint equations with the rigid body constraint conditions, the displacement of a spatial mechanism with these four points' Cartesian coordinates can be directly established. The velocities and accelerations of the manipulator can be expressed in a form of the first-order and second-order derivatives of these Cartesian coordinates with respect to time which are linear combinations of these coordinates. Therefore, numerical iterative methods such as the Newton–Raphson method, the Newton–GMRES method, and so on, are only needed to calculate the positions for real-time applications. The analysis of velocity, acceleration, and jerk is immediately gained by solving liner equations. In the widely used method based on the transformation matrix, the numerical iterative method will have to be used not only in position, but also in velocity, acceleration, and jerk.

Although Kim and Ryu [8] presented a dimensionally homogeneous Jacobian matrix formulation by three end effector points for optimal design of parallel manipulators, the method could not be used for non-coplanar points. We, therefore, propose a kinematic model with four non-coplanar points and further address the sufficient and necessary criteria for using four non-coplanar points for any spatial parallel manipulators via a mathematics proof in this paper.

Assume

$$\boldsymbol{x} = \begin{bmatrix} x_1 & y_1 & z_1 & x_2 & y_2 & z_2 & x_3 & y_3 & z_3 & x_4 & y_4 & z_4 \end{bmatrix}^T \tag{7.1}$$

where $x_i, y_i, z_i\,(i = 1,2,3,4)$ are the Cartesian coordinates of the specified four points M_1, M_2, M_3, and M_4, which are not coplanar.

The relationship constraint equations can be denoted by:

$$\boldsymbol{P}(\boldsymbol{x}) = \boldsymbol{0}. \tag{7.2}$$

The Jacobian matrix of equation (7.2) is obtained:

$$\boldsymbol{P}'(\boldsymbol{x}) = \frac{\partial \boldsymbol{P}(\boldsymbol{x})}{\partial \boldsymbol{x}}. \tag{7.3}$$

Immediately, the singularity criteria of the mechanism should be expressed as:

$$\left|\boldsymbol{P}'(\boldsymbol{x})\right| = 0 \tag{7.4}$$

where $|\cdot|$ represents the determinant of "$\cdot$". When equation (7.4) does not hold, the forward displacement solutions can be obtained through numerical methods, such as the Newton–Raphson method [1], which is not the focus of this chapter:

$$\boldsymbol{x}_{j+1} = \boldsymbol{x}_j - \left[\boldsymbol{P}'(\boldsymbol{x}_j)\right]^{-1} \boldsymbol{P}(\boldsymbol{x}_j) \tag{7.5}$$

where $j = 0, 1, \mathrm{L}$ denotes the sequence of iterating steps.

After gaining x via iterating formula (7.5), we can analyze the angular displacement of the end effector. Obviously, the unit normal vector of the end effector plane can be expressed with the Cartesian coordinates of three of the prescribed points M_1, M_2, and M_3 on the end effector plane:

$$\boldsymbol{n} = \begin{bmatrix} n_x \\ n_y \\ n_z \end{bmatrix} = \frac{\overrightarrow{M_1M_2} \times \overrightarrow{M_2M_3}}{\left\|\overrightarrow{M_1M_2}\right\| \left\|\overrightarrow{M_2M_3}\right\| \sin \angle M_1M_2M_3}. \tag{7.6}$$

Therefore, the angles from the end effector plane to the three coordinate axes can be immediately obtained from equation (7.6):

$$\begin{cases} \alpha_x = \arcsin\left(n_x\right) \\ \alpha_y = \arcsin\left(n_y\right) \\ \alpha_z = \arcsin\left(n_z\right) \end{cases} \tag{7.7}$$

where α_x,α_y, and α_z denote the angles from the end effector plane to x-, y- and z-axes, individually.

The derivatives of equation (7.2) with respect to time are expressed in the following matrix form:

$$\mathbf{A}\dot{\boldsymbol{x}} = \boldsymbol{b} \tag{7.8}$$

where $\mathbf{A} = \mathbf{A}(\boldsymbol{x})$ which is a known matrix made up of the Cartesian coordinates of the prescribed four points, $\boldsymbol{b}$ is a known column vector containing the actuations and

$$\dot{\boldsymbol{x}} = \begin{bmatrix} \dot{x}_1 & \dot{y}_1 & \dot{z}_1 & \dot{x}_2 & \dot{y}_2 & \dot{z}_2 & \dot{x}_3 & \dot{y}_3 & \dot{z}_3 & \dot{x}_4 & \dot{y}_4 & \dot{z}_4 \end{bmatrix}^T. \tag{7.9}$$

After gaining $\boldsymbol{x}$ with equation (7.5), we can easily obtain $\dot{\boldsymbol{x}}$ by solving equation (7.8) for the coefficient matrix $\mathbf{A}(\boldsymbol{x})$ is a linear function of $\boldsymbol{x}$. Similarly, the acceleration $\ddot{\boldsymbol{x}}$ can also be deduced from (7.8):

$$\mathbf{A}\ddot{x} = \boldsymbol{d} \tag{7.10}$$

where $\ddot{x} = \begin{bmatrix} \ddot{x}_1 & \ddot{y}_1 & \ddot{z}_1 & \ddot{x}_2 & \ddot{y}_2 & \ddot{z}_2 & \ddot{x}_3 & \ddot{y}_3 & \ddot{z}_3 & \ddot{x}_4 & \ddot{y}_4 & \ddot{z}_4 \end{bmatrix}^T$ and $\boldsymbol{d} = \boldsymbol{d}\left(\boldsymbol{x},\dot{\boldsymbol{x}}\right)$ is a function of the general displacements and velocities.

Generally, we can select the Cartesian coordinates of any four non-coplanar vertices, M_1, M_2, eM_3, and M_4, on the end effector as the natural coordinates. As shown in Figure 7.1, the natural coordinates are denoted by (7.1).

(A) Lemma. The nature coordinates of any point within a space can be uniquely expressed as the linear combination of those of four non-coplanar points; the velocity and acceleration of the point also have the same linear combination of those of the four non-coplanar points.

(B) Proof. Suppose that the coordinates of the four non-coplanar points in the absolute coordinate system shown in Figure 7.1 are

$$\boldsymbol{r}_{M_1} = \begin{bmatrix} x_1 \\ y_1 \\ z_1 \end{bmatrix}, \quad \boldsymbol{r}_{M_2} = \begin{bmatrix} x_2 \\ y_2 \\ z_2 \end{bmatrix}, \quad \boldsymbol{r}_{M_3} = \begin{bmatrix} x_3 \\ y_3 \\ z_3 \end{bmatrix}, \quad \boldsymbol{r}_{M_4} = \begin{bmatrix} x_4 \\ y_4 \\ z_4 \end{bmatrix}.$$

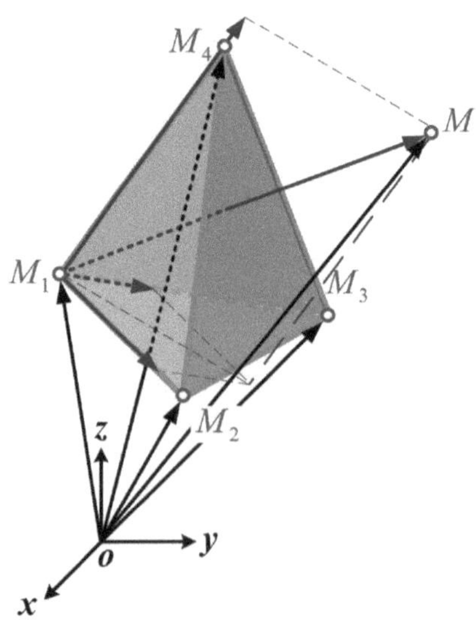

FIGURE 7.1 A space determined by four non-coplanar points.

Immediately, we obtain

$$\boldsymbol{r}_{M_1M_2} = \boldsymbol{r}_{M_2} - \boldsymbol{r}_{M_1} = \begin{bmatrix} x_2 - x_1 \\ y_2 - y_1 \\ z_2 - z_1 \end{bmatrix}, \quad \boldsymbol{r}_{M_1M_3} = \boldsymbol{r}_{M_3} - \boldsymbol{r}_{M_1} = \begin{bmatrix} x_3 - x_1 \\ y_3 - y_1 \\ z_3 - z_1 \end{bmatrix},$$

$$\boldsymbol{r}_{M_1M_4} = \boldsymbol{r}_{M_4} - \boldsymbol{r}_{M_1} = \begin{bmatrix} x_4 - x_1 \\ y_4 - y_1 \\ z_4 - z_1 \end{bmatrix}.$$

For any point, $M(x,y,z)$, within the space, there should be

$$\boldsymbol{r}_{M_1M} = \boldsymbol{r}_{M} - \boldsymbol{r}_{M_1} = \begin{bmatrix} x - x_1 \\ y - y_1 \\ z - z_1 \end{bmatrix}.$$

Assume

$$\boldsymbol{r}_{M_1M} = k_1\boldsymbol{r}_{M_1M_2} + k_2\boldsymbol{r}_{M_1M_3} + k_3\boldsymbol{r}_{M_1M_4} \tag{7.11}$$

where $k_i\,(i = 1,2,3)$ are three coefficients that are not zeros simultaneously.

Suppose there are another set of coefficients $c_i\,(i = 1,2,3)$ such that

$$\boldsymbol{r}_{M_1M} = c_1\boldsymbol{r}_{M_1M_2} + c_2\boldsymbol{r}_{M_1M_3} + c_3\boldsymbol{r}_{M_1M_4}. \tag{7.12}$$

Associating (7.11) and (7.12), we obtain

$$\left(k_1 - c_1\right)\boldsymbol{r}_{M_1M_2} + \left(k_2 - c_2\right)\boldsymbol{r}_{M_1M_3} + \left(k_3 - c_3\right)\boldsymbol{r}_{M_1M_4} = 0. \tag{7.13}$$

Because the four points, M_1, M_2, M_3, and M_4, are four non-coplanar points, $\boldsymbol{r}_{M_1M_2}, \boldsymbol{r}_{M_1M_3}$ and $\boldsymbol{r}_{M_1M_4}$ are three non-coplanar vectors and they are independent in the three-dimensional space. As a result, equation (7.13) only results in

$$\begin{cases} k_1 - c_1 = 0 \\ k_2 - c_2 = 0 \\ k_3 - c_3 = 0. \end{cases} \tag{7.14}$$

Equation (7.14) indicates that the three coefficients in (7.11) or (7.12) are uniquely determined by the position of point $M(x,y,z)$ with respect to the space determined by the four non-coplanar points. In other words, the three coefficients, k_i $(i = 1,2,3)$, are unique when point $M(x,y,z)$ is specified. Expanding (7.11) presents

$$\boldsymbol{r}_M - \boldsymbol{r}_{M_1} = k_1\left(\boldsymbol{r}_{M_2} - \boldsymbol{r}_{M_1}\right) + k_2\left(\boldsymbol{r}_{M_3} - \boldsymbol{r}_{M_1}\right) + k_3\left(\boldsymbol{r}_{M_4} - \boldsymbol{r}_{M_1}\right) \tag{7.15}$$

then we have

$$\boldsymbol{r}_M = \phi\left(\boldsymbol{x}\right) = \mathbf{X}\boldsymbol{K} \tag{7.16}$$

where the linear operation $\phi(\cdot)$ is uniquely determined by the geometry relationship between point M and the specified four points, and $\mathbf{X} = [\boldsymbol{r}_{M_1}\ \ \boldsymbol{r}_{M_2}\ \ \boldsymbol{r}_{M_3}\ \ \boldsymbol{r}_{M_4}]$,$\boldsymbol{K} = [1 - k_1 - k_2 - k_3\ \ k_1\ \ k_2\ \ k_3]^T$.

As a result, the coordinates of any point $M(x,y,z)$ within the space determined by the four non-coplanar points can be expressed as the linear combination of those of four points. Suppose that the coordinates of any point $M_i\ (i = 5,6,\ldots,n)$ on the end effector are denoted by

$$\boldsymbol{r}_{M_i} = \begin{bmatrix} x_i & y_i & z_i \end{bmatrix}^T \tag{7.17}$$

and we can denote the linear combination by:

$$\boldsymbol{r}_{M_i} = \phi_i(\boldsymbol{x}) \tag{7.18}$$

where the linear operation $\phi_i(\cdot)$ is uniquely determined by the geometry relationship between point M_i and the specified four points, and therefore $\phi_i(\cdot)$ is invariable with respect to time. As a result, the velocity and the acceleration of point M_i can be directly obtained, respectively:

$$\begin{cases} \boldsymbol{r}_{\dot{M}_i} = \phi_i(\dot{\boldsymbol{x}}) \\ \boldsymbol{r}_{\ddot{M}_i} = \phi_i(\ddot{\boldsymbol{x}}) \end{cases} \tag{7.19}$$

where $\boldsymbol{r}_{\dot{M}_i}$ and $\boldsymbol{r}_{\ddot{M}_i}$ denote the velocity and acceleration of point M_i, respectively.

Therefore, the velocity and acceleration of any point $M_i\ (x_i,y_i,z_i)$ within the space determined by the four non-coplanar points are also expressed as the same linear combination of those of the prescribed four points.

For the prescribed four points, M_1,M_2,M_3, and M_4, there will be the following velocity relationship:

$$\boldsymbol{v}_{M_i} = \boldsymbol{v}_C + \boldsymbol{\omega} \times \boldsymbol{r}_{CM_i} \tag{7.20}$$

where $i = 1,2,3,4$, $\boldsymbol{v}_{M_i} = \begin{bmatrix} \dot{x}_i & \dot{y}_i & \dot{z}_i \end{bmatrix}^T$, v_C denotes the velocity of the reference point on the end effector, $\boldsymbol{\omega}$ denotes the angular velocity of the end effector, and $\boldsymbol{r}_{CM_i}$ denotes the absolute vector from the reference point C to point M_i.

From equation (7.20), we can deduce that

$$\boldsymbol{v}_{M_2} - \boldsymbol{v}_{M_1} = \boldsymbol{\omega} \times \left(\boldsymbol{r}_{CM_2} - \boldsymbol{r}_{CM_1}\right) = \boldsymbol{\omega} \times \boldsymbol{r}_{M_1M_2}. \tag{7.21}$$

Cross-multiplying vector $\boldsymbol{r}_{M_1M_2}$ at both sides of equation (7.21) yields:

$$\boldsymbol{r}_{M_1M_2} \times \left(\boldsymbol{v}_{M_2} - \boldsymbol{v}_{M_1}\right) = \boldsymbol{r}_{M_1M_2} \times \left(\boldsymbol{\omega} \times \boldsymbol{r}_{M_1M_2}\right). \tag{7.22}$$

The right side of equation (7.22) can be further simplified:

$$\boldsymbol{r}_{M_1M_2} \times \left(\boldsymbol{\omega} \times \boldsymbol{r}_{M_1M_2}\right) = \left(\left\|\boldsymbol{r}_{M_1M_2}\right\|^2 \mathbf{I} - \boldsymbol{r}_{M_1M_2}\boldsymbol{r}^T_{M_1M_2}\right)\boldsymbol{\omega} \tag{7.23}$$

where $\mathbf{I}$ denotes a 3×3 identity matrix.

From equations (7.22) and (7.23), we can obtain:

$$\left(\left\|\boldsymbol{r}_{M_1M_2}\right\|^2 \mathbf{I} - \boldsymbol{r}_{M_1M_2}\boldsymbol{r}_{M_1M_2}^T\right)\boldsymbol{\omega} = \boldsymbol{r}_{M_1M_2} \times \left(\boldsymbol{v}_{M_2} - \boldsymbol{v}_{M_1}\right). \tag{7.24}$$

Assume

$$\begin{cases} \mathbf{B}_{12} = \left\|\boldsymbol{r}_{M_1M_2}\right\|^2 \mathbf{I} - \boldsymbol{r}_{M_1M_2}\boldsymbol{r}_{M_1M_2}^T \\ \boldsymbol{e}_{12} = \boldsymbol{r}_{M_1M_2} \times \left(\boldsymbol{v}_{M_2} - \boldsymbol{v}_{M_1}\right). \end{cases} \tag{7.25}$$

Similarly, with the Cartesian coordinates of points M_1 and M_3, we can conceive that

$$\begin{cases} \mathbf{B}_{13} = \left\|\boldsymbol{r}_{M_1M_3}\right\|^2 \mathbf{I} - \boldsymbol{r}_{M_1M_3}\boldsymbol{r}_{M_1M_3}^T \\ \boldsymbol{e}_{13} = \boldsymbol{r}_{M_1M_3} \times \left(\boldsymbol{v}_{M_3} - \boldsymbol{v}_{M_1}\right). \end{cases} \tag{7.26}$$

We next prove the following two theorems:

Theorem 7.1. $\mathbf{B}_{12}$ *and* $\mathbf{B}_{13}$ *are individually singular by nature;*

Theorem 7.2. $\mathbf{B}$ *will not be singular when we let*

$$\mathbf{B} = \mathbf{B}_{12} + \mathbf{B}_{13} \tag{7.27}$$

provided that the three different points M_1, M_2, and M_3 do not remain collinear.

It is not difficult to prove Theorem 7.1, and so the proof is omitted here. Following, we only prove Theorem 7.2.

Proof. Assume that matrix $\mathbf{B}$ is singular, and we immediately have

$$\det\left(\mathbf{B}\right) = 0 \tag{7.28}$$

which yields

$$\left|\left(\left\|\boldsymbol{r}_{M_1M_2}\right\|^2 + \left\|\boldsymbol{r}_{M_1M_3}\right\|^2\right)\mathbf{I} - \left(\boldsymbol{r}_{M_1M_2}\boldsymbol{r}_{M_1M_2}^T + \boldsymbol{r}_{M_1M_3}\boldsymbol{r}_{M_1M_3}^T\right)\right| = 0 \tag{7.29}$$

where $|\cdot|$ denotes the determinant of a matrix "$\cdot$".

Equation (7.29) shows that $\left\|\boldsymbol{r}_{M_1M_2}\right\|^2 + \left\|\boldsymbol{r}_{M_1M_3}\right\|^2$ is one eigenvalue of matrix $\boldsymbol{r}_{M_1M_2}\boldsymbol{r}_{M_1M_2}^T + \boldsymbol{r}_{M_1M_3}\boldsymbol{r}_{M_1M_3}^T$. Therefore, $\exists \boldsymbol{x} \in \mathbb{R}^3$, and $x \neq 0$, such that

$$\left(\boldsymbol{r}_{M_1M_2}\boldsymbol{r}_{M_1M_2}^T + \boldsymbol{r}_{M_1M_3}\boldsymbol{r}_{M_1M_3}^T\right)\boldsymbol{x} = \left(\left\|\boldsymbol{r}_{M_1M_2}\right\|^2 + \left\|\boldsymbol{r}_{M_1M_3}\right\|^2\right)\boldsymbol{x}. \tag{7.30}$$

Dot multiplying $\boldsymbol{x}^T$ at both sides of (7.30) yields

$$\boldsymbol{x}^T\left(\boldsymbol{r}_{M_1M_2}\boldsymbol{r}_{M_1M_2}^T + \boldsymbol{r}_{M_1M_3}\boldsymbol{r}_{M_1M_3}^T\right)\boldsymbol{x} = \left(\left\|\boldsymbol{r}_{M_1M_2}\right\|^2 + \left\|\boldsymbol{r}_{M_1M_3}\right\|^2\right)\boldsymbol{x}^T\boldsymbol{x} \tag{7.31}$$

which results in

$$\left(\boldsymbol{r}_{M_1M_2},\boldsymbol{x}\right)^2 + \left(\boldsymbol{r}_{M_1M_3},\boldsymbol{x}\right)^2 = \left\|\boldsymbol{r}_{M_1M_2}\right\|^2 \|\boldsymbol{x}\|^2 + \left\|\boldsymbol{r}_{M_1M_3}\right\|^2 \|\boldsymbol{x}\|^2 \tag{7.32}$$

where $\left(\boldsymbol{r}_{M_1M_2},\boldsymbol{x}\right)$ denotes the inner product of two vectors, $\boldsymbol{r}_{M_1M_2}$ and $\boldsymbol{x}$.

However, according to the Cauchy–Schwartz inequality, we have

$$\begin{cases} \left(\boldsymbol{r}_{M_1M_2},\boldsymbol{x}\right)^2 \le \left\|\boldsymbol{r}_{M_1M_2}\right\|^2 \|\boldsymbol{x}\|^2 \\ \left(\boldsymbol{r}_{M_1M_3},\boldsymbol{x}\right)^2 \le \left\|\boldsymbol{r}_{M_1M_3}\right\|^2 \|\boldsymbol{x}\|^2. \end{cases} \tag{7.33}$$

From equation (7.33), we have

$$\left(\boldsymbol{r}_{M_1M_2},\boldsymbol{x}\right)^2 + \left(\boldsymbol{r}_{M_1M_3},\boldsymbol{x}\right)^2 \le \left\|\boldsymbol{r}_{M_1M_2}\right\|^2 \|\boldsymbol{x}\|^2 + \left\|\boldsymbol{r}_{M_1M_3}\right\|^2 \|\boldsymbol{x}\|^2. \tag{7.34}$$

Notice that the equality holds if and only if $\boldsymbol{r}_{M_1M_2}$ and $\boldsymbol{r}_{M_1M_3}$ are both parallel to $\boldsymbol{x}$, which is in conflict with the preconditions that the three different points M_1, M_2, and M_3 do not remain collinear. Therefore, equation (7.32) does not hold. And as a result, det $(B) \neq 0$.

From Theorems 7.1 and 7.2, we can find that the angular velocity of the end effector can be solved with a linear equation set whose coefficient matrix is constructed only with the Cartesian coordinates of three non-collinear points. This selection of Cartesian coordinates is also the sufficient and necessary criterion for the kinematics model.

Therefore, the angular velocity of the end effector can be expressed as:

$$\boldsymbol{\omega} = \mathbf{B}^{-1}\boldsymbol{e} \tag{7.35}$$

where

$$\boldsymbol{e} = \boldsymbol{e}_{12} + \boldsymbol{e}_{13}. \tag{7.36}$$

The derivatives of equation (7.21) with respect to time can be expressed as:

$$\dot{\boldsymbol{v}}_{M_2} - \dot{\boldsymbol{v}}_{M_1} = \dot{\boldsymbol{\omega}} \times \boldsymbol{r}_{M_1M_2} + \boldsymbol{\omega} \times \left(\boldsymbol{\omega} \times \boldsymbol{r}_{M_1E_2}\right). \tag{7.37}$$

Considering

$$\begin{cases} \dot{\boldsymbol{v}}_{M_i} = \begin{bmatrix} \ddot{x}_i & \ddot{y}_i & \ddot{z}_i \end{bmatrix} \\ \boldsymbol{\varepsilon} = \dot{\boldsymbol{\omega}}. \end{cases} \tag{7.38}$$

We can transform equation (7.37) into

$$\boldsymbol{\varepsilon} \times \boldsymbol{r}_{M_1M_2} = \left(\dot{\boldsymbol{v}}_{M_2} - \dot{\boldsymbol{v}}_{M_1}\right) - \boldsymbol{\omega} \times \left(\boldsymbol{\omega} \times \boldsymbol{r}_{M_1M_2}\right). \tag{7.39}$$

Left cross multiplying $\boldsymbol{r}_{M_1M_2}$ at both sides of equation (7.39) yields

$$\mathbf{B}_{12}\boldsymbol{\varepsilon} = \boldsymbol{r}_{M_1M_2} \times \left(\dot{\boldsymbol{v}}_{M_2} - \dot{\boldsymbol{v}}_{M_1}\right) + \left(\boldsymbol{\omega} \cdot \boldsymbol{r}_{M_1M_2}\right) \boldsymbol{\omega} \times \boldsymbol{r}_{M_1M_2}. \tag{7.40}$$

Similarly, we obtain

$$\mathbf{B}_{13}\boldsymbol{\varepsilon} = \boldsymbol{r}_{M_1M_3} \times \left(\dot{\boldsymbol{v}}_{M_3} - \dot{\boldsymbol{v}}_{M_1}\right) + \left(\boldsymbol{\omega} \cdot \boldsymbol{r}_{M_1M_3}\right) \boldsymbol{\omega} \times \boldsymbol{r}_{M_1M_3} \tag{7.41}$$

and

$$\mathbf{B}\boldsymbol{\varepsilon} = \boldsymbol{p} \tag{7.42}$$

where $\boldsymbol{p} = \boldsymbol{r}_{M_1M_2} \times (\dot{\boldsymbol{v}}_{M_2} - \dot{\boldsymbol{v}}_{M_1}) + (\boldsymbol{\omega} \cdot \boldsymbol{r}_{M_1M_2})\, \boldsymbol{\omega} \times \boldsymbol{r}_{M_1M_2} + \boldsymbol{r}_{M_1M_3} \times (\dot{\boldsymbol{v}}_{M_3} - \dot{\boldsymbol{v}}_{M_1}) + (\boldsymbol{\omega} \cdot \boldsymbol{r}_{M_1M_3})\, \boldsymbol{\omega} \times \boldsymbol{r}_{M_1M_3}$. Therefore, the angular acceleration of the end effector is obtained from equation (7.42):

$$\boldsymbol{\varepsilon} = \mathbf{B}^{-1}\boldsymbol{p}. \tag{7.43}$$

Hereto, the forward displacement and kinematic problems have been settled only with the three points' coordinates. The fourth point's coordinates are only used to linearly express any other point off the plane determined by the previous three points on the end effector. However, there are many tricks in establishing the relationship constraint equations of a spatial parallel manipulator, which will be investigated in the next section.

7.2 FORWARD KINEMATICS OF SPATIAL PARALLEL MANIPULATORS

Now we will illustrate the efficiency and simplicity of the kinematics model with the four non-coplanar points' coordinates method.

7.2.1 Forward Kinematics of a Gough–Stewart Manipulator

The mobile end effector of a Gough–Stewart manipulator shown in Figure 7.2 is linked with the fixed base through six separate *SPS* kinematic chains.

For the sake of simplicity, we select the absolute coordinates of a point P_0 on the spindle and joints P_1, P_2, P_3 on the end effector as the natural coordinates.

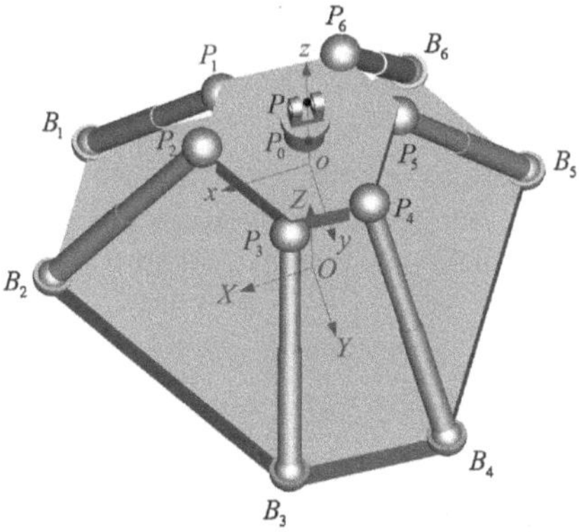

FIGURE 7.2 Structure of a Gough–Stewart manipulator.

Namely,

$$\boldsymbol{x} = \begin{bmatrix} x_0 & y_0 & z_0 & x_1 & y_1 & z_1 & x_2 & y_2 & z_2 & x_3 & y_3 & z_3 \end{bmatrix}^T. \tag{7.44}$$

As proved in the above section, the coordinates of the other three vertices of the hexagon, $P_4(x_4 \quad y_4 \quad z_4), P_5(x_5 \quad y_5 \quad z_5)$ and $P_6(x_6 \quad y_6 \quad z_6)$, and any other points attached to the end effector can be expressed as the linear combinations of the selected non-coplanar four points. According to equation (7.15), we presume:

$$\boldsymbol{r}_{P_i} - \boldsymbol{r}_{P_0} = \begin{bmatrix} \boldsymbol{r}_{P_1} - \boldsymbol{r}_{P_0} & \boldsymbol{r}_{P_2} - \boldsymbol{r}_{P_0} & \boldsymbol{r}_{P_3} - \boldsymbol{r}_{P_0} \end{bmatrix} \boldsymbol{k}_i \tag{7.45}$$

where $\boldsymbol{k}_i = \begin{bmatrix} k_{i1} & k_{i2} & k_{i3} \end{bmatrix}^T (i = 4,5,6)$ are known constants determined uniquely by the geometry of the end effector. Particularly, we can investigate the position, velocity and acceleration of the distal point $P(x \quad y \quad z)$ on the end effector in the same way.

From equations (7.45), we first obtain the coordinates of the other three vertices of the end effector:

$$\boldsymbol{r}_{P_i} = \mathbf{X}\boldsymbol{K}_i \tag{7.46}$$

where
$\mathbf{X} = \begin{bmatrix} \boldsymbol{r}_{P_0} & \boldsymbol{r}_{P_1} & \boldsymbol{r}_{P_2} & \boldsymbol{r}_{P_3} \end{bmatrix}, \boldsymbol{K}_i = \begin{bmatrix} 1 - k_{i1} - k_{i2} - k_{i3} & k_{i1} & k_{i2} & k_{i3} \end{bmatrix}^T,$
$(i = 4,5,6)$.

In addition, six quadratic constraint equations of actuations can be obtained:

$$\left(\boldsymbol{r}_{P_i} - \boldsymbol{r}_{B_i}\right)^T \left(\boldsymbol{r}_{P_i} - \boldsymbol{r}_{B_i}\right) - l_i^2 = 0 \tag{7.47}$$

where $l_i\,(i = 1,2,\ldots,6)$ are the lengths of the legs, which reflect the actuations.

Now, let's consider the geometry relationship for the vertices of the manipulator $P_1P_2P_3P_4P_5P_6$ shown in Figure 7.2. In the rigid body $P_1P_2P_3P_4P_5P_6$, the tetrahedron $P_0P_1P_2P_3$ will never deform during moving. Therefore we obtain a set of rigid body constraints:

$$\left(\boldsymbol{r}_{P_i} - \boldsymbol{r}_{P_j}\right)^T \left(\boldsymbol{r}_{P_i} - \boldsymbol{r}_{P_j}\right) - l_{P_iP_j}^2 = 0 \tag{7.48}$$

where $i, j = 0,1,2,3$ and $i \neq j$.

With equation (7.1), we have the following constraint equations by combining equations (7.47) and (7.48):

$$P(x) = \begin{bmatrix} (x_1 - x_{B_1})^2 + (y_1 - y_{B_1})^2 + (z_1 - z_{B_1})^2 - l_1^2 \\ (x_2 - x_{B_2})^2 + (y_2 - y_{B_2})^2 + (z_2 - z_{B_2})^2 - l_2^2 \\ (x_3 - x_{B_3})^2 + (y_3 - y_{B_3})^2 + (z_3 - z_{B_3})^2 - l_3^2 \\ (x_4 - x_{B_4})^2 + (y_4 - y_{B_4})^2 + (z_4 - z_{B_4})^2 - l_4^2 \\ (x_5 - x_{B_5})^2 + (y_5 - y_{B_5})^2 + (z_5 - z_{B_5})^2 - l_5^2 \\ (x_6 - x_{B_6})^2 + (y_6 - y_{B_6})^2 + (z_6 - z_{B_6})^2 - l_6^2 \\ (x_1 - x_2)^2 + (y_1 - y_2)^2 + (z_1 - z_2)^2 - L_{M_1M_2}^2 \\ (x_2 - x_3)^2 + (y_2 - y_3)^2 + (z_2 - z_3)^2 - L_{M_2M_3}^2 \\ (x_3 - x_1)^2 + (y_3 - y_1)^2 + (z_3 - z_1)^2 - L_{M_3M_1}^2 \\ (x_1 - x_O)^2 + (y_1 - y_O)^2 + (z_1 - z_O)^2 - L_{M_OM_1}^2 \\ (x_2 - x_O)^2 + (y_2 - y_O)^2 + (z_2 - z_O)^2 - L_{M_OM_2}^2 \\ (x_3 - x_O)^2 + (y_3 - y_O)^2 + (z_3 - z_O)^2 - L_{M_OM_2}^2 \end{bmatrix} = 0. \tag{7.49}$$

The Jacobian matrix of equation (7.49) can be gained:

$$P'(x) = \frac{\partial P(x)}{\partial x} = 2\begin{bmatrix} \mathbf{A}_{11} & \mathbf{A}_{12} & \mathbf{A}_{13} & \mathbf{A}_{14} \\ \mathbf{A}_{21} & \mathbf{A}_{22} & \mathbf{A}_{23} & \mathbf{A}_{24} \\ \mathbf{A}_{31} & \mathbf{A}_{32} & \mathbf{A}_{33} & \mathbf{A}_{34} \\ \mathbf{A}_{41} & \mathbf{A}_{42} & \mathbf{A}_{43} & \mathbf{A}_{44} \end{bmatrix} \tag{7.50}$$

where

$$\mathbf{A}_{11} = \mathbf{0}_{3\times3},$$

$$\mathbf{A}_{12} = \begin{bmatrix} \boldsymbol{r}_{P_1} - \boldsymbol{r}_{B_1} & \mathbf{0}_{3\times2} \end{bmatrix}^T,$$

$$\mathbf{A}_{13} = \begin{bmatrix} \mathbf{0}_{3\times1} & \boldsymbol{r}_{P_2} - \boldsymbol{r}_{B_2} & \mathbf{0}_{3\times1} \end{bmatrix}^T,$$

$$\mathbf{A}_{14} = \begin{bmatrix} \mathbf{0}_{3\times2} & \boldsymbol{r}_{P_3} - \boldsymbol{r}_{B_3} \end{bmatrix}^T,$$

$$\mathbf{A}_{21} = \begin{bmatrix} (1-k_{41}-k_{42}-k_{43})(x_4-x_{B_4}) & (1-k_{41}-k_{42}-k_{43})(y_4-y_{B_4}) & (1-k_{41}-k_{42}-k_{43})(z_4-z_{B_4}) \\ (1-k_{51}-k_{52}-k_{53})(x_5-x_{B_5}) & (1-k_{51}-k_{52}-k_{53})(y_5-y_{B_5}) & (1-k_{51}-k_{52}-k_{53})(z_5-z_{B_5}) \\ (1-k_{61}-k_{62}-k_{63})(x_6-x_{B_6}) & (1-k_{61}-k_{62}-k_{63})(y_6-y_{B_6}) & (1-k_{61}-k_{62}-k_{63})(z_6-z_{B_6}) \end{bmatrix},$$

$$\mathbf{A}_{22} = \begin{bmatrix} k_{41}(x_4-x_{B_4}) & k_{41}(y_4-y_{B_4}) & k_{41}(z_4-z_{B_4}) \\ k_{51}(x_5-x_{B_5}) & k_{51}(y_5-y_{B_5}) & k_{51}(z_5-z_{B_5}) \\ k_{61}(x_6-x_{B_6}) & k_{61}(y_6-y_{B_6}) & k_{61}(z_6-z_{B_6}) \end{bmatrix},$$

$$\mathbf{A}_{23} = \begin{bmatrix} k_{42}(x_4 - x_{B_4}) & k_{42}(y_4 - y_{B_4}) & k_{42}(z_4 - z_{B_4}) \\ k_{52}(x_5 - x_{B_5}) & k_{52}(y_5 - y_{B_5}) & k_{52}(z_5 - z_{B_5}) \\ k_{62}(x_6 - x_{B_6}) & k_{62}(y_6 - y_{B_6}) & k_{62}(z_6 - z_{B_6}) \end{bmatrix},$$

$$\mathbf{A}_{24} = \begin{bmatrix} k_{43}(x_4 - x_{B_4}) & k_{43}(y_4 - y_{B_4}) & k_{43}(z_4 - z_{B_4}) \\ k_{53}(x_5 - x_{B_5}) & k_{53}(y_5 - y_{B_5}) & k_{53}(z_5 - z_{B_5}) \\ k_{63}(x_6 - x_{B_6}) & k_{63}(y_6 - y_{B_6}) & k_{63}(z_6 - z_{B_6}) \end{bmatrix},$$

$$\mathbf{A}_{31} = \mathbf{0}_{3\times 3},$$

$$\mathbf{A}_{32} = \begin{bmatrix} \boldsymbol{r}_{P_1} - \boldsymbol{r}_{P_2} & \mathbf{0}_{3\times 1} & \boldsymbol{r}_{P_1} - \boldsymbol{r}_{P_3} \end{bmatrix}^T,$$

$$\mathbf{A}_{33} = \begin{bmatrix} \boldsymbol{r}_{P_2} - \boldsymbol{r}_{P_1} & \boldsymbol{r}_{P_2} - \boldsymbol{r}_{P_3} & \mathbf{0}_{3\times 1} \end{bmatrix}^T,$$

$$\mathbf{A}_{34} = \begin{bmatrix} \mathbf{0}_{3\times 1} & \boldsymbol{r}_{P_3} - \boldsymbol{r}_{P_2} & \boldsymbol{r}_{P_3} - \boldsymbol{r}_{P_1} \end{bmatrix}^T,$$

$$\mathbf{A}_{41} = \begin{bmatrix} \boldsymbol{r}_{P_0} - \boldsymbol{r}_{P_1} & \boldsymbol{r}_{P_0} - \boldsymbol{r}_{P_2} & \boldsymbol{r}_{P_0} - \boldsymbol{r}_{P_1} \end{bmatrix}^T,$$

$$\mathbf{A}_{42} = \begin{bmatrix} \boldsymbol{r}_{P_1} - \boldsymbol{r}_{B_0} & \mathbf{0}_{3\times 2} \end{bmatrix}^T,$$

$$\mathbf{A}_{43} = \begin{bmatrix} \mathbf{0}_{3\times 1} & \boldsymbol{r}_{P_2} - \boldsymbol{r}_{B_0} & \mathbf{0}_{3\times 1} \end{bmatrix}^T,$$

$$\mathbf{A}_{44} = \begin{bmatrix} \mathbf{0}_{3\times 2} & \boldsymbol{r}_{P_3} - \boldsymbol{r}_{B_0} \end{bmatrix}^T.$$

The singularity criteria of the manipulator will be naturally expressed as:

$$\left|\boldsymbol{P}'(\boldsymbol{x})\right| = 0. \tag{7.51}$$

When equation (7.51) does not hold, the forward displacement solutions can be obtained through the Newton–Raphson method [9] or the Newton–GMRES method [10]. However, what must be pointed out is that the iterating results might be different because of the multi-solutions of equation (7.49) if we use different start points for numerical iterating. As to all solutions of a Gough–Stewart forward kinematics, Merlet [11] presented an algorithm based on interval analysis, which allows us to consider the physical and technological constraints.

The derivative of equation (7.49) with respect to time can be expressed as:

$$\mathbf{A}\dot{\boldsymbol{x}} = \boldsymbol{b} \tag{7.52}$$

where

$$\mathbf{A} = \begin{bmatrix} \mathbf{A}_{11} & \mathbf{A}_{12} & \mathbf{A}_{13} & \mathbf{A}_{14} \\ \mathbf{A}_{21} & \mathbf{A}_{22} & \mathbf{A}_{23} & \mathbf{A}_{24} \\ \mathbf{A}_{31} & \mathbf{A}_{32} & \mathbf{A}_{33} & \mathbf{A}_{34} \\ \mathbf{A}_{41} & \mathbf{A}_{42} & \mathbf{A}_{43} & \mathbf{A}_{44} \end{bmatrix},$$

$$\boldsymbol{b} = \begin{bmatrix} l_1\dot{l}_1 & l_2\dot{l}_2 & l_3\dot{l}_3 & l_4\dot{l}_4 & l_5\dot{l}_5 & l_6\dot{l}_6 & 0 & 0 & 0 & 0 & 0 & 0 \end{bmatrix}^T.$$

After $\boldsymbol{x}$ is iterated with equation (7.5), $\dot{x}$ will be immediately solved with (7.8) for the coefficient matrix $\mathbf{A}$ are known constant matrix.

Therefore, the positions of the manipulator's geometry center C can be obtained according to equation (7.15). Suppose that

$$(\boldsymbol{r}_{P_0} - \boldsymbol{r}) = k_1(\boldsymbol{r}_{P_1} - \boldsymbol{r}) + k_2(\boldsymbol{r}_{P_2} - \boldsymbol{r}) + k_3(\boldsymbol{r}_{P_3} - \boldsymbol{r})$$

where $k_i\,(i = 1,2,3)$ are specified by

$$\begin{cases} k_1x_1 + k_2x_2 + k_3x_3 = x_0 \\ k_1y_1 + k_2y_2 + k_3y_3 = y_0 \\ k_1z_1 + k_2z_2 + k_3z_3 = z_0. \end{cases}$$

We have

$$\boldsymbol{r} = \frac{k_1\boldsymbol{r}_{P_1} + k_2\boldsymbol{r}_{P_2} + k_3\boldsymbol{r}_{P_3} - \boldsymbol{r}_{P_0}}{k_1 + k_2 + k_3 - 1}. \tag{7.53}$$

Therefore,

$$\begin{bmatrix} x \\ y \\ z \end{bmatrix} = \begin{bmatrix} K_0 & 0 & 0 & K_1 & 0 & 0 & K_2 & 0 & 0 & K_3 & 0 & 0 \\ 0 & K_0 & 0 & 0 & K_1 & 0 & 0 & K_2 & 0 & 0 & K_3 & 0 \\ 0 & 0 & K_0 & 0 & 0 & K_1 & 0 & 0 & K_2 & 0 & 0 & K_3 \end{bmatrix} \boldsymbol{x} \tag{7.54}$$

where $K_0 = \frac{-1}{k_1+k_2+k_3-1}$, $K_1 = \frac{k_1}{k_1+k_2+k_3-1}$, $K_2 = \frac{k_2}{k_1+k_2+k_3-1}$, $K_3 = \frac{k_3}{k_1+k_2+k_3-1}$.

Therefore, the velocities of the manipulator's geometry center can be obtained:

$$\begin{bmatrix} \dot{x} \\ \dot{y} \\ \dot{z} \end{bmatrix} = \begin{bmatrix} K_0 & 0 & 0 & K_1 & 0 & 0 & K_2 & 0 & 0 & K_3 & 0 & 0 \\ 0 & K_0 & 0 & 0 & K_1 & 0 & 0 & K_2 & 0 & 0 & K_3 & 0 \\ 0 & 0 & K_0 & 0 & 0 & K_1 & 0 & 0 & K_2 & 0 & 0 & K_3 \end{bmatrix} \dot{\boldsymbol{x}}. \tag{7.55}$$

Consequently, the angular velocities of the manipulator can be solved according to equation (7.35).

The derivative of equation (7.52) with respect to time are expressed as:

$$\mathbf{A}\ddot{\boldsymbol{x}} = \boldsymbol{d} \tag{7.56}$$

where $\ddot{\boldsymbol{x}} = \begin{bmatrix} \ddot{x}_0 & \ddot{y}_0 & \ddot{z}_0 & \ddot{x}_1 & \ddot{y}_1 & \ddot{z}_1 & \ddot{x}_2 & \ddot{y}_2 & \ddot{z}_2 & \ddot{x}_3 & \ddot{y}_3 & \ddot{z}_3 \end{bmatrix}^T$,

$$\boldsymbol{d} = \begin{bmatrix} l_1\ddot{l}_1 + (\dot{l}_1)^2 - (\dot{x}_1)^2 - (\dot{y}_1)^2 - (\dot{z}_1)^2 \\ l_2\ddot{l}_2 + (\dot{l}_2)^2 - (\dot{x}_2)^2 - (\dot{y}_2)^2 - (\dot{z}_2)^2 \\ l_3\ddot{l}_3 + (\dot{l}_3)^2 - (\dot{x}_3)^2 - (\dot{y}_3)^2 - (\dot{z}_3)^2 \\ l_4\ddot{l}_4 + (\dot{l}_4)^2 - [(1-k_{41}-k_{42}-k_{43})\dot{x}_0 + k_{41}\dot{x}_1 + k_{42}\dot{x}_2 + k_{43}\dot{x}_3]^2 \\ -[(1-k_{41}-k_{42}-k_{43})\dot{y}_0 + k_{41}\dot{y}_1 + k_{42}\dot{y}_2 + k_{43}\dot{y}_3]^2 - [(1-k_{41}-k_{42}-k_{43})\dot{z}_0 + k_{41}\dot{z}_1 + k_{42}\dot{z}_2 + k_{43}\dot{z}_3]^2 \\ l_5\ddot{l}_5 + (\dot{l}_5)^2 - [(1-k_{51}-k_{52}-k_{53})\dot{x}_0 + k_{51}\dot{x}_1 + k_{52}\dot{x}_2 + k_{53}\dot{x}_3]^2 \\ -[(1-k_{51}-k_{52}-k_{53})\dot{y}_0 + k_{51}\dot{y}_1 + k_{52}\dot{y}_2 + k_{53}\dot{y}_3]^2 - [(1-k_{51}-k_{52}-k_{53})\dot{z}_0 + k_{51}\dot{z}_1 + k_{52}\dot{z}_2 + k_{53}\dot{z}_3]^2 \\ l_6\ddot{l}_6 + (\dot{l}_6)^2 - [(1-k_{61}-k_{62}-k_{63})\dot{x}_0 + k_{61}\dot{x}_1 + k_{62}\dot{x}_2 + k_{63}\dot{x}_3]^2 \\ -[(1-k_{61}-k_{62}-k_{63})\dot{y}_0 + k_{61}\dot{y}_1 + k_{62}\dot{y}_2 + k_{63}\dot{y}_3]^2 - [(1-k_{61}-k_{62}-k_{63})\dot{z}_0 + k_{61}\dot{z}_1 + k_{62}\dot{z}_2 + k_{63}\dot{z}_3]^2 \\ -(\dot{x}_1-\dot{x}_2)^2 - (\dot{y}_1-\dot{y}_2)^2 - (\dot{z}_1-\dot{z}_2)^2 \\ -(\dot{x}_2-\dot{x}_3)^2 - (\dot{y}_2-\dot{y}_3)^2 - (\dot{z}_2-\dot{z}_3)^2 \\ -(\dot{x}_3-\dot{x}_1)^2 - (\dot{y}_3-\dot{y}_1)^2 - (\dot{z}_3-\dot{z}_1)^2 \\ -(\dot{x}_1-\dot{x}_0)^2 - (\dot{y}_1-\dot{y}_0)^2 - (\dot{z}_1-\dot{z}_0)^2 \\ -(\dot{x}_2-\dot{x}_0)^2 - (\dot{y}_2-\dot{y}_0)^2 - (\dot{z}_2-\dot{z}_0)^2 \\ -(\dot{x}_3-\dot{x}_0)^2 - (\dot{y}_3-\dot{y}_0)^2 - (\dot{z}_3-\dot{z}_0)^2 \end{bmatrix}.$$

According to equation (7.56), the absolute accelerations of the manipulator's geometry center can be described as:

$$\begin{bmatrix} \ddot{x} \\ \ddot{y} \\ \ddot{z} \end{bmatrix} = \begin{bmatrix} K_0 & 0 & 0 & K_1 & 0 & 0 & K_2 & 0 & 0 & K_3 & 0 & 0 \\ 0 & K_0 & 0 & 0 & K_1 & 0 & 0 & K_2 & 0 & 0 & K_3 & 0 \\ 0 & 0 & K_0 & 0 & 0 & K_1 & 0 & 0 & K_2 & 0 & 0 & K_3 \end{bmatrix} \ddot{\boldsymbol{x}}. \tag{7.57}$$

Accordingly, the angular velocity and angular acceleration of the end effector can be obtained in accordance with equations (7.35) and (7.43).

The parameters of a Gough–Stewart manipulator are specified in Table 7.1.

The initial configuration of the Stewart platform is that the orientation angles of the end effector equal to zero and the coordinate of the center point is $\boldsymbol{t} = \begin{bmatrix} 0 & 0 & 1800 \end{bmatrix}^T$.

TABLE 7.1 The Parameters of a Gough–Stewart Manipulator

Notations	Description
$l_i \in$ [1800mm,3200mm]	Length of each leg, $i = 1,2,3,4,5,6$
$r = 300$ mm	Radius of the manipulator circumcircle
$R = 1000$ mm	Radius of the base circumcircle
$h = 20$ mm	Offset distance of different joints

The coordinates of the joints on the fixed base in coordinate system $OXYZ$ are

$$B_1\begin{pmatrix} R\cos 45^\circ & -R\sin 45^\circ & 0 \end{pmatrix},\quad B_2\begin{pmatrix} R\cos 15^\circ & -R\sin 15^\circ & 0 \end{pmatrix},$$
$$B_3\begin{pmatrix} R\sin 15^\circ & R\cos 15^\circ & 0 \end{pmatrix},\quad B_4\begin{pmatrix} -R\sin 15^\circ & R\cos 15^\circ & 0 \end{pmatrix},$$
$$B_5\begin{pmatrix} -R\cos 15^\circ & -R\sin 15^\circ & 0 \end{pmatrix},\quad B_6\begin{pmatrix} -R\cos 45^\circ & -R\sin 45^\circ & 0 \end{pmatrix}.$$

In the local coordinate system $oxyz$, the coordinates of the joints on the end effector are

$$P_1\begin{pmatrix} r\sin 10^\circ & -r\cos 10^\circ & 0 \end{pmatrix}, P_2\begin{pmatrix} r\cos 20^\circ & r\sin 20^\circ & h \end{pmatrix},$$
$$P_3\begin{pmatrix} r\cos 40^\circ & r\sin 40^\circ & 0 \end{pmatrix}, P_4\begin{pmatrix} -r\sin 50^\circ & r\cos 50^\circ & h \end{pmatrix},$$
$$P_5\begin{pmatrix} -r\sin 70^\circ & r\cos 70^\circ & 0 \end{pmatrix}, P_6\begin{pmatrix} -r\sin 10^\circ & -r\cos 10^\circ & h \end{pmatrix}.$$

And the local coordinates of point P_0 are $P_0\begin{pmatrix} 0 & 0 & 3h \end{pmatrix}$. According to (7.45), we obtain:

$$\begin{cases} k_{41}(r\sin 10^\circ - 0) + k_{42}(r\cos 20^\circ - 0) + k_{43}(r\cos 40^\circ - 0) = r\cos 20^\circ - 0 \\ k_{41}(-r\cos 10^\circ - 0) + k_{42}(r\sin 20^\circ - 0) + k_{43}(r\sin 40^\circ - 0) = r\sin 20^\circ - 0 \\ k_{41}(0 - 3h) + k_{42}(h - 3h) + k_{43}(0 - 3h) = h - 3h. \end{cases} \tag{7.58}$$

As a result, $\boldsymbol{k}_4$ can be solved from (7.58):

$$\boldsymbol{k}_4 = \begin{bmatrix} 0.0247 & -2.9420 & 2.6033 \end{bmatrix}^T. \tag{7.59}$$

With similar process, we can obtain $\boldsymbol{k}_5$ and $\boldsymbol{k}_6$ by solving the linear constant equations (7.45).

$$\begin{cases} \boldsymbol{k}_5 = \begin{bmatrix} 0.3690 & -3.4665 & 2.9420 \end{bmatrix}^T \\ \boldsymbol{k}_6 = \begin{bmatrix} 0.8880 & -0.3690 & 0.0247 \end{bmatrix}^T \end{cases} \tag{7.60}$$

Substituting $\boldsymbol{k}_4$,$\boldsymbol{k}_5$, and $\boldsymbol{k}_6$ into equation (7.49), the $\boldsymbol{x}$ can be gained through the Newton–Raphson method [9], and then the velocity, acceleration, angular velocity, and angular acceleration of the end effector can be obtained according to equations (7.55), (7.57), (7.35), and (7.43), respectively.

To verify the effectiveness and efficiency of the proposed method, another two different methods [12,13] are introduced to solve the same kinematic problems. The position and orientation of the end effector were described by six variables in the rotation transformation matrix method [12]. They are

$$\boldsymbol{x} = \begin{bmatrix} X & Y & Z & \phi & \theta & \psi \end{bmatrix}^T$$

where $(X\ \ Y\ \ Z)$ indicate the position of the end effector and the Euler angles $(\phi\ \ \theta\ \ \psi)$ are used to describe the orientation of the end effector. The rotation transformation matrix can be obtained as following.

$$\mathbf{R} = \begin{bmatrix} c\psi c\phi - c\theta s\phi s\psi & -s\psi c\phi - c\theta s\phi c\psi & s\theta s\phi \\ c\psi s\phi + c\theta c\phi s\psi & -s\psi s\phi + c\theta c\phi c\psi & -s\theta c\phi \\ s\psi s\theta & c\psi s\theta & c\theta \end{bmatrix}$$

where s and c denote sine and cosine, respectively.

Considering the six constraints equations of the six legs, the forward kinematics of the Stewart platform can express a system of six equations with six unknown variables.

The idea of the method [13] is to use the elements of the direction cosine matrix, rather than the angle value, to represent orientation:

$$\mathbf{R} = \begin{bmatrix} n_x & o_x & a_x \\ n_y & o_y & a_y \\ n_z & o_z & a_z \end{bmatrix}.$$

The elements of the rotation transformation matrix are interdependent in the following fashion:

$$\begin{cases} \boldsymbol{n} \cdot \boldsymbol{n} = n_x^2 + n_y^2 + n_z^2 = 1 \\ \boldsymbol{o} \cdot \boldsymbol{o} = o_x^2 + o_y^2 + o_z^2 = 1 \\ \boldsymbol{n} \cdot \boldsymbol{o} = n_x o_x + n_y o_y + n_z o_z = 0 \\ \boldsymbol{a} = \boldsymbol{n} \times \boldsymbol{o}. \end{cases}$$

Then the forward kinematics can be expressed as a system of nine equations with nine unknown variables.

$$\boldsymbol{x} = \begin{bmatrix} X & Y & Z & n_x & n_y & n_z & o_x & o_y & o_z \end{bmatrix}^T$$

Assume that the velocity of each leg of the Gough–Stewart manipulator changes with respect to time is shown in Figure 7.3, $t_1 = 4$, $t_2 = 6$, $t_3 = 10$, and the maximum velocity of each leg is shown in Table 7.2.

The algorithm applied in solving the forward kinematic problems is the Newton–Raphson method. The iterating step is equal to 0.1 s and the last solution is chosen as the initial start point in the next iterating cycle. Then the computing times of three different methods are listed in Table 7.3 and the displacements of the end effector with respect to time are shown in Figure 7.4. It should be pointed out that the computing time is the time to solve the $\boldsymbol{x}$ by the Newton–Raphson method.

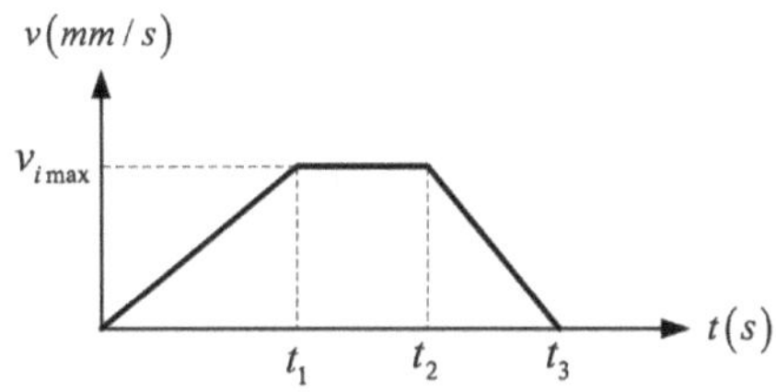

FIGURE 7.3 The velocity of each leg changes with respect to time.

TABLE 7.2 The Maximum Velocity of Each Leg

Leg Number	$v_{i\max}(mm/s)$
1	80
2	40
3	70
4	65
5	60
6	50

TABLE 7.3 Computing Time of Different Methods

Method	Computing Time (s)
Rotation transformation matrix	498.1775
Direction cosine	489.4073
Four points	83.8176

From Figure 7.4, we can find that the direction cosine method and the method with four points' Cartesian coordinates can effectively accomplish real-time calculations, but the rotation transformation matrix method failed. Moreover, the four points' Cartesian coordinates method is the most efficient, as shown by comparing the computing time in Table 7.3.

The displacement, velocity, angular velocity, acceleration, and angular acceleration with respect to time are shown in Figures 7.4, 7.5, 7.6, 7.7, and 7.8, respectively.

From the above analysis, we find that the method with four points' Cartesian coordinates can effectively and efficiently accomplish real-time calculations through numerical simulations no matter how many extra Cartesian Coordinates of points are introduced.

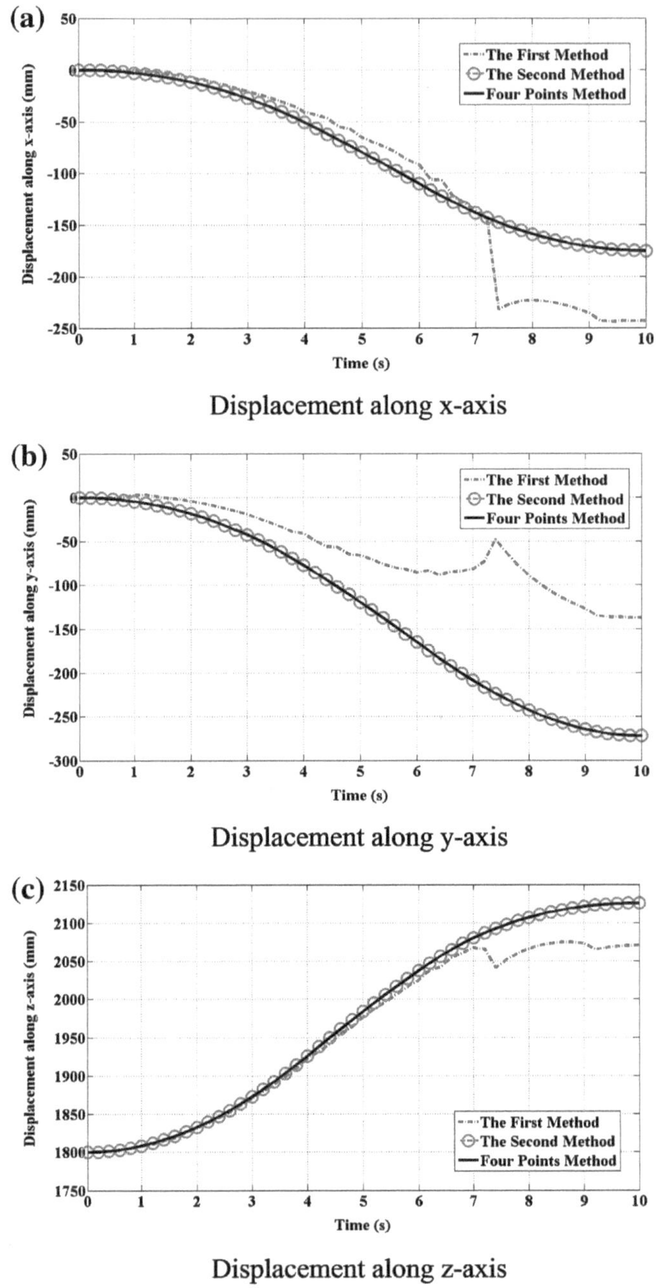

FIGURE 7.4 Displacement-time curves of the Gough–Stewart manipulator.

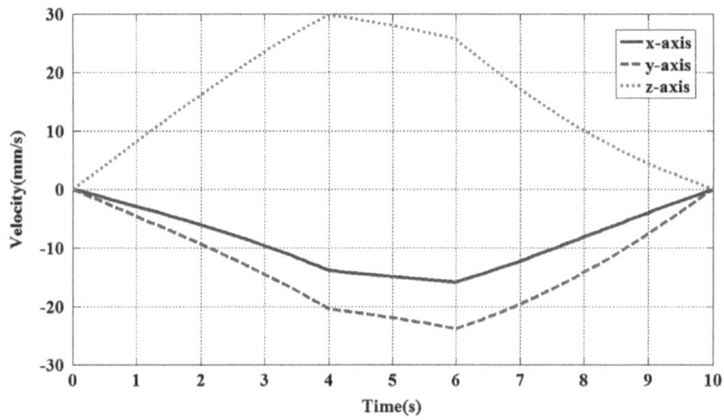

FIGURE 7.5 Velocity–time curves of the Gough–Stewart manipulator.

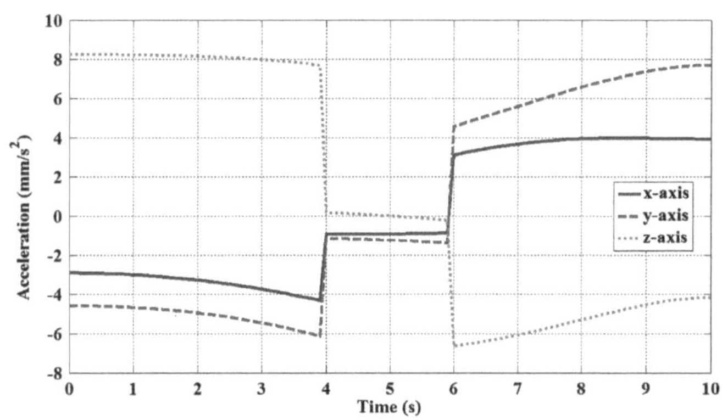

FIGURE 7.6 Acceleration–time curves of the Gough–Stewart manipulator.

7.2.2 Kinematic Problem of a Spatial Parallel Manipulator with One Translation and Three Rotational Degrees of Freedom

As mentioned in Section 7.1, the forward displacement and kinematics could be settled only with three points' coordinates. As a result, for some spatial parallel manipulators with fewer degrees of freedom (DOFs), the forward displacement and kinematic problems could be solved only with three points' coordinates. As Figure 7.9 shows, the moving end effector of a spatial parallel manipulator is linked with the fixed base through one central *PS* kinematic chain and three *PUS* chains. Its end effector has one translational DOF and three rotational DOFs, which can be analyzed according to [14].

The top view of the end effector is shown in Figure 7.10, the three vertexes of M_1, M_2, and M_3 are equally distributed in a same circle with the center

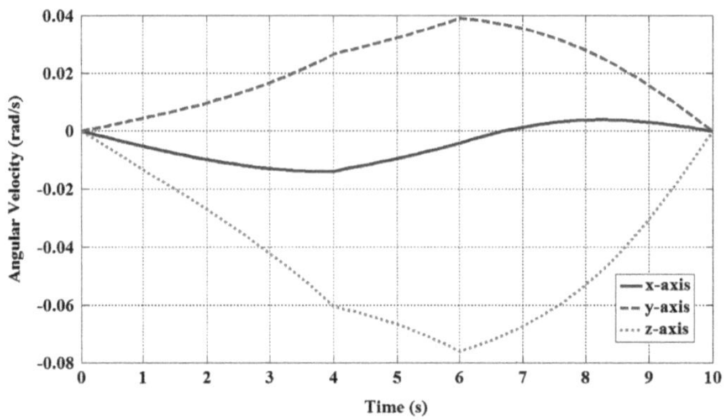

FIGURE 7.7 Angular velocity–time curves of the Gough–Stewart manipulator.

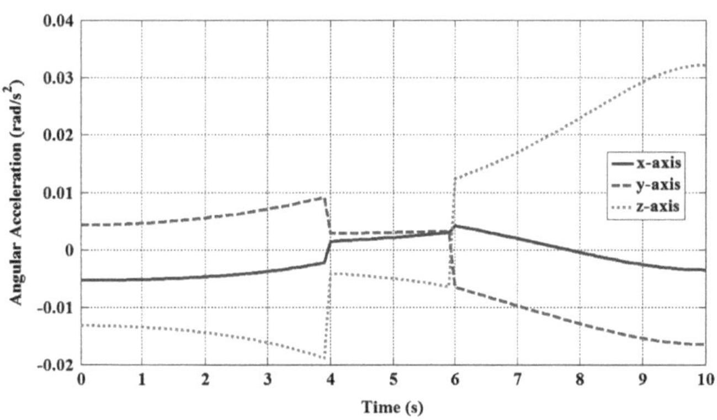

FIGURE 7.8 Angular acceleration–time curves of the Gough–Stewart manipulator.

on vertex M_4. Therefore, if we particularly choose the Cartesian coordinates of points M_1, M_2, and M_3 as the natural coordinates, the positions of the manipulator's geometry center, M_4, can be immediately obtained according to equation (7.18):

$$\begin{bmatrix} x_4 \\ y_4 \\ z_4 \end{bmatrix} = \frac{1}{3}\begin{bmatrix} 1 & 0 & 0 & 1 & 0 & 0 & 1 & 0 & 0 \\ 0 & 1 & 0 & 0 & 1 & 0 & 0 & 1 & 0 \\ 0 & 0 & 1 & 0 & 0 & 1 & 0 & 0 & 1 \end{bmatrix} \boldsymbol{x} \tag{7.61}$$

where $\boldsymbol{x} = \begin{bmatrix} x_1 & y_1 & z_1 & x_2 & y_2 & z_2 & x_3 & y_3 & z_3 \end{bmatrix}^T$.

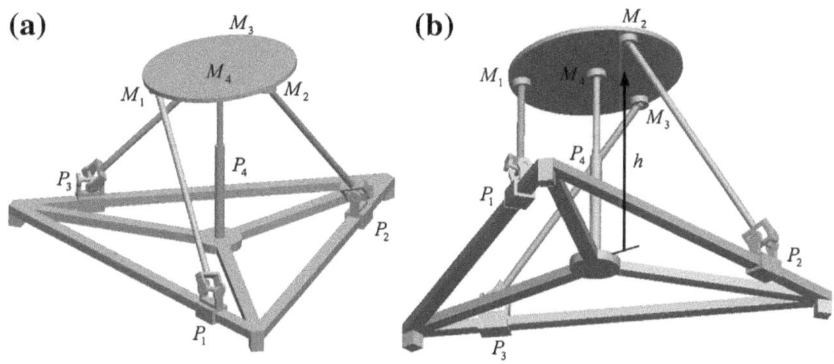

FIGURE 7.9 Architecture of a parallel manipulator with one translation and three rotations.

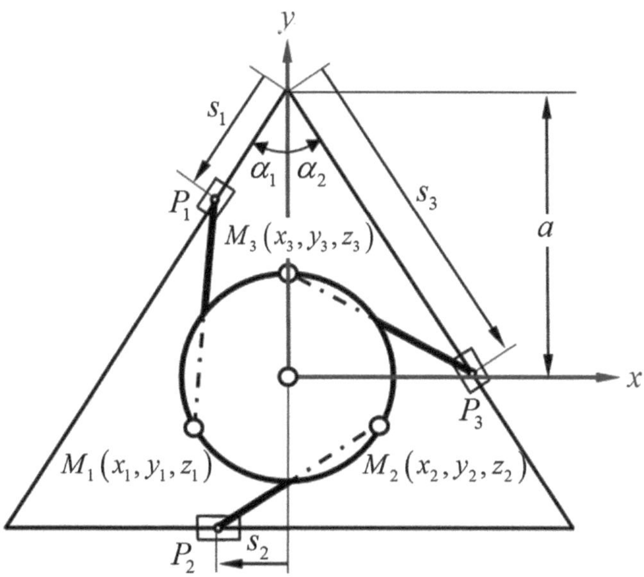

FIGURE 7.10 Top view of the mechanism.

Expanding equation (7.61) yields:

$$\begin{bmatrix} x_4 \\ y_4 \\ z_4 \end{bmatrix} = \frac{1}{3} \begin{bmatrix} x_1 + x_2 + x_3 \\ y_1 + y_2 + y_3 \\ z_1 + z_2 + z_3 \end{bmatrix}. \tag{7.62}$$

According to the column structure's rigid body constraints, we can find that the displacement of vertex M_4 satisfies

$$\begin{bmatrix} x_4 \\ y_4 \\ z_4 \end{bmatrix} = \begin{bmatrix} 0 \\ 0 \\ h \end{bmatrix} \tag{7.63}$$

where h denotes the vertical displacement induced by actuator P_4.

As a result, we obtain a set of actuation constraints from equations (7.62) and (7.63):

$$\begin{cases} x_1 + x_2 + x_3 = 0 \\ y_1 + y_2 + y_3 = 0 \\ z_1 + z_2 + z_3 - 3h = 0. \end{cases} \tag{7.64}$$

The other three-actuation constraints are

$$\begin{cases} (x_1 - x_{P_1})^2 + (y_1 - y_{P_1})^2 + (z_1 - z_{P_1})^2 - l_1^2 = 0 \\ (x_2 - x_{P_2})^2 + (y_2 - y_{P_2})^2 + (z_2 - z_{P_2})^2 - l_2^2 = 0 \\ (x_3 - x_{P_3})^2 + (y_3 - y_{P_3})^2 + (z_3 - z_{P_3})^2 - l_3^2 = 0 \end{cases} \tag{7.65}$$

where l_1, l_2, and l_3 denote the fixed lengths of the limbs P_1M_1, P_2M_2, and P_3M_3.

If we assume the displacements s_1, s_2, and s_3 to be a new set of actuation variables and the positive directions to be shown by their arrows, the Cartesian coordinates of the three sliders can be expressed:

$$\begin{cases} x_{P_1} = -s_1 \sin\alpha_1 \\ y_{P_1} = a - s_1 \cos\alpha_1 \\ z_{P_1} = 0 \end{cases}, \quad \begin{cases} x_{P_2} = -s_2 \\ y_{P_2} = 0 \\ z_{P_2} = 0 \end{cases} \quad \text{and} \quad \begin{cases} x_{P_3} = s_3 \sin\alpha_2 \\ y_{P_3} = a - s_3 \cos\alpha_1 \\ z_{P_3} = 0 \end{cases}$$

where α is a constant value and $\alpha_1 = \alpha_2 = \frac{\pi}{6}$ here. Therefore, equation (7.65) can be transformed to

$$\begin{cases} (x_1 + \frac{1}{2}s_1)^2 + (y_1 - a + \frac{\sqrt{3}}{2}s_1)^2 + z_1^2 - l_1^2 = 0 \\ (x_2 + s_2)^2 + y_2^2 + z_2^2 - l_2^2 = 0 \\ (x_3 - \frac{1}{2}s_3)^2 + (y_3 - a + \frac{\sqrt{3}}{2}s_3)^2 + z_3^2 - l_3^2 = 0. \end{cases} \tag{7.66}$$

The rigid body constraints are:

$$\begin{cases} (x_1 - x_2)^2 + (y_1 - y_2)^2 + (z_1 - z_2)^2 - L_{M_1M_2}^2 = 0 \\ (x_2 - x_3)^2 + (y_2 - y_3)^2 + (z_2 - z_3)^2 - L_{M_2M_3}^2 = 0 \\ (x_3 - x_1)^2 + (y_3 - y_1)^2 + (z_3 - z_1)^2 - L_{M_3M_1}^2 = 0. \end{cases} \tag{7.67}$$

Therefore, the relationship constraint equations can be obtained from equations (7.64), (7.66), and (7.67):

$$\boldsymbol{P}\left(\boldsymbol{x}\right) = \begin{bmatrix} (x_1 + \frac{1}{2}s_1)^2 + (y_1 - a + \frac{\sqrt{3}}{2}s_1)^2 + z_1^2 - l_1^2 \\ (x_2 + s_2)^2 + y_2^2 + z_2^2 - l_2^2 \\ (x_3 - \frac{1}{2}s_3)^2 + (y_3 - a + \frac{\sqrt{3}}{2}s_3)^2 + z_3^2 - l_3^2 \\ (x_1 - x_2)^2 + (y_1 - y_2)^2 + (z_1 - z_2)^2 - L_{M_1M_2}^2 \\ (x_2 - x_3)^2 + (y_2 - y_3)^2 + (z_2 - z_3)^2 - L_{M_2M_3}^2 \\ (x_3 - x_1)^2 + (y_3 - y_1)^2 + (z_3 - z_1)^2 - L_{M_3M_1}^2 \\ x_1 + x_2 + x_3 \\ y_1 + y_2 + y_3 \\ z_1 + z_2 + z_3 - 3h \end{bmatrix} = \boldsymbol{0}. \tag{7.68}$$

The fourth to eighth relationship constraint equations of (7.68) are obtained via rigid body constraint conditions and the others are four actuations' relationship constraint conditions.

The Jacobian matrix of equation (7.68) can be gained:

$$\boldsymbol{P}'(\boldsymbol{x}) = \frac{\partial \boldsymbol{P}(\boldsymbol{x})}{\partial \boldsymbol{x}} = 2\begin{bmatrix} x_1 + \frac{1}{2}s_1 & y_1 - a + \frac{\sqrt{3}}{2}s_1 & z_1 & 0 & 0 & 0 & 0 & 0 & 0 \\ 0 & 0 & 0 & x_2 + s_2 & y_2 & z_2 & 0 & 0 & 0 \\ 0 & 0 & 0 & 0 & 0 & 0 & x_3 - \frac{1}{2}s_3 & y_3 - a + \frac{\sqrt{3}}{2}s_3 & z_3 \\ x_1 - x_2 & y_1 - y_2 & z_1 - z_2 & x_2 - x_1 & y_2 - y_1 & z_2 - z_1 & 0 & 0 & 0 \\ 0 & 0 & 0 & x_2 - x_3 & y_2 - y_3 & z_2 - z_3 & x_3 - x_2 & y_3 - y_2 & z_3 - z_2 \\ x_1 - x_3 & y_1 - y_3 & z_1 - z_3 & 0 & 0 & 0 & x_3 - x_1 & y_3 - y_1 & z_3 - z_1 \\ \frac{1}{2} & 0 & 0 & \frac{1}{2} & 0 & 0 & \frac{1}{2} & 0 & 0 \\ 0 & \frac{1}{2} & 0 & 0 & \frac{1}{2} & 0 & 0 & \frac{1}{2} & 0 \\ 0 & 0 & \frac{1}{2} & 0 & 0 & \frac{1}{2} & 0 & 0 & \frac{1}{2} \end{bmatrix} \tag{7.69}$$

The singularity criteria of the manipulator will be naturally expressed as:

$$\left|\boldsymbol{P}'(\boldsymbol{x})\right| = 0. \tag{7.70}$$

When equation (7.70) does not hold, the forward displacement solutions can be obtained with the Newton–Raphson method [9] or the Newton–GMRES method [10]. The derivative of equation set (7.68) with respect to time can be expressed as:

$$\begin{bmatrix} (x_1 + \frac{1}{2}s_1)(\dot{x_1} + \frac{1}{2}\dot{s_1}) + (y_1 - a + \frac{\sqrt{3}}{2}s_1)(\dot{y_1} + \frac{\sqrt{3}}{2}\dot{s_1}) + z_1\dot{z_1} \\ (x_2 + s_2)(\dot{x_2} + \dot{s_2}) + y_2\dot{y_2} + z_2\dot{z_2} \\ (x_3 - \frac{1}{2}s_3)(\dot{x_3} - \frac{1}{2}\dot{s_3}) + (y_3 - a + \frac{\sqrt{3}}{2}s_3)(\dot{y_3} + \frac{\sqrt{3}}{2}\dot{s_3}) + z_3\dot{z_3} \\ (x_1 - x_2)(\dot{x_1} - \dot{x_2}) + (y_1 - y_2)(\dot{y_1} - \dot{y_2}) + (z_1 - z_2)(\dot{z_1} - \dot{z_2}) \\ (x_2 - x_3)(\dot{x_2} - \dot{x_3}) + (y_2 - y_3)(\dot{y_2} - \dot{y_3}) + (z_2 - z_3)(\dot{z_2} - \dot{z_3}) \\ (x_3 - x_1)(\dot{x_3} - \dot{x_1}) + (y_3 - y_1)(\dot{y_3} - \dot{y_1}) + (z_3 - z_1)(\dot{z_3} - \dot{z_1}) \\ \dot{x_1} + \dot{x_2} + \dot{x_3} \\ \dot{y_1} + \dot{y_2} + \dot{y_3} \\ \dot{z_1} + \dot{z_2} + \dot{z_3} - 3\dot{h} \end{bmatrix} = \mathbf{0}. \tag{7.71}$$

In matrix form, equation (7.71) can be rewritten as:

$$\mathbf{A}\dot{\boldsymbol{x}} = \boldsymbol{c} \tag{7.72}$$

where $\dot{\boldsymbol{x}} = \begin{bmatrix} \dot{x_1} & \dot{y_1} & \dot{z_1} & \dot{x_2} & \dot{y_2} & \dot{z_2} & \dot{x_3} & \dot{y_3} & \dot{z_3} \end{bmatrix}^T$,

$$\mathbf{A} = \begin{bmatrix} x_1 + \frac{1}{2}s_1 & y_1 - a + \frac{\sqrt{3}}{2}s_1 & z_1 & 0 & 0 & 0 & 0 & 0 & 0 \\ 0 & 0 & 0 & x_2 + s_2 & y_2 & z_2 & 0 & 0 & 0 \\ 0 & 0 & 0 & 0 & 0 & 0 & x_3 - \frac{1}{2}s_3 & y_3 - a + \frac{\sqrt{3}}{2}s_3 & z_3 \\ x_1 - x_2 & y_1 - y_2 & z_1 - z_2 & x_2 - x_1 & y_2 - y_1 & z_2 - z_1 & 0 & 0 & 0 \\ 0 & 0 & 0 & x_2 - x_3 & y_2 - y_3 & z_2 - z_3 & x_3 - x_2 & y_3 - y_2 & z_3 - z_2 \\ x_1 - x_3 & y_1 - y_3 & z_1 - z_3 & 0 & 0 & 0 & x_3 - x_1 & y_3 - y_1 & z_3 - z_1 \\ 1 & 0 & 0 & 1 & 0 & 0 & 1 & 0 & 0 \\ 0 & 1 & 0 & 0 & 1 & 0 & 0 & 1 & 0 \\ 0 & 0 & 1 & 0 & 0 & 1 & 0 & 0 & 1 \end{bmatrix},$$

$$\boldsymbol{c} = \begin{bmatrix} -\left[\frac{1}{2}(x_1 + \frac{1}{2}s_1) + \frac{\sqrt{3}}{2}(y_1 - a + \frac{\sqrt{3}}{2}s_1)\right]\dot{s_1} \\ -(x_2 + s_2)\dot{s_2} \\ \left[\frac{1}{2}(x_3 - \frac{1}{2}s_3) - \frac{\sqrt{3}}{2}(y_3 - a + \frac{\sqrt{3}}{2}s_3)\right]\dot{s_3} \\ 0 \\ 0 \\ 0 \\ 0 \\ 0 \\ 3\dot{h} \end{bmatrix}.$$

After $\boldsymbol{x}$ is iterated with equation (7.5), $\dot{\boldsymbol{x}}$ will be easily solved via equation (7.72) for the coefficient matrix $\mathbf{A}$ are known matrix function of $\boldsymbol{x}$.

Therefore, the velocities of the manipulator's geometry center, M_4, can be obtained by associating equations (7.19) and (7.63):

$$\begin{bmatrix} \dot{x_4} \\ \dot{y_4} \\ \dot{z_4} \end{bmatrix} = \begin{bmatrix} 0 \\ 0 \\ \dot{h} \end{bmatrix}. \tag{7.73}$$

Consequently, the angular velocities of the manipulator can be solved according to equation (7.35).

The derivative of equation set (7.71) with respect to time can be expressed as:

$$\mathbf{A}\ddot{\boldsymbol{x}} = \boldsymbol{d} \tag{7.74}$$

where

$$\boldsymbol{d} = -\begin{bmatrix} (\dot{x}_1 + \frac{1}{2}\dot{s}_1)^2 + \frac{1}{2}(x_1 + \frac{1}{2}s_1)\ddot{s}_1 + \frac{\sqrt{3}}{2}(y_1 - a + \frac{\sqrt{3}}{2}s_1)\ddot{s}_1 + (\dot{y}_1 + \frac{\sqrt{3}}{2}\dot{s}_1)^2 + (\dot{z}_1)^2 \\ (x_2 + s_2)\ddot{s}_2 + (\dot{x}_2 + \dot{s}_2)^2 + (\dot{y}_2)^2 + (\dot{z}_2)^2 \\ -\frac{1}{2}(x_3 - \frac{1}{2}s_3)\ddot{s}_3 + \frac{\sqrt{3}}{2}(y_3 - a + \frac{\sqrt{3}}{2}s_3)\ddot{s}_3 + (\dot{x}_3 - \frac{1}{2}\dot{s}_3)^2 + (\dot{y}_3 + \frac{\sqrt{3}}{2}\dot{s}_3)^2 + (\dot{z}_3)^2 \\ (\dot{x}_1 - \dot{x}_2)^2 + (\dot{y}_1 - \dot{y}_2)^2 + (\dot{z}_1 - \dot{z}_2)^2 \\ (\dot{x}_2 - \dot{x}_3)^2 + (\dot{y}_2 - \dot{y}_3)^2 + (\dot{z}_2 - \dot{z}_3) \\ (\dot{x}_3 - \dot{x}_1)^2 + (\dot{y}_3 - \dot{y}_1)^2 + (\dot{z}_3 - \dot{z}_1) \\ 0 \\ 0 \\ -3\ddot{h} \end{bmatrix}.$$

The accelerations of the end effector can also be solved with a similar process as those in Section 7.2.1.

According to the above analysis, we conduct numerical simulation with the following parameters.

The length range of each leg:

$$0 \le s_1 \le 500\text{mm}, \quad 0 \le s_2 \le 500\text{mm}, \quad 500\text{mm} \le s_3 \le 1000\text{mm},$$
$$800\text{mm} \le h \le 1000\text{mm}, \quad l_i = 1000\text{mm}\left(i = 1{,}2{,}3\right).$$

And the side length of the base triangle is 1000 mm. The side length of the manipulator:

$$L_{M_1M_2} = L_{M_2M_3} = L_{M_3M_1} = 600\sqrt{3}$$

If we presume that the displacement of each actuator with respect to time are shown in Table 7.4, the velocities, accelerations, angular velocities, and angular accelerations with respect to time are shown in Figures 7.11, 7.12, 7.13, and 7.14, respectively. The total time of simulation is 20 s. Numerical simulations show that this method in natural coordinates can effectively and efficiently accomplish real-time calculations.

TABLE 7.4 Displacement of each Actuator with Respect to Time

Actuators	Displacements (mm)
P_1	$300 + 60\cos t$
P_2	$240 + 60\left(\frac{t}{4} + \frac{\pi}{3}\right)$
P_3	$680 + 60\cos\left(\frac{t}{2} + \frac{\pi}{6}\right)$
P_4	$870 + 60\cos\frac{t}{2}$

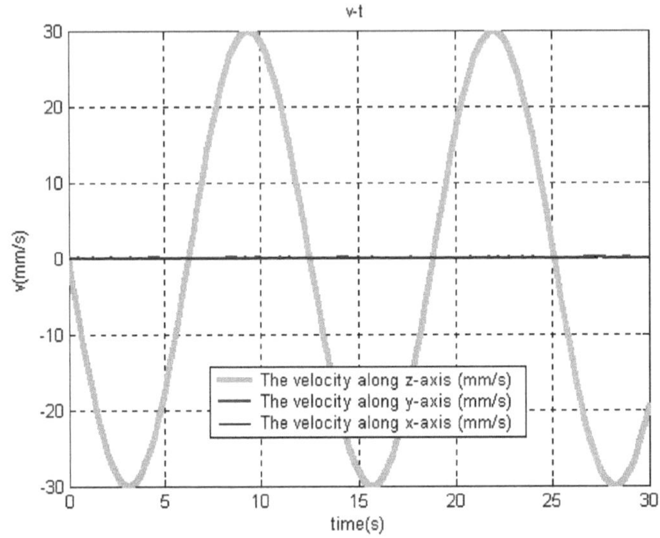

FIGURE 7.11 Velocity–time curves of the mechanism.

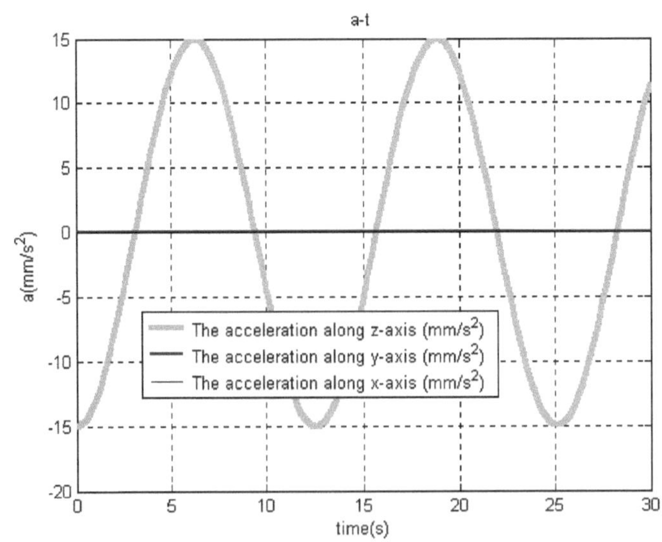

FIGURE 7.12 Acceleration–time curves of the mechanism.

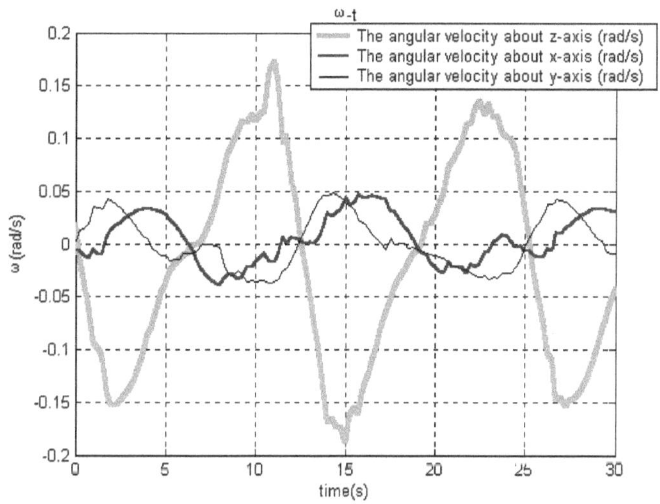

FIGURE 7.13 Angular velocity–time curves of the mechanism.

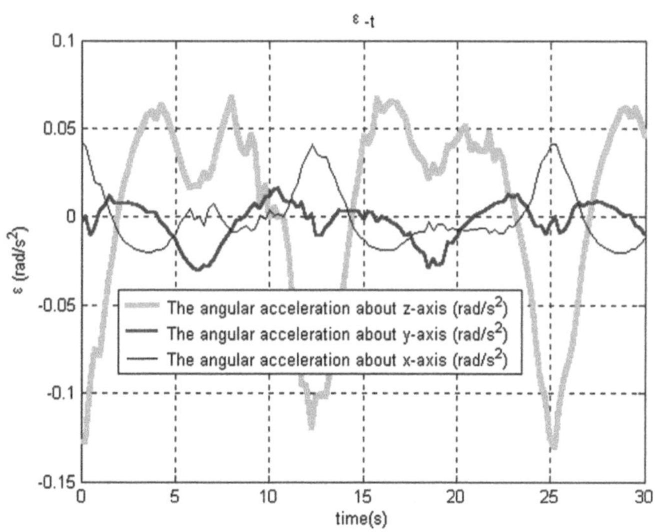

FIGURE 7.14 Angular acceleration–time curves of the mechanism.

7.3 CONCLUSIONS

In this chapter, the kinematics model of the spatial parallel robot mechanism is discussed. On the basis of the natural coordinate method, the sufficiency and necessity are proved that the kinematics model can be established for a type of spatial parallel mechanism only with four non-coplanar points' Cartesian coordinates. This method has a significant advantage in dealing with kinematics issues of spatial parallel mechanisms with more than three DOFs. And the established kinematics model shares the same advantages of ordinary natural coordinate methods and can also minimize the number of independent variables. This can particularly apply to solving the kinematics modeling of multi-DOF spatial parallel robot mechanisms. Two different manipulators are employed to illustrate the efficiency and simplicity of the kinematics model with four non-coplanar points' coordinates method. The advantages of the four-point coordinate method have brought great convenience to kinematic and dynamic analysis.

REFERENCES

[1] M.D. Lioa, Vittore Cossalter and Roberto Lot, On the use of natural coordinates in optimal synthesis of mechanisms, Mechanism and Machine Theory 35 (2000) 1367–1389.
[2] J.G. de Jalón, E. Bayo, Kinematic and Dynamic Simulation of Multibody Sytems: The Real-Time Challenge, Springer-Verlag, New York, 1994.
[3] A. Hernández, O. Altuzarra, R. Avilés, V. Petuya, Kinematic analysis of mechanisms via a velocity equation based in a geometric matrix, Mechanism and Machine Theory 38 (2003) 1413–1429.
[4] T. Szkodny, Forward and inverse kinematics of IRb-6 manipulator, Mechanism and Machine Theory 30 (1995) 1039–1056.
[5] M. Ceccarelli, P.M. Decio Fino, J.M. Jimenez, Dynamic performance of CaPaMan by numerical simulations, Mechanism and Machine Theory 37 (2002) 241–266.
[6] T. Szkodny, Dynamics of industrial robot manipulators, Mechanism and Machine Theory 30 (1995) 1057–1072.
[7] R.V. Schwerin, Multibody System Simulation—Numerical Methods, Algorithms, and Software, Springer-Verlag, Berlin Heidelberg, 1999.
[8] S.G. Kim, J. Ryu, New dimensionally homogeneous jacobian matrix formulation by three end effector points for optimal design of parallel manipulators, IEEE Transactions on Robotics and Automation 19 (2003) 731–737.
[9] C.T. Kelley, Iterative Methods for Linear and Nonlinear Equations, North Carolina State University, Society for Industrial and Applied Mathematics, Philadelphia, 1995.
[10] J.S. Zhao, X. Yang, L. Zhu, Z.J. Feng, K. Zhou, On the forward and inverse displacement of spatial parallel manipulators, The International Journal of Advanced Manufacturing Technology 29 (2006) 1284–1294.
[11] J.P. Merlet, Solving the forward kinematics of a Gough-type parallel manipulator with interval analysis, The International Journal of Robotics Research 23 (2004) 221–235.
[12] K. Harib, K. Srinivasan, Kinematic and dynamic analysis of stewart platform-based machine tool structures, Robotica 21 (2003) 541–554.
[13] D. Jakobovic, L. Budin, Forward Kinematics of a Stewart Platform Mechanism, in: IEEE Proc. Of the 6th International Conference on Intelligent Engineering Systems (INES), Opatija, Croatia, 2002.
[14] J.S. Zhao, K. Zhou, D.Z. Mao, Y.F. Gao, Y.F. Fang, A new method to study the degree of freedom of spatial parallel mechanisms, The International Journal of Advanced Manufacturing Technology 23 (2004) 288–294.

Chapter 8

Kinematics and Statics of Manipulators

As the kinematics and statics play a very important role in determining the actuating inputs and the effective loads that the end effector sustains, this chapter focuses on this issue and proposes an analytical process to study the forward and inverse kinematics and statics of spatial manipulators. As series manipulators and parallel manipulators show different features in kinematics and statics, this chapter discusses them separately. Firstly, the forward and inverse velocity problems of the manipulator linkages are investigated with reciprocal screw theory. Then, the static balance conditions together with forward and inverse statics of the manipulator linkages are established through virtual power theory. In the kinematics analysis, the primary conditions for feasible motions of an end effector are addressed through velocity screws. Illustrative examples indicate that the method proposed in this chapter can be used to guide the singularity identification, path planning and feasible motion determination.

8.1 KEY PROBLEMS IN KINEMATICS AND STATICS ANALYSIS

The forward and inverse kinematics and statics analysis play a very important role in robotics research. Shigley and Uicher [1], Ficher [2], McCarthy [3], and Yan [4] investigated the primary kinematics of general manipulators. Kumar [5] presented the instantaneous kinematics of parallel robotic mechanisms. The kinematics of a manipulator, as a matter of fact, is one of the primary considerations in robot control analysis. Phillips and Hunt [6] extended the Aronhold–Kennedy theorem from planar cases to the spatial motions of general three rigid bodies, which expands the applications of screw theory in kinematics and singularity analysis of spatial robot mechanisms. Etemadi-Zanganeh and Angeles [7] investigated the instantaneous kinematics of hybrid parallel manipulators. For the applications of kinematics of spatial mechanisms, Kim and Chung [8] proposed an analytical formulation of reciprocal screws and discussed its applications for non-redundant robots.

Advanced Theory of Constraint and Motion Analysis for Robot Mechanisms.
http://dx.doi.org/10.1016/B978-0-12-420162-0.00008-4

The most popular methods for controlling robotic arms are mainly based on manual tables that are obtained by offline kinematic models [9–11]. However, the fact that the inverse kinematics of most series robotic arms is multivalued increases the difficulty of the method in practice [12]. Similarly, the forward and inverse statics analysis is also of primary importance in the design of an autonomous robot. Estremera and Waldron [13] investigated the powered autonomous running robots and developed an efficient actuator system for generating the radial thrust in the legs. Fischer [14] presented a method to partition the forces and torques exerted on the joints of a mechanism of static equilibrium. Agrawal and Roth [15] analyzed the statics of parallel manipulators to provide an explicit relationship between the loads and actuations. Pennock and Oncu [16] applied screw theory to the inverse statics of a six-degrees-of-freedom cylindrical robot. They analyzed the dual actuator forces exerted on each of the three cylindrical joints and the power required by each rotary–linear actuator to support the specified external wrenches acting on the end effector.

The forward and inverse statics are also the primary considerations to evaluate the stiffness of the mechanisms under different conditions. Juan and Mirats [17] presented a review of the basic issues about the statics of metamorphic structures and provided a mathematical framework to analyze the rigidity and stability properties of tensegrity structures from the motions, forces, and energy approaches. Williams and Gallina [18] discussed a technique to improve the serious cable interference problems and to avoid configurations where negative cable tensions are required to exert general forces on the planar cable-direct-driven robot. Gosselin [19] investigated the kinematics, statics, and dynamics of a kind of lightweight and deployable tensegrity mechanism. Elnagar and Hussein [20] proposed an approach to generate acceleration-based optimal piecewise smooth trajectories by using the minimal energy criterion in the trajectory planning. Arsenault and Shai [21] investigated the dualism relationship between mechanisms and determinate trusses and established the connection between statics and kinematics. However, the feasible motions and loads of an end effector are seldom systematically discussed in previous literatures. Therefore, this chapter focuses on the kinematics and statics of general manipulators. The primary conditions for feasible motions of an end effector are obtained by analyzing its velocities. Besides, forward and inverse velocity problems of the manipulator linkages are discussed with reciprocal screw theory. Furthermore, static balance conditions together with forward and inverse statics solutions of the manipulator linkages are presented by using virtual power theory. As manipulator kinematics and statics mainly discuss the motions and forces between the end effector and the actuating inputs, the major work of this chapter is to establish the relationship between these factors.

Besides, the method proposed in this chapter can also be utilized to synthesize a mechanism with specified kinematic specification. Arsenault and Boudreau [22] discussed the synthesis of planar parallel mechanisms by considering the workspace, dexterity, stiffness, and singularity avoidance problems.

Merlet [23] addressed the singular configurations of parallel manipulators by Grassmann geometry. Gosselin and Angeles [24], and Park and Kim [25] investigated the singularity problems of closed kinematic chain mechanisms. Bhattacharya [26] compared the exact and approximate methods of singularity avoidance problems in platform type parallel manipulators. As a matter of fact, the workspace boundary and singularity search can be accomplished in the kinematics and statics analysis model proposed in this chapter.

In this chapter, kinematics is first discussed and the statics follows. The forward velocity problem of a manipulator is used to calculate the velocity of the end effector after knowing a set of joint velocities; and the inverse velocity problem of the manipulator linkages is used to analyze all possible sets of joint velocities that could be used to attain the velocity of a prescribed end effector. Similarly, forward statics of the manipulator linkages are used to calculate the affordable forces of the end effector under the condition that a set of joint driving forces are known, while the inverse statics of a manipulator are used to calculate all possible sets of joint driving forces under the condition that the affordable forces of the end effector are known.

8.2 KINEMATICS AND STATICS OF SERIES MANIPULATORS

Nowadays, series manipulators are widely used in industrial robots. Compared with parallel manipulators, forward kinematics of the series manipulators is relatively easy. As a result, this chapter will first discuss the kinematics of series manipulators.

8.2.1 Velocity Analysis of Series Manipulators

Suppose that the end effector of a series manipulator has m $(1 \le m \le 6)$ DOFs and the sum of the relative DOFs of all joints is $n (n \ge m)$. Also suppose that the generalized angular speeds of the joints in the series kinematic chain are represented by a column vector $\dot{\theta} = \begin{bmatrix} \dot{\theta}_1 & \dot{\theta}_2 & \cdots & \dot{\theta}_n \end{bmatrix}^T \in \mathbb{R}^{n\times 1}$. Then, the velocity of the end effector can be expressed as

$$\boldsymbol{v}_E = \mathbf{J}\dot{\boldsymbol{\theta}} \tag{8.1}$$

where $\mathbf{J} = \begin{bmatrix} \$_1 & \$_2 & \cdots & \$_n \end{bmatrix}$. $\mathbf{J}$ is a $6 \times n$ Jacobian matrix that is made up of kinematic screws of each joint, $\$_i$ $(i = 1,2,\cdots,n)$ is the Plücker column vector of the ith unit kinematic screw in a fixed coordinate frame, $\dot{\boldsymbol{\theta}} = \begin{bmatrix} \dot{\theta}_1 & \dot{\theta}_2 & \cdots & \dot{\theta}_n \end{bmatrix}^T$, and $\dot{\theta}_i$ is the generalized angular speed corresponding to $\$_i$.

Equation (8.1) shows that the velocity screws of the end effector connected in series by n generalized revolute joints can be represented by a linear combination of the kinematic screws of the chain in a fixed coordinate frame. If all kinematic screws are unit screws, then each coefficient in the linear combination describes the generalized angular velocity of each joint in the kinematic chain. Rewrite equation (8.1) in the following form

$$\boldsymbol{v}_E = \dot{\theta}_1 \$_1 + \dot{\theta}_2 \$_2 + \cdots + \dot{\theta}_n \$_n. \tag{8.2}$$

If $n = m$, then the Jacobian matrix of the end effector is non-singular in columns, which means that the column vectors $\$_i \left(i = 1,2,\cdots,n\right)$ are linearly independent. Assume $\boldsymbol{v}_S$ is an n-dimension subspace spanned by $\left\{ \$_1 \quad \$_2 \quad \cdots \quad \$_n \right\}$, that is

$$\mathbb{V}_s = \text{span}\left\{ \$_1 \quad \$_2 \quad \cdots \quad \$_n \right\} \tag{8.3}$$

From equations (8.2) and (8.3), we can conclude that vector $\boldsymbol{v}_E$ belongs to the linear space $\boldsymbol{v}_S$, that is

$$\boldsymbol{v}_E \in \mathbb{V}_s. \tag{8.4}$$

As a result, $\boldsymbol{V}_S$ is called feasible motion space of the end effector of the series manipulator. In most cases, the Jacobian matrix, $\mathbf{J}$, is not a square one, so the inverse velocity problem of the manipulator linkage is a bit difficult. However, for any real velocity, $\boldsymbol{v}_E$, we can work out the solution via pseudo-inverse techniques [27,28]. Left multiplying $\mathbf{J}^T$ on both sides of equation (8.1) yields

$$\mathbf{J}^T \boldsymbol{v}_E = \mathbf{J}^T \mathbf{J} \dot{\boldsymbol{\theta}}. \tag{8.5}$$

If $n = m$, which means $\mathbf{J}$ is non-singular in column, then $\mathbf{J}^T\mathbf{J}$ is an $m \times m$ non-singular square matrix. As a result, left multiplying $(\mathbf{J}^T\mathbf{J})^{-1}$ on both sides of equation (8.5) yields the inverse kinematic solution of the series manipulator

$$\dot{\boldsymbol{\theta}} = \left(\mathbf{J}^T\mathbf{J}\right)^{-1} \mathbf{J}^T \boldsymbol{v}_E \tag{8.6}$$

where $(\mathbf{J}^T\mathbf{J})^{-1}\mathbf{J}^T$ is called the pseudo inverse of $\mathbf{J}$. Under the condition of $n = m$, the solution of equation (8.6) is unique.

If $n > m$, the solution is not unique which is discussed in the related documents such as [28]. Now the forward and inverse velocity problems of series manipulator linkages will be discussed using reciprocal screw theory.

Definition 8.1. The reciprocal product of two screws, $\$_i \left(i = 1,2,\cdots\right)$ and $\$_j \left(j = 1,2,\cdots\right)$, is denoted by

$$M_{ij} = \$_i^T \boldsymbol{\Delta} \$_j \tag{8.7}$$

where $\boldsymbol{\Delta} = \begin{bmatrix} \mathbf{0}_{3\times3} & \mathbf{I}_{3\times3} \\ \mathbf{I}_{3\times3} & \mathbf{0}_{3\times3} \end{bmatrix}$ and $\mathbf{0}_{3\times3} = \begin{bmatrix} 0 & 0 & 0 \\ 0 & 0 & 0 \\ 0 & 0 & 0 \end{bmatrix}$, and $\mathbf{I}_{3\times3} = \begin{bmatrix} 1 & 0 & 0 \\ 0 & 1 & 0 \\ 0 & 0 & 1 \end{bmatrix}$.

M_{ij} denote the instant power when the screws $\$_i$ and $\$_j$ represent velocity and force, individually. And the screws $\$_i$ and $\$_j$ are called reciprocal when $M_{ij} = 0$. Dai and Jones [29,30] presented an explicit interrelationship between screw systems and the reciprocal screws.

Theorem 8.1. *Assume the columns of a screw matrix* $\mathbf{J} = \begin{bmatrix} \$_1 & \$_2 & \cdots & \$_n \end{bmatrix}$ are n $(1 \le n \le 6)$ *linearly independent screws,* $\$_1, \$_2, \cdots, \$_n$, *then for each screw* $\$_i$ $(i = 1,2,\cdots,n)$, *there is at least one screw* $\$^r_{J_j}$ *such that* $\$^r_{J_j}$ *is reciprocal to all screws in* $\mathbf{J}_j = \begin{bmatrix} \$_1 & \cdots & \$_{j-1} & \$_{j+1} & \cdots & \$_n \end{bmatrix}$ *but the reciprocal product of* $\$^r_{J_j}$ *and* $\$_j$ *is not zero, that is,* $\left(\$^r_{J_j}\right)^T \boldsymbol{\Delta}\$_j \ne 0$.

Proof. When $n = 6$, according to equation (8.7), the solution of $\mathbf{J}^T\boldsymbol{\Delta x} = 0$ is $\boldsymbol{x} = 0$. Since Rank $(\mathbf{J}_j) = n - 1 = 5$, there must be a non-zero solution $\boldsymbol{x} = \$^r_{J_j} \ne 0$ such that $\mathbf{J}_j^T\boldsymbol{\Delta}\$^r_{J_j} = 0$. Assume $\boldsymbol{x} = \$^r_{J_j}$ also satisfies $\$_j^T\boldsymbol{\Delta x} = 0$, then $\boldsymbol{x} = \$^r_{J_j}$ must also be the solution of $\mathbf{J}^T\boldsymbol{\Delta x} = 0$, the only solution of which is $\boldsymbol{x} = 0$ $\boldsymbol{x} - 0$. Thus, this induces the conflicts with the assumption that $\boldsymbol{x} = \$^r_{J_j} \ne 0$. So the assumption of $\$_j^T\boldsymbol{\Delta}\$^r_{J_j} = 0$ is wrong. As a result, $\$_j^T\boldsymbol{\Delta}\$^r_{J_j} \ne 0$.

When $1 \le n \le 6$, according to equation (8.7), the basic solution system of $\mathbf{J}^T\boldsymbol{\Delta x} = 0$ can be represented as

$$\boldsymbol{x} = k_1\$^r_1 + k_2\$^r_2 + \cdots + k_{(6-n)}\$^r_{(6-n)} \tag{8.8}$$

where k_i $(i = 1,2,\cdots,6-n)$ are any constants but not all equal to zero simultaneously, and $\$^r_1, \$^r_2, \cdots, \$^r_{6-n}$ are linearly independent column vectors.

Similarly, the basic solution system of $\mathbf{J}_j^T\boldsymbol{\Delta x} = 0$ can be represented by

$$\boldsymbol{x} = \$^r_{J_j} = k_1\$^r_1 + k_2\$^r_2 + \cdots + k_{(6-n)}\$^r_{(6-n)} + k_{(7-n)}\$^r_{(7-n)} \tag{8.9}$$

where k_i $(i = 1,2,\cdots,6-n,7-n)$ are any constants but not all equal to zero at the same time, $\$^r_1, \$^r_2, \cdots, \$^r_{6-n}, \$^r_{7-n}$ are linearly independent column vectors, too.

According to equation (8.9), there is at least one group of coefficients $k_1 = k_2 = \cdots = k_{6-n} = 0, k_{7-n} \ne 0$, such that $\boldsymbol{x} = \$^r_{J_j} = k_{(7-n)}\$^r_{(7-n)} \ne 0$ is the solution of equation $\mathbf{J}_j^T\boldsymbol{\Delta x} = 0$.

Similarly, we can prove that $\$_j^T\boldsymbol{\Delta}\$^r_{J_j} \ne 0$.

Assume $\$_j^T\boldsymbol{\Delta}\$^r_{J_j} = 0$, then $\boldsymbol{x} = \$^r_{J_j}$ is also the solution of equation $\mathbf{J}^T\boldsymbol{\Delta x} = 0$. Thus, the basic solution systems (8.8) and (8.9) are equivalent. So there must be $7 - n = 6 - n$, which is obviously wrong. Therefore, the assumption of $\$_j^T\boldsymbol{\Delta}\$^r_{J_j} = 0$ is wrong, and as a result, the equation $\$_j^T\boldsymbol{\Delta}\$^r_{J_j} \ne 0$ must hold.

Based on the above analysis, Theorem 8.1 is proved.

Now, left multiplying $\left(\$^r_{J_j}\right)^T \boldsymbol{\Delta}$ on both sides of equation (8.2) yields

$$\left[\left(\$^r_{J_j}\right)^T \boldsymbol{\Delta}\right]\boldsymbol{v}_E = \dot{\theta}_1\left[\left(\$^r_{J_j}\right)^T \boldsymbol{\Delta}\$_1\right] + \cdots + \dot{\theta}_j\left[\left(\$^r_{J_j}\right)^T \boldsymbol{\Delta}\$_j\right] + \cdots \dot{\theta}_n\left[\left(\$^r_{J_j}\right)^T \boldsymbol{\Delta}\$_n\right] = \dot{\theta}_j\left[\left(\$^r_{J_j}\right)^T \boldsymbol{\Delta}\$_j\right]. \quad (8.10)$$

According to Theorem 8.1, $\left(\$^r_{J_j}\right)^T \boldsymbol{\Delta}\$_j \neq 0$. So, to attain a real velocity $\boldsymbol{v}_E$ of the end effector of a series manipulator, the angular velocity of the jth $(j = 1,2,\cdots,n)$ joint can be calculated from equation (8.10)

$$\dot{\theta}_j = \frac{\left(\$^r_{J_j}\right)^T \boldsymbol{\Delta}\boldsymbol{v}_E}{\left(\$^r_{J_j}\right)^T \boldsymbol{\Delta}\$_j}. \quad (8.11)$$

Thus, to obtain the velocity, $\boldsymbol{v}_E$, the angular velocity of each joint can be represented in the form of matrix multiplications as

$$\dot{\boldsymbol{\theta}} = \mathbf{D}\mathbf{r}^T \boldsymbol{\Delta}\boldsymbol{v}_E \quad (8.12)$$

where

$$\mathbf{D} = \operatorname{diag}\left(\frac{1}{\left(\$^r_{J_1}\right)^T \boldsymbol{\Delta}\$_1} \quad \frac{1}{\left(\$^r_{J_2}\right)^T \boldsymbol{\Delta}\$_2} \quad \cdots \quad \frac{1}{\left(\$^r_{J_n}\right)^T \boldsymbol{\Delta}\$_n}\right),$$

$$\mathbf{r} = \left[\$^r_{J_1} \quad \$^r_{J_2} \quad \cdots \quad \$^r_{J_n}\right].$$

According to the constructing process of $\$^r_{J_j}$, it is not difficult to prove that Rank $(\mathbf{r}) = n$. Substituting equation (8.12) into equation (8.1), we can find

$$\left(\mathbf{J}\mathbf{D}\mathbf{r}^T \boldsymbol{\Delta} - \mathbf{I}_{6\times 6}\right)\boldsymbol{v}_E = 0 \quad (8.13)$$

where $\mathbf{I}_{6\times 6}$ is a 6×6 unit matrix.

This means that any feasible velocity $\boldsymbol{v}_E$ of the end effector must satisfy equation (8.13). In other words, all velocities that do not satisfy equation (8.13) are infeasible. Piccardo and Honderd [31] studied the on-line path planning problems in a non-static environment. In practice, equation (8.13) can be directly used as an associated constraint equation in the trajectory planning of an end effector which will be discussed in the following example.

Example 8.1. In a series manipulator shown in Figure 8.1, the number of DOFs of the end effector *CD* is 3. *CD* is connected to the fixed frame through a *RRR*-kinematic chain. Try to analyze the angular velocity relationship between each joint and the feasible motions of the end effector when *D* tracks the trajectory shown by the dashed arrow.

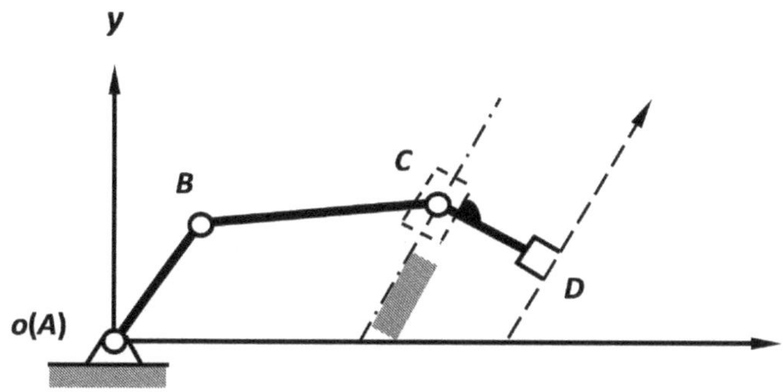

FIGURE 8.1 A planar serial manipulator of three DOFs.

Solution. As shown in Figure 8.1, letting the hinge axis of the constant rotation joint A be the z-axis, the crossing point of the z-axis and motion plane be the origin of the coordinate frame, a horizontal ray to the right be the x-axis, we can build up a right-handed coordinate frame with the y-axis upwards.

In this fixed coordinate frame, the kinematic screw matrix of the end effector CD can be represented by

$$\mathbf{J} = \begin{bmatrix} \$_A & \$_B & \$_C \end{bmatrix}$$

where $\$_A = \begin{bmatrix} 0 & 0 & 1 & 0 & 0 & 0 \end{bmatrix}^T$, $\$_B = \begin{bmatrix} 0 & 0 & 1 & y_B & -x_B & 0 \end{bmatrix}^T$, $\$_C = \begin{bmatrix} 0 & 0 & 1 & y_C & -x_C & 0 \end{bmatrix}^T$.
Then we obtain that

$$\mathbf{r} = \begin{bmatrix} \$^r_{J_A} & \$^r_{J_B} & \$^r_{J_C} \end{bmatrix}$$

where $\$^r_{J_A} = \begin{bmatrix} x_C - x_B & y_C - y_B & 0 & 0 & 0 & x_B y_C - x_C y_B \end{bmatrix}^T$, $\$^r_{J_B} = \begin{bmatrix} x_C & y_C & 0 & 0 & 0 & 0 \end{bmatrix}^T$, $\$^r_{J_C} = \begin{bmatrix} x_B & y_B & 0 & 0 & 0 & 0 \end{bmatrix}^T$.

Since the end effector CD moves along a line shown by the dashed arrow in Figure 8.1, we may assume that the feasible velocity screw of the end effector is

$$\boldsymbol{v}_E = \begin{bmatrix} 0 & 0 & 0 & a & b & c \end{bmatrix}^T$$

where $[a \quad b \quad c]^T$ represents the direction vector of the dashed arrow line. If $c = 0$, then $\boldsymbol{v}_E$ expresses the series manipulator shown by the dash-dotted line in Figure 8.1, that is, a planar slider-crank mechanism.

According to equation (8.11), the angular velocity of revolute joint A can be worked out as

$$\dot{\theta_A} = \frac{\left(\$^r_{J_A}\right)^T \boldsymbol{\Delta} \boldsymbol{v}_E}{\left(\$^r_{J_A}\right)^T \boldsymbol{\Delta} \$_A} = \frac{a\left(x_C - x_B\right) + b\left(y_C - y_B\right)}{x_B y_C - x_C y_B}. \tag{8.14}$$

It is not difficult to prove that the result of equation (8.14) is an inverse velocity solution of the planar slider-crank mechanism. Furthermore, we can obtain that

$$\mathbf{D} = \begin{bmatrix} \dfrac{1}{x_B y_C - x_C y_B} & 0 & 0 \\ 0 & \dfrac{1}{x_C y_B - x_B y_C} & 0 \\ 0 & 0 & \dfrac{1}{x_B y_C - x_C y_B} \end{bmatrix}.$$

Therefore, the angular velocities of each revolute joint can also be expressed by equation (8.12).

$$\dot{\boldsymbol{\theta}} = \mathbf{D}\mathbf{r}^T \boldsymbol{\Delta} \boldsymbol{v}_E = \begin{bmatrix} \dfrac{a\left(x_C - x_B\right) + b\left(y_C - y_B\right)}{x_B y_C - x_C y_B} \\ \dfrac{a x_C + b y_C}{x_C y_B - x_B y_C} \\ \dfrac{a x_B + b y_B}{x_B y_C - x_C y_B} \end{bmatrix} \tag{8.15}$$

Substituting $\boldsymbol{v}_E = \begin{bmatrix} 0 & 0 & 0 & a & b & c \end{bmatrix}^T$ into equation (8.13) yields

$$\left(\mathbf{J}\mathbf{D}\mathbf{r}^T \boldsymbol{\Delta} - \mathbf{I}\right) \boldsymbol{v}_E$$
$$= \left(\begin{bmatrix} 0 & 0 & 0 \\ 0 & 0 & 0 \\ 1 & 1 & 1 \\ 0 & y_B & y_C \\ 0 & -x_B & -x_C \\ 0 & 0 & 0 \end{bmatrix} \begin{bmatrix} \dfrac{1}{x_B y_C - x_C y_B} & 0 & 0 \\ 0 & \dfrac{1}{x_C y_B - x_B y_C} & 0 \\ 0 & 0 & \dfrac{1}{x_B y_C - x_C y_B} \end{bmatrix} \right.$$

$$\times \begin{bmatrix} x_C - x_B & x_C & x_B \\ y_C - y_B & y_C & y_B \\ 0 & 0 & 0 \\ 0 & 0 & 0 \\ 0 & 0 & 0 \\ x_B y_C - x_C y_B & 0 & 0 \end{bmatrix}^T \boldsymbol{\Delta} - \mathbf{I} \Bigg) \begin{bmatrix} 0 \\ 0 \\ 0 \\ a \\ b \\ c \end{bmatrix}$$

$$= \left(\begin{bmatrix} 0 & 0 & 0 \\ 0 & 0 & 0 \\ \dfrac{1}{x_B y_C - x_C y_B} & \dfrac{1}{x_C y_B - x_B y_C} & \dfrac{1}{x_B y_C - x_C y_B} \\ 0 & \dfrac{y_B}{x_C y_B - x_B y_C} & \dfrac{y_C}{x_B y_C - x_C y_B} \\ 0 & \dfrac{-x_B}{x_C y_B - x_B y_C} & \dfrac{-x_C}{x_B y_C - x_C y_B} \\ 0 & 0 & 0 \end{bmatrix} \right.$$

$$\left. \times \begin{bmatrix} x_C - x_B & y_C - y_B & 0 & 0 & 0 & x_B y_C - x_C y_B \\ x_C & y_C & 0 & 0 & 0 & 0 \\ x_B & y_B & 0 & 0 & 0 & 0 \end{bmatrix} \boldsymbol{\Delta} - \mathbf{I} \right) \begin{bmatrix} 0 \\ 0 \\ 0 \\ a \\ b \\ c \end{bmatrix}$$

$$= \left(\begin{bmatrix} 0 & 0 & 0 & 0 & 0 & 0 \\ 0 & 0 & 0 & 0 & 0 & 0 \\ 0 & 0 & 0 & 0 & 0 & 1 \\ 1 & 0 & 0 & 0 & 0 & 0 \\ 0 & 1 & 0 & 0 & 0 & 0 \\ 0 & 0 & 0 & 0 & 0 & 0 \end{bmatrix} \boldsymbol{\Delta} - \mathbf{I} \right) \begin{bmatrix} 0 \\ 0 \\ 0 \\ a \\ b \\ c \end{bmatrix}$$

$$= \begin{bmatrix} -1 & 0 & 0 & 0 & 0 & 0 \\ 0 & -1 & 0 & 0 & 0 & 0 \\ 0 & 0 & 0 & 0 & 0 & 0 \\ 0 & 0 & 0 & 0 & 0 & 0 \\ 0 & 0 & 0 & 0 & 0 & 0 \\ 0 & 0 & 0 & 0 & 0 & -1 \end{bmatrix} \begin{bmatrix} 0 \\ 0 \\ 0 \\ a \\ b \\ c \end{bmatrix}$$

$$= \begin{bmatrix} 0 & 0 & 0 & 0 & 0 & -c \end{bmatrix}^T.$$

According to equation (8.13), when $c \neq 0$, the motion is infeasible, which can be clearly seen from Figure 8.1. The end effector can only move in *xoy*-plane while the prescribed motion is along the direction of $[\,a \quad b \quad c\,]^T$ in space, so the given motion is infeasible, which verifies the correction of equation (8.13). The path planning problem studied by Piccardo and Honderd [31] can be modeled by the above process.

What is more, for the end effector of the planar series manipulator with 3 DOFs shown in Figure 8.1, all motions that have the form of $\boldsymbol{v}_E = [\,0 \quad 0 \quad n \quad p \quad q \quad 0\,]^T$ (n, p, and q are arbitrary parameters so long as they are not zeros simultaneously) are its feasible motions. Gan *et al.* [12] researched the inverse kinematics of a 5-DOF robot to control the arm to get any reachable position, which can also be well predicted online by the model proposed here.

8.2.2 Statics Analysis of Series Manipulators

Now the static forces of series manipulators will be discussed in this section. It is a primary problem to calculate the driving forces/torques of each joint under the condition that the loads on the end effector are known. Assume that the end effector of a series manipulator is under the control of a force screw, $\boldsymbol{w}_E$, then to keep the balance of the manipulator, the driving torques of each joint are $\tau_1, \tau_2, \cdots, \tau_n$. Introduce a vector of joint driving torques:

$$\boldsymbol{\tau} = \begin{bmatrix} \tau_1 & \tau_2 & \cdots & \tau_n \end{bmatrix}^T. \tag{8.16}$$

According to equation (8.7), the power consumption at any instant of the end effector under the given loads is

$$P_L = \boldsymbol{v}_E^T \boldsymbol{\Delta} \boldsymbol{w}_E \tag{8.17}$$

where P_L denotes the instant power of the loads exerted on the end effector.

Substituting equation (8.1) into equation (8.17) yields

$$P_L = \dot{\boldsymbol{\theta}}^T \mathbf{J}^T \boldsymbol{\Delta} \boldsymbol{w}_E. \tag{8.18}$$

At this instant, the power supplied by the driving forces of all actuators is

$$P_A = \dot{\boldsymbol{\theta}}^T \boldsymbol{\tau} \tag{8.19}$$

where P_A denotes the instant driving power of the actuators.

According to the principle of virtual power, to keep the balance of a manipulator, the power provided by the driving forces of all actuators at any instant should be equal to the power consumed by the end effector at the same instant, that is

$$P_A = P_L. \tag{8.20}$$

Substituting equations (8.18) and (8.19) into equation (8.20), we obtain

$$\dot{\boldsymbol{\theta}}^T\left(\boldsymbol{\tau} - \mathbf{J}^T\boldsymbol{\Delta}\boldsymbol{w}_E\right) = 0. \tag{8.21}$$

Because $\dot{\boldsymbol{\theta}}$ is an arbitrary vector, the necessary and sufficient condition for equation (8.21) is

$$\boldsymbol{\tau} = \mathbf{J}^T\boldsymbol{\Delta}\boldsymbol{w}_E. \tag{8.22}$$

According to equation (8.22), the revolute moments of each joint can be represented by

$$\tau_i = \$_i^T\boldsymbol{\Delta}\boldsymbol{w}_E. \tag{8.23}$$

Equations (8.22) and (8.23) represent the inverse statics solutions of series manipulators. According to Theorem 8.1, $\boldsymbol{w}_E$ can be expressed as the linear combinations of $\$^r_{J_1}, \$^r_{J_2}, \cdots, \$^r_{J_n}$:

$$\boldsymbol{w}_E = a_1\$^r_{J_1} + a_2\$^r_{J_2} + \cdots + a_n\$^r_{J_n}. \tag{8.24}$$

Substituting equation (8.24) into equation (8.23) yields

$$\tau_i = \$_i^T\boldsymbol{\Delta}\left(a_1\$^r_{J_1} + a_2\$^r_{J_2} + \cdots + a_n\$^r_{J_n}\right) = a_i\$_i^T\boldsymbol{\Delta}\$^r_{J_i}.$$

Therefore,

$$a_i = \frac{\tau_i}{\$_i^T\boldsymbol{\Delta}\$^r_{J_i}}. \tag{8.25}$$

Substituting equations (8.25) and (8.24) into equation (8.22) yields

$$\boldsymbol{w}_E = \sum_{i=1}^{n}\frac{\tau_i}{\$_i^T\boldsymbol{\Delta}\$^r_{J_i}}\$^r_{J_i}. \tag{8.26}$$

Equation (8.26) can be rewritten in the form of matrix:

$$\boldsymbol{w}_E = \mathbf{rD}\boldsymbol{\tau}. \tag{8.27}$$

Equation (8.27) presents the forward statics solutions of series manipulators. Substituting equation (8.22) into equation (8.27), we gain

$$\left(\mathbf{rDJ}^T\boldsymbol{\Delta} - \mathbf{I}\right)\boldsymbol{w}_E = 0. \tag{8.28}$$

This means that all controllable loads, $\boldsymbol{w}_E$, should satisfy equation (8.28). In other words, any loads that do not satisfy equation (8.28) are uncontrollable. Nevertheless, these uncontrollable loads, in reality, are completely resisted by the rigid structures of the mechanism.

8.3 KINEMATICS AND STATICS OF PARALLEL MANIPULATORS

Firstly, we can discuss the motion of the end effectors of parallel manipulators. Since kinematic screw systems of a hybrid kinematic chain can be equivalently substituted by kinematic screw systems of a series kinematic chain according to reciprocal screw theory [32], we only discuss the situation that the kinematic chains connected with the end effector are all serial. Assume that the number of driving inputs of the mechanism is n and the number of kinematic chains connected with the end effector is p. According to equation (8.2), for the kth $(k = 1,2,\cdots,p)$ series kinematic chain of the end effector, the instantaneous velocity of the end effector can be represented by the linear combinations of the column vectors of its kinematic screw matrix

$$\boldsymbol{v}_E = \dot{\theta}_{k1}\ \$_{k1} + \dot{\theta}_{k2}\ \$_{k2} + \cdots + \left(\dot{q}_{kd_1}\ \$_{kd_1} + \cdots + \dot{q}_{kd_i}\ \$_{kd_i}\right) + \cdots \dot{\theta}_{kc}\ \$_{kc} \tag{8.29}$$

where q_{kd_i} $\left(0 \le i < n\right)$ are the displacements (angular displacements or linear displacements) of the joints chosen as actuators in the kth kinematic chain, c is the sum of the relative DOFs of all joints in the kth chain.

According to equation (8.10), the angular velocities or linear velocities of the driving joints in the kth kinematic chain can be represented as

$$\dot{q}_{kd_i} = \frac{\left(\$^r_{Jkd_i}\right)^T \boldsymbol{\Delta v}_E}{\left(\$^r_{Jkd_i}\right)^T \boldsymbol{\Delta}\$_{kd_i}}. \tag{8.30}$$

So, if we let

$$\dot{\boldsymbol{q}} = \begin{bmatrix} \dot{q}_{1d_1} & \cdots & \dot{q}_{jd_i} & \cdots & \dot{q}_{pd_n} \end{bmatrix}^T \tag{8.31}$$

then the inverse velocity solution of parallel mechanisms can be expressed by

$$\dot{\boldsymbol{q}} = A\boldsymbol{\Delta v}_E \tag{8.32}$$

where

$$\mathbf{A} = \begin{bmatrix} \dfrac{\$^r_{Jkd_1}}{\left(\$^r_{Jkd_1}\right)^T \boldsymbol{\Delta}\$_{kd_1}} & \dfrac{\$^r_{Jkd_2}}{\left(\$^r_{Jkd_2}\right)^T \boldsymbol{\Delta}\$_{kd_2}} & \cdots & \dfrac{\$^r_{Jkd_n}}{\left(\$^r_{Jkd_n}\right)^T \boldsymbol{\Delta}\$_{kd_n}} \end{bmatrix}^T.$$

Assuming equation (8.7) to be zero, we immediately obtain the instantaneous velocity screw of the end effector

$$\boldsymbol{v}_E = b_1\$_1 + b_2\$_2 + \cdots + b_m\$_m \tag{8.33}$$

where $\$_1,\$_2,\cdots,\$_m$ denote a group of bases of the velocity screw system of the end effector, m represents the number of the base screws, b_i $\left(i = 1,2,\cdots,m\right)$ denote the linear combination coefficients.

Substituting equation (8.33) into equation (8.32) yields

$$\dot{q} = (\mathbf{A}\boldsymbol{\Delta}\$_1)\, b_1 + (\mathbf{A}\boldsymbol{\Delta}\$_2)\, b_2 + \cdots + (\mathbf{A}\boldsymbol{\Delta}\$_m)\, b_m. \tag{8.34}$$

Equation (8.34) can be represented in the form of matrix products as

$$\dot{q} = (\mathbf{A}_{n\times 6}\boldsymbol{\Delta}\mathbf{J}_{6\times m})\, \boldsymbol{b} \tag{8.35}$$

where $\mathbf{J} = \begin{bmatrix} \$_1 & \$_2 & \cdots & \$_m \end{bmatrix}$, $b = \begin{bmatrix} b_1 & b_2 & \cdots & b_m \end{bmatrix}^T$.

If $n = m$, Matrix $\mathbf{A}\boldsymbol{\Delta}\mathbf{J}$ must be a reversible square matrix, as a result, $\boldsymbol{b}$ can be solved from equation (8.35)

$$\boldsymbol{b} = (\mathbf{A}\boldsymbol{\Delta}\mathbf{J})^{-1}\, \dot{q}. \tag{8.36}$$

Substituting equation (8.36) into equation (8.33) yields

$$v_E = \mathbf{J}\,(\mathbf{A}\boldsymbol{\Delta}\mathbf{J})^{-1}\, \dot{q}. \tag{8.37}$$

Equation (8.37) represents the forward velocity solution of parallel manipulators. Assume that the end effector can keep in balance under the loads of $\boldsymbol{w}_E$, then the power consumption of the end effector at any instant can be obtained with equation (8.7)

$$P_L = \left(\mathbf{J}\,(\mathbf{A}\boldsymbol{\Delta}\mathbf{J})^{-1}\, \dot{q}\right)^T \boldsymbol{\Delta}\boldsymbol{w}_E = \dot{\boldsymbol{q}}^T \left(\mathbf{J}^T \boldsymbol{\Delta}\mathbf{A}^T\right)^{-1} \mathbf{J}^T \boldsymbol{\Delta}\boldsymbol{w}_E. \tag{8.38}$$

At the same instant, the power supplied by the driving forces of all actuators is

$$P_A = \dot{\boldsymbol{q}}^T \boldsymbol{\tau}. \tag{8.39}$$

According to the virtual power principle, to keep the balance of the manipulator, the power supplied by driving forces of all actuators should be equal to the power consumption of the end effector, that is

$$\dot{\boldsymbol{q}}^T \left[\boldsymbol{\tau} - \left(\mathbf{J}^T \boldsymbol{\Delta}\mathbf{A}^T\right)^{-1} \mathbf{J}^T \boldsymbol{\Delta}\boldsymbol{w}_E\right] = 0. \tag{8.40}$$

Since $\dot{\boldsymbol{q}}$ might be arbitrary, the necessary and sufficient condition for the establishment of equation (8.40) is

$$\boldsymbol{\tau} = \left(\mathbf{J}^T \boldsymbol{\Delta}\mathbf{A}^T\right)^{-1} \mathbf{J}^T \boldsymbol{\Delta}\boldsymbol{w}_E. \tag{8.41}$$

Equation (8.41) expresses the inverse static solution of parallel manipulators. According to Theorem 8.1, $\boldsymbol{w}_E$ can also be represented by the linear combination of a group of bases $\mathbb{W} = \begin{Bmatrix} w_1 & w_2 & \cdots & w_m \end{Bmatrix}$ which belongs to the controllable load space

$$\boldsymbol{w}_E = c_1\boldsymbol{w}_1 + c_2\boldsymbol{w}_2 + \cdots + c_m\boldsymbol{w}_m \tag{8.42}$$

where $c_1, c_2, \cdots, c_m$ are not zeros simultaneously, $\boldsymbol{w}_j\ (j = 1,2,\cdots,\boldsymbol{m})$ can be given out from the velocity screw space of the end effector. For an example, if the velocity screw space of an end effector is denoted by $\mathbb{V}_s = \begin{Bmatrix} s_1 & s_2 & s_3 & s_1 & s_2 & s_3 \\ 0 & 0 & 0 & 0 & 0 & 0 \end{Bmatrix}$, the controllable load space can be represented by $\mathbb{V}_s = \begin{Bmatrix} 0 & 0 & 0 & s_1 & s_2 & s_3 \\ s_1 & s_2 & s_3 & 0 & 0 & 0 \end{Bmatrix}$.

Substituting equation (8.42) into equation (8.41), we can gain

$$\left(\mathbf{J}^T \boldsymbol{\Delta} \mathbf{A}^T\right) \boldsymbol{\tau} = \left(\mathbf{J}^T \boldsymbol{\Delta} \boldsymbol{w}_1\right) c_1 + \left(\mathbf{J}^T \boldsymbol{\Delta} \boldsymbol{w}_2\right) c_2 + \cdots + \left(\mathbf{J}^T \boldsymbol{\Delta} \boldsymbol{w}_m\right) c_m \quad (8.43)$$

$$\boldsymbol{c} = \left(\mathbf{J}^T \boldsymbol{\Delta} \boldsymbol{w}\right)^{-1} \left(\mathbf{J}^T \boldsymbol{\Delta} \mathbf{A}^T\right) \boldsymbol{\tau} \quad (8.44)$$

where $\mathbf{w} = \begin{bmatrix} w_1 & w_2 & \cdots & w_m \end{bmatrix}$, $c = \begin{bmatrix} c_1 & c_2 & \cdots & c_m \end{bmatrix}^T$.

Substituting equation (8.44) into equation (8.42) and rewriting the result in the form of a matrix, we obtain

$$\boldsymbol{w}_E = \boldsymbol{w} \left(\mathbf{J}^T \boldsymbol{\Delta} \boldsymbol{w}\right)^{-1} \left(\mathbf{J}^T \boldsymbol{\Delta} \mathbf{A}^T\right) \boldsymbol{\tau}. \quad (8.45)$$

Equation (8.45) represents the forward statics solutions of parallel manipulators.

Example 8.2. Analyze the kinematics and statics of the Schoenflies parallel manipulator [33] shown in Figure 8.2.

Solution. Assume that the coordinates of $B_i\ (i = 1,2,3,4)$ are known in the Cartesian coordinate frame shown in Figure 8.2, the subtended angles of each kinematic chain plane and the x-axis are α_i, and the Cartesian coordinates of M_i are undetermined. Then, the analyses are as follows:

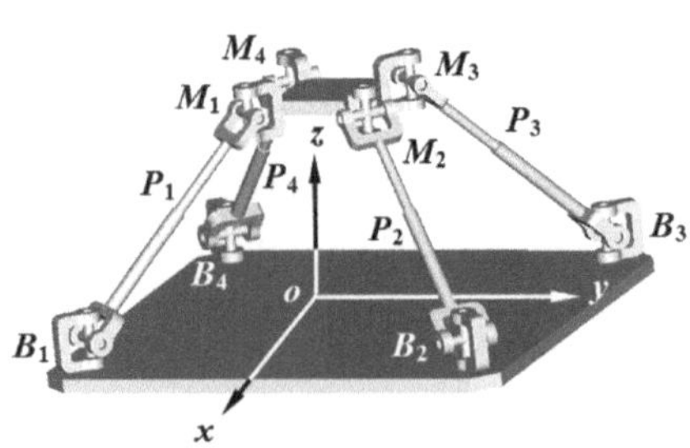

FIGURE 8.2 A Schoenflies parallel manipulator consisting of 4-UPU kinematic chains.

(1) Analyze the kinematic screws of each chain.

$$\$_{i1} = \begin{bmatrix} \cos\alpha_i & \sin\alpha_i & 0 & 0 & 0 & x_{B_i}\sin\alpha_i - y_{B_i}\cos\alpha_i \end{bmatrix}^T$$

$$\$_{i2} = \begin{bmatrix} 0 & 0 & 1 & y_{B_i} & -x_{B_i} & 0 \end{bmatrix}^T$$

$$\$_{i3} = \begin{bmatrix} 0 & 0 & 0 & x_{M_i} - x_{B_i} & y_{M_i} - y_{B_i} & z_{M_i} \end{bmatrix}^T$$

$$\$_{i4} = \begin{bmatrix} \cos\alpha_i & \sin\alpha_i & 0 & -z_{M_i}\sin\alpha_i & z_{M_i}\cos\alpha_i & x_{M_i}\sin\alpha_i - y_{M_i}\cos\alpha_i \end{bmatrix}^T$$

$$\$_{i5} = \begin{bmatrix} 0 & 0 & 1 & y_{M_i} & -x_{M_i} & 0 \end{bmatrix}^T$$

where $\$_{i3}$ acts as an actuator.

(2) Solve the forward and inverse velocity solutions of the manipulator. As

$$\$^r_{J_{i3}} = \begin{bmatrix} x_{M_i} - x_{B_i} & y_{M_i} - y_{B_i} & z_{M_i} & y_{B_i}z_{M_i} & -x_{B_i}z_{M_i} & x_{B_i}y_{M_i} - x_{M_i}y_{B_i} \end{bmatrix}^T$$

then, we can obtain the matrix **A**

$$\boldsymbol{A} = \begin{bmatrix}
\frac{x_{M_1}-x_{B_1}}{(x_{M_1}-x_{B_1})^2+(y_{M_1}-y_{B_1})^2+z_{M_1}^2} & \frac{x_{M_2}-x_{B_2}}{(x_{M_2}-x_{B_2})^2+(y_{M_2}-y_{B_2})^2+z_{M_2}^2} & \times\frac{x_{M_3}-x_{B_3}}{(x_{M_3}-x_{B_3})^2+(y_{M_3}-y_{B_3})^2+z_{M_3}^2} & \frac{x_{M_4}-x_{B_4}}{(x_{M_4}-x_{B_4})^2+(y_{M_4}-y_{B_4})^2+z_{M_4}^2} \\
\frac{y_{M_1}-y_{B_1}}{(x_{M_1}-x_{B_1})^2+(y_{M_1}-y_{B_1})^2+z_{M_1}^2} & \frac{y_{M_2}-y_{B_2}}{(x_{M_2}-x_{B_2})^2+(y_{M_2}-y_{B_2})^2+z_{M_2}^2} & \times\frac{y_{M_3}-y_{B_3}}{(x_{M_3}-x_{B_3})^2+(y_{M_3}-y_{B_3})^2+z_{M_3}^2} & \frac{y_{M_4}-y_{B_4}}{(x_{M_4}-x_{B_4})^2+(y_{M_4}-y_{B_4})^2+z_{M_4}^2} \\
\frac{z_{M_1}}{(x_{M_1}-x_{B_1})^2+(y_{M_1}-y_{B_1})^2+z_{M_1}^2} & \frac{z_{M_2}}{(x_{M_2}-x_{B_2})^2+(y_{M_2}-y_{B_2})^2+z_{M_2}^2} & \times\frac{z_{M_3}}{(x_{M_3}-x_{B_3})^2+(y_{M_3}-y_{B_3})^2+z_{M_3}^2} & \frac{z_{M_4}}{(x_{M_4}-x_{B_4})^2+(y_{M_4}-y_{B_4})^2+z_{M_4}^2} \\
\frac{y_{B_1}z_{M_1}}{(x_{M_1}-x_{B_1})^2+(y_{M_1}-y_{B_1})^2+z_{M_1}^2} & \frac{y_{B_2}z_{M_2}}{(x_{M_2}-x_{B_2})^2+(y_{M_2}-y_{B_2})^2+z_{M_2}^2} & \times\frac{y_{B_3}z_{M_3}}{(x_{M_3}-x_{B_3})^2+(y_{M_3}-y_{B_3})^2+z_{M_3}^2} & \frac{y_{B_4}z_{M_4}}{(x_{M_4}-x_{B_4})^2+(y_{M_4}-y_{B_4})^2+z_{M_4}^2} \\
\frac{-x_{B_1}z_{M_1}}{(x_{M_1}-x_{B_1})^2+(y_{M_1}-y_{B_1})^2+z_{M_1}^2} & \frac{-x_{B_2}z_{M_2}}{(x_{M_2}-x_{B_2})^2+(y_{M_2}-y_{B_2})^2+z_{M_2}^2} & \times\frac{-x_{B_3}z_{M_3}}{(x_{M_3}-x_{B_3})^2+(y_{M_3}-y_{B_3})^2+z_{M_3}^2} & \frac{-x_{B_4}z_{M_4}}{(x_{M_4}-x_{B_4})^2+(y_{M_4}-y_{B_4})^2+z_{M_4}^2} \\
\frac{x_{B_1}y_{M_1}-x_{M_1}y_{B_1}}{(x_{M_1}-x_{B_1})^2+(y_{M_1}-y_{B_1})^2+z_{M_1}^2} & \frac{x_{B_2}y_{M_2}-x_{M_2}y_{B_2}}{(x_{M_2}-x_{B_2})^2+(y_{M_2}-y_{B_2})^2+z_{M_2}^2} & \times\frac{x_{B_3}y_{M_3}-x_{M_3}y_{B_3}}{(x_{M_3}-x_{B_3})^2+(y_{M_3}-y_{B_3})^2+z_{M_3}^2} & \frac{x_{B_4}y_{M_4}-x_{M_4}y_{B_4}}{(x_{M_4}-x_{B_4})^2+(y_{M_4}-y_{B_4})^2+z_{M_4}^2}
\end{bmatrix}^T$$

Presuming equation (8.7) to be zero, the instantaneous velocity screw of the end effector can be represented by

$$\boldsymbol{v}_E = b_1\$_1 + b_2\$_2 + b_3\$_3 + b_4\$_4 \tag{8.46}$$

where b_i $(i=1,2,3,4)$ denote a set of arbitrary real numbers that are not zeros simultaneously, $\$_1 = \begin{bmatrix} 0 & 0 & 0 & 1 & 0 & 0 \end{bmatrix}^T$, $\$_2 = \begin{bmatrix} 0 & 0 & 0 & 0 & 1 & 0 \end{bmatrix}^T$, $\$_3 = \begin{bmatrix} 0 & 0 & 0 & 0 & 0 & 1 \end{bmatrix}^T$, and $\$_4 = \begin{bmatrix} 0 & 0 & 1 & 0 & 0 & 0 \end{bmatrix}^T$.

Thus,

$$\mathbf{J} = \begin{bmatrix} 0 & 0 & 0 & 0 \\ 0 & 0 & 0 & 0 \\ 0 & 0 & 0 & 1 \\ 1 & 0 & 0 & 0 \\ 0 & 1 & 0 & 0 \\ 0 & 0 & 1 & 0 \end{bmatrix}.$$

Any velocity that has the form of equation (8.46) is a feasible motion for the end effector:

$$\boldsymbol{v}_E = \begin{bmatrix} 0 & 0 & b_4 & b_1 & b_2 & b_3 \end{bmatrix}^T.$$

Equation (8.46) indicates that the end effector, $M_1M_2M_3M_4$, shown in Figure 8.2 has three translational DOFs and one rotational DOF about the z-direction. And therefore, the z-coordinates of any points on the end effector are always identical except for at the singular pose or position. For the sake of convenience, we may assume that the instantaneous velocity of the end effector is

$$\boldsymbol{v}_E = \sqrt{3}\begin{bmatrix} 0 & 0 & 0 & \frac{\sqrt{3}}{3} & \frac{\sqrt{3}}{3} & \frac{\sqrt{3}}{3} \end{bmatrix}^T.$$

Based on the above analysis, we can suppose that the z-coordinate of any point on the end effector is denoted by z. According to equation (8.32), the inverse velocity solutions of the given parallel manipulator can be obtained.

$$\dot{q} = \mathbf{AE}\boldsymbol{v}_E = \begin{bmatrix} \dfrac{x_{M_1}+y_{M_1}+z_{M_1}-x_{B_1}-y_{B_1}}{\left(x_{M_1}-x_{B_1}\right)^2+\left(y_{M_1}-y_{B_1}\right)^2+z^2} \\ \dfrac{x_{M_2}+y_{M_2}+z_{M_2}-x_{B_2}-y_{B_2}}{\left(x_{M_2}-x_{B_2}\right)^2+\left(y_{M_2}-y_{B_2}\right)^2+z^2} \\ \dfrac{x_{M_3}+y_{M_3}+z_{M_3}-x_{B_3}-y_{B_3}}{\left(x_{M_3}-x_{B_3}\right)^2+\left(y_{M_3}-y_{B_3}\right)^2+z^2} \\ \dfrac{x_{M_4}+y_{M_4}+z_{M_4}-x_{B_4}-y_{B_4}}{\left(x_{M_4}-x_{B_4}\right)^2+\left(y_{M_4}-y_{B_4}\right)^2+z^2} \end{bmatrix}$$

Given the driving velocity $\dot{q}$, we can also obtain the velocity of the end effector in accordance with equation (8.37).

(3) Solve the forward and inverse statics solutions of the parallel manipulator: According to the velocity of the end effector represented by equation (8.46), we can gain a controllable load space of the end effector

$$w = \left\{ w_1 \;\; w_2 \;\; w_3 \;\; w_4 \right\}$$

where $w_1 = \begin{bmatrix} 1 & 0 & 0 & 0 & 0 & 0 \end{bmatrix}^T$, $w_2 = \begin{bmatrix} 0 & 1 & 0 & 0 & 0 & 0 \end{bmatrix}^T$, $w_3 = \begin{bmatrix} 0 & 0 & 1 & 0 & 0 & 0 \end{bmatrix}^T$, and $w_4 = \begin{bmatrix} 0 & 0 & 0 & 0 & 0 & 1 \end{bmatrix}^T$.

Thus, we can let $w_{J_1} = \begin{bmatrix} 0 & 0 & 0 & 1 & 0 & 0 \end{bmatrix}^T$, $w_{J_2} = \begin{bmatrix} 0 & 0 & 0 & 0 & 1 & 0 \end{bmatrix}^T$, $w_{J_3} = \begin{bmatrix} 0 & 0 & 0 & 0 & 0 & 1 \end{bmatrix}^T$, $w_{J_4} = \begin{bmatrix} 0 & 0 & 1 & 0 & 0 & 0 \end{bmatrix}^T$.

If we assume the column vector of the driving forces to be

$$\tau = \begin{bmatrix} 1 & 2 & -3 & 4 \end{bmatrix}^T$$

then according to equation (8.45), we can obtain

$$w_E = \begin{bmatrix} \dfrac{x_{M_1} - x_{B_1}}{\left(x_{M_1} - x_{B_1}\right)^2 + \left(y_{M_1} - y_{B_1}\right)^2 + z^2} + \dfrac{2\left(x_{M_2} - x_{B_2}\right)}{\left(x_{M_2} - x_{B_2}\right)^2 + \left(y_{M_2} - y_{B_2}\right)^2 + z^2} \\ - \dfrac{3\left(x_{M_3} - x_{B_3}\right)}{\left(x_{M_3} - x_{B_3}\right)^2 + \left(y_{M_3} - y_{B_3}\right)^2 + z^2} + \dfrac{4\left(x_{M_4} - x_{B_4}\right)}{\left(x_{M_4} - x_{B_4}\right)^2 + \left(y_{M_4} - y_{B_4}\right)^2 + z^2} \\ \dfrac{y_{M_1} - y_{B_1}}{\left(x_{M_1} - x_{B_1}\right)^2 + \left(y_{M_1} - y_{B_1}\right)^2 + z^2} + \dfrac{2\left(y_{M_2} - y_{B_2}\right)}{\left(x_{M_2} - x_{B_2}\right)^2 + \left(y_{M_2} - y_{B_2}\right)^2 + z^2} \\ - \dfrac{3\left(y_{M_3} - y_{B_3}\right)}{\left(x_{M_3} - x_{B_3}\right)^2 + \left(y_{M_3} - y_{B_3}\right)^2 + z^2} + \dfrac{4\left(y_{M_4} - y_{B_4}\right)}{\left(x_{M_4} - x_{B_4}\right)^2 + \left(y_{M_4} - y_{B_4}\right)^2 + z^2} \\ \dfrac{z}{\left(x_{M_1} - x_{B_1}\right)^2 + \left(y_{M_1} - y_{B_1}\right)^2 + z^2} + \dfrac{2z}{\left(x_{M_2} - x_{B_2}\right)^2 + \left(y_{M_2} - y_{B_2}\right)^2 + z^2} \\ - \dfrac{3z}{\left(x_{M_3} - x_{B_3}\right)^2 + \left(y_{M_3} - y_{B_3}\right)^2 + z^2} + \dfrac{4z}{\left(x_{M_4} - x_{B_4}\right)^2 + \left(y_{M_4} - y_{B_4}\right)^2 + z^2} \\ 0 \\ 0 \\ \dfrac{x_{B_1} y_{M_1} - x_{M_1} y_{B_1}}{\left(x_{M_1} - x_{B_1}\right)^2 + \left(y_{M_1} - y_{B_1}\right)^2 + z^2} + \dfrac{2\left(x_{B_2} y_{M_2} - x_{M_2} y_{B_2}\right)}{\left(x_{M_2} - x_{B_2}\right)^2 + \left(y_{M_2} - y_{B_2}\right)^2 + z^2} \\ - \dfrac{3\left(x_{B_3} y_{M_3} - x_{M_3} y_{B_3}\right)}{\left(x_{M_3} - x_{B_3}\right)^2 + \left(y_{M_3} - y_{B_3}\right)^2 + z^2} + \dfrac{4\left(x_{B_4} y_{M_4} - x_{M_4} y_{B_4}\right)}{\left(x_{M_4} - x_{B_4}\right)^2 + \left(y_{M_4} - y_{B_4}\right)^2 + z^2} \end{bmatrix}. \tag{8.47}$$

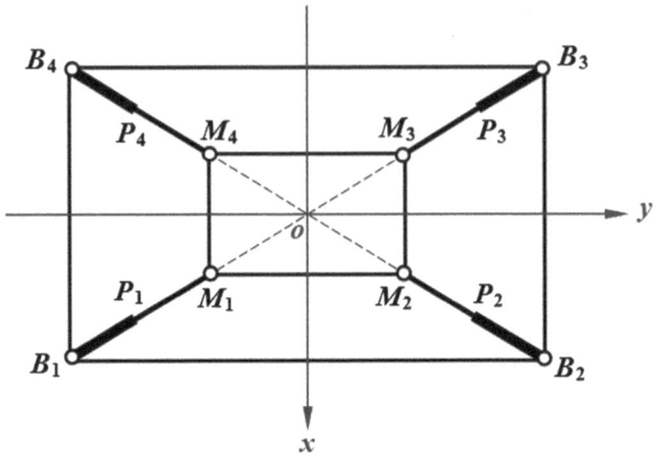

FIGURE 8.3 A special position of the manipulator.

Equation (8.47) denotes the controllable loads of the manipulator under the present position and pose. The inverse statics solution of the given parallel manipulator can be obtained from equation (8.41). Obviously, the forward and inverse statics solutions of the manipulator are both influenced by its position and pose, so the analysis should be dynamical.

Next, we can make a further investigation on the forward statics solution (8.47). If we presume that the lengths of M_1M_2 and M_1M_4 are $2a$ and $2b$, the lengths of B_1B_2 and B_1B_4 are $2c$ and $2d$, respectively, then the coordinates of the four vertexes of the end effector and the four fixed vertexes in the projected *oxyz*-coordinate system shown in Figure 8.3 can be expressed as

$$\begin{cases} \boldsymbol{r}_{M_1} = [b \quad -a \quad z]^T \\ \boldsymbol{r}_{M_2} = [b \quad a \quad z]^T \\ \boldsymbol{r}_{M_3} = [-b \quad a \quad z]^T \\ \boldsymbol{r}_{M_4} = [-b \quad -a \quad z]^T \end{cases} \tag{8.48}$$

$$\begin{cases} \boldsymbol{r}_{B_1} = [d \quad -c \quad 0]^T \\ \boldsymbol{r}_{B_2} = [d \quad c \quad 0]^T \\ \boldsymbol{r}_{B_3} = [-d \quad c \quad 0]^T \\ \boldsymbol{r}_{B_4} = [-d \quad -c \quad 0]^T. \end{cases} \tag{8.49}$$

Substituting equations (8.48) and (8.49) into equation (8.47) yields

$$w_E = \frac{1}{(b-d)^2 + (a-c)^2 + z^2} \begin{bmatrix} 2(b-d) \\ 6(a-c) \\ 4z \\ 0 \\ 0 \\ 4(da-bc) \end{bmatrix}. \tag{8.50}$$

Obviously, we can find that W_E will degenerate when any of the following equations hold:

1. $b - d = 0$,
2. $a - c = 0$,
3. $z = 0$, and
4. $da - bc = 0$.

Now, we can discuss these four cases, individually.

1. When $b - d = 0$, there will be $b = d$.
   In this case, the end effector will gain one rotational DOF about the x-direction shown in Figure 8.4.
2. When $a - c = 0$, there will be $a = c$.
   In this case, the end effector will gain one rotational DOF about the y-direction shown in Figure 8.5.
3. When $z = 0$, the manipulator degenerates to a planar one.
   In this case, the instantaneous translation of the end effector in the z-direction will be out of control.
4. When $da - bc = 0$, there will be $\frac{a}{b} = \frac{c}{d}$.

In this case, the rotation about the z-axis of the end effector shown by the double arrows in Figure 8.6 will be out of control. This singularity is also called similar singularity [33] for the end effector and the fixed base are similar in geometry which is roughly shown by Figure 8.6.

All of the four singularities can be avoided by adjusting the structure parameters and the working dimension in the z-direction. The four cases discussed above only illustrate power of the singularity identification with the method when the geometry center of the end effector moves to the z-axis and the rotational angle about the z-axis is zero. Of course, the singularities within the whole reachable workspace can also be investigated online with the model proposed in this chapter. Therefore, the method can be utilized not only in kinematics and statics analysis but also in singularity identification.

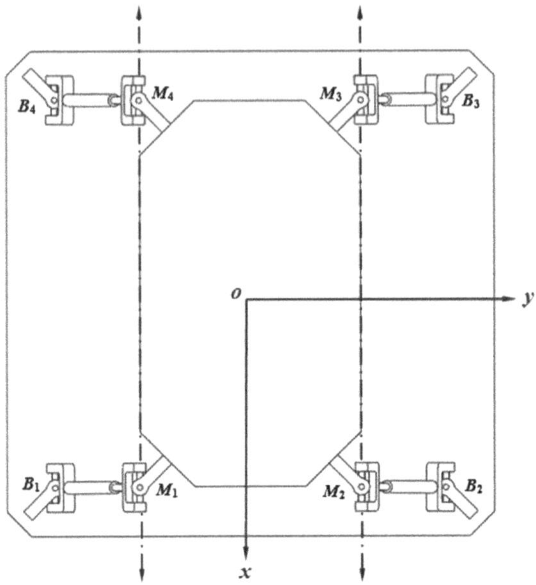

FIGURE 8.4 Top view of the first singularity pose.

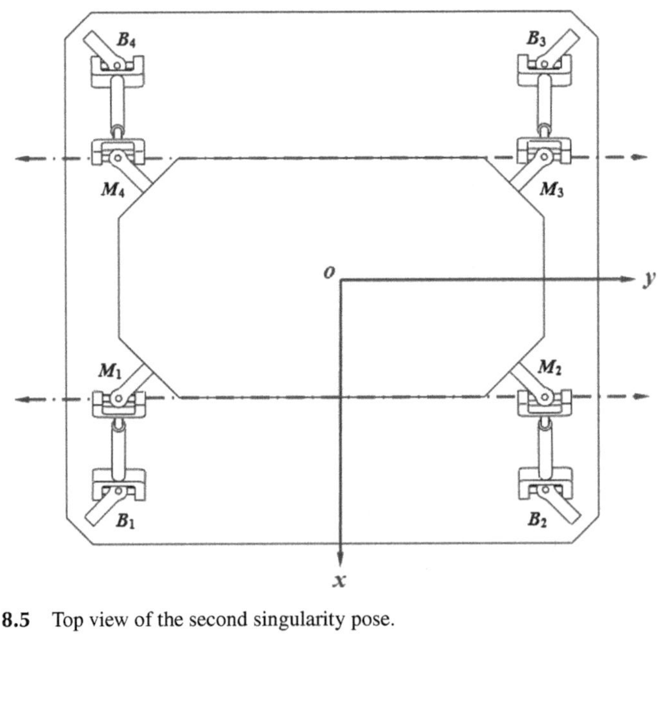

FIGURE 8.5 Top view of the second singularity pose.

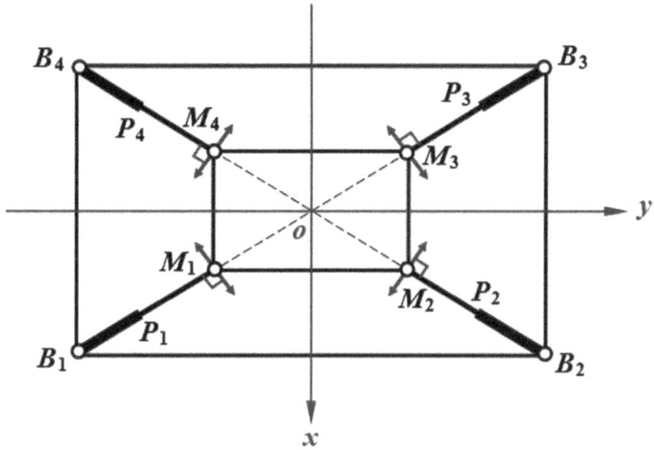

FIGURE 8.6 The fourth singularity case.

Schmalz and Agrawal [34] address the issues of control and workspace determination of planar active tensegrity or tensegrity-like structures. The motion of such structures is generally produced by actuated cables, which cannot tolerate compressive forces. As a matter of fact, all these problems can be well predicted and solved online by the models proposed in this chapter.

8.4 CONCLUSIONS

This chapter investigates the forward and inverse kinematics and statics of spatial manipulators by using reciprocal screw theory. The primary conditions for feasible motions of an end effector are firstly addressed by analyzing its velocity screws. The forward and inverse velocity problems are discussed both for series and for parallel manipulators. And then, the static balance conditions together with forward and inverse statics of series and parallel manipulators are exploited by using virtual power theory. Both the forward and the inverse velocity models of the manipulators provide explicit relationships between the possible joint velocities and the end effector velocities. Similarly, the forward and the inverse statics of the manipulators represent the relationship between the affordable forces of the end effectors and driving forces of the joints. The method proposed in this chapter is particularly useful in singularity identification, path planning and the controllability determination of series manipulators as well as parallel manipulators.

REFERENCES

[1] J.E. Shigley, J.J. Uicher, Theory of Machines and Mechanisms, McGraw-Hill Companies, Inc., New York, 1980.

[2] E.F. Ficher, A Stewart platform-based manipulator: general theory and practical construction, The International Journal of Robotics Research 5 (1986) 157–182.

[3] J.M. McCarthy, An Introduction to Theoretical Kinematics, MIT Press, Cambridge, MA, 1990.

[4] H.-S. Yan, C.-H. Kuo, Topological representations and characteristics of variable kinematic joints, Journal of Mechanical Design 128 (2006) 384–391.

[5] V. Kumar, Instantaneous kinematics of parallel-chain robotic mechanisms, Journal of Mechanical Design 114 (1992) 349–358.

[6] J.R. Phillips, K.H. Hunt, On the theorem of three axes in the spatial motion of three bodies, Australian Journal of Exerted Science 15 (1964) 267–287.

[7] K. Etemadi-Zanganeh, J. Angeles, Instantaneous kinematics of general hybrid parallel manipulators, Journal of Mechanical Design 117 (1995) 581–588.

[8] D. Kim, W.K. Chung, Analytic formulation of reciprocal screws and its application to nonredundant robot manipulators, Journal of Mechanical Design 125 (2003) 158–164.

[9] R.D. Klafter, T.A. Chmielewski, M. Negin, Robotic Engineering: An Integrated Approach, Prentice Hall, 1989.

[10] P.J. McKerrow, Introduction to Robotics, Addison-Wesley, New York, 1991.

[11] S.B. Niku, Introduction to Robotics: Analysis, Systems, Applications, Prentice Hall, 2001.

[12] J.Q. Gan, E. Oyama, E.M. Rosales, H.S. Hu, A complete analytical solution to the inverse kinematics of the pioneer 2 robotic arm, Robotica 23 (2005) 123–129.

[13] J. Estremera, K.J. Waldron, Thrust Control Stabilization and Energetics of a Quadruped Running Robot, The International Journal of Robotics Research 27 (2008) 1135–1151.

[14] I.S. Fischer, Static equilibrium of spatial mechanisms with cylindrical joints, Journal of Mechanical Design 130 (2008) 062305-1–062305-9.

[15] S.K. Agrawal, B. Roth, Statics of in-parallel manipulator systems, Journal of Mechanical Design 114 (1992) 564.

[16] G.R. Pennock, B.A. Oncu, Static force analysis of a three-cylindric robot using the theory of screws, Robotics and Autonomous Systems 9 (1992) 201–211.

[17] S.H. Juan, J.M. Mirats Tur, Tensegrity frameworks: static analysis review, Mechanism and Machine Theory 43 (2008) 859–881.

[18] R.L. Williams II, P. Gallina, Translational planar cable-direct-driven robots, Journal of Intelligent and Robotic Systems 37 (2003) 69–96.

[19] M. Arsenault, C.M. Gosselin, Kinematic, static and dynamic analysis of a planar 2-DOF tensegrity mechanism, Mechanism and Machine Theory (2006) 1072–1089.

[20] A. Elnagar, A. Hussein, On optimal constrained trajectory planning in 3D environments, Robotics and Autonomous Systems 33 (2000) 195–206.

[21] O. Shai, Utilization of the dualism between determinate trusses and mechanisms, Mechanism and Machine Theory 37 (2002) 1307–1323.

[22] M. Arsenault, R. Boudreau, Synthesis of planar parallel mechanisms while considering workspace, dexterity, stiffness and singularity avoidance, Journal of Mechanical Design 128 (2006) 69–78.

[23] J.-P. Merlet, Singular configurations of parallel manipulators and Grassmann geometry, The International Journal of Robotics Research 8 (1989) 45–56.

[24] C. Gosselin, J. Angeles, Singularity analysis of closed-loop kinematic chains, IEEE Transactions on Robotics and Automation 6 (1990) 281–290.

[25] F.C. Park, J.W. Kim, Singularity analysis of closed kinematic chains, Journal of Mechanical Design 121 (1999) 32–38.

[26] S. Bhattacharya, H. Hatwal, A. Ghosh, Comparison of an exact and an approximate method of singularity avoidance in platform type parallel manipulators, Mechanism and Machine Theory 33 (1998) 965–974.

[27] C.T. Kelley, Iterative Methods for Linear and Nonlinear Equations, North Carolina State University, Society for Industrial and Applied Mathematics, Philadelphia, 1995.

[28] Y. Nakamura, Advanced Robotics: Redundancy and Optimization, Addison-Wesley: Addison-Wesley Publishing Company, 1991.
[29] J.S. Dai, J.R. Jones, Interrelationship between Screw Systems and Corresponding Reciprocal Systems and Applications, Mechanism and Machine Theory 6 (2001) 633–651.
[30] J.S. Dai, J.R. Jones, Null-space construction using cofactors from a screw-algebra context, Proceedings of the Royal Society of London, Series A 458 (2002) 1845–1866.
[31] H.R. Piccardo, G. Honderd, A new approach to on-line path planning and generation for robots in non-static environments, Robotics and Autonomous Systems 8 (1991) 187–201.
[32] J.-S. Zhao, K. Zhou, et al., The singularity study of spatial hybrid mechanism based on screw theory, The International Journal of Advanced Manufacturing Technology 25 (2005) 1053–1061.
[33] J.-S. Zhao, Y.-Z. Fu, K. Zhou, Z.J. Feng, Mobility properties of a schoenflies type parallel mechanism, Robotics and Computer-Integrated Manufacturing 22 (2006) 124–133.
[34] A.P. Schmalz, S.K. Agrawal, Dynamic workspace and control of planar active tensegritylike structures, Journal of Mechanical Design 130 (2008) 122301-1–122301-12.

Chapter 9

Fundamental Factors to Investigating the Motions and Actuations of a Mechanism

The degree of freedom (DOF) of a mechanism is of primary importance in the theory of mechanisms and has been extensively discussed either in analytical methods or in quick calculation approaches based on the algebra summations of the numbers of the links, joints and the constraints induced by the joints in the previous literature during the past 150 years [1]. This lengthy development process on the one hand demonstrates the complexity and difficulty of this project, and on the other hand indicates that it is vital important both in theory and in engineering applications.

As mentioned above, all of the previous methods are generally classified into two groups, one is analytical and the other is based on the principle of algebra summations of the numbers of the links, joints, and the constraints induced by the joints. No matter which group the method or formula belongs to, the critical steps to correctly analyze the DOF of a mechanism are the same. As to the development of the DOF of a mechanism, Gogu [1] carried out the latest review. Therefore, we do not intend to make a full review about the history of DOF of a mechanism in this chapter. On the contrary, we will only investigate the critical steps to correctly analyze the DOF of a mechanism based on our previous work [2–5].

According to the physical meaning of the reciprocal screws, we can solve the constraint spaces spanned by the reciprocal screws, and then the free motion spaces of the end effector and its DOF. With this methodology, not only can we differentiate the system from its members, but we can also extend the concept of DOF, which should have the attributes of number, type, and direction.

Obviously, the DOF of an end effector is not larger than 6, but the independent number of actuators required to uniquely control the end effector might be any non-negative integer. Bearing the above two definitions in mind, we can fall into two steps to investigate the mobility of a mechanism—the DOF of the end effector and the configuration degree of freedom (CDOF) of the mechanism with the end effector. The former definition indicates the

Advanced Theory of Constraint and Motion Analysis for Robot Mechanisms.
http://dx.doi.org/10.1016/B978-0-12-420162-0.00009-6

full instantaneous motion properties of the end effector through a mathematic concept of free mobility space, while the later one tells us the instantaneous controllability of the mechanism system. Therefore, we introduce the following two definitions:

Definition 9.1. The DOF of an end effector totally characterizes the motions of the end effector including the number, type and direction of the independent motions.

Definition 9.2. The CDOF of a mechanism with an end effector indicates the independent number of actuations required to uniquely control the end effector under any configurations.

9.1 THE BASIC STEPS FOR CALCULATING THE DOF OF THE MECHANISM WITH A PRESCRIBED END EFFECTOR

Now let us consider the motions of a free rigid body in a three-dimensional physical space. The three independent translations and three independent rotations can be immediately expressed with a screw matrix, **S**, in a six-dimensional Euclidean space:

$$\mathbf{S} = \begin{bmatrix} \$_{\omega_x} & \$_{\omega_y} & \$_{\omega_z} & \$_{v_x} & \$_{v_y} & \$_{v_z} \end{bmatrix} = \begin{bmatrix} 1 & 0 & 0 & 0 & 0 & 0 \\ 0 & 1 & 0 & 0 & 0 & 0 \\ 0 & 0 & 1 & 0 & 0 & 0 \\ 0 & 0 & 0 & 1 & 0 & 0 \\ 0 & 0 & 0 & 0 & 1 & 0 \\ 0 & 0 & 0 & 0 & 0 & 1 \end{bmatrix} \tag{9.1}$$

where $\$_{\omega_x}$ indicates the unit rotation about the x-axis, $\$_{v_x}$ indicates the unit translation along the x-axis; the meanings of the other symbols can be similarly deduced.

Equation (9.1) just represents a basis of the six-dimensional Euclidean space. Reciprocally, the constraint matrix of a fully restricted rigid body, $\$^C_{Fix}$, can also be expressed in the same six-dimensional Euclidean space:

$$\$^C_{Fix} = \begin{bmatrix} \$_{f_x} & \$_{f_y} & \$_{f_z} & \$_{\tau_x} & \$_{\tau_y} & \$_{\tau_z} \end{bmatrix} = \begin{bmatrix} 1 & 0 & 0 & 0 & 0 & 0 \\ 0 & 1 & 0 & 0 & 0 & 0 \\ 0 & 0 & 1 & 0 & 0 & 0 \\ 0 & 0 & 0 & 1 & 0 & 0 \\ 0 & 0 & 0 & 0 & 1 & 0 \\ 0 & 0 & 0 & 0 & 0 & 1 \end{bmatrix} \tag{9.2}$$

where $\$_{f_x}$ indicates the unite force along the x-axis, $\$_{\tau_x}$ indicates the unite moment about the x-axis; and the meanings of the other symbols can also be similarly deduced.

Consequently, the free motion(s) of an end effector must be a subspace of the six-dimensional Euclidean space denoted by its basis (9.1); and correspondingly, the constraint(s) of the end effector must also be a subspace of the six-dimensional Euclidean space denoted by its basis (9.2). These two subspaces are tightly connected by the reciprocities of motions and constraints.

$$\$^T \Delta \$^r = 0 \tag{9.3}$$

where $\$$ is called the screw matrix of the kinematic chain; $\$^r$ is the reciprocal screws of the kinematic chain, which is determined by the maximum independent set of the kinematic screws, $\$$.

Consider a rigid body spherically jointed to a fixed frame shown in Figure 9.1. The constraints exerted to the rigid body through the point by the fixed frame are three orthogonal forces (bold arrows), which should be easily expressed with a subspace of the Euclidean space denoted by equation (9.2), $\$^C_{body}$.

$$\$^C_{body} = \begin{bmatrix} \$_{f_x} & \$_{f_y} & \$_{f_z} \end{bmatrix} = \begin{bmatrix} 1 & 0 & 0 \\ 0 & 1 & 0 \\ 0 & 0 & 1 \\ 0 & 0 & 0 \\ 0 & 0 & 0 \\ 0 & 0 & 0 \end{bmatrix} \tag{9.4}$$

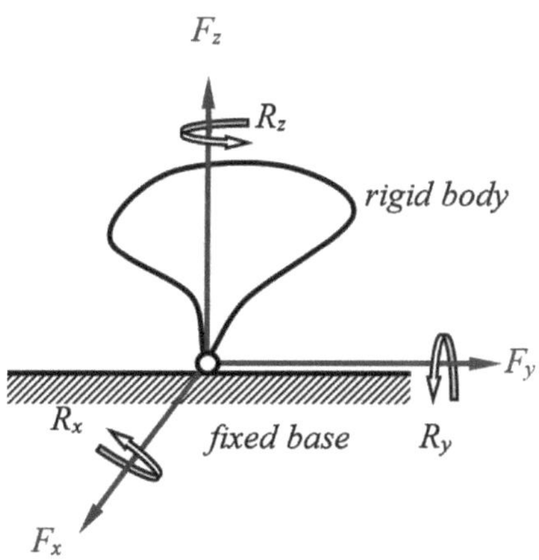

FIGURE 9.1 A rigid body spherically jointed by the fixed frame.

The motions restricted by $\$_{body}^{C}$ can be expressed with the constraint matrix:

$$\mathbf{S}_{body}^{C}=\begin{bmatrix}\$_{v_x} & \$_{v_y} & \$_{v_z}\end{bmatrix}=\begin{bmatrix}0&0&0\\0&0&0\\0&0&0\\1&0&0\\0&1&0\\0&0&1\end{bmatrix}. \tag{9.5}$$

Abstractly, $\$_{body}^{C}$ and $\mathbf{S}_{body}^{C}$ are mutually orthogonal supplementary in a six-dimensional Euclidean space. Considering the subspace of the free motions of any rigid body, $\mathbf{S}_{body}^{F}$, we have the relationship [3]:

$$\mathbf{S}_{body}^{F}\oplus\mathbf{S}_{body}^{C}=\mathbf{S} \tag{9.6}$$

where $\oplus$ denotes the direct sum of two Euclidean spaces.

Therefore, $\mathbf{S}_{body}^{F}$ will be immediately obtained from equations (9.5) and (9.6):

$$\mathbf{S}_{body}^{F}=\begin{bmatrix}\$_{\omega_x} & \$_{\omega_y} & \$_{\omega_z}\end{bmatrix}=\begin{bmatrix}1&0&0\\0&1&0\\0&0&1\\0&0&0\\0&0&0\\0&0&0\end{bmatrix}. \tag{9.7}$$

The free motions are also shown in Figure 9.1 (hollow bow arrows). It is not difficult to find that equations (9.4) and (9.7) satisfy reciprocal screw theory. Therefore, if we know the constraints of the end effector, the free motions can be equivalently found through reciprocal screw theory. In this regard, the constraints exerted on the end effector and the free motions of the end effector under these constraints are reciprocally equivalent. As a result, we only need to analyze the constraints exerted to the end effector or its free motions.

For the spatial mechanism is shown in Figure 9.2, we first establish the Cartesian coordinate system as Figure 9.3 shows by setting the origin at the geometry center of the fixed frame and the x-axis parallel to B_3B_1 and the y-axis perpendicular to B_3B_1, the kinematic screws of the PS kinematic chain OP_4O_1 are:

$$\$_{OP_4O_1}=\begin{bmatrix}\$_{O_11} & \$_{O_12} & \$_{O_13} & \$_{O_14}\end{bmatrix} \tag{9.8}$$

where

$$\$_{O_11}=\begin{bmatrix}0&0&0&0&0&1\end{bmatrix}^T,\quad \$_{O_12}=\begin{bmatrix}1&0&0&0&z_{O_1}&0\end{bmatrix}^T,$$

$$\$_{O_13}=\begin{bmatrix}0&1&0&-z_{O_1}&0&0\end{bmatrix}^T,\quad \$_{O_14}=\begin{bmatrix}0&0&1&0&0&0\end{bmatrix}^T$$

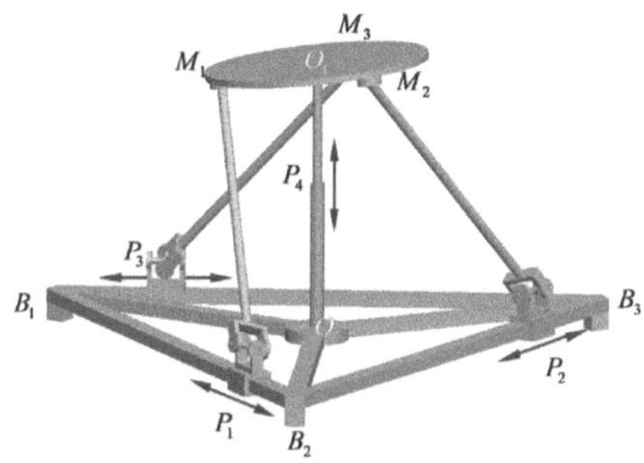

FIGURE 9.2 A spatial parallel manipulator with three PUS and one PS chains.

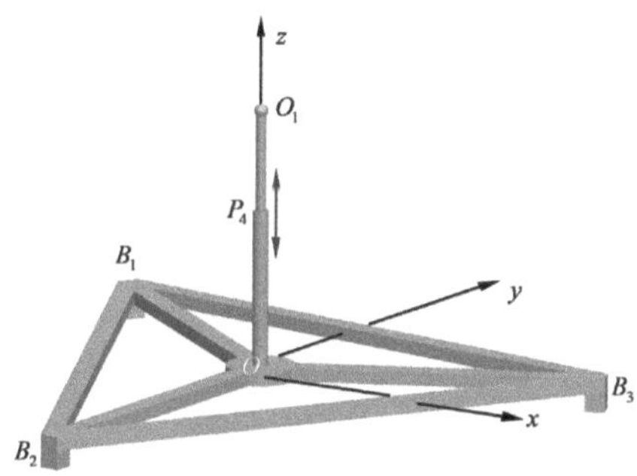

FIGURE 9.3 PS kinematic chain.

where $\$_{O_1 1}$ denotes the homogeneous Plücker coordinates of the prismatic joint P_4; $\$_{O_1 2}$,$\$_{O_1 3}$, and $\$_{O_1 4}$ denote the homogeneous Plücker coordinates of the top spherical joint.

Substituting equation (9.8) into equation (9.3) yields

$$\begin{bmatrix} 0 & 0 & 1 & 0 & 0 & 0 \\ 0 & z_{O_1} & 0 & 1 & 0 & 0 \\ -z_{O_1} & 0 & 0 & 0 & 1 & 0 \\ 0 & 0 & 0 & 0 & 0 & 1 \end{bmatrix} \begin{bmatrix} l^r \\ m^r \\ n^r \\ p^r \\ q^r \\ r^r \end{bmatrix} = 0. \tag{9.9}$$

The solutions for equation (9.9) are:

$$\begin{cases} \$^r_{O_1 1} = \begin{bmatrix} 1 & 0 & 0 & 0 & z_{O_1} & 0 \end{bmatrix}^T \\ \$^r_{O_1 2} = \begin{bmatrix} 0 & 1 & 0 & -z_{O_1} & 0 & 0 \end{bmatrix}^T. \end{cases} \tag{9.10}$$

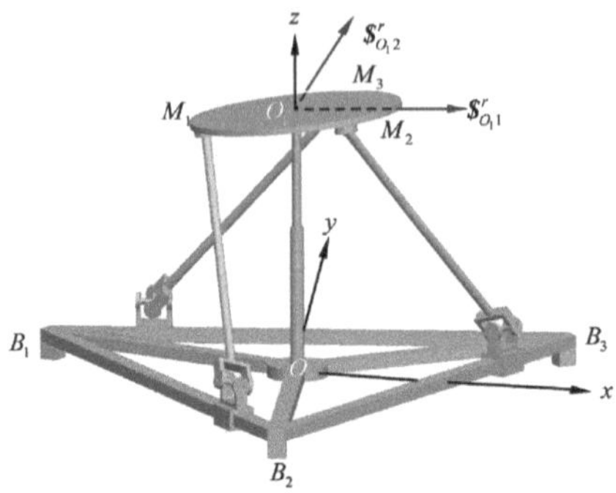

FIGURE 9.4 Terminal constraints of the PS kinematic chain.

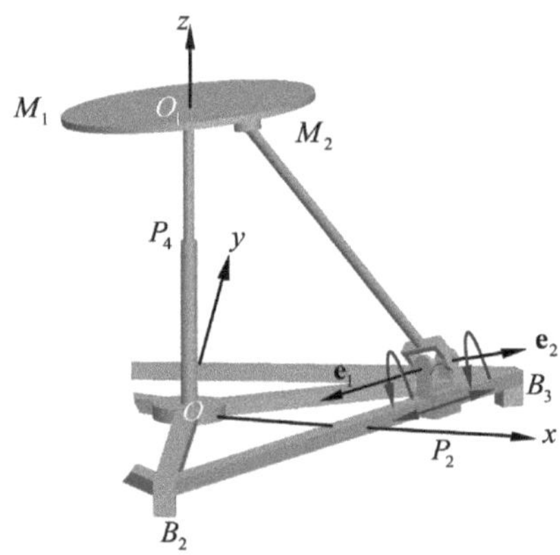

FIGURE 9.5 A PUS kinematic chain.

Equation (9.10) represents the terminal constraints exerted to the end effector from the PS kinematic chain. $\$^r_{O_11}$ indicates a pure force along the line passing through the spherical joint with a direction of $[1 \quad 0 \quad 0]^T$, and $\$^r_{O_12}$ indicates a pure force along the line passing through the spherical joint with a direction of $[0 \quad 1 \quad 0]^T$, which are shown in Figure 9.4.

Next, we consider the terminal constraints of the other three identical kinematic chains. For the sake of simplicity, we only need to study any one of the three, for instance, kinematic chain P_2M_2 shown in Figure 9.5.

According to the setting principle of the Cartesian coordinate system, we can find the axis direction vector $\boldsymbol{e}_1$ is parallel to the side, B_3B_2, of the equilateral triangle $B_1B_2B_3$. Therefore, the direction vector $\boldsymbol{e}_1$ can be denoted as:

$$\boldsymbol{e}_1 = \begin{bmatrix} -\cos\frac{\pi}{3} & -\sin\frac{\pi}{3} & 0 \end{bmatrix}^T. \tag{9.11}$$

The direction ratio of $\boldsymbol{e}_2$ can be expressed by:

$$\boldsymbol{e}_2 = \frac{\overrightarrow{P_2M_2}}{\left\|\overrightarrow{P_2M_2}\right\|} \times \boldsymbol{e}_1. \tag{9.12}$$

Therefore, the twist matrix of this chain are represented by:

$$\boldsymbol{\$}_{P_2M_2} = \begin{bmatrix} \boldsymbol{\$}_{M_21} & \boldsymbol{\$}_{M_22} & \boldsymbol{\$}_{M_23} & \boldsymbol{\$}_{M_24} & \boldsymbol{\$}_{M_25} & \boldsymbol{\$}_{M_26} \end{bmatrix} \tag{9.13}$$

where

$$\boldsymbol{\$}_{M_21} = \begin{bmatrix} \boldsymbol{0} \\ \boldsymbol{e}_1 \end{bmatrix}, \quad \boldsymbol{\$}_{M_22} = \begin{bmatrix} \boldsymbol{e}_1 \\ \boldsymbol{r}_{P_2} \times \boldsymbol{e}_1 \end{bmatrix}, \quad \boldsymbol{\$}_{M_23} = \begin{bmatrix} \boldsymbol{e}_2 \\ \boldsymbol{r}_{P_2} \times \boldsymbol{e}_2 \end{bmatrix},$$

$$\boldsymbol{\$}_{M_24} = \begin{bmatrix} 1 & 0 & 0 & 0 & z_{M_2} & -y_{M_2} \end{bmatrix}^T,$$

$$\boldsymbol{\$}_{M_25} = \begin{bmatrix} 0 & 1 & 0 & -z_{M_2} & 0 & x_{M_2} \end{bmatrix}^T,$$

$$\boldsymbol{\$}_{M_26} = \begin{bmatrix} 0 & 0 & 1 & y_{M_2} & -x_{M_2} & 0 \end{bmatrix}^T.$$

It is not difficult to solve the reciprocal screw of equation (9.13)

$$\boldsymbol{\$}^r_{P_2M_2} = \begin{bmatrix} 0 & 0 & 0 & 0 & 0 & 0 \end{bmatrix}^T. \tag{9.14}$$

Therefore, the kinematic chain P_2M_2 will exert no constraint to the end effector, so are the chains P_1M_1 and P_3M_3. As a result, the constraints exerted to the end effector are two forces provided by kinematic chain OP_4O_1 only.

Now, let us consider the DOF of the end effector shown in Figure 9.2. We have known that the constraints exerted to the end effector are expressed with equation (9.10). Assume

$$\boldsymbol{\$}^r_M = \begin{bmatrix} 1 & 0 \\ 0 & 1 \\ 0 & 0 \\ 0 & -z_{O_1} \\ z_{O_1} & 0 \\ 0 & 0 \end{bmatrix}. \tag{9.15}$$

According to equation (9.3), we can get the following homogeneous equations:

$$\left(\boldsymbol{\$}^r_M\right)^T \boldsymbol{\Delta\$}_M = \boldsymbol{0}. \tag{9.16}$$

So, the free motions of the end effector can be found from equation (9.16):

$$\$_M = \begin{bmatrix} \$_M^1 \\ \$_M^2 \\ \$_M^3 \\ \$_M^4 \end{bmatrix} = \begin{bmatrix} 1 & 0 & 0 & 0 & z_{O_1} & 0 \\ 0 & 1 & 0 & -z_{O_1} & 0 & 0 \\ 0 & 0 & 1 & 0 & 0 & 0 \\ 0 & 0 & 0 & 1 & 0 & 0 \end{bmatrix}^T \tag{9.17}$$

where $\$_M^1, \$_M^2, \$_M^3$ represent three rotations about the spherical joint O_1 and $\$_M^4$ represents a translation along the z-axis.

Therefore, $\$_M$ in equation (9.17) fully characterizes the DOF of the end effector at any instant. The number of DOFs is Rank $(\$_M) = 4$, and the types and directions are expressed by the corresponding screws in (9.17).

9.2 ACTUATION SCHEMES TO CONTROL OF THE MECHANISM

9.2.1 Possible Actuation Scheme Analysis

As we know, the mobility of a mechanism and the actuations needed to control the end effector might be different if the actuation schemes are selected differently. Consequently, it is urgently necessary to analyze the number of actuations to control the end effector after knowing the DOF of the end effector.

Assume X to be the target set and G is a group to operate on X. For each $a \in X$, there is

$$X_a = \left\{ g(a)|_{g \in G} \right\}$$

where X_a is called the obit of X under the operation of G, and a is the representative element of the orbit. Therefore, the order of the group X can be denoted as $|X| = \sum_{a \in X} |X_a|$, where $|\cdot|$ indicates the order of group "·". Also assume $g \in G$ and $a \in X$, a is the fixed point of g if $g(a) = a$. The set of all groups whose fixed point is a can be represented by $G_a = \{g | g \in G, g(a) = a\}$. According to the Burnside lemma [6], the number of the orbits of a finite set X under the operations of the finite group G is

$$N_{12} = \frac{1}{|G|} \sum_{g \in G} \chi(g)$$

where $\chi(g)$ represents the number of fixed points of X under the operation of g, the summation, $\sum$, operates on every element of the group, and $\chi(g) = |\{a | a \in X, g(a) = a\}|$. For each a that satisfies the relationship $g(a) = a$, the actuation schemes corresponding to the same cycle of g must be the same, therefore, the total number of possible actuation schemes of each cycle is n. The number of cycles that g contains is $\lambda_1 + \lambda_2 + \cdots + \lambda_m$, so the number of possible selections of the actuation schemes that satisfy the needs of $g(a) = a$

is $n^{\lambda_1+\lambda_2+\cdots+\lambda_m}$. Therefore, the number of different actuation schemes can be obtained when $\chi(g) = n^{\lambda_1+\lambda_2+\cdots+\lambda_m}$ is substituted into the formula of Burnside formula [6]:

$$N_{12} = \frac{1}{|D_m|} \sum_{g \in D_m} n^{\lambda_1+\lambda_2+\cdots+\lambda_m} \tag{9.18}$$

where g is expressed in $1^{\lambda_1} 2^{\lambda_2} \cdots m^{\lambda_m}$ format, D_m is a dihedral group, and the summation is operated on every permutation of D_m.

After the DOF of the end effector is obtained, we have to make it clear whether it is enough to control the end effector with the equal number of actuations as that of DOF of the end effector. In order to answer this question, we firstly investigate a planar manipulator with a single open chain shown in Figure 9.6. If we set up a Cartesian coordinate system as Figure 9.6 shows, the screw matrix of the open chain is:

$$\$ = \begin{bmatrix} \$_A & \$_B & \$_C & \$_D \end{bmatrix} = \begin{bmatrix} 0 & 0 & 0 & 0 \\ 0 & 0 & 0 & 0 \\ 1 & 1 & 1 & 1 \\ 0 & y_B & y_C & y_D \\ 0 & -x_B & -x_C & -x_D \\ 0 & 0 & 0 & 0 \end{bmatrix}. \tag{9.19}$$

The rank of (9.19) is three under general configurations, and the terminal constraints of the chain exerted to the end effector DE can be quickly solved by substituting equation (9.19) into equation (9.3), and the terminal constraint matrix is:

$$\$^r_{DE} = \begin{bmatrix} 0 & 0 & 0 \\ 0 & 0 & 0 \\ 1 & 0 & 0 \\ 0 & 1 & 0 \\ 0 & 0 & 1 \\ 0 & 0 & 0 \end{bmatrix} \tag{9.20}$$

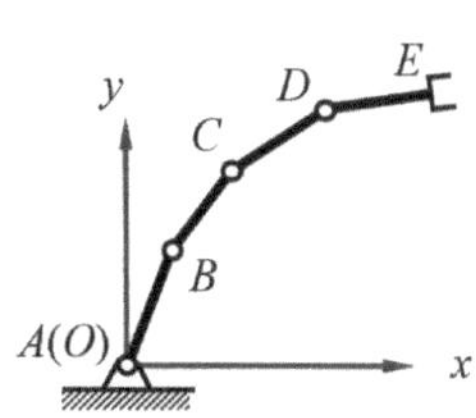

FIGURE 9.6 A planar manipulator with single open chain.

The free motion space of this end effector, DE, is the null space of matrix $\mathbf{A} = \left(\$^r_{DE}\right)^T \boldsymbol{\Delta}$. The standard basis of the free motion space can also be quickly found by solving equation (9.3) again:

$$\$_{DE} = \begin{bmatrix} 0 & 0 & 0 \\ 0 & 0 & 0 \\ 1 & 0 & 0 \\ 0 & 1 & 0 \\ 0 & 0 & 1 \\ 0 & 0 & 0 \end{bmatrix}. \tag{9.21}$$

Therefore, the number of DOFs of the end effector under general configurations is $\text{Rank}(\$_{DE}) = 3$. However, if the coordinates of the kinematic pairs satisfy the criteria:

$$\frac{y_B - y_A}{x_B - x_A} = \frac{y_C - y_A}{x_C - x_A} = \frac{y_D - y_A}{x_D - x_A} \tag{9.22}$$

which is shown in Figure 9.7, the rank of equation (9.19) will be two and the terminal constraint matrix now becomes:

$$\$^r_{DE} = \begin{bmatrix} 0 & 0 & 0 & x_B \\ 0 & 0 & 0 & y_B \\ 1 & 0 & 0 & 0 \\ 0 & 1 & 0 & 0 \\ 0 & 0 & 1 & 0 \\ 0 & 0 & 0 & 0 \end{bmatrix}. \tag{9.23}$$

A basis for the free motion space of the end effector DE under this configuration can be obtained by substituting equation (9.23) into equation (9.3):

$$\$^{\omega}_{DE} = \begin{bmatrix} 0 & 0 & 1 & 0 & 0 & 0 \\ 0 & 0 & 0 & -y_B & x_B & 0 \end{bmatrix}^T. \tag{9.24}$$

These two free motions of the end effector under such a configuration are individually represented by the hollow arrow (rotation) and the bi-directional arrow (translation) shown in Figure 9.7. From equations (9.21) and (9.24), it is not difficult to find that the number of DOFs of the end effector DE is three or two. However, to uniquely control the end effector DE, four actuations are necessary which is indicated by equation (9.19). So, the number of the actuations required to uniquely control an end effector are not always equal to the number of the DOFs of the end effector.

Therefore, the general mobility of a mechanism sometimes gives no real meanings for both the DOF of the end effector and the actuations required to drive the end effector. And consequently, determining the number of actuations

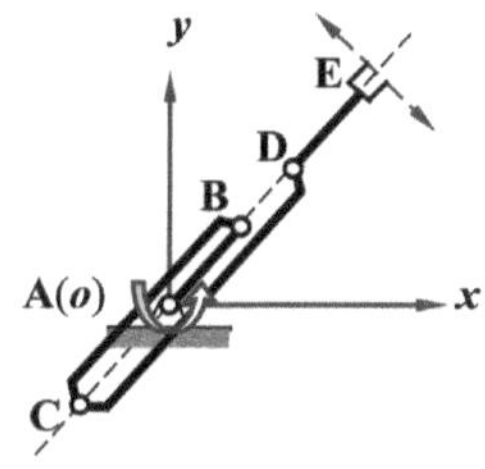

FIGURE 9.7 Special configuration of the serial mechanism.

needed to uniquely control the end effector is another problem after obtaining the DOF of the end effector. The following section will focus on this problem.

From the above analysis, we have known that the independent actuations required to uniquely control the end effector are not less than the DOF of the end effector in number. Therefore, it is sufficient to check whether the DOF is zero after exerting a number of actuations, which are equal to the former DOF of the end effector in number. If yes, the number of actuations required to uniquely control the end effector under this configuration is that of the DOF; otherwise, we should either reselect the actuation scheme or repeat the above analysis process until the end effector is completely controlled. Hence, the number of actuations required to uniquely control the end effector is the sum of these actuations.

We now investigate the number of actuations required to control the end effector shown in Figure 9.2. The DOF of the end effector are three rotations about one moving point O_1 and one translation along the vertical axis OO_1 of the support column. Therefore, a general actuation scheme is to assign these four actuators to the three sliders P_1, P_2, P_3, and the vertical prismatic joint P_4.

After applying these actuations, the parallel manipulator becomes a "new one," which comprises one S chain and three US kinematic chains.

For the OP_4O_1 chain (only consists of a single S joint), the twists can be similarly obtained from equation (9.8):

$$\$_{OO_1} = \begin{bmatrix} \$_{O_12} & \$_{O_13} & \$_{O_14} \end{bmatrix}. \tag{9.25}$$

Substituting equation (9.25) into equation (9.3) yields the following terminal constraints of this actuated kinematic chain:

$$\begin{cases} \$^r_{O_11} = \begin{bmatrix} 1 & 0 & 0 & 0 & z_{O_1} & 0 \end{bmatrix}^T \\ \$^r_{O_12} = \begin{bmatrix} 0 & 1 & 0 & -z_{O_1} & 0 & 0 \end{bmatrix}^T \\ \$^r_{O_13} = \begin{bmatrix} 0 & 0 & 1 & 0 & 0 & 0 \end{bmatrix}^T. \end{cases} \tag{9.26}$$

For the P_2M_2 chain (US kinematic chain), the twist matrix of the chain can also be similarly obtained from equation (9.13):

$$\$_{P_2M_2} = \begin{bmatrix} \$_{M_22} & \$_{M_23} & \$_{M_24} & \$_{M_25} & \$_{M_26} \end{bmatrix}. \tag{9.27}$$

Substituting equation (9.27) into equation (9.3) yields a unit terminal constraint of this actuated kinematic chain:

$$\$^r_{P_2M_2} = \frac{\begin{bmatrix} x_{M_2}-x_{P_2} & y_{M_2}-y_{P_2} & z_{M_2}-z_{P_2} & z_{M_2}y_{P_2}-y_{M_2}z_{P_2} & x_{M_2}z_{P_2}-z_{M_2}x_{P_2} & y_{M_2}x_{P_2}-x_{M_2}y_{P_2} \end{bmatrix}^T}{\sqrt{(x_{M_2}-x_{P_2})^2+(y_{M_2}-y_{P_2})^2+(z_{M_2}-z_{P_2})^2}}. \tag{9.28}$$

Similarly, the terminal constraints of the other kinematic chains, P_1M_1 and P_3M_3 can be obtained in the same Cartesian coordinate system, respectively:

$$\$^r_{P_1M_1} = \frac{\begin{bmatrix} x_{M_1}-x_{P_1} & y_{M_1}-y_{P_1} & z_{M_1}-z_{P_1} & z_{M_1}y_{P_1}-y_{M_1}z_{P_1} & x_{M_1}z_{P_1}-z_{M_1}x_{P_1} & y_{M_1}x_{P_1}-x_{M_1}y_{P_1} \end{bmatrix}^T}{\sqrt{(x_{M_1}-x_{P_1})^2+(y_{M_1}-y_{P_1})^2+(z_{M_1}-z_{P_1})^2}}. \tag{9.29}$$

$$\$^r_{P_3M_3} = \frac{\begin{bmatrix} x_{M_3}-x_{P_3} & y_{M_3}-y_{P_3} & z_{M_3}-z_{P_3} & z_{M_3}y_{P_3}-y_{M_3}z_{P_3} & x_{M_3}z_{P_3}-z_{M_3}x_{P_3} & y_{M_3}x_{P_3}-x_{M_3}y_{P_3} \end{bmatrix}^T}{\sqrt{(x_{M_3}-x_{P_3})^2+(y_{M_3}-y_{P_3})^2+(z_{M_3}-z_{P_3})^2}}. \tag{9.30}$$

Considering the homogeneousness of Plücker coordinates, the coefficient constants can be eliminated for the sake of simplicity. Therefore, from equations (9.26), (9.28), (9.29) and (9.30), the constraints exerted on the end effector under these four actuations are expressed by the terminal constraint matrix:

$$\$^r_{M_1M_2M_3} = \begin{bmatrix} 1 & 0 & 0 & x_{M_1}-x_{P_1} & x_{M_2}-x_{P_2} & x_{M_3}-x_{P_3} \\ 0 & 1 & 0 & y_{M_1}-y_{P_1} & y_{M_2}-y_{P_2} & y_{M_3}-y_{P_3} \\ 0 & 0 & 1 & z_{M_1}-z_{P_1} & z_{M_2}-z_{P_2} & z_{M_3}-z_{P_3} \\ 0 & -z_{o1} & 0 & z_{M_1}y_{P_1}-y_{M_1}z_{P_1} & z_{M_2}y_{P_2}-y_{M_2}z_{P_2} & z_{M_3}y_{P_3}-y_{M_3}z_{P_3} \\ z_{o1} & 0 & 0 & x_{M_1}z_{P_1}-z_{M_1}x_{P_1} & x_{M_2}z_{P_2}-z_{M_2}x_{P_2} & x_{M_3}z_{P_3}-z_{M_3}x_{P_3} \\ 0 & 0 & 0 & y_{M_1}x_{P_1}-x_{M_1}y_{P_1} & y_{M_2}x_{P_2}-x_{M_2}y_{P_2} & y_{M_3}x_{P_3}-x_{M_3}y_{P_3} \end{bmatrix}. \tag{9.31}$$

Substituting equation (9.31) into equation (9.3), we can obtain a basis of the free motion space of the actuated end effector. Obviously, the necessary and sufficient criteria for non-zero solution of equation (9.3) are that the rank of $\$^r_{M_1M_2M_3}$ is less than six. That is, the determinate of matrix $\$^r_{M_1M_2M_3}$ is zero:

$$\left|\$^r_{M_1M_2M_3}\right| = 0. \tag{9.32}$$

When equation (9.32) holds under certain configurations, the number of actuations needed instantaneously to control the end effector will increase. Traditionally, these configurations are often called singularities.

However, what must be pointed out is that we should avoid selecting the DOFs in a kinematic chain that do not do real work to the end effector as actuations. Otherwise, such actuations will be assumed to be valid idle DOFs to the mechanism.

According to the above analysis, the computation cycle to obtain the number of actuations required to uniquely control the end effector of a spatial mechanism can be roughly divided into two major phases. One is to investigate the DOF of the end effector; the other is to analyze the actuation(s) needed to control the end effector. The crucial steps to investigate the CDOF of a mechanism can also be shown in Figure 9.8.

It is not difficult to find that the whole analysis process is quite adapted to programming at a computer. As an example, let us consider a spatial parallel manipulator whose actuations required are more than the number of DOFs of its end effector. This kind of mechanism is often called a mechanism with redundant mobility and therefore has quite good controlling flexibility and often has a larger workspace. The structure of such a mechanism is shown in Figure 9.9. The end effector $M_1M_2M_3$ is connected with the fixed frame $B_1B_2B_3$ through three PURS kinematic chains and one PS chain. By a similar

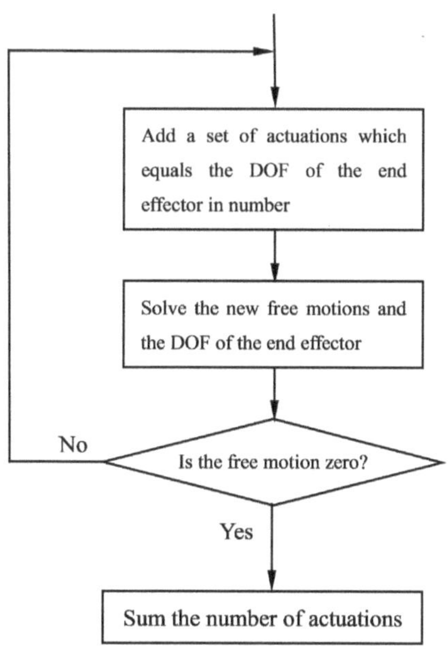

FIGURE 9.8 Process to investigate the CDOF of a mechanism.

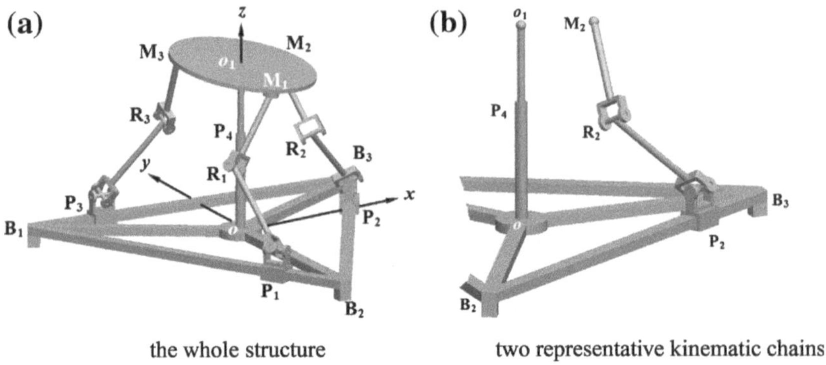

FIGURE 9.9 The structure of a spatial mechanism with redundant DOFs.

investigation process, we can find that the end effector has the same DOFs as the one shown in Figure 9.2.

It is not difficult to find that none of the kinematic chains of the end effector comprises local coaxial rotation. Therefore, we can directly implement as Figure 9.8 shows.

1. Assign four actuators to the three sliders P_1, P_2, P_3, and one to the vertical prismatic joint P_4.

After exerting these actuations, the parallel manipulator becomes a "new one," which comprises one S-chain and three URS kinematic chains.

For the OP_4O_1 chain (S kinematic chain), the kinematic screws are also expressed with equation (9.25), and therefore the new terminal constraints of this actuated kinematic chain are equation (9.26). For the other three kinematic chains (currently become URS kinematic chains), $P_1R_1M_1$, $P_2R_2M_2$, and $P_3R_3M_3$, the terminal constraints of them are all zeros.

Therefore, the terminal constraints yielded by these four actuations are just those expressed by equation (9.26). And as a result, the free motion space of the end effector is expressed with equation (9.25). For Rank $\left(\$_{OO_1}\right) = 3 \neq 0$, we have to implement another computation cycle.

2. Add another Rank $\left(\$_{OO_1}\right) = 3$ actuators to the three rotational joints R_1, R_2, and R_3

After applying these three more actuations, the parallel manipulator becomes a new one, which comprises one S chain and three US kinematic chains. After solving the terminal constraints exerted to the end effector by these three US kinematic chains, we can find the free motion(s) of the end effector again.

If the free motion of the end effector is zero under these three more actuations, the number of actuations required to control the end effector at this

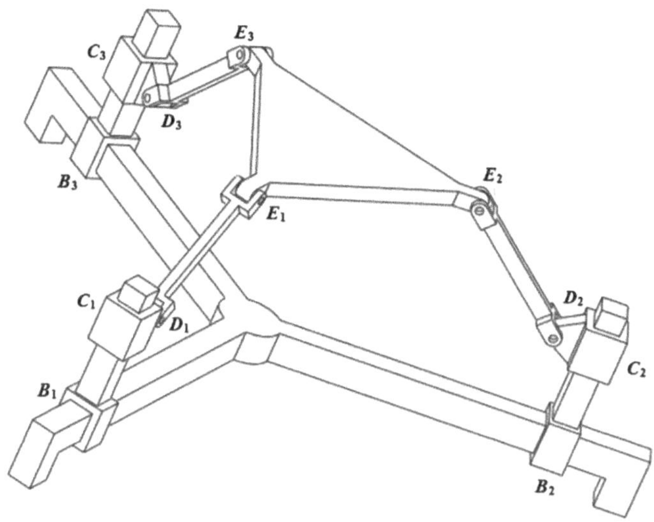

FIGURE 9.10 A 3-PPRR spatial parallel mechanism.

configuration for the mechanism with the architecture shown in Figure 9.9 is $4+3=7$, and therefore, the CDOF of the mechanism at this configuration is 7.

And the DOF of the end effector and the number of actuations needed to control the end effector are really different concepts and should not be ambiguously expressed by one concept of general mobility. In order to get a complete analysis of the mobility of a mechanism, the programmable analytical methodology shown in Figures 9.5 and 9.7 should have to be applied individually.

For another example, the primary architecture characteristics of the mechanism shown in Figure 9.10 are that the end effector, $E_1E_2E_3$, is connected with the fixed base, $B_1B_2B_3$, through three PPRR kinematic chains, and the two prismatic pairs B_i, $C_i (i = 1,2,3)$ in each chain are orthogonal to each other. These three PPRR kinematic chains, $B_iC_iD_iE_i (i = 1,2,3)$, are identical in topology [7]. In what follows, the degree of freedom of the end effector, $E_1E_2E_3$, will be firstly investigated by exploiting the geometry characteristics of the mechanism; and then the position of the end effector relative to the fixed base will be established in one Cartesian coordinate frame, after which the mobility of the mechanism will be discussed from the viewpoint of the number of independent parameters of position equations.

Before developing the first topic, a further observation into the structure geometry of the mechanism might shed a new light on the cracking of the problem. As is shown in Figure 9.11, in each PPRR kinematic chain of the end effector, the links will always remain in the same plane. For instance, kinematic chain $B_1C_1D_1E_1$ is always in the π_1-plane, $B_2C_2D_2E_2$ is always in the π_2-plane, and $B_3C_3D_3E_3$ is always in π_3-plane. Obviously, each PPRR series kinematic chain only allows its terminal connector to have three independent planar motions within the plane the kinematic chain locates, including two translations and one rotation about the normal of the plane. That is, any rotations about the direction parallel to the plane are impossible. Consequently, the end effector, $E_1E_2E_3$, constrained by three such kinematic chains only has one translation along the common intersecting line, O_BO_E, of the three planes π_1, π_2, π_2.

Therefore, it is explicit that the end effector of the mechanism only has one translational DOF along the common intersecting line of the three planes of its kinematic chains. The common intersecting line is also the centralized line of the mechanism, and therefore, the mechanism shown in Figure 9.10 is also called a centralized moving parallel mechanism.

As is analyzed, the end effector, $E_1E_2E_3$, can only move up and down along the centralized line, O_BO_E. Therefore, we can assume that the z-axis is along this centralized line, the x-axis is along the sliding direction of prismatic pair B_1, and the origin of the coordinate frame locates at the cross point, O_B, of the three fixed guide lines. Therefore, the y-axis is determined in accordance with the right-hand rule. The Cartesian coordinate frame is shown in Figure 9.12.

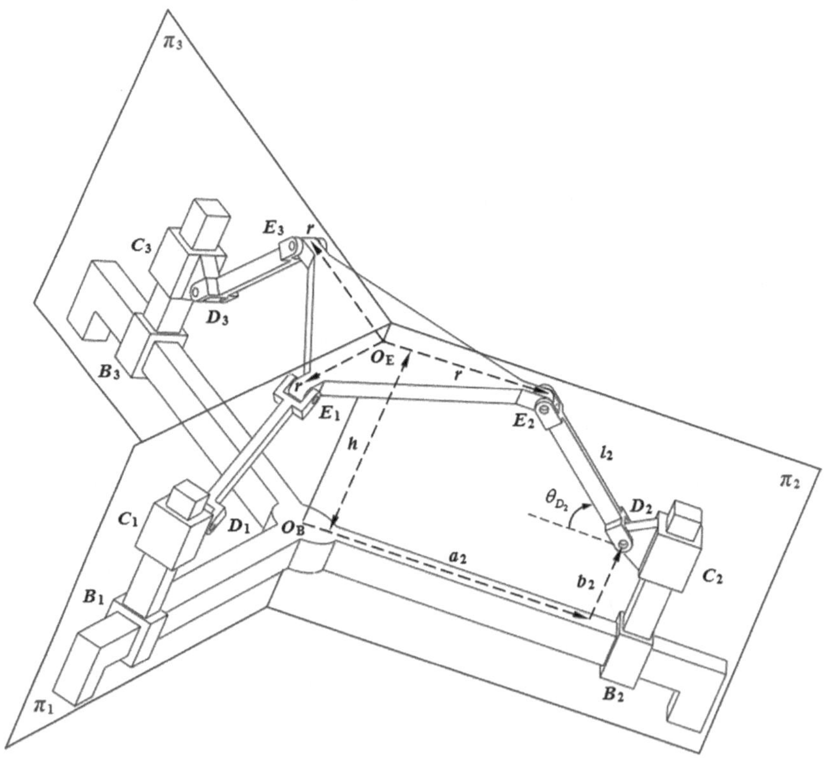

FIGURE 9.11 Geometry constraints for the parallel mechanism.

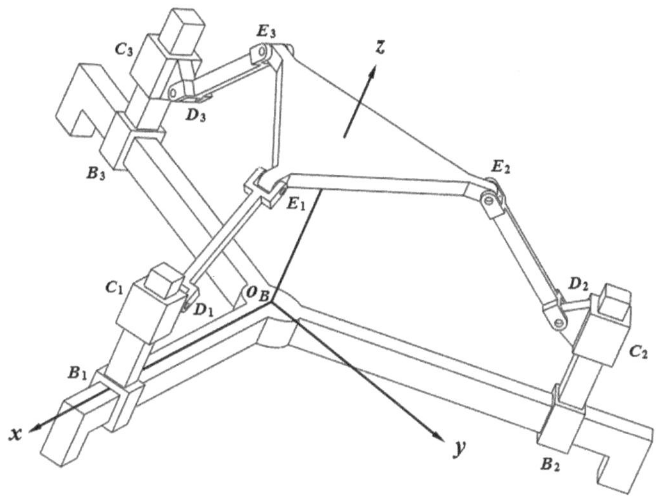

FIGURE 9.12 Coordinate frame established in the mechanism.

Consequently, the position constraints can be built up for each kinematic chain when the displacement of the end effector in z-direction is assumed to be h. Also presume that the position of the prismatic pair $B_i (i = 1,2,3)$ relative to the geometry center of the fixed base, O_B, is a_i, the position of the prismatic pair C_i relative to pair B_i is b_i, the length of the link $D_i E_i$ is l_i and the rotational angle relative to the horizontal line is θ_{D_i}, the radius of the circumcircle of the end effector is r. The position constraint equations for the mechanism can be established.

In the π_2-plane the second kinematic chain locates, for instance, the constraint equations can be expressed as

$$\begin{cases} a_2 - r = l_2 \cos \theta_{D_2} \\ h - b_2 = l_2 \sin \theta_{D_2}. \end{cases} \tag{9.33}$$

Similarly, the constraint equations for the other two kinematic chains are

$$\begin{cases} a_1 - r = l_1 \cos \theta_{D_1} \\ h - b_1 = l_1 \sin \theta_{D_1} \end{cases} \tag{9.34}$$

$$\begin{cases} a_3 - r = l_3 \cos \theta_{D_3} \\ h - b_3 = l_3 \sin \theta_{D_3}. \end{cases} \tag{9.35}$$

Therefore, the position constraint equations for the mechanism can be obtained by associating equation (9.33) with equations (9.34) and (9.35).

$$\begin{cases} a_1 - l_1 \cos\theta_{D_1} - r = 0 \\ b_1 + l_1 \sin\theta_{D_1} - h = 0 \\ a_2 - l_2 \cos\theta_{D_2} - r = 0 \\ b_2 + l_2 \sin\theta_{D_2} - h = 0 \\ a_3 - l_3 \cos\theta_{D_3} - r = 0 \\ b_3 + l_3 \sin\theta_{D_3} - h = 0 \end{cases} \tag{9.36}$$

where r is a known structure parameter, and h, $a_i, b_i, \theta_{D_i} (i = 1,2,3)$ are unknown variables.

From equation (9.36), it is not difficult to find that the mechanism system has 10 variables and six constraint equations. Therefore, the number of independent variables in the equation system is four. As is well known, the mobility of a mechanism is often defined as "the number of independent coordinates required to define its position" in current textbooks [8]. Therefore, the mobility of the mechanism shown in Figure 9.10 should be 4 according to the above definition. In other words, the mechanism will be completely controlled if we add four actuations to the mechanism. Is this true in reality?

For the mechanism shown in Figure 9.10, the total number of kinematic pairs is 12, each one of which can be utilized as a possible actuation. Therefore, the number of possible actuation schemes can be calculated with combination formula

$$N = C_{12}^4 = \frac{12 \times 11 \times 10 \times 9}{4 \times 3 \times 2 \times 1} = 495. \tag{9.37}$$

However, among these 495 possible actuation schemes, there are some schemes that are the same when the symmetries of the mechanism structure are considered. For the sake of convenience in this chapter, each actuation scheme is represented by the symbols of the actuated pairs in the mechanism. For an example, actuation scheme $B_1B_2B_3C_1$ represents that kinematic pairs B_1, B_2, B_3 and C_1 are assigned one actuation, individually.

This actuation scheme can also be denoted by $B_1C_1B_2B_3$ which is the same as $B_1B_2C_2B_3$ and $B_1B_2B_3C_3$ in essence. That is to say, actuation schemes $B_1C_1B_2B_3$, $B_1B_2C_2B_3$, and $B_1B_2B_3C_3$ are four different schemes in equation (9.37), but should be grouped into one actuation scheme because they are identical in topology when the symmetries are taken into consideration. Consequently, in what follows the different number of actuation schemes in essence should be investigated first.

According to the position constraint equation, it is explicit that the mobility of the mechanism is 4. Therefore, amongst the three kinematic chains of the end effector, there must be one kinematic chain in which there are two kinematic pairs that are selected as the actuations. As a result, the actuation schemes for this three-chain mechanism can be roughly divided into three groups, namely: (1) the maximum number of actuations for each kinematic chain is two; (2) the maximum number of actuations for each kinematic chain is three; and (3) the maximum number of actuations for each kinematic chain is four.

For the first group, we can also split it into two classes. One is where the other two actuations are evenly allocated to the other two kinematic chains, and the second is where the other two actuations are all allocated to one of the other two kinematic chains.

The number of actuation schemes for the first class can be divided into two steps: one is to find the number of possible means to select two to be the actuations from the four pairs in a kinematic chain, and the second step is to find the different choices of selecting one pair from the other two kinematic chains. For the first step, the number of possible selections can be easily expressed by the combination formula, $N_{11} = C_4^2 = \frac{4\times 3}{2\times 1} = 6$. For the second step, the operations of a group on a finite set should be first recalled for the sake of simplicity.

Now, if the end effector, $E_1E_2E_3$, shown in Figure 9.10, is thought of as an equilateral triangle and has three identical kinematic chains in topology, $B_iC_iD_iE_i (i = 1,2,3)$, the symmetries for the two kinematic chains the other two actuations will be assigned and can be expressed by a dihedral group $G = D_2 = \{1\ 12\}$. If every kinematic pair of each chain is considered to be different and to be entitled different numbers to indicate their differences, the possible actuations in each kinematic chain forms a finite set, $X = \{1\,2\,3\,4\}$. According to equation (9.18), the different number of possible actuation schemes in essence is the number of orbit of the set X under the operations of group G:

$$N_{12} = \frac{1}{|G|}\sum_{g\in G}\chi(g) = \frac{1}{2}\left(1\times 4^2 + 1\times 4\right) = 10. \tag{9.38}$$

Therefore, the total number of actuation schemes for the first class in the first group can be expressed as

$$N_1 = N_{11}N_{12} = 6\times 10 = 60. \tag{9.39}$$

For the second class of the first group, the symmetries for the two kinematic chains can also be expressed by a dihedral group $G = D_2 = \{1\ 12\}$. In this type, there are two kinematic chains each one of which has two pairs that will be selected to be the actuators. Therefore, there are $C_4^2 = 6$ selections for each chain. In this type, there are two chains each one of which contains two actuators. Consequently, according to the Burnside lemma [6], the different number of possible actuation schemes in essence is the number of orbit of a six-element set under the operations of group G. Therefore, the number of actuation schemes will be $N_{13} = \frac{1}{|G|}\sum_{g\in G}\chi(g) = \frac{1}{2}\left(1\times 6^2 + 1\times 6\right) = 21$ when the symmetries of the two kinematic chains are considered to be identical in topology.

Therefore, the total number of actuation schemes for the first group can be obtained:

$$N_1 = N_{11}N_{12} + N_{13} = 6 \times 10 + 21 = 81. \tag{9.40}$$

For the second group, the number of possible actuation schemes can be directly calculated by combination formula:

$$N_2 = C_4^3 C_4^1 = 4 \times 4 = 16. \tag{9.41}$$

Similarly, the number of possible actuation schemes of the third group can also be directly computed by combination formula:

$$N_3 = C_4^4 = 1. \tag{9.42}$$

The number of different actuation schemes in essence by considering the symmetries of the mechanism architecture in topology can be obtained from equations (9.40), (9.41), and (9.42):

$$N = N_1 + N_2 + N_3 = 81 + 16 + 1 = 98. \tag{9.43}$$

Equation (9.43) indicates that the possible actuation schemes still amount to 98 although the number is greatly decreased compared with equation (9.37). Is each possible actuation scheme feasible in reality? This is a new problem raised after knowing the mobility of the mechanism.

9.2.2 Feasible Actuation Scheme Analysis

This section will investigate the possible actuation schemes of the mechanism from the aspect of statics with the mechanism example of Figure 9.9. As is analyzed in the above sections, the end effector only has one translational DOF along the z-direction shown in Figure 9.12. In other words, the end effector only has the ability to output power for the loads in the z-direction. Therefore, the load of the end effector in z-direction can be denoted by a variable, F_z. Now, the statics equations of the mechanism might be established for each one of the above 98 actuation schemes.

Suppose the mechanism is actuated under the actuation scheme $B_1C_1B_2B_3$. Therefore, the four actuations are assigned to the kinematic pairs B_1, C_1, B_2, and B_3 respectively.

Assume that the loads in z-direction exerted to the terminals of the three kinematic chains of the end effector are F_1, F_2, and F_3, respectively, the following equation holds:

$$F_1 + F_2 + F_3 = F_z. \tag{9.44}$$

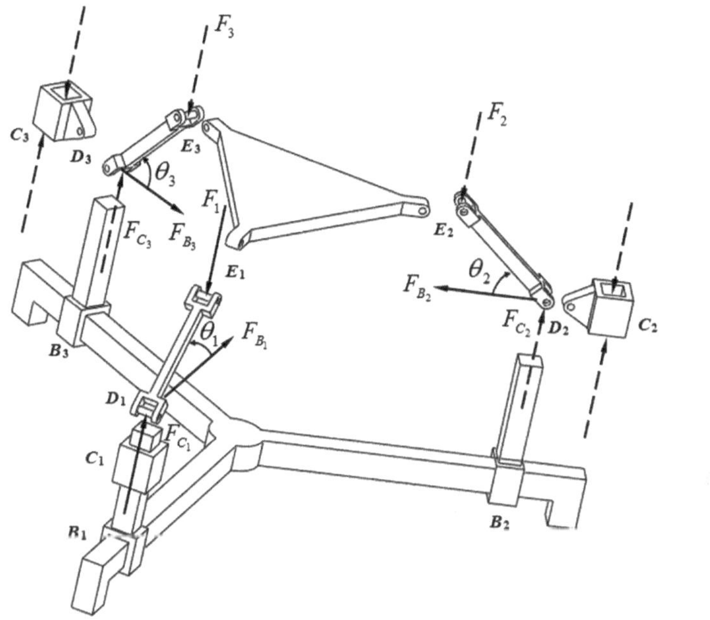

FIGURE 9.13 Statics of the mechanism under the $B_1C_1B_2B_3$ actuation scheme.

Suppose that the angle from the *xoy*-plane to link D_iE_i ($i = 1,2,3$) is denoted by θ_i. In the first kinematic chain shown in Figure 9.13, there are the following equations

$$\begin{cases} F_{C_1} - F_1 = 0 \\ F_{B_1} - F_1 \cot\theta_1 = 0. \end{cases} \tag{9.45}$$

For $\cot\theta_1 = \frac{x_{E_1}-x_{D_1}}{y_{E_1}-y_{D_1}}$, the actuating forces can be expressed as

$$\begin{cases} F_{C_1} = F_1 \\ F_{B_1} = \frac{x_{E_1}-x_{D_1}}{y_{E_1}-y_{D_1}} F_1. \end{cases} \tag{9.46}$$

Similarly, there are the following equations in the second kinematic chain:

$$\begin{cases} F_{C_2} = F_2 \\ F_{B_2} = \frac{x_{E_2}-x_{D_2}}{y_{E_2}-y_{D_2}} F_2. \end{cases} \tag{9.47}$$

Because there is no actuation at pair C_1, there must be $F_{C_2} = 0$. As a result, there must be

$$F_2 = 0 \tag{9.48}$$

$$F_{B_2} = 0. \tag{9.49}$$

Similarly, the following equations should also hold

$$F_3 = 0 \tag{9.50}$$

$$F_{B_3} = 0. \tag{9.51}$$

Substituting equations (9.48) and (9.50) into equation (9.44) yields

$$F_1 = F_z. \tag{9.52}$$

Substituting equation (9.52) into equation (9.46) yields

$$\begin{cases} F_{C_1} = F_z \\ F_{B_1} = \frac{x_{E_1} - x_{D_1}}{y_{E_1} - y_{D_1}} F_z. \end{cases} \tag{9.53}$$

Associating equation (9.53) with equations (9.49) and (9.51) yields the inverse static solutions

$$\begin{cases} F_{B_1} = \frac{x_{E_1} - x_{D_1}}{y_{E_1} - y_{D_1}} F_z \\ F_{C_1} = F_z \\ F_{B_2} = 0 \\ F_{B_3} = 0. \end{cases} \tag{9.54}$$

The theoretical result represented by equation (9.54) indicates that only the actuators B_1 and C_1 act while the other two actuators B_2 and B_3 do not bear any load exerted to the end effector. Can these two actuators, B_2 and B_3, be cancelled?

Before answering this question, a further analysis on the geometry characteristics should be addressed. As the end effector, $E_1E_2E_3$, only has one translational DOF along the z-direction, the mechanism structure shown in Figure 9.10 can be equivalently substituted by the one shown in Figure 9.14.

In any plane of π_1, π_2, and π_3, the planar kinematic chain $o_B B_i C_i D_i E_i o_E$ (i = 1,2,3) forms a planar PPRRP five-link mechanism. Obviously, all the links in the five-link mechanism,$o_B B_1 C_1 D_1 E_1 o_E$, have determinate motions when prismatic pairs B_1 and C_1 are assigned actuators. Therefore, the end effector, $E_1E_2E_3$, has a determinate position along the z-direction. As a result, the relative motions of the links in the five-link mechanism, $o_B B_2 C_2 D_2 E_2 o_E$, will be also be determinate when the prismatic pair, C_2, is assigned an actuator. The same is true of the five-link mechanism, $o_B B_3 C_3 D_3 E_3 o_E$, when an actuation is allotted to the prismatic pair, C_3.

Obviously, these two actuators, C_2 and C_3, are just utilized to keep the configurations of the planar five-link mechanisms, $o_B B_2 C_2 D_2 E_2 o_E$ and $o_B B_3 C_3 D_3 E_3 o_E$, but could not input any useful power to the end effector. Therefore, these two actuators can be canceled in propelling the end effector to output power in theory. That is, the $B_1C_1B_2B_3$ actuated scheme is equivalent

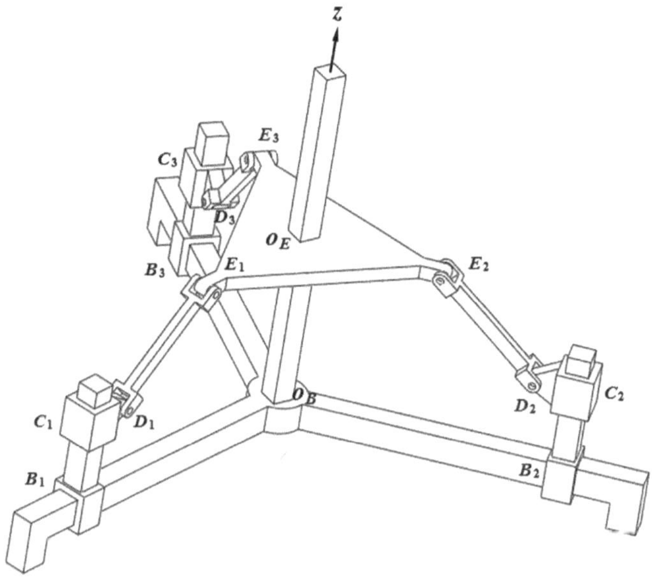

FIGURE 9.14 The equivalent mechanism structure in mobility.

to the B_1C_1 actuated scheme from the viewpoint of propelling the end effector. The only differences between these two schemes are that the relative motions of the links in the $B_2C_2D_2E_2$ and $B_3C_3D_3E_3$ chains are determinate in the former scheme while the relative motions of the links in the $B_2C_2D_2E_2$ and $B_3C_3D_3E_3$ chains are not determined in the latter one.

With the advent of spatial parallel mechanisms, the primary considerations of the designers have been focused on nothing but the degree of freedom of the end effector and its controllability. In this regard, it is also very necessary to investigate the actuation schemes after knowing the DOF of an end effector.

With a similar process to the above, the controllability of the other 97 actuation schemes can also be analyzed. The possible actuation schemes of class (1) are listed in Table 9.1, and all of the possible actuation schemes of class (2) and class (3) are listed in Table 9.2.

All of the possible actuation schemes for the mechanism shown in Figure 9.10 are listed in Tables 9.1 and 9.2. Among these 98 actuation schemes, 30 kinds are feasible if we require that every link of the mechanism should be controlled. As a matter of fact, only two actuations are effectively utilized to output power to the end effector even in these 30 feasible actuation schemes which can be similarly analyzed with the above process in this section. The remained two actuations in any of these 30 feasible actuation schemes only keep the configurations of the other two kinematic chains. Among these 98 actuation schemes, there are also 28 kinds that are feasible to control the end effector but unfeasible to control every link of the mechanism, while there are

TABLE 9.1 Possible Actuation Schemes of Class (1)

Actuated Kinematic Pairs				Possible Schemes	Feasible or Not to Control		Comments	Sequence
					End Effector	Every Link		
B_i	C_i	B_j	B_k	$B_iC_iB_jB_k$	Feasible	Feasible	All the links are controlled.	1
			C_k	$B_iC_iB_jC_k$	Feasible	Feasible	All the links are controlled.	2
			D_k	$B_iC_iB_jD_k$	Feasible	Feasible	All the links are controlled.	3
			E_k	$B_iC_iB_jE_k$	Feasible	Feasible	All the links are controlled.	4
		C_j	C_k	$B_iC_iC_jC_k$	Feasible	Feasible	All the links are controlled.	5
			D_k	$B_iC_iC_jD_k$	Feasible	Feasible	All the links are controlled.	6
			E_k	$B_iC_iC_jE_k$	Feasible	Feasible	All the links are controlled.	7
		D_j	D_k	$B_iC_iD_jD_k$	Feasible	Feasible	All the links are controlled.	8
			E_k	$B_iC_iD_jE_k$	Feasible	Feasible	All the links are controlled.	9
		E_j	E_k	$B_iC_iE_jE_k$	Feasible	Feasible	All the links are controlled.	10
B_i	D_i	B_j	B_k	$B_iD_iB_jB_k$	Unfeasible	Unfeasible	All the links are not controlled.	11
			C_k	$B_iD_iB_jC_k$	Unfeasible	Unfeasible	All the links are not controlled.	12
			D_k	$B_iD_iB_jD_k$	Unfeasible	Unfeasible	All the links are not controlled.	13
			E_k	$B_iD_iB_jE_k$	Unfeasible	Unfeasible	All the links are not controlled.	14
		C_j	C_k	$B_iD_iC_jC_k$	Unfeasible	Unfeasible	All the links are not controlled.	15
			D_k	$B_iD_iC_jD_k$	Unfeasible	Unfeasible	All the links are not controlled.	16
			E_k	$B_iD_iC_jE_k$	Unfeasible	Unfeasible	All the links are not controlled.	17
		D_j	D_k	$B_iD_iD_jD_k$	Unfeasible	Unfeasible	All the links are not controlled.	18
			E_k	$B_iD_iD_jE_k$	Unfeasible	Unfeasible	All the links are not controlled.	19
		E_j	E_k	$B_iD_iE_jE_k$	Unfeasible	Unfeasible	All the links are not controlled.	20

(continued)

TABLE 9.1 (Continued)

Actuated Kinematic Pairs				Possible Schemes	Feasible or Not to Control		Comments	Sequence
					End Effector	Every Link		
B_i	E_i	B_j	B_k	$B_iE_iB_jB_k$	Unfeasible	Unfeasible	All the links are not controlled.	21
			C_k	$B_iE_iB_jC_k$	Unfeasible	Unfeasible	All the links are not controlled.	22
			D_k	$B_iE_iB_jD_k$	Unfeasible	Unfeasible	All the links are not controlled.	23
			E_k	$B_iE_iB_jE_k$	Unfeasible	Unfeasible	All the links are not controlled.	24
		C_j	C_k	$B_iE_iC_jC_k$	Unfeasible	Unfeasible	All the links are not controlled.	25
			D_k	$B_iE_iC_jD_k$	Unfeasible	Unfeasible	All the links are not controlled.	26
			E_k	$B_iE_iC_jE_k$	Unfeasible	Unfeasible	All the links are not controlled.	27
		D_j	D_k	$B_iE_iD_jD_k$	Unfeasible	Unfeasible	All the links are not controlled.	28
			E_k	$B_iE_iD_jE_k$	Unfeasible	Unfeasible	All the links are not controlled.	29
		E_j	E_k	$B_iE_iE_jE_k$	Unfeasible	Unfeasible	All the links are not controlled.	30
C_i	D_i	B_j	B_k	$C_iD_iB_jB_k$	Feasible	Feasible	All the links are controlled.	31
			C_k	$C_iD_iB_jC_k$	Feasible	Feasible	All the links are controlled.	32
			D_k	$C_iD_iB_jD_k$	Feasible	Feasible	All the links are controlled.	33
			E_k	$C_iD_iB_jE_k$	Feasible	Feasible	All the links are controlled.	34
		C_j	C_k	$C_iD_iC_jC_k$	Feasible	Feasible	All the links are controlled.	35
			D_k	$C_iD_iC_jD_k$	Feasible	Feasible	All the links are controlled.	36
			E_k	$C_iD_iC_jE_k$	Feasible	Feasible	All the links are controlled.	37
		D_j	D_k	$C_iD_iD_jD_k$	Feasible	Feasible	All the links are controlled.	38
			E_k	$C_iD_iD_jE_k$	Feasible	Feasible	All the links are controlled.	39
		E_j	E_k	$C_iD_iE_jE_k$	Feasible	Feasible	All the links are controlled.	40

(continued)

TABLE 9.1 (Continued)

Actuated Kinematic Pairs				Possible Schemes	Feasible or Not to Control		Comments	Sequence
					End Effector	Every Link		
C_i	E_i	B_j	B_k	$C_iE_iB_jB_k$	Feasible	Feasible	All the links are controlled.	41
			C_k	$C_iE_iB_jC_k$	Feasible	Feasible	All the links are controlled.	42
			D_k	$C_iE_iB_jD_k$	Feasible	Feasible	All the links are controlled.	43
			E_k	$C_iE_iB_jE_k$	Feasible	Feasible	All the links are controlled.	44
		C_j	C_k	$C_iE_iC_jC_k$	Feasible	Feasible	All the links are controlled.	45
			D_k	$C_iE_iC_jD_k$	Feasible	Feasible	All the links are controlled.	46
			E_k	$C_iE_iC_jE_k$	Feasible	Feasible	All the links are controlled.	47
		D_j	D_k	$C_iE_iD_jD_k$	Feasible	Feasible	All the links are controlled.	48
			E_k	$C_iE_iD_jE_k$	Feasible	Feasible	All the links are controlled.	49
		E_j	E_k	$C_iE_iE_jE_k$	Feasible	Feasible	All the links are controlled.	50
D_i	E_i	B_j	B_k	$D_iE_iB_jB_k$	Unfeasible	Unfeasible	All the links are not controlled.	51
			C_k	$D_iE_iB_jC_k$	Unfeasible	Unfeasible	All the links are not controlled.	52
			D_k	$D_iE_iB_jD_k$	Unfeasible	Unfeasible	All the links are not controlled.	53
			E_k	$D_iE_iB_jE_k$	Unfeasible	Unfeasible	All the links are not controlled.	54
		C_j	C_k	$D_iE_iC_jC_k$	Unfeasible	Unfeasible	All the links are not controlled.	55
			D_k	$D_iE_iC_jD_k$	Unfeasible	Unfeasible	All the links are not controlled.	56
			E_k	$D_iE_iC_jE_k$	Unfeasible	Unfeasible	All the links are not controlled.	57
		D_j	D_k	$D_iE_iD_jD_k$	Unfeasible	Unfeasible	All the links are not controlled.	58
			E_k	$D_iE_iD_jE_k$	Unfeasible	Unfeasible	All the links are not controlled.	59
		E_j	E_k	$D_iE_iE_jE_k$	Unfeasible	Unfeasible	All the links are not controlled.	60

(*continued*)

TABLE 9.1 (Continued)

Actuated Kinematic Pairs				Possible Schemes	Feasible or Not to Control		Comments	Sequence
					End Effector	Every Link		
B_i	C_i	B_j	C_j	$B_iC_iB_jC_j$	Feasible	Unfeasible	There is one redundant actuation in the actuation scheme while the kinematic chain $B_kC_kD_kE_k$ is not controlled.	61
		B_j	D_j	$B_iC_iB_jD_j$	Feasible	Unfeasible	There is one redundant actuation in the kinematic chain $B_jC_jD_jE_j$ while the kinematic chain $B_kC_kD_kE_k$ is not controlled.	62
		B_j	E_j	$B_iC_iB_jE_j$	Feasible	Unfeasible	There is one redundant actuation in the kinematic chain $B_jC_jD_jE_j$ while the kinematic chain $B_kC_kD_kE_k$ is not controlled.	63
		C_j	D_j	$B_iC_iC_jD_j$	Feasible	Unfeasible	There is one redundant actuation in the actuation scheme while the kinematic chain $B_kC_kD_kE_k$ is not controlled.	64
		C_j	E_j	$B_iC_iC_jE_j$	Feasible	Unfeasible	There is one redundant actuation in the actuation scheme while the kinematic chain $B_kC_kD_kE_k$ is not controlled.	65
		D_j	E_j	$B_iC_iD_jE_j$	Feasible	Unfeasible	There is one redundant actuation in the kinematic chain $B_jC_jD_jE_j$ while the kinematic chain $B_kC_kD_kE_k$ is not controlled.	66

(*continued*)

TABLE 9.1 (Continued)

Actuated Kinematic Pairs				Possible Schemes	Feasible or Not to Control		Comments	Sequence
					End Effector	Every Link		
B_i	D_i	B_j	D_j	$B_iD_iB_jD_j$	Unfeasible	Unfeasible	All the links are not controlled.	67
		B_j	E_j	$B_iD_iB_jE_j$	Unfeasible	Unfeasible	All the links are not controlled.	68
		C_j	D_j	$B_iD_iC_jD_j$	Feasible	Unfeasible	There is one redundant actuation in the kinematic chain $B_iC_iD_iE_i$ while the kinematic chain $B_kC_kD_kE_k$ is not controlled.	69
		C_j	E_j	$B_iD_iC_jE_j$	Feasible	Unfeasible	There is one redundant actuation in the kinematic chain $B_iC_iD_iE_i$ while the kinematic chain $B_kC_kD_kE_k$ is not controlled.	70
		D_j	E_j	$B_iD_iD_jE_j$	Unfeasible	Unfeasible	All the links are not controlled.	71
B_i	E_i	B_j	E_j	$B_iE_iB_jE_j$	Unfeasible	Unfeasible	All the links are not controlled.	72
		C_j	D_j	$B_iE_iC_jD_j$	Feasible	Unfeasible	There is one redundant actuation in the kinematic chain $B_iC_iD_iE_i$ while the kinematic chain $B_kC_kD_kE_k$ is not controlled.	73
		C_j	E_j	$B_iE_iC_jE_j$	Feasible	Unfeasible	There is one redundant actuation in the kinematic chain $B_iC_iD_iE_i$ while the kinematic chain $B_kC_kD_kE_k$ is not controlled.	74
		D_j	E_j	$B_iE_iD_jE_j$	Unfeasible	Unfeasible	All the links are not controlled.	75

(*continued*)

TABLE 9.1 (Continued)

Actuated Kinematic Pairs				Possible Schemes	Feasible or Not to Control		Comments	Sequence
					End Effector	Every Link		
C_i	D_i	C_j	D_j	$C_iD_iC_jD_j$	Feasible	Unfeasible	There is one redundant actuation in the actuation scheme while the kinematic chain $B_kC_kD_kE_k$ is not controlled.	76
		C_j	E_j	$C_iD_iC_jE_j$	Feasible	Unfeasible	There is one redundant actuation in the actuation scheme while the kinematic chain $B_kC_kD_kE_k$ is not controlled.	77
		D_j	E_j	$C_iD_iD_jE_j$	Feasible	Unfeasible	There is one redundant actuation in the kinematic chain $B_jC_jD_jE_j$ while the kinematic chain $B_kC_kD_kE_k$ is not controlled.	78
C_i	E_i	C_j	E_j	$C_iE_iC_jE_j$	Feasible	Unfeasible	There is one redundant actuation in the actuation scheme while the kinematic chain $B_kC_kD_kE_k$ is not controlled.	79
		D_j	E_j	$C_iE_iD_jE_j$	Feasible	Unfeasible	There is one redundant actuation in the kinematic chain $B_jC_jD_jE_j$ while the kinematic chain $B_kC_kD_kE_k$ is not controlled.	80
D_i	E_i	D_j	E_j	$D_iE_iD_jE_j$	Unfeasible	Unfeasible	All the links are not controlled.	81

TABLE 9.2 Possible Actuation Schemes of Class (2) and Class (3)

Actuated Kinematic Pairs		Possible Schemes	Feasible or Not to Control		Comments	Sequence
			End Effector	Every Link		
$B_iC_iD_i$	B_j	$B_iC_iD_iB_j$	Feasible	Unfeasible	There is one redundant actuation in the kinematic chain $B_iC_iD_iE_i$ while the kinematic chain $B_kC_kD_kE_k$ is not controlled.	82
	C_j	$B_iC_iD_iC_j$	Feasible	Unfeasible	There is one redundant actuation in the kinematic chain $B_iC_iD_iE_i$ while the kinematic chain $B_kC_kD_kE_k$ is not controlled.	83
	D_j	$B_iC_iD_iD_j$	Feasible	Unfeasible	There is one redundant actuation in the kinematic chain $B_iC_iD_iE_i$ while the kinematic chain $B_kC_kD_kE_k$ is not controlled.	84
	E_j	$B_iC_iD_iE_j$	Feasible	Unfeasible	There is one redundant actuation in the kinematic chain $B_iC_iD_iE_i$ while the kinematic chain $B_kC_kD_kE_k$ is not controlled.	85
$B_iC_iE_i$	B_j	$B_iC_iE_iB_j$	Feasible	Unfeasible	There is one redundant actuation in the kinematic chain $B_iC_iD_iE_i$ while the kinematic chain $B_kC_kD_kE_k$ is not controlled.	86
	C_j	$B_iC_iE_iC_j$	Feasible	Unfeasible	There is one redundant actuation in the kinematic chain $B_iC_iD_iE_i$ while the kinematic chain $B_kC_kD_kE_k$ is not controlled.	87
	D_j	$B_iC_iE_iD_j$	Feasible	Unfeasible	There is one redundant actuation in the kinematic chain $B_iC_iD_iE_i$ while the kinematic chain $B_kC_kD_kE_k$ is not controlled.	88
	E_j	$B_iC_iE_iE_j$	Feasible	Unfeasible	There is one redundant actuation in the kinematic chain $B_iC_iD_iE_i$ while the kinematic chain $B_kC_kD_kE_k$ is not controlled.	89

(continued)

TABLE 9.2 (Continued)

Actuated Kinematic Pairs		Possible Schemes	Feasible or Not to Control		Comments	Sequence
			End Effector	Every Link		
$C_iD_iE_i$	B_j	$C_iD_iE_iB_j$	Feasible	Unfeasible	There is one redundant actuation in the kinematic chain $B_iC_iD_iE_i$ while the kinematic chain $B_kC_kD_kE_k$ is not controlled.	90
	C_j	$C_iD_iE_iC_j$	Feasible	Unfeasible	There is one redundant actuation in the kinematic chain $B_iC_iD_iE_i$ while the kinematic chain $B_kC_kD_kE_k$ is not controlled.	91
	D_j	$C_iD_iE_iD_j$	Feasible	Unfeasible	There is one redundant actuation in the kinematic chain $B_iC_iD_iE_i$ while the kinematic chain $B_kC_kD_kE_k$ is not controlled.	92
	E_j	$C_iD_iE_iE_j$	Feasible	Unfeasible	There is one redundant actuation in the kinematic chain $B_iC_iD_iE_i$ while the kinematic chain $B_kC_kD_kE_k$ is not controlled.	93
$B_iD_iE_i$	B_j	$B_iD_iE_iB_j$	Unfeasible	Unfeasible	All the links are not controlled.	94
	C_j	$B_iD_iE_iC_j$	Unfeasible	Unfeasible	All the links are not controlled.	95
	D_j	$B_iD_iE_iD_j$	Unfeasible	Unfeasible	All the links are not controlled.	96
	E_j	$B_iD_iE_iE_j$	Unfeasible	Unfeasible	All the links are not controlled.	97
$B_iC_iD_iE_i$		$B_iC_iD_iE_i$	Feasible	Unfeasible	There are two redundant actuations in the kinematic chain $B_iC_iD_iE_i$ while the two kinematic chains $B_jC_jD_jE_j$ and $B_kC_kD_kE_k$ are not controlled.	98

40 kinds that can neither control the end effector nor control the links of the mechanism.

Therefore, the controllability of the mechanism is affected by the actuation schemes. In this regard, it is very necessary to investigate the actuation schemes after knowing the DOF of the end effector. In fact, this problem can be easily avoided if we focus on the DOF of the end effector and its actuation.

Because of the elegance in depicting the relationship between the motions and the constraints, reciprocal screw theory is a good selection to accomplish this work [9]. Therefore, the reciprocal screw theory is utilized to analyze the DOF of the end effector of the mechanism and the number of actuators required to control the end effector shown in Figure 9.9.

Firstly, the degree of freedom of the end effector of the mechanism, shown in Figure 9.10, should be analyzed. In this mechanism, the end effector, $E_1E_2E_3$, has three PPRR kinematic chains connected with the fixed base. For the sake of convenience, we can still use the Cartesian coordinate frame shown in Figure 9.12 for the mechanism.

The direction vectors of the prismatic pairs, B_1, B_2, and B_3, are $\boldsymbol{e}_{B_1} = (1\ \ 0\ \ 0)^T$, $\boldsymbol{e}_{B_2} = \left(\cos\frac{2\pi}{3}\ \ \sin\frac{2\pi}{3}\ \ 0\right)^T$, $\boldsymbol{e}_{B_3} = \left(\cos\frac{4\pi}{3}\ \ \sin\frac{4\pi}{3}\ \ 0\right)^T$, respectively; the direction vectors of the prismatic pairs, C_1, C_2, and C_3, are identical and can be uniformly expressed as $\boldsymbol{e}_{C_i} = (0\ \ 0\ \ 1)^T$, the rotational vector of the revolute pair C_i can therefore be expressed by $\boldsymbol{e}_{D_i} = \boldsymbol{e}_{C_i} \times \boldsymbol{e}_{B_i}$, the rotational vector of the revolute pair E_i ($i = 1,2,3$) is identical to that of D_i which can be denoted by $\boldsymbol{e}_{E_i} = \boldsymbol{e}_{D_i}$.

The DOF of the end effector of the mechanism can be investigated in accordance with the process proposed in [2]. The twist matrix for the kinematic chain, $B_1C_1D_1E_1$, can be expressed as

$$\$_{B_1C_1D_1E_1} = [\$_{B_1}\ \ \$_{C_1}\ \ \$_{D_1}\ \ \$_{E_1}] \tag{9.55}$$

where $\$_{B_1} = (0\ \ 0\ \ 0\ \ 1\ \ 0\ \ 0)^T$, $\$_{C_1} = (0\ \ 0\ \ 0\ \ 0\ \ 0\ \ 1)^T$, $\$_{D_1} = (0\ \ 1\ \ 0\ \ -z_{D_1}\ \ 0\ \ x_{D_1})^T$, $\$_{E_1} = (0\ \ 1\ \ 0\ \ -z_{E_1}\ \ 0\ \ x_{E_1})^T$.

The terminal constraints of the kinematic chain can be obtained with reciprocal screw theory and the constraint matrix is represented by:

$$\$^r_{B_1C_1D_1E_1} = \left[\$_1^{r1}\ \ \$_1^{r2}\ \ \$_1^{r3}\right] \tag{9.56}$$

where $\$_1^{r1} = (0\ \ 0\ \ 0\ \ 1\ \ 0\ \ 0)^T$, $\$_1^{r2} = (0\ \ 0\ \ 0\ \ 0\ \ 0\ \ 1)^T$, $\$_1^{r3} = (0\ \ 1\ \ 0\ \ -z_{E_1}\ \ 0\ \ x_{E_1})^T$.

Similarly, we can obtain the terminal constraints of the other two kinematic chains.

$$\$^r_{B_2C_2D_2E_2} = \left[\$_2^{r1}\ \ \$_2^{r2}\ \ \$_2^{r3}\right] \tag{9.57}$$

where $\$_2^{r1} = \begin{pmatrix} 0 & 0 & 0 & -\frac{1}{2} & \frac{\sqrt{3}}{2} & 0 \end{pmatrix}^T$, $\$_2^{r2} = (0\ 0\ 0\ 0\ 0\ 1)^T$, $\$_2^{r3} = \begin{pmatrix} -\frac{\sqrt{3}}{2} & -\frac{1}{2} & 0 & \frac{1}{2}z_{E_2} & -\frac{\sqrt{3}}{2}z_{E_2} & -\frac{1}{2}x_{E_2} + \frac{\sqrt{3}}{2}y_{E_2} \end{pmatrix}^T$.

$$\$^r_{B_3C_3D_3E_3} = \begin{bmatrix} \$_3^{r1} & \$_3^{r2} & \$_3^{r3} \end{bmatrix} \tag{9.58}$$

where $\$_3^{r1} = \begin{pmatrix} 0 & 0 & 0 & -\frac{1}{2} & -\frac{\sqrt{3}}{2} & 0 \end{pmatrix}^T$, $\$_3^{r2} = (0\ 0\ 0\ 0\ 0\ 1)^T$, $\$_3^{r3} = \begin{pmatrix} \frac{\sqrt{3}}{2} & -\frac{1}{2} & 0 & \frac{1}{2}z_{E_3} & \frac{\sqrt{3}}{2}z_{E_3} & -\frac{1}{2}x_{E_3} - \frac{\sqrt{3}}{2}y_{E_3} \end{pmatrix}^T$.

It is obvious that $\$_1^{r2}$, $\$_2^{r2}$ and $\$_3^{r2}$ are common constraints for these three kinematic chains. Therefore, the terminal constraints exerted on the end effector by the three kinematic chains can be denoted as

$$\$^r_{E_1E_2E_3} = \begin{bmatrix} \$_1^{r1} & \$_2^{r1} & \$_3^{r1} & \$_1^{r2} & \$_1^{r3} & \$_2^{r3} & \$_3^{r3} \end{bmatrix}. \tag{9.59}$$

It is not difficult to find that the rank of $\$^r_{E_1E_2E_3}$ expressed by equation (9.59) is 5, and the free motions of the end effector, $E_1E_2E_3$, can be again solved with reciprocal screw theory:

$$\$^F_{E_1E_2E_3} = \begin{pmatrix} 0 & 0 & 0 & 0 & 0 & 1 \end{pmatrix}^T. \tag{9.60}$$

Equation (9.60) indicates that the end effector has one translational DOF along the direction vector $\boldsymbol{e} = (0\ 0\ 1)^T$ which is consistent with that obtained in Section 9.2.1. Of course, the number of DOFs of the end effector is $\mathrm{Rank}(\$^F_{E_1E_2E_3}) = 1$, the direction is $\boldsymbol{e} = (0\ 0\ 1)^T$ and the type is translation, which is fully represented by the screw expression (9.60).

Now, we can analyze the number of actuation(s) required to control the end effector and the actuation scheme. Because the number of DOFs of the end effector is 1, it is reasonable for one to expect that the end effector can be fully controlled only with one actuation. When one actuation is added to any pair of the mechanism, B_1 for an example, we can find that the end effector still has one translational DOF along the direction $\boldsymbol{e} = (0\ 0\ 1)^T$ when we repeat the above analysis. Therefore, we have to add another actuation to the mechanism. Of course, we can add the second actuation to any one of the other pairs. However, it is not difficult to find from Tables 9.1 and 9.2 that the end effector will not be controlled unless the second actuation is exerted to the prismatic pair C_1 under the condition that the first actuation is exerted to B_1. This, as a matter of fact, forms the B_1C_1 actuation scheme which is analyzed in Section 9.2.1

Of course, we can also exert the second actuation to the prismatic pair B_2 after assigning the first actuation to the prismatic pair B_1. Again, we can find that the end effector still has one free translation along the direction $\boldsymbol{e} = (0\ 0\ 1)^T$ when we repeat the above process. So, we have to continue to add the third actuation to the prismatic pair B_3. However, the end effector will not be controlled until a fourth actuation is applied to one of the prismatic pairs, C_1, C_2, vand C_3. This forms actuation scheme 1 in Table 9.1. So, under this actuation scheme, the number of actuations needed to control the end effector is 4. As mentioned in Section 9.2.1, there are two actuations among these four actuations that are only utilized to keep the configuration of the two non-actuated chains. Therefore, the effects of the actuation scheme $B_iC_iB_jB_k$ $(i,j,k = 1,2,3)$ and the actuation scheme B_iC_i are identical in propelling the end effector to output power. With a similar process, any feasible actuation schemes shown in Tables 9.1 and 9.2 can be obtained. Moreover, any actuation schemes that might control the end effector but not control all the links of the mechanism can also be obtained with the above process.

What must be pointed out is that two actuations at least are needed in order to completely control the one DOF end effector shown in Figure 9.9. The result that the mobility of the mechanism is 4 only indicates the least number of actuations to keep the configurations of every link in the mechanism. Therefore, the controllability of the end effector is different from the mobility of the mechanism. The controllability of the end effector is affected by the DOF and the actuation scheme but not the mobility of the mechanism. The controllability of the end effector and the actuation scheme can be investigated with the process proposed in this section after knowing the DOF of the end effector.

9.3 CONCLUSIONS

This chapter focuses on the actuation schemes of mechanisms with a representative 3-PPRR parallel linkage. Firstly, the mechanism architecture is presented, and then the mobility of the mechanism is analyzed from the aspect of the number of independent parameters in position constraint equations. Based on the mobility obtained through position equations, the number of possible actuation schemes is calculated with a combination formula and the Burnside lemma. Through statics analysis of each actuation scheme, the characteristics of every possible actuation scheme are represented in two tables. This research indicates that the mobility of a mechanism can only provide the number of actuations needed to control all the links in the mechanism, but not safeguard that any actuation scheme is available to control the end effector.

REFERENCES

[1] G. Gogu, Mobility of mechanisms: a critical review, Mechanism and Machine Theory 40 (2005) 1068–1097.

[2] J.S. Zhao, K. Zhou, et al., A new method to study the degree of freedom of spatial parallel mechanisms, The International Journal of Advanced Manufacturing Technology 23 (2004) 288–294.

[3] J.S. Zhao, Z.J. Feng, K. Zhou, D.W. Jin, Re-analysis on the degree-of-freedom configuration of the platforms in spatial parallel mechanisms with constraints spaces, The International Journal of Advanced Manufacturing Technology 28 (2006) 190–196.
[4] J.S. Zhao, K. Zhou, Z.J. Feng, A theory of degrees of freedom for mechanisms, Mechanism and Machine Theory 39 (2004) 621–643.
[5] J.S. Zhao, Z.J. Feng, J.X. Dong, Computation of the configuration degree of freedom of a spatial parallel mechanism by using reciprocal screw theory, Mechanism and Machine Theory 41 (2006) 148–1504.
[6] J.J. Rotman, Advanced Modern Algebra, Pearson Education, Inc., 2002, Publishing as Prentice Hall.
[7] J.S. Dai, J.R. Jones, Matrix representation of topological changes in metamorphic mechanisms, ASME Journal of Mechanical Design 127 (2005) 837–840.
[8] R.L. Norton, Design of Machinery—An Introduction to the Synthesis and Analysis of Mechanisms and Machines, McGraw-Hill Companies, Inc., 2001.
[9] J.S. Dai, An historical review of the theoretical development of rigid body displacements from Rodrigues parameters to the finite twist, Mechanism and Machine Theory 41 (2006) 41–52.

Chapter 10

Motion Characteristics of a Robotic Mechanism

Chapter 9 discussed the fundamental factors to investigate the motions and actuations of a mechanism; this chapter will analyze the free motions of the end effector and the configuration degree of freedom of a spatial mechanism.

The degree of freedom (DOF) of a mechanism is the first consideration in the kinematics and dynamics of the mechanism [1]. As a matter of fact, it should also be specified before designing the mechanism, and then the drive and control can be determined. Many scholars have presented various theories about the analysis and calculation of the DOFs of mechanisms [1–7]. However, the previous theories and formulas are right only under certain circumstances. In some special cases, (for example, the 4-PUU spatial mechanism shown in Figure 10.1 and the 3-PUU spatial mechanism shown in Figure 10.2), the existing theories may lead to a completely wrong result, which has been discussed in Chapters 1 and 4.

In this chapter, therefore, we will investigate a stricter theory system of DOF that can be used to analyze and compute the DOF of the end effector and the actuators to drive the mechanisms. Traditionally, the DOF of a mechanism was thought of as the number of independent parameters to define the configuration of a mechanism. However, with the advent of spatial parallel mechanisms, the primary considerations of the designers are focused on the DOF of the end effector. In this regard, it is urgently necessary to differentiate the DOF of the end effector from the independent number of actuations to completely control the end effector under any configurations.

10.1 PROBLEMS EXISTING IN THE TRADITIONAL THEORIES OF DOF

In the following, we will first analyze the DOF of the mechanisms shown in Figures 10.1 and 10.2 with the traditional theories of DOF.

Advanced Theory of Constraint and Motion Analysis for Robot Mechanisms.
http://dx.doi.org/10.1016/B978-0-12-420162-0.00010-2

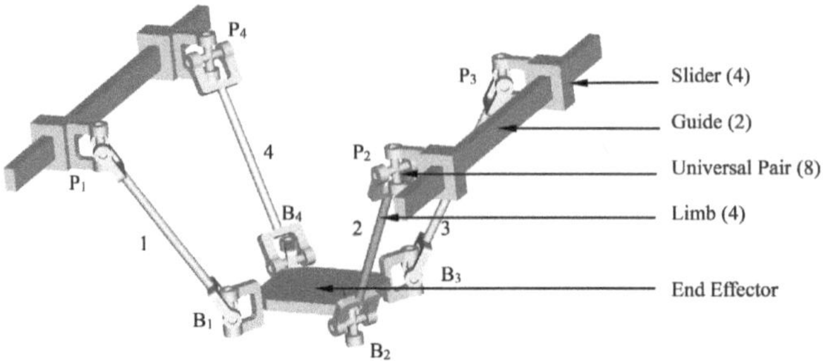

FIGURE 10.1 A spatial parallel mechanism made up of 4-PUU.

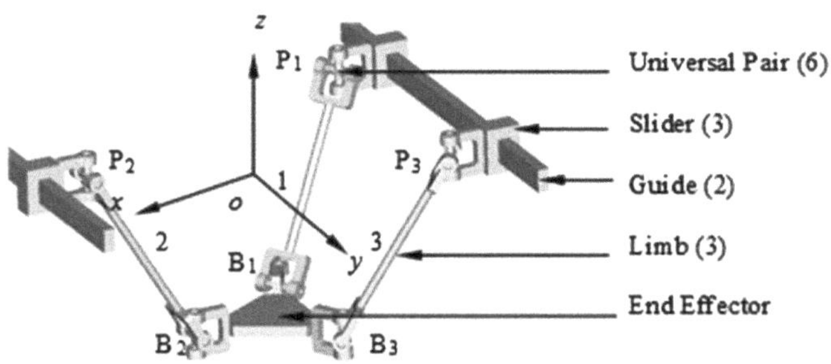

FIGURE 10.2 A spatial parallel mechanism made up of 3-PUU.

10.1.1 Representative Formula of Kutzbach–Grübler

In many monographs [1,2,4], there are various forms of Kutzbach–Grübler formula, which are often divided into two kinds—one is for planar and spherical mechanisms, the other is for spatial mechanisms.

For planar and spherical mechanisms, we have the following Kutzbach–Grübler formula:

$$M = 3(n - j - 1) + \sum_{i=1}^{j} f_i.$$

For spatial mechanisms, we have the following Kutzbach–Grübler formula:

$$M = 6(n - j - 1) + \sum_{i=1}^{j} f_i$$

where M represents the DOF of the mechanism, n represents the number of the total members of the system, j denotes the number of the total kinematic pairs, and f_i indicates the DOF number of the ith kinematic pair.

Now, we will analyze the example in Figure 10.1 with Kutzbach–Grübler's formula.

Here, $n = 10$, $j = 12$, $\sum_{i=1}^{j} f_i = 1 \times 4 + 2 \times 8 = 20$, therefore, we have

$$M = 6(n - j - 1) + \sum_{i=1}^{j} f_i = 6(10 - 12 - 1) + 20 = 2.$$

10.1.2 Dobrovolski's Formula

In monograph [5], Dobrovolski's formula is given as follows:

$$M = (6 - f)n - \sum_{j=f+1}^{5} (j - f)c_j$$

where f is the number of DOFs that are eliminated from all of the links of the system, n represents the number of total movable links, j indicates the class of the kinematic pair, and c_j denotes the number of kinematic pairs in class j.

We adopt Dobrovolski's formula to analyze the example in Figure 10.1 again. According to Dobrovolski's Formula, for all links, $f = 0$ holds.

$$\begin{aligned} M &= (6 - f)n - \sum_{j=f+1}^{5} (j - f)c_j = 6n - (c_1 + 2c_2 + 3c_3 + 4c_4 + 5c_5) \\ &= 6 \times 9 - (4 \times 8 + 5 \times 4) = 2. \end{aligned}$$

The result is the same as that of Kutzbach–Grübler's formula.

10.1.3 Methods of Hunt *et al.*

Hunt [2], Lung-Wen Tsai [1], *et al.* presented the following formula:

$$M = d(n - g - 1) + \sum_{i=1}^{g} f_i \text{ or } M = \sum_{i=1}^{g} f_i - dl$$

where d represents the order of the mechanism, $d = 6 - \lambda$ and λ is the number of the common constraints of the mechanism, n denotes the number of the members in the mechanism, g indicates the number of the kinematic pairs of the mechanism f_i is the DOF of the ith kinematic pair, and l is the number of the independent loops of a mechanism.

Hunt also gives a more general formula:

$$M = \sum_{i=1}^{g} f_i - \sum_{j=1}^{l} d_j$$

where $d_j (j = 1, \ldots, l)$ is the order of the $jth (j = 1, \ldots, l)$ loop. However, he also points out that this formula will pose further questions, such as

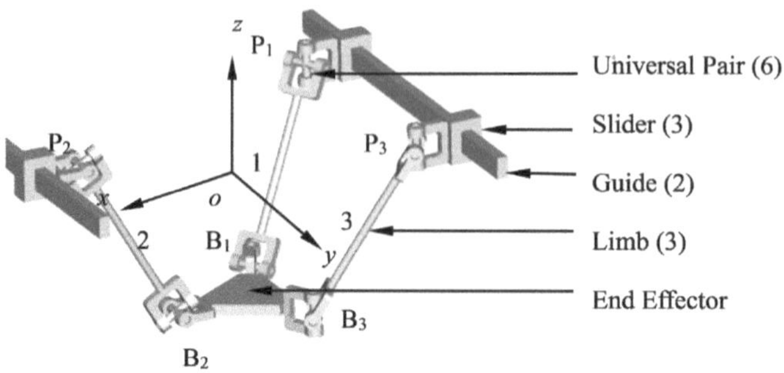

FIGURE 10.3 Another spatial parallel mechanism made up of 3-PUU.

how to identify suitable independent loops and how to assign the right d-values.

According to the method proposed above, the result of DOF for the mechanism shown in Figure 10.1 is also 2, which equals the results of traditional formulas. But in fact, the DOF of *the mechanism with a prescribed end effector* in Figure 10.1 is four rather than two. In Section 10.4.3, we will discuss its DOF thoroughly.

Similarly, when the traditional formulae are used to analyze the mechanisms shown in Figure 10.2, there are the same problems. The mechanism shown in Figure 10.2 and the mechanism in Figure 10.3 appear to be the same, but they are quite different. The DOF of the end effector shown in Figure 10.3 is three, including one translation along the x-axis, one translation along the y-axis, and one translation along the z-axis; but the DOF of the end effector shown in Figure 10.2 is four rather than three, in which there is one more rotation about the z-axis. Therefore, the traditional theories of DOF often lead to serious problems when they are applied to certain unusual mechanisms. Below, we first analyze the reasons that give rise to the above problems.

In the traditional theories of DOF, the mechanisms are not distinguished according to the end effectors. For example, when we consider the DOF of the mechanism shown in Figure 10.4, conflicts arise. If we consider the DOF of the whole spatial 4-bar mechanism (spatial parallel mechanism), the members AB and CD obviously have zero mobility because the directional view along the z-axis is a determinate structure shown in Figure 10.5. When considering the independent mobility of the link BC, however, we can get a different result shown in Figure 10.6. The link can execute an independent rotation about the z-axis freely.

Therefore, the DOF of a mechanism might be different if the end effector is selected differently. As the hybrid kinematic chains can be equivalently transformed into serial chains according to Theorem 4.1, for the sake of uniformity, we make a stipulation in this chapter that the calculations of DOF are all for *the mechanism with a prescribed end effector*, and the kinematic

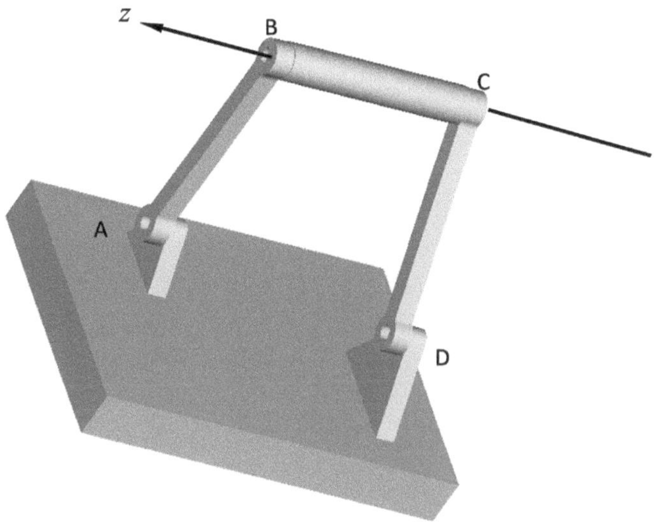

FIGURE 10.4 A spatial 4-bar mechanism.

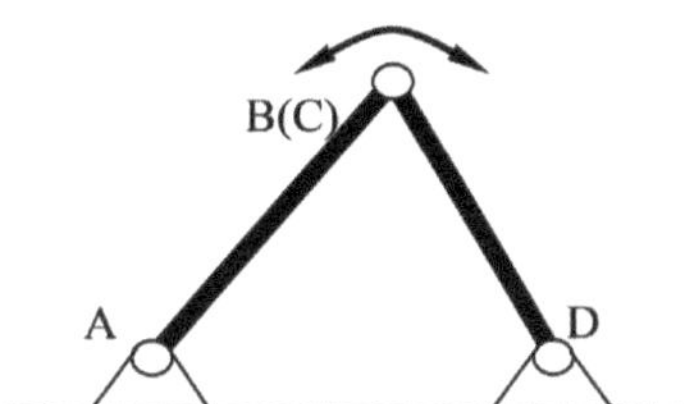

FIGURE 10.5 Directional view along the z-axis for the spatial 4-bar mechanism.

chains connecting the *end effector* and the base should be pure series kinematic pairs. Of course, in many cases, the results are often identical with those of the traditional methods except for those that cannot be solved with the traditional formulae.

As is well known, in order to obtain the DOF of a mechanism, we only need to calculate the values of the conventional formulae with the numbers of the links and pairs. Through intuitive deduction, it is safe to say that these methods must have problems with certain spatial mechanisms because the DOF of a mechanism should not only be the function of the numerical relationships between the links and pairs, it should also be the function of the constraints and geometry configurations of the mechanism.

For instance, according to the conventional methods, the DOF of the mechanism shown in Figure 10.2 and the mechanism in Figure 10.3 should be identical because both the numbers of the links and the numbers of pairs are

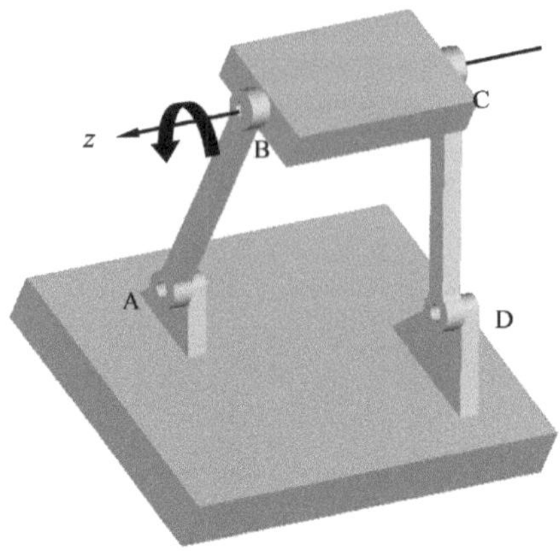

FIGURE 10.6 DOF of the spatial 4-bar mechanism with end effector BC.

identical, respectively. In fact, they are different. The DOF of *the mechanism with the prescribed end effector* in Figure 10.2 is 4 while the DOF of *the mechanism with the prescribed end effector* in Figure 10.3 is 3.

Although many scholars [2,3] have investigated the constraint relationships of members and amended the formulae, these methods still have their own limits and therefore are not widely applied.

10.2 THEORY OF DOF FOR THE MECHANISM WITH AN END EFFECTOR

In order to avoid the problems existing in the traditional methods, we will first introduce the conception of *constraint spaces* in this chapter. The definition is:

The constraint spaces, S_C, of a rigid body are the linear spaces spanned by all of the constraints exerted by its neighboring bodies.

For example, the *constraint spaces* of a fixed rigid body can be expressed:

$$\mathbf{S}^C_{fixed} = \{T_x \quad T_y \quad T_z \quad R_x \quad R_y \quad R_z\}$$

where S^C_{fixed} represents the *constraint spaces* of a fixed rigid body, T_x denotes the unit vector of a translational constraint along the x-axis, R_x indicates the unit vector of a rotational constraint about the x-axis; the meanings of the other symbols can be deduced.

After the introduction of the conception of *constraint spaces*, we can find that the DOF of any rigid body, F, can be obtained if we presume $d = \dim \mathbf{S}_C$:

$$F = 6 - d. \tag{10.1}$$

10.2.1 Basic Formula for Calculating the DOF of the Mechanism with a Prescribed End Effector

The formula for calculating the DOF of *the mechanism with a prescribed end effector* is represented below:

$$M = F + \sum_{i=1}^{n} r_i = 6 - d + \sum_{i=1}^{n} r_i \tag{10.2}$$

where M denotes the DOF of the mechanism with an output member, d indicates the number of dimensions that all of the reciprocal screws can be spanned in the constraint spaces, $r_i, i = 1, 2, \ldots, n$ represents the rank deficiency of the kinematic screws of the ith chain.

Through the formula, we can depict the independent actuations and avoid the trivial classifications of mechanisms, the order of a mechanism, the motion parameter and so on.

10.2.2 Concept of Configuration DOF and Its Formula

In most cases, after analyzing M, we have to compute the *configuration degree of freedom* (CDOF) of the system, which uniformly depicts the instantaneous mobility, singularity and stability of the mechanism system. The CDOF, denoted by M_{conf}, means the DOF under the configuration at the instant. The number of CDOFs is equal to the number of independent coordinates needed to uniquely specify the position and orientation of the mechanism under the configuration of the system at the instant.

As is well known, when the machine works at a singular position, the DOFs will be changed suddenly. When the number of DOFs decreases, over-input phenomena may occur; and when the number of DOFs increases, out-of-control phenomena might also occur. When the number of DOFs decreases, the independent actuations needed will be less than the inputs, therefore over-input phenomena must occur; and *vice versa*, when the number of DOFs increases, the independent drives needed will be more than the inputs, and therefore out-of-control phenomena may also occur. Both of these two cases can bring about tremendous destruction in engineering applications. Hence, only if the number of inputs is equal to the number of the CDOFs, can the system be stably controlled and work according to the trajectory designed in advance, without over-input phenomena, and without out-of-control phenomena at any instant.

As a result, the analysis of singularity of a mechanism is naturally transformed into the calculation of the CDOFs. Similarly, if the number of inputs is equal to the number of CDOFs at any instant, the system can be stably controlled. Therefore, the calculation of DOFs, the analysis of singularity, and the stability of the mechanism are solved at the same time after the introduction of CDOF: *Only if the number of the independent inputs is equal to the number of CDOFs can the whole system be stable.*

Assume that the whole system is stable under the control of F independent actuations at a configuration; then when the F independent actuations are specified at this configuration, the newly computed DOFs at the $jth(j = 1,2,\ldots)$ time, namely F_j, at the configuration must be zero, or else, another newly increased F_j independent actuations must be added to ensure the stability of the system at the configuration. That is, the CDOFs of the system at the configuration is:

$$M_{conf} = M + \sum_{j=1}^{m-1} F_j = 6 - d + \sum_{i=1}^{n} r_i + \sum_{j=1}^{m-1} F_j \tag{10.3}$$

where $F_j, j = 1,2,\ldots,m$ denotes the DOFs of *the mechanism with a prescribed end effector* when we analyze the mechanism at the jth time, m indicates the times repeated to analyze the DOF of the end effector, and there must be $F_m = 0$.

10.2.3 The General Steps to Analyze the DOF of the Mechanism with a Prescribed End Effector

The following are the general steps:

(1) Decompose the mechanism
In general, we can decompose the mechanism into n kinematic chains connecting the end effector with the base. The criteria of decomposition are:

Each kinematic chain can connect once and only once with the fixed base. That is, each kinematic chain is in pure series.

(2) Investigate the $i^{th}(1 \le i \le n)$ kinematic chain and obtain the Plücker coordinates of its kinematic pairs
To facilitate the description, a Cartesian coordinate system (local or absolute) often needs to be created for each kinematic chain. After the establishment of the coordinate system, the Plücker coordinates of every kinematic pairs can be obtained.

(3) Solve the reciprocal screws $\$_i^r$ of the $i^{th}(1 \le i \le n)$ kinematic chain
According to reciprocal screw theory, we can get the following equation:

$$\$_i^T \boldsymbol{\Delta} \$_i^r = 0$$

where $\boldsymbol{\Delta} = \begin{bmatrix} 0 & \mathrm{I} \\ \mathrm{I} & 0 \end{bmatrix}$ and $\mathbf{I}$ denotes a 3×3 identity matrix and 0 denotes a 3×3 zero matrix.

The reciprocal screws, $\$_i^r$, can be solved with linear algebra methods.

(4) Solve the terminal constraints of the $(i+1)^{\text{th}}(1 \le i \le n-1)$ kinematic chain in turn
If the twists of the $(\boldsymbol{i}+\mathbf{1})^{\text{th}}(\mathbf{1} \le \boldsymbol{i} \le \boldsymbol{n}-\mathbf{1})$ kinematic chain are all the same as those of any of the kinematic chains solved, then go on with the next step; otherwise return and repeat steps 2, 3, and 4, until all of the terminal constraints of every kinematic chain of the end effector have been solved.

(5) Decide the dimension d that the terminal constraints exerted on the end effector can be spanned in the constraint spaces
According to the physical meaning of reciprocal screws of twists, we can obtain the constraint spaces that should be spanned by all of the terminal constraints. Then we can decide the dimension d of the constraint spaces. Therefore, the DOF of the end effector, F, can be denoted as:

$$F = 6 - d.$$

(6) Investigate the rank deficiency of the twists of the $i^{\text{th}}(1 \le i \le n)$ kinematic chain
In the kinematic chains, there may be many kinematic pairs that will always remain collinear. If the collinear rotational pairs do not act as actuations, we should make every possible combination of two of them and simplify their corresponding rotational twists with the maximum linearly independent screws. Replace the twists, $\$_i$, of the $ith(i = 1,2,\ldots,n)$ kinematic chain with the simplified screws. If the total number of the simplified screws is denoted by n_i, and the rank of $\$_i$ is denoted by Rank($\$_i$), the rank deficiency of the twists $\$_i$ of the kinematic chain, r_i, can be obtained:

$$r_i = n_i - \text{Rank}(\$_i).$$

For example, a kinematic chain consists of SSS pairs shown in Figure 10.7, should be simplified because they include two collinear rotational idle DOFs. The coordinate system is created as Figure 10.7 shows; its origin is superimposed with spherical pair C and the z-axis is along the axis of CB. We presume the coordinates of B and A are $(0 \quad 0 \quad l)$ and $(x_A \quad y_A \quad z_A)$.

First, let us simplify the twists of pair C and B:
The twists of pair C are

$$\$_C^1 = \begin{bmatrix} 1 & 0 & 0 & 0 & 0 & 0 \end{bmatrix}^T, \quad \$_C^2 = \begin{bmatrix} 0 & 1 & 0 & 0 & 0 & 0 \end{bmatrix}^T,$$
$$\$_C^3 = \begin{bmatrix} 0 & 0 & 1 & 0 & 0 & 0 \end{bmatrix}^T.$$

The twists of pair B are

$$\$_B^1 = \begin{bmatrix} 0 & 0 & 0 & 1 & 0 & 0 \end{bmatrix}^T, \quad \$_B^2 = \begin{bmatrix} 0 & 0 & 1 & 0 & 0 & 0 \end{bmatrix}^T,$$
$$\$_B^3 = \begin{bmatrix} 0 & 0 & 1 & 0 & 0 & 0 \end{bmatrix}^T.$$

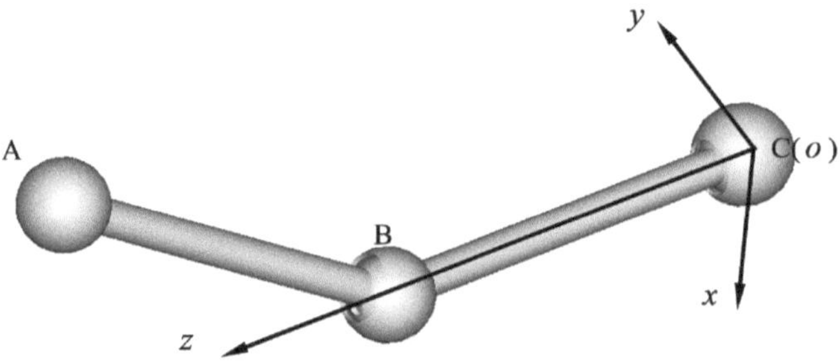

FIGURE 10.7 SSS pairs and the coordinate system.

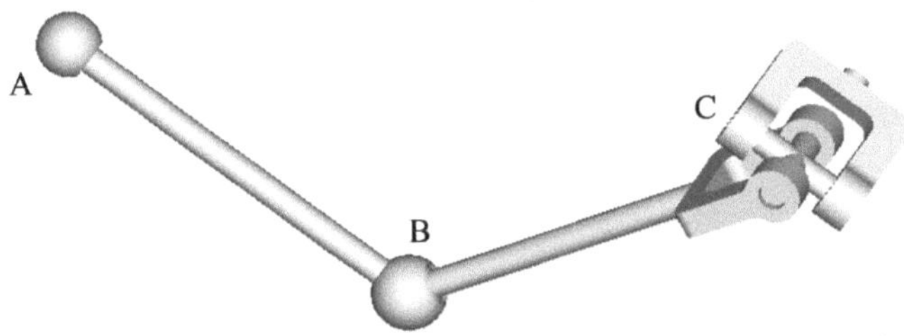

FIGURE 10.8 SSU chain.

If we presume that $\$_{CB}$ denote the twists of pairs B and C, we can obtain

$$\$_{CB} = \begin{bmatrix} \$_C^1 & \$_C^2 & \$_C^3 & \$_B^1 & \$_B^2 & \$_B^3 \end{bmatrix}.$$

With algebra method, we can obtain

$$\$_{CB}^m = \begin{bmatrix} \$_C^1 & \$_C^2 & \$_B^1 & \$_B^2 & \$_B^3 \end{bmatrix}$$

where $\$_{CB}^m$ represents the maximum linearly independent set of $\$_{CB}$. Therefore, with the combination of pairs C and B, we can obtain the first simplified kinematic chain shown in Figure 10.8.

Similarly, with the combination of pairs B and A in Figure 10.8, we can obtain the second and the last simplified kinematic chain shown in Figure 10.10.

We presume that the simplified twist set in this example is $\$_{CBA}^s$. Here, the number of the twists of the kinematic chain—CBA in Figure 10.9, is 7. $r = 7 - \text{Rank}(\$_{CBA}^s) = 7 - 6 = 1$.

The DOF of the mechanism with the end effector can be obtained with the following formula:

$$M = F + \sum_{i=1}^{n} r_i = 6 - d + \sum_{i=1}^{n} r_i.$$

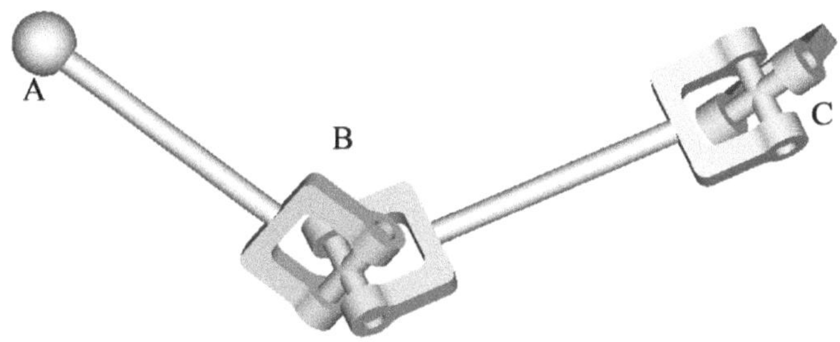

FIGURE 10.9 SUU chain.

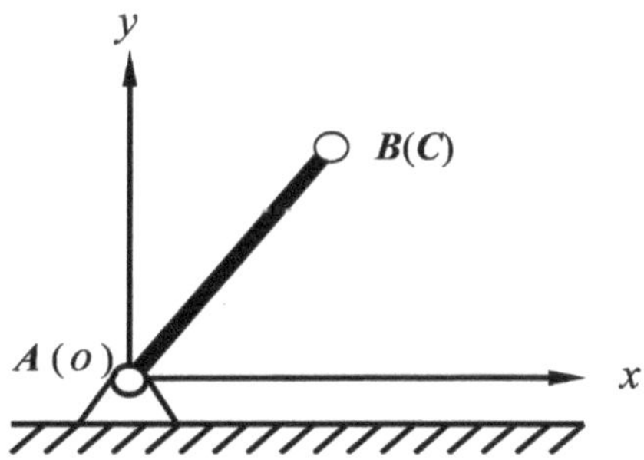

FIGURE 10.10 Cartesian coordinate System of limb AB.

Accordingly, we can presume that the formula for calculating the DOF of the mechanism with the same end effector at the jth time is:

$$M_j = F_j + \sum_{i=1}^{n} r_i^j = 6 - d_j + \sum_{i=1}^{n} r_i^j, \quad (j = 0,1,2,\ldots)$$

where d_j represents the dimension of the constraints spaces, M_j denotes the DOF of the output member, r_i^j indicates the rank deficiency of kinematic screws of the ith kinematic chain. Of course, $M_0 = F$.

(7) **Add $M_j,(j = 0,1,2,\ldots,m-1)$ actuations to the kinematic chains and repeat the first six steps and obtain a newly computed F_{j+1}**
Add M_j actuations to the kinematic chains and repeat the first six steps and obtain a newly computed $F_{j+1},(j = 0,1,2,\ldots,m-1)$. If the newly computed $F_{j+1} > 0$, it denotes that the actuations we selected are not linearly independent. We can reselect or add another F_{j+1} actuations and repeat the first six steps until $F_m = 0$. Fortunately, in most cases, m ≤ 2.

The best layout of the actuations is to directly assign the corresponding controls to the shortest chain connecting the end effector.

(8) Compute the configuration degrees of freedom M_{conf} and analyze the instantaneous stability of the system with the formula below

$$M_{conf} = \sum_{j=0}^{m-1} M_j = F + \sum_{i=1}^{n} r_i + \sum_{j=1}^{m-1} M_j = 6 - d + \sum_{i=1}^{n} r_i + \sum_{j=1}^{m-1} F_j \tag{10.4}$$

10.3 PROOF OF THE THEORY

In order to accurately depict the position and orientation of any spatial rigid body, only six independent variables are needed. That is, the DOFs of any spatial rigid body is six. According to the screw theory [2,7], the reciprocal screws of twists are a set of general forces, including three orthogonal forces and three orthogonal moments. Therefore, the constraint spaces of the reciprocal screws applied to the end effector by the n kinematic chains may be obtained. As a result, the dimension d of the constraint spaces is the DOFs lost by the end effector. So, the DOFs of the end effector is:

$$F = 6 - d.$$

That is, F independent actuations should be given to stably control the end effector.

With the simplified twists, the collinear rotational *idle degree(s) of freedom* can be deleted. The rank deficiency of the simplified twists $\$_i$ of the ith chain, r_i, depicts the dependent actuations of the ith kinematic chain, which increase the flexibility of the kinematic chain. That is, the dependent actuations are a family of different dependent means to control the ith kinematic chain.

In fact, every kinematic chain can be treated as an open one with certain terminal constraints, and every simplified twist corresponds to a kinematic input. When the twists are not linearly $\sum_{i=1}^{n} r_i$ independent, the actuation(s) corresponding to the linearly dependent twist(s) as the iterate inputs cannot affect the terminal constraints. So, the number of these actuations must be added to the above formula, the number of the similar DOF of all of the kinematic chains in total is $\sum_{i=1}^{n} r_i$; hence, we have the following formula:

$$M_0 = M = F + \sum_{i=1}^{n} r_i = 6 - d + \sum_{i=1}^{n} r_i.$$

Adding $M_j (j = 0,1,2,\ldots,m-1)$ actuations to the kinematic chains, we can calculate $F_{j+1} (j = 0,1,2,\ldots,m-1)$ by repeating the first six steps. If the output member is stable under the control of M_j independent actuations at any configuration, the newly computed DOFs $F_{j+1}(j = 0,1,2,\ldots,m-1)$ at the configuration must be zero when the M_j independent actuations are provided; otherwise, the newly increased F_{j+1} independent actuations have to be added to ensure the stability of the system at the configuration. That is, the actuations we selected are not linearly independent if $F_{j+1} > 0$. However, the actuations we selected are still the possible means to control the mechanism, so, $\sum_{j=1}^{m-1} F_j$ depicts the like items. Of course, we can also reselect the actuations to ensure $F_{j+1} = 0$.

Therefore, the CDOF of the mechanism with the end effector is

$$M_{conf} = \sum_{j=0}^{m-1} M_j = F + \sum_{i=1}^{n} r_i + \sum_{j=1}^{m-1} M_j = 6 - d + \sum_{i=1}^{n} r_i + \sum_{j=1}^{m-1} F_j.$$

10.4 APPLICATIONS AND DISCUSSIONS

10.4.1 Investigate the DOF of a Spatial 4-Bar Mechanism with Link BC as the End Effector

A spatial 4-bar mechanism, shown in Figure 10.4, is made up of 2-RR (2 Rotational pairs) kinematic chains. To study the DOF of *the mechanism with a prescribed end effector*—link BC, we can first decompose the spatial parallel mechanism to form two kinematic chains connecting the end effector BC with the base and then study the terminal constraints of each kinematic chain.

Because the two kinematic chains connecting the link with the base are identical in isomorphism, we can only select one of them, say, member AB to analyze the terminal constraints applied to link BC.

The Cartesian coordinate system $oxyz$ is established as Figure 10.10 shows, the origin of the coordinate system is superimposed with pair A, the x-axis is parallel to the base plane and the y-axis is vertical to the base plane, and therefore, the z-axis is naturally parallel to the axis of link BC. If we presume the coordinates of pair B are $(x_B \;\; y_B \;\; 0)$, the Plücker coordinates of pairs A and B can be denoted as:

$$\$_A = \begin{bmatrix} 0 & 0 & 1 & 0 & 0 & 0 \end{bmatrix}^T, \quad \$_B = \begin{bmatrix} 0 & 0 & 1 & y_B & -x_B & 0 \end{bmatrix}^T.$$

Therefore, the twist matrix of kinematic chain AB can be denoted:

$$\$_{AB} = \begin{bmatrix} \$_A & \$_B \end{bmatrix}.$$

The terminal constraint matrix of $\$_{AB}$ are:

$$\$^r_{AB} = \begin{bmatrix} \$^{r1}_{AB} & \$^{r2}_{AB} & \$^{r3}_{AB} & \$^{r4}_{AB} \end{bmatrix}$$

where

$$\$_{AB}^{r1} = \begin{bmatrix} \frac{x_B}{\sqrt{x_B^2 + y_B^2}} & \frac{y_B}{\sqrt{x_B^2 + y_B^2}} & 0 & 0 & 0 & 0 \end{bmatrix}^T,$$

$$\$_{AB}^{r2} = \begin{bmatrix} 0 & 0 & 1 & 0 & 0 & 0 \end{bmatrix}^T, \quad \$_{AB}^{r3} = \begin{bmatrix} 0 & 0 & 0 & 1 & 0 & 0 \end{bmatrix}^T,$$

$$\$_{AB}^{r4} = \begin{bmatrix} 0 & 0 & 0 & 0 & 1 & 0 \end{bmatrix}^T.$$

In order to analyze the terminal constraints exerted on the end effector BC by chain DC, we can also use the coordinate system $oxyz$. The coordinates of pair D and C are $D(x_D \ \ 0 \ \ z_D)$,$C(x_B \ \ y_B \ \ z_C)$. The Plücker coordinates of kinematic pair D and C can be denoted as:

$$\$_D = \begin{bmatrix} 0 & 0 & 1 & 0 & -x_D & 0 \end{bmatrix}^T, \quad \$_C = \begin{bmatrix} 0 & 0 & 1 & y_B & -x_B & 0 \end{bmatrix}^T.$$

Therefore, the twist matrix of kinematic chain AB can be denoted:

$$\$_{DC} = \begin{bmatrix} \$_D & \$_C \end{bmatrix}.$$

The terminal constraint matrix of $\$_{DC}$ are:

$$\$_{DC}^r = \begin{bmatrix} \$_{DC}^{r1} & \$_{DC}^{r2} & \$_{DC}^{r3} & \$_{DC}^{r4} \end{bmatrix},$$

where

$$\$_{DC}^{r1} = \begin{bmatrix} \frac{x_B - x_D}{\sqrt{(x_B - x_D)^2 + y_B^2}} & \frac{y_B}{\sqrt{(x_B - x_D)^2 + y_B^2}} & 0 & 0 & 0 & 0 \end{bmatrix}^T,$$

$$\$_{DC}^{r2} = \begin{bmatrix} 0 & 0 & 1 & 0 & 0 & 0 \end{bmatrix}^T, \$_{DC}^{r3} = \begin{bmatrix} 0 & 0 & 0 & 1 & 0 & 0 \end{bmatrix}^T,$$

$$\$_{DC}^{r4} = \begin{bmatrix} 0 & 0 & 0 & 0 & 1 & 0 \end{bmatrix}^T.$$

Therefore, the terminal constraint matrix exerted on the end effector BC can be obtained:

$$\$_{BC}^C = \begin{bmatrix} \frac{x_B}{\sqrt{x_B^2 + y_B^2}} & \frac{x_B - x_D}{\sqrt{(x_B - x_D)^2 + y_B^2}} & 0 & 0 & 0 \\ \frac{y_B}{\sqrt{x_B^2 + y_B^2}} & \frac{y_B}{\sqrt{(x_B - x_D)^2 + y_B^2}} & 0 & 0 & 0 \\ 0 & 0 & 1 & 0 & 0 \\ 0 & 0 & 0 & 1 & 0 \\ 0 & 0 & 0 & 0 & 1 \\ 0 & 0 & 0 & 0 & 0 \end{bmatrix}$$

The constraint spaces $\$^C_{BC}$ can be denoted as:

$$\$^C_{BC} = \left\{ T_x \quad T_y \quad T_z \quad R_x \quad R_y \right\}.$$

The dimension of the constraint space, $\$^C_{BC}$, is:

$$d = \text{Rank}(\$^C_{BC}) = 5$$

$$F = 6 - d = 6 - 5 = 1, \quad M_{conf} = M = F + \sum_{i=1}^{2} r_i = 1 + 0 = 1.$$

According to step (7) in Section 10.2.3, the end effector only has one rotational DOF. Therefore, we should first select pair B or C as the actuation.

If we apply one rotational control to pair B or C, the link BC will be stably controlled. Therefore,

$$M_{conf} = 6 - d + \sum_{i=1}^{n} r_i + \sum_{j=1}^{m-1} F_j = 6 - 5 + 0 + 0 = 1.$$

However, if we first select pair A (or D) as the actuation, we can find that $F_1 = 1$; then if we continue to select pair D (or A) as the actuation, $F_2 = 1$; the link will not be stably controlled until we select pair B or C as the actuation. That is,

$$M_{conf} = 6 - d + \sum_{i=1}^{n} r_i + \sum_{j=1}^{2} F_j = 6 - 5 + 0 + 2 = 3.$$

In order to study the reason, we can only select kinematic chain AB to analyze. The movement transmitted by link AB from the actuation A to pair B is no longer a rotational movement about the coaxes of pairs B and C while the solely possible motion of the end effector BC is a rotation about the co-axes of pairs B and C. Therefore, F_1 and F_2 are meaningless and unfeasible inputs.

It is important to note that in this special mechanism, the input and output members are superimposed, which might be used as adjustable bearing in reality.

10.4.2 Investigate the DOF of a Planar Parallel Mechanism with a Prescribed End Effector—Link *EF*

A planar nine-bar mechanism is shown in Figure 10.11. If the end effector is assigned to member EF, this planar mechanism can be thought of as a parallel mechanism that is made up of two kinematic chains, $ABCDE$ and $IHJF$. The origin of the coordinate system is superimposed with the revolute pair A, the x-axis is along the horizontal direction and the y-axis is along the vertical direction shown in Figure 10.11.

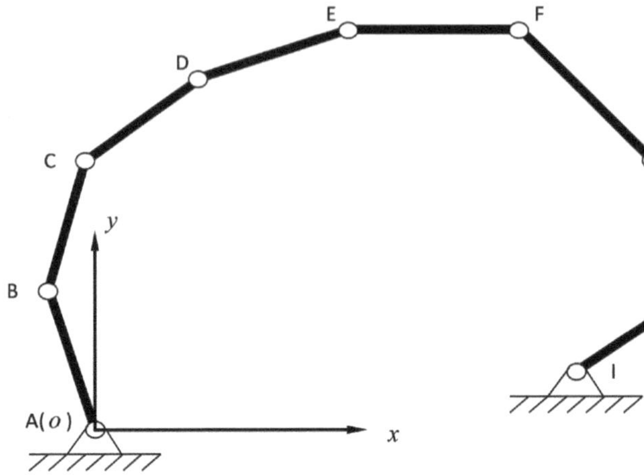

FIGURE 10.11 The coordinate system of a 9-bar planar mechanism.

The twist matrix of the kinematic chain $ABCDE$ is represented by:

$$\$_{ABCDE} = \begin{bmatrix} \$_A & \$_B & \$_C & \$_D & \$_E \end{bmatrix}$$

where

$$\$_{\mathrm{A}} = \begin{bmatrix} 0 & 0 & 1 & 0 & 0 & 0 \end{bmatrix}^T, \$_{\mathrm{B}} = \begin{bmatrix} 0 & 0 & 1 & y_B & -x_B & 0 \end{bmatrix}^T,$$
$$\$_{\mathrm{C}} = \begin{bmatrix} 0 & 0 & 1 & y_C & -x_C & 0 \end{bmatrix}^T, \$_{\mathrm{D}} = \begin{bmatrix} 0 & 0 & 1 & y_D & -x_D & 0 \end{bmatrix}^T,$$
$$\$_{\mathrm{E}} = \begin{bmatrix} 0 & 0 & 1 & y_E & -x_E & 0 \end{bmatrix}^T.$$

The rank of the twist matrix of the chain can be obtained with the algebra method:

$$\mathrm{Rank}(\$_{ABCDE}) = 3.$$

The rank deficiency of the chain is:

$$r_{ABCDE} = n_{ABCDE} - \mathrm{Rank}(\$_{ABCDE}) = 5 - 3 = 2.$$

The terminal constraint matrix of the kinematic chain can be obtained:

$$\$^r_{ABCDE} = \begin{bmatrix} \$^r_1 & \$^r_2 & \$^r_3 \end{bmatrix}$$

where

$$\$^r_1 = \begin{bmatrix} 0 & 0 & 1 & 0 & 0 & 0 \end{bmatrix}^T, \ \$^r_2 = \begin{bmatrix} 0 & 0 & 0 & 1 & 0 & 0 \end{bmatrix}^T,$$
$$\$^r_3 = \begin{bmatrix} 0 & 0 & 0 & 0 & 1 & 0 \end{bmatrix}^T.$$

$\$^r_{ABCDE}$ are just the planar constraints.

The twist matrix of the kinematic chain $IHGF$ is:

$$\$_{IHGF} = \begin{bmatrix} \$_I & \$_H & \$_G & \$_F \end{bmatrix}$$

where

$$\$_{\mathrm{I}} = \begin{bmatrix} 0 & 0 & 1 & y_I & -x_I & 0 \end{bmatrix}^T, \$_{\mathrm{H}} = \begin{bmatrix} 0 & 0 & 1 & y_H & -x_H & 0 \end{bmatrix}^T,$$
$$\$_{\mathrm{G}} = \begin{bmatrix} 0 & 0 & 1 & y_G & -x_G & 0 \end{bmatrix}^T, \$_{\mathrm{F}} = \begin{bmatrix} 0 & 0 & 1 & y_F & -x_F & 0 \end{bmatrix}^T.$$
$$\mathrm{Rank}(\$_{ABCDE}) = 3.$$

The rank deficiency of the chain is:

$$r_{IHGF} = n_{IHGF} - \mathrm{Rank}(\$_{IHGF}) = 4 - 3 = 1.$$

The terminal constraint matrix of the kinematic chain is:

$$\$^r_{IHGF} = \begin{bmatrix} \$^r_1 & \$^r_2 & \$^r_3 \end{bmatrix}$$

where

$$\$^r_1 = \begin{bmatrix} 0 & 0 & 1 & 0 & 0 & 0 \end{bmatrix}^T, \quad \$^r_2 = \begin{bmatrix} 0 & 0 & 0 & 1 & 0 & 0 \end{bmatrix}^T,$$
$$\$^r_3 = \begin{bmatrix} 0 & 0 & 0 & 0 & 1 & 0 \end{bmatrix}^T.$$

$\$^r_{IHGF}$ also denotes the same planar constraints.

The constraint spaces that the terminal constraints of the end effector can span are

$$S_C = \left\{ T_z \quad R_x \quad R_y \right\}.$$

Therefore, the CDOF of the mechanism with a prescribed end effector EF is:

$$M_{conf} = 6 - d + \sum_{i=1}^{2} r_i = 6 - 3 + (2 + 1) = 6.$$

10.4.3 Analyze the DOF of a 4-PUU Spatial Parallel Mechanism with a Prescribed End Effector

The spatial parallel mechanism, shown in Figure 10.1, is made up of 4-PUU (one prismatic pair and two universal pairs) kinematic chains. To study the DOF of the mechanism with the end effector, we can first decompose the spatial parallel mechanism to be four kinematic chains connecting the end effector with the base and then study the terminal constraints of each kinematic chain.

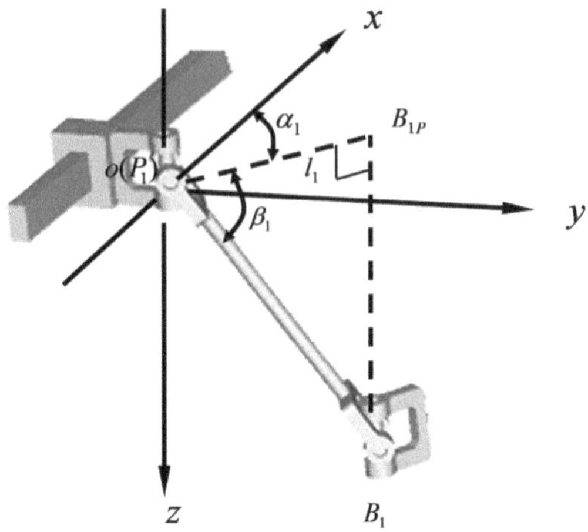

FIGURE 10.12 PUU kinematic chain.

Because all the kinematic chains connecting the end effector with the base are identical, we can simplify the process of analysis. For example, we can analyze chain P_1B_1 and create local Cartesian coordinates shown in Figure 10.12; B_{1P} is the projecting point of B_1 in the xoy plane, and OB_{1P} is the projecting line of OB_1 in the xoy plane. β_1 is the angle from line OB_{1P} to line OB_1, α_1 is the angle from the x-axis to line OB_{1P}.

The Plücker coordinates of slider P_1 are:

$$\$_1 = \begin{bmatrix} 0 & 0 & 0 & 1 & 0 & 0 \end{bmatrix}^T.$$

The universal pair P_1 can be decomposed as two orthogonal revolute pairs. The individual Plücker coordinates are:

$$\$_2 = \begin{bmatrix} \sin\alpha_1 & -\cos\alpha_1 & 0 & 0 & 0 & 0 \end{bmatrix}^T, \quad \$_3 = \begin{bmatrix} 0 & 0 & 1 & 0 & 0 & 0 \end{bmatrix}^T.$$

And the same is true for universal pair B_1. The coordinates of B_1 are:

$$B_1(l\cos\beta_1\cos\alpha_1 \quad l\cos\beta_1\sin\alpha_1 \quad l\sin\beta_1).$$

The Plücker coordinates can be denoted as:

$$\$_4 = \begin{bmatrix} s_4^1 \\ s_4^0 \end{bmatrix}, \quad \$_5 = \begin{bmatrix} s_5^1 \\ s_5^0 \end{bmatrix}.$$

where

$$s_4^1 = \begin{bmatrix} \sin\alpha_1 & -\cos\alpha_1 & 0 \end{bmatrix}^T, \quad s_5^1 = \begin{bmatrix} 0 & 0 & 1 \end{bmatrix}^T,$$

$$s_4^0 = \boldsymbol{r}_{B_1} \times s_4^1 = \begin{bmatrix} l\sin\beta_1\cos\alpha_1 & l\sin\beta_1\sin\alpha_1 & -l\cos\beta_1 \end{bmatrix}^T,$$

$$\$_5^0 = \boldsymbol{r}_{B_1} \times \boldsymbol{s}_5^1 = \begin{bmatrix} l\cos\beta_1 \sin\alpha_1 & -l\cos\beta_1\cos\alpha_1 & 0 \end{bmatrix}^T,$$

$$\therefore\ \$_4 = \begin{bmatrix} \sin\alpha_1 & -\cos\alpha_1 & 0 & l\sin\beta_1\cos\alpha_1 & l\sin\beta_1\sin\alpha_1 & -l\cos\beta_1 \end{bmatrix}^T,$$

$$\$_5 = \begin{bmatrix} 0 & 0 & 1 & l\cos\beta_1\sin\alpha_1 & -l\cos\beta_1\cos\alpha_1 & 0 \end{bmatrix}^T.$$

Therefore, the twists of the kinematic chain P_1B_1 can be expressed as:

$$\$_{P_1B_1} = \begin{bmatrix} \$_1 & \$_2 & \$_3 & \$_4 & \$_5 \end{bmatrix}.$$

The terminal constraints of kinematic chain P_1B_1 can be obtained:

$$\$^r_{P_1B_1} = \begin{bmatrix} 0 & 0 & 0 & \cos\alpha_1 & \sin\alpha_1 & 0 \end{bmatrix}^T.$$

According to the physical meaning of the reciprocal screws of twists, $\$^r_{P_1B_1}$ denotes a pure moment that is perpendicular to the universal pair plane.

The motion restricted by $\$^r_{P_1B_1}$ is $[\cos\alpha_1 \ \ \sin\alpha_1 \ \ 0 \ \ 0 \ \ 0 \ \ 0]^T$. This is a pure rotation about the direction $[\cos\alpha_1 \ \ \sin\alpha_1 \ \ 0]^T$, which is perpendicular to the planes of the universal pairs P_1 and B_1 shown in Figure 10.13. That is, the end effector can not execute any rotation about the direction $[\cos\alpha_1 \ \ \sin\alpha_1 \ \ 0]^T$. The same is true for the other three kinematic chains of the end effector. Therefore, the terminal constraints exerted on the end effector are four pure moments, which are all perpendicular to the direction of the z-axis, shown in Figure 10.14. The motions restricted by the terminal constraints exerted on the end effector are:

$$\$_E^C = \begin{bmatrix} \$^C_{P_1B_1} & \$^C_{P_2B_2} & \$^C_{P_3B_3} & \$^C_{P_4B_4} \end{bmatrix}$$

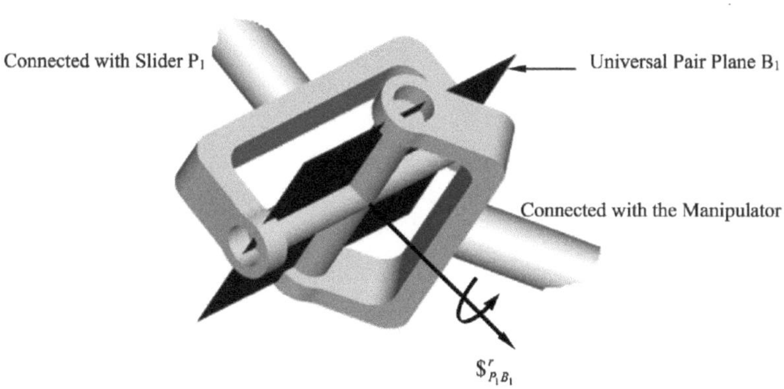

FIGURE 10.13 Terminal constraint of kinematic chain P_1B_1 exerted on the end effector.

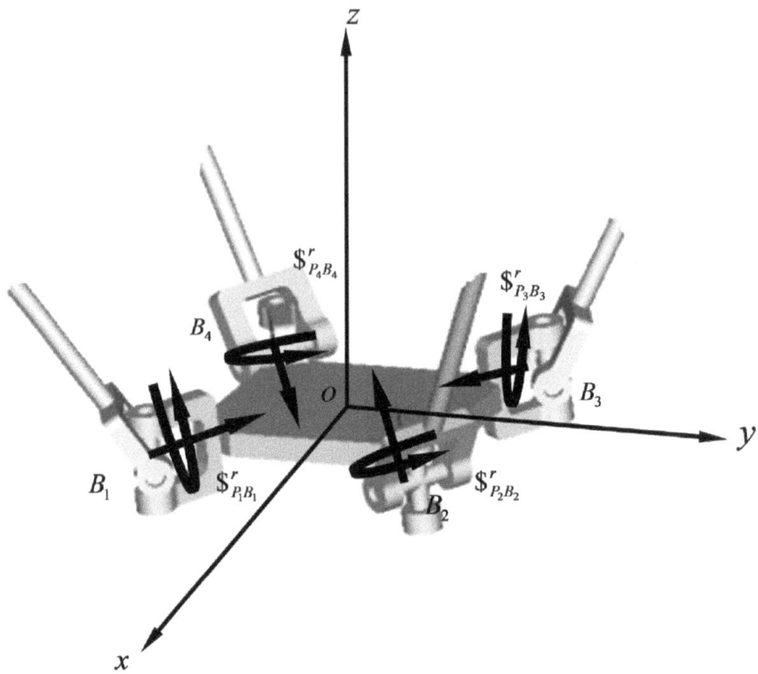

FIGURE 10.14 Terminal constraints exerted on the end effector $B_1B_2B_3B_4$.

where

$$\$^C_{P_1B_1} = \begin{bmatrix} \cos\alpha_1 & \sin\alpha_1 & 0 & 0 & 0 & 0 \end{bmatrix}^T,$$
$$\$^C_{P_2B_2} = \begin{bmatrix} \cos\alpha_2 & \sin\alpha_2 & 0 & 0 & 0 & 0 \end{bmatrix}^T,$$
$$\$^C_{P_3B_3} = \begin{bmatrix} \cos\alpha_3 & \sin\alpha_3 & 0 & 0 & 0 & 0 \end{bmatrix}^T,$$
$$\$^C_{P_4B_4} = \begin{bmatrix} \cos\alpha_4 & \sin\alpha_4 & 0 & 0 & 0 & 0 \end{bmatrix}^T.$$

$\alpha_i (i = 1,2,3,4)$ is the angle from the x-axis of the absolute coordinate system to the projecting line of the ith kinematic chain in the plane xoy.

Obviously, $1 \le \text{Rank}\left(\$^C_E\right) \le 2$, when $\text{Rank}\left(\$^C_E\right) = 1$, there must be $\alpha_1 = \alpha_4 = \frac{\pi}{2}$ and $\alpha_2 = \alpha_3 = \frac{3\pi}{2}$. This case corresponds to one kind of traditional singularity and can be avoided by adjusting the structure parameters.

In general, $\text{Rank}\left(\$^C_E\right) = 2$ and $\$^C_E$ can be fully spanned in a subspace S_C:

$$S_C = \begin{Bmatrix} R_x & R_y \end{Bmatrix}.$$

Therefore, we have:

$$M_0 = F = 6 - \text{Rank}\left(\$_E^C\right) = 6 - 2 = 4.$$

If we select the four sliders shown in Figure 10.1 as the actuations, F_1 can be analyzed as follows:

When the actuations are given, the four sliders can be seen as fixed pairs. Therefore, the twists of kinematic chain P_1B_1 can be rewritten as:

$$\$_{P_1B_1}^{Act} = \begin{bmatrix} \$_2 & \$_3 & \$_4 & \$_5 \end{bmatrix}.$$

Therefore, the terminal constraints of kinematic chain P_1B_1 can be obtained:

$$\$_{P_1B_1}^{r1} = \begin{bmatrix} 0 & 0 & 0 & \cos\alpha_1 & \sin\alpha_1 & 0 \end{bmatrix}^T,$$

$$\$_{P_1B_1}^{r2} = \begin{bmatrix} \cos\beta_1\cos\alpha_1 & \cos\beta_1\sin\alpha_1 & -\sin\beta_1 & 0 & 0 & 0 \end{bmatrix}^T.$$

According to the physical meaning of the reciprocal screws of twists, $\$_{P_1B_1}^{r1}$, denotes a pure moment that is perpendicular to the universal pair plane, and $\$_{P_1B_1}^{r2}$ is a pure force along the axis of limb P_1B_1 shown in Figure 10.15.

The same is true for the other three kinematic chains of the end effector. Therefore, the terminal constraints exerted on the end effector are four pure forces along the individual axes of the kinematic chains and four pure moments, which are all perpendicular to the direction of the z-axis in the absolute coordinate system shown in Figure 10.16.

From the above analysis, we know that the four pure moments can prevent the end effector from rolling about the x-axis and the y-axis in the absolute

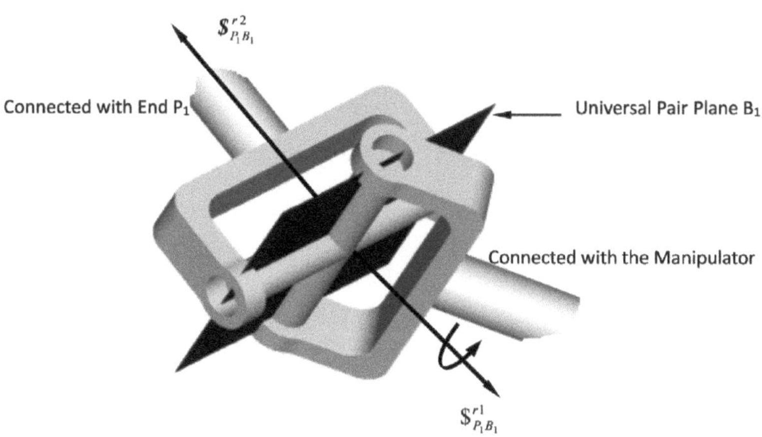

FIGURE 10.15 Terminal constraints of the kinematic chain.

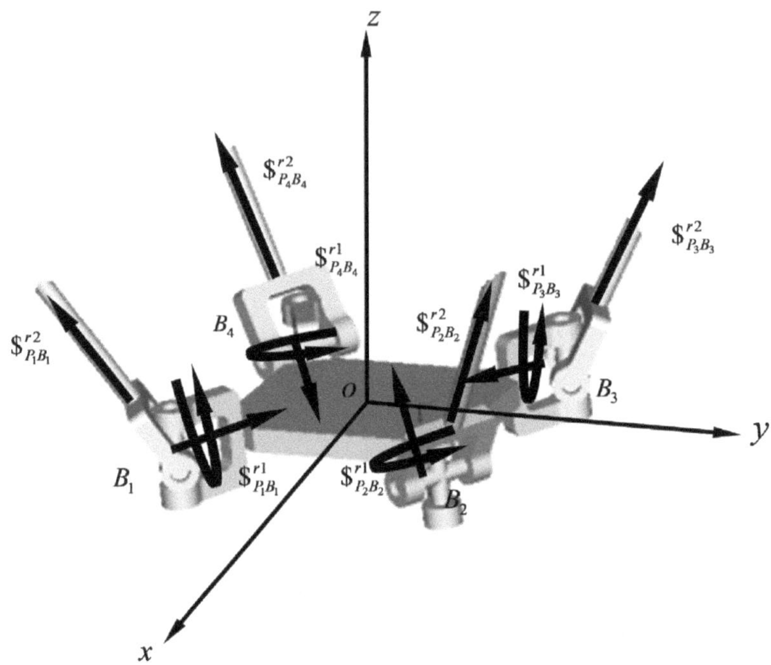

FIGURE 10.16 Terminal constraints exerted on the end effector $B_1B_2B_3B_4$ by the kinematic chains.

coordinate system and can be spanned to a subspace S_C. Therefore, we can only analyze whether the four pure forces $\$_E^{Act}$ (reciprocal screws) applied to the end effector can be spanned fully to the annihilated complementary spaces of S_C, denoted by $S_C^{\perp}$, if so, the end effector must be stable.

$$\$_E^{Act} = \begin{bmatrix} \$^{r2}_{P_1B_1} & \$^{r2}_{P_2B_2} & \$^{r2}_{P_3B_3} & \$^{r2}_{P_4B_4} \end{bmatrix}$$

where

$$\begin{aligned}
\$^{r2}_{P_1B_1} &= \begin{bmatrix} \cos\beta_1\cos\alpha_1 & \cos\beta_1\sin\alpha_1 & -\sin\beta_1 & 0 & 0 & 0 \end{bmatrix}^T, \\
\$^{r2}_{P_2B_2} &= \begin{bmatrix} \cos\beta_2\cos\alpha_2 & \cos\beta_2\sin\alpha_2 & -\sin\beta_2 & 0 & 0 & 0 \end{bmatrix}^T, \\
\$^{r2}_{P_3B_3} &= \begin{bmatrix} \cos\beta_3\cos\alpha_3 & \cos\beta_3\sin\alpha_3 & -\sin\beta_3 & 0 & 0 & 0 \end{bmatrix}^T, \\
\$^{r2}_{P_4B_4} &= \begin{bmatrix} \cos\beta_4\cos\alpha_4 & \cos\beta_4\sin\alpha_4 & -\sin\beta_4 & 0 & 0 & 0 \end{bmatrix}^T.
\end{aligned}$$

Therefore, the end effector must be stable if $\$_E^{Act}$ is spanned fully to $S_C^{\perp}$, the reciprocal screws $\$^{r2}_{P_1B_1}$, $\$^{r2}_{P_2B_2}$, $\$^{r2}_{P_3B_3}$ and $\$^{r2}_{P_4B_4}$ should prevent the end effector from executing translational movements along the three orthogonal axes and one rotational movement about the z-axis in the absolute coordinate system under any foreign forces. Assume that $k_i (i = 1,2,3,4)$ denote the intensity of the four reciprocal screws, F_x, F_v, and F_z to be the three component foreign forces exerted on the end effector and M_z to be the component foreign torque about the z-axis exerted on the end effector. Also assuming $B_1B_4 = 2a$, $B_1B_2 = 2b$, we get:

$$\begin{cases} \sum F_x = 0 & (k_1 \cos\theta_1 + k_2 \cos\theta_2 + k_3 \cos\theta_3 + k_4 \cos\theta_4)\cos\beta + F_x = 0 \\ \sum F_y = 0 & (k_1 \sin\theta_1 + k_2 \sin\theta_2 + k_3 \sin\theta_3 + k_4 \sin\theta_4)\cos\beta + F_y = 0 \\ \sum F_z = 0 & (k_1 + k_2 + k_3 + k_4)\sin\beta + F_z = 0 \\ \sum M_z = 0 & (-a\sin\theta_1 + b\cos\theta_1)k_1 + (a\sin\theta_2 + b\cos\theta_2)k_2 \\ & +(a\sin\theta_3 - b\cos\theta_3)k_3 + (-a\sin\theta_4 - b\cos\theta_4)k_4 + T_z = 0 \end{cases}$$

where $\theta_i (i = 1,2,3,4)$ denotes the angles from the projecting lines of the four limbs to the x-axis in the absolute coordinate system. The above equations can be transformed into:

$$\boldsymbol{A}\boldsymbol{F}_i = \boldsymbol{F}_f$$

where

$$\mathbf{A} = \begin{bmatrix} \cos\theta_1 & \cos\theta_2 & \cos\theta_3 & \cos\theta_4 \\ \sin\theta_1 & \sin\theta_2 & \sin\theta_3 & \sin\theta_4 \\ 1 & 1 & 1 & 1 \\ a\sin\theta_1 + b\cos\theta_1 & a\sin\theta_2 - b\cos\theta_2 & -a\sin\theta_3 - b\cos\theta_3 & -a\sin\theta_4 + b\cos\theta_4 \end{bmatrix},$$

$$\boldsymbol{F}_i = [k_1 \quad k_2 \quad k_3 \quad k_4]^T,$$
$$\boldsymbol{F}_f = [-F_x/\cos\beta \quad -F_y/\cos\beta \quad -F_z/\sin\beta \quad -T_z]^T.$$

The rank of matrix **A** is the dimension that the terminal constraints $\$_E^{Act}$ can be spanned. Obviously, if the equation below holds, the end effector should be uncontrollable:

$$|\mathbf{A}| = 0.$$

That is,

$$\begin{aligned} &a[(\sin\theta_1 - \sin\theta_2)\sin(\theta_4 - \theta_3) + (\sin\theta_4 - \sin\theta_3)\sin(\theta_2 - \theta_1)] + \\ &b[(\cos\theta_4 - \cos\theta_1)\sin(\theta_2 - \theta_3) + (\cos\theta_3 - \cos\theta_2)\sin(\theta_4 - \theta_1)] = 0 \end{aligned}.$$

When the above equation holds,

$$\text{Rank}\left(\$_E^{Act}\right) < 4.$$

$$\alpha_i \neq \frac{\pi}{2}, \alpha_i \neq \frac{3\pi}{2}$$

When and $|\mathbf{A}| \neq 0$, Rank $\left(\$_E^{Act}\right) = 4$, the four actuations can stably control the end effector.In this case,

$$F_1 = 0, \quad \text{and,} \quad M_{conf} = 6 - d + \sum_{i=1}^{n} r_i + \sum_{j=1}^{m} M_j = 6 - 2 + 0 + 0 = 4.$$

Therefore, if we select the four sliders as actuations, the end effector shall not be controllable except for $|\mathbf{A}| \neq 0$. This is because the four actuations are not linearly independent at these positions. In this case, $F_1 > 0$, and we can add another F_1 actuation or reselect the initial four actuations to control the end effector. In fact, these positions are the traditional singularities and we have:

$$M_{conf} = 6 - d + \sum_{i=1}^{n} r_i + \sum_{j=1}^{m-1} F_j = 6 - 2 + 0 + \sum_{j=1}^{m-1} F_j = 4 + \sum_{j=1}^{m-1} F_j > 4.$$

If $\$_E^{Act}$ denotes the reciprocal screws applied to the end effector, in ordinary cases, we also have:

$$\text{Rank}\left(\$_E^{Act}\right) = 2.$$

Therefore, in the ordinary cases and we have

$$F = 6 - \text{Rank}\left(\$_E^{Act}\right) = 6 - 2 = 4.$$

If we select the three sliders and any one vertical pivot of the six universal pairs as the actuations, the end effector is controllable and $F_1 = 0$. Therefore, we have:

$$M_{conf} = 6 - d + \sum_{i=1}^{n} r_i + \sum_{j} F_j = 6 - 2 + 0 + 0 = 4.$$

However, if the structure's parameters satisfy the following equations:

$$\alpha_1 = \frac{\pi}{2} \quad \text{and} \quad \alpha_2 = \alpha_3 = \frac{3\pi}{2},$$

We get:

$$\text{Rank}\left(\$_E^{Act}\right) = 1. \quad M = 6 - \text{Rank}\left(\$_E^{Act}\right) = 6 - 1 = 5.$$

If we appropriately select another two actuations (for example, select any one vertical and any one horizontal pivot from the six Universal pairs) besides the three sliders, we can keep $F_1 = 0$ and the CDOF of the mechanism with the end effector will be:

$$M_{conf} = \sum_{j} M_j = 5 + 0 + 0 = 5.$$

With the above procedure, we can also obtain the DOF of any other *mechanism with a prescribed end effector*. As mentioned in Chapter 9, the configuration DOF is partly dependent on the actuation schemes. Therefore, the actuation scheme should be discussed first after obtaining the DOF of the end effector of a mechanism.

10.5 CONCLUSIONS

By uniformly depicting the kinematic and geometrical constraints of mechanism with screw theory, this chapter probes a system method that should be widely used to study the DOF of *spatial parallel mechanisms with an end effector*. For the complex spatial mechanisms, the kinematic chains of the specified end effector might not be pure series which often contain sub-parallel chains; however, the twists of the sub-parallel chains can be equivalently substituted with the method discussed in Section 4.4 of Chapter 4. Therefore, the method investigated in this chapter can solve the analysis and calculation problems of the DOF in many complex spatial mechanisms.

REFERENCES

[1] L.W. Tsai, Mechanism Design: Enumeration of Kinematic Structures According to Function, CRC Press LLC, 200 N.W. Corporate Blvd., Boca Raton, Florida 33431, 2001, pp. 66–94.
[2] K.H. Hunt, Kinematic Geometry of Mechanisms, Oxford University Press, Oxford, 1978.
[3] J. Phillips, Freedom in machinery, Introducing Screw Theory, vol. 1, Cambridge University Press, Cambridge, 1984.
[4] K.J. Waldron, G.L. Kinzel, Kinematics, Dynamics, and Design of Machinery, Hamilton Printing, John Wiley & Sons, New York, 1999.
[5] D.B. Marghitu, M.J. Crocker, Analytical Elements of Mechanism, Cambridge University Press, Cambridge, 2001.
[6] J.M. McCarthy, Geometrical Design of Linkages, Springer-Verlag, New York, 2000.
[7] R.S. Ball, A Treatise on The Theory of Screws, Cambridge University Press, Cambridge, 1900, (First Published 1900, First Paperback Edition 1998, Printed in the United States of America).

Chapter 11

Mechanism Theory and Application of Deployable Structures Based on Scissor-Like Elements

This chapter aims at investigating the mechanism theory and application of deployable structures based on scissor-like elements (SLEs). The deployable structure has a wide application both in space structures and in general mechanical engineering. However, the mechanism theory of such deployable structures based on SLEs is seldom probed systematically. Therefore, this chapter starts from the mobility analysis of simple planar linkages, and then explores the mechanism theory of spatial deployable units utilized in flat, cylindrical, and spherical deployable structures. The theoretical results obtained in this chapter have a general adaptation in conceiving any kinds of deployable structures.

11.1 DEPLOYABLE STRUCTURES

Deployable structural systems are most often found in some biological organisms. Hachem et al. [1] surveyed biological systems in terms of their morphology, kinematics, and structural characteristics, and presented the similarity of the biological systems with existing deployable structures and potential applications. In the 1960s, a Spanish architect, E. P. Pienro, initially used the deployable concept to the movable theatre [2,3]. Deployable structures were widely surveyed and utilized in space structures with the increasing needs of complex space missions in the 1970s [4]. After that, deployable structures received great attention not only in space structures but also in ordinary familiar mechanical engineering. Deployable structures are generally used in cases requiring minimal damage to structural components during repeated deployment/undeployment [5]. From the viewpoint of theory of mechanism, the deployable structures were probed as early as the 19th century by Sarrus, Bricard, and in the early of 20th century by Bennett and et al. [6]. Deployable structures consist of units, which are composed of pairs of bars connected at

Advanced Theory of Constraint and Motion Analysis for Robot Mechanisms.
http://dx.doi.org/10.1016/B978-0-12-420162-0.00011-4

joint that allows a compact and deployed configuration [7]. Many applications can be found from small-scale structures [8], or swimming pool covers [9], to aerospace structures [4,7].

However, the most widely adopted deployable structures are those based on SLEs. On the basis of the basic module, various deployable trusses can be developed, such as equilateral triangles, squares, or normal hexagons [3]. From a geometric point of view, the whole idea of this type of deployable structure is that pairs of bars are connected to each other at an intermediate point through a revolute joint which allows them to rotate freely about an axis perpendicular to their common plane, but restrains all other degrees of freedom (DOFs), while at the same time, their end points are hinged to the end points of other SLEs [10]. For general three-dimensional configurations, geometric design is performed according to a set of geometric constraints resulting from the requirement of zero stresses at the two extreme configurations [10] and throughout the deployment process [7], which can be used to analyze and synthesize any type of spatial deployable structure. Nevertheless, it is very difficult to use this index to directly guide the deployable structure's design and analysis.

This chapter, therefore, addresses the mechanism theory that can guarantee the zero stress both at the two extreme configurations and throughout the whole deployment process. To begin with the mobility analysis of a rigid SLE, it deals with the rigid body model and kinematic principles of deployable scissor-like mechanisms.

Deployable structure units herein are restricted to pantographic foldable structures, which are often called SLEs. The simplest flat deployable structure such as the one shown in Figure 11.1 is most often found in deployable gates and so on, while the flat deployable structure shown on the left and the cylindrical deployable structure shown on the right of Figure 11.2 are also widely utilized in space and civil engineering. The most common characteristics of all these deployable structures are that each consists of a number of identical deployable structure units and the structure has one DOF during folding and unfolding

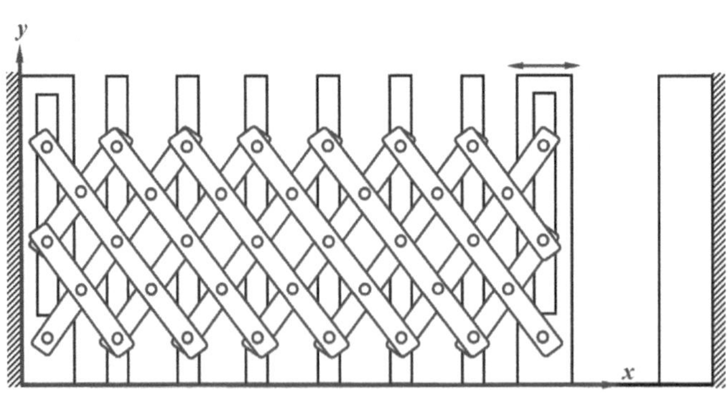

FIGURE 11.1 A deployable gate mechanism.

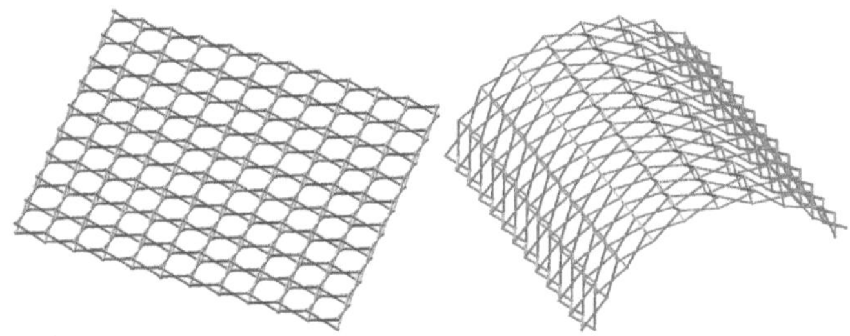

FIGURE 11.2 Deployable structures used in space and civil engineering.

when a reference point is given on the structure; this will be investigated in the following sections.

Therefore, from the viewpoint of mechanism theory, the mobility analysis of the basic unit plays a very important role in the synthesis and analysis of deployable structures. Prior to probing the mechanism theory of the deployable structure unit, this chapter will first present a brief review of reciprocal screw theory based on exponential transformation [11].

11.2 BRIEF REVIEW OF RECIPROCAL SCREW THEORY BASED ON EXPONENTIAL PRODUCTION

It is very convenient to express the kinematics of a serial kinematic chain by exponential production of the revolute angles of the joints in the chain [11]. Therefore, this section will briefly investigate the mobility of an SLE with reciprocal screw theory [12–17] based on exponential production [11].

Assume that $SO(3) = \{\mathbf{R} \in \mathbb{R}^{3\times3} | \mathbf{R}\mathbf{R}^T = \mathbf{I}, \det(\mathbf{R}) = 1\}$ denotes a 3×3 special orthogonal matrix group. As a matter of fact, group $SO(3)$ denotes the three-dimensional rotations, and therefore, it is also called configuration space.

The exponential transformation for a point on a link rotated θ about an axis $\boldsymbol{\omega}$ can be expressed with:

$$\mathbf{R} = \exp\left(\hat{\boldsymbol{\omega}}\theta\right) \tag{11.1}$$

where

$$\hat{\boldsymbol{\omega}} = \left(\begin{bmatrix} \omega_x \\ \omega_y \\ \omega_z \end{bmatrix}\right)^{\wedge} = \begin{bmatrix} 0 & -\omega_z & \omega_y \\ \omega_z & 0 & -\omega_x \\ -\omega_y & \omega_x & 0 \end{bmatrix} \tag{11.2}$$

and θ denotes the rotational angle about the joint axis $\boldsymbol{\omega}$.

Let $so(3) = \{\mathbf{S} \in \mathbb{R}^{n\times n} | \mathbf{S}^T = -\mathbf{S}\}$, if $\hat{\boldsymbol{\omega}} \in so(3)$ and $\theta \in R$ which is a real number, there must be $\exp\left(\hat{\boldsymbol{\omega}}\,\theta\right) \in SO(3)$. It is not difficult to prove that the mapping from $so(3)$ to $SO(3)$ is surjective.

Define groups:

$$se(3) = \left\{ (\hat{\omega}, v) \middle| \hat{\omega} \in so(3), v \in \mathbb{R}^3 \right\},$$

and

$$SE(3) = \left\{ (\mathbf{R}, p) | \mathbf{R} \in SO(3), p \in \mathbb{R}^3 \right\} = \mathbb{R}^3 \times SO(3).$$

Obviously, the element $(\mathbf{R}, p) \in SE(3)$ can be used to express the transformation from one coordinate system to another.

Suppose the coordinates of a twist are denoted by $\xi = \begin{bmatrix} \omega \\ v \end{bmatrix}$. Obviously, if a Vee operation, $\vee$, and a wedge operation, $\wedge$, are defined such that $\hat{\xi} = \begin{bmatrix} \hat{\omega} & v \\ 0 & 0 \end{bmatrix}$ and $(\hat{\xi})^{\vee} = \begin{bmatrix} \hat{\omega} & v \\ 0 & 0 \end{bmatrix}^{\vee} = \begin{bmatrix} \omega \\ v \end{bmatrix} = \xi$, then $e^{\hat{\xi}\theta} \in SE(3)$. If we let $\mathbf{g}(0)$ represent the initial configuration of a rigid body relative to a frame, then the final configuration, still with respect to the same frame, is given by:

$$\mathbf{g}(\theta) = \exp\left(\hat{\xi}\theta\right) \mathbf{g}(0). \tag{11.3}$$

Thus, the exponential map for a twist gives the relative motion of a rigid body. Therefore, the exponential of a twist is a mapping from initial configuration to the final one. It can be proven that given $\mathbf{g} \in SE(3), \hat{\xi} \in se(3)$ exists, and $\theta \in R$ such that $\mathbf{g} = \exp(\hat{\xi}\theta)$. The forward kinematics map for a series kinematic chain is given by:

$$\mathbf{g}_{ab}(\theta) = \prod_{i=1}^{n} \exp\left(\hat{\xi}_i \theta_i\right) \mathbf{g}_{ab}(0) \tag{11.4}$$

where $\mathbf{g}_{ab}(\theta)$ represents the rigid motion of the end effector connected by the series kinematic chain, and $g_{ab}(0)$ represents transformation from b to a when the manipulator is in its reference configuration.

Equation (11.4) can be expanded to

$$\mathbf{g}_{ab}(\theta) = \begin{bmatrix} \mathbf{R}_{ab}(\theta) & p_{ab}(\theta) \\ 0 & 1 \end{bmatrix}. \tag{11.5}$$

The coordinates of a point q attached to the rigid end effector in spatial coordinates are given by

$$q_a(\theta) = \mathbf{g}_{ab}(\theta) q_b. \tag{11.6}$$

The velocity of the point can be derived by directly differentiating the vector with respect to time:

$$v_{q_a} = \dot{q}_a = \dot{\mathbf{g}}_{ab}(\theta) q_b = \dot{\mathbf{g}}_{ab}(\theta) \mathbf{g}_{ab}^{-1}(\theta) q_a. \tag{11.7}$$

The velocity of the end effector can be specified by defining wedge and Vee operations in the absolute coordinate system:

$$\hat{v}_{ab}^{a} = \dot{\mathbf{g}}_{ab}\mathbf{g}_{ab}^{-1} = \begin{bmatrix} \dot{\mathbf{R}}_{ab}\mathbf{R}_{ab}^{T} & -\dot{\mathbf{R}}_{ab}\mathbf{R}_{ab}^{T}\boldsymbol{p}_{ab} + \dot{\boldsymbol{p}}_{ab} \\ 0 & 0 \end{bmatrix} \tag{11.8}$$

$$\boldsymbol{v}_{ab}^{a} = \left(\hat{v}_{ab}^{a}\right)^{\vee} = \begin{bmatrix} \boldsymbol{\omega}_{ab}^{a} \\ \boldsymbol{v}_{ab}^{a} \end{bmatrix} = \begin{bmatrix} \left(\dot{\mathbf{R}}_{ab}\mathbf{R}_{ab}^{T}\right)^{\vee} \\ -\dot{\mathbf{R}}_{ab}\mathbf{R}_{ab}^{T}\boldsymbol{p}_{ab} + \dot{\boldsymbol{p}}_{ab} \end{bmatrix}. \tag{11.9}$$

Therefore, from equations (11.4), (11.8), and (11.9), we can obtain the velocity of the end effector. From equation (11.8), we might find that:

$$\hat{v}_{ab}^{a} = \dot{\mathbf{g}}_{ab}\mathbf{g}_{ab}^{-1} = \left(\sum_{i=1}^{n} \frac{\partial \mathbf{g}_{ab}}{\partial \theta_i}\dot{\theta}_i\right)\mathbf{g}_{ab}^{-1} = \sum_{i=1}^{n}\left(\frac{\partial \mathbf{g}_{ab}}{\partial \theta_i}\mathbf{g}_{ab}^{-1}\right)\dot{\theta}_i. \tag{11.10}$$

In twist coordinates, equation (11.10) can be rewritten as:

$$\boldsymbol{v}_{ab}^{a} = \mathbf{J}_{ab}^{a}\dot{\theta} \tag{11.11}$$

where

$$\mathbf{J}_{ab}^{a} = \left[\left(\frac{\partial \mathbf{g}_{ab}}{\partial \theta_1}\mathbf{g}_{ab}^{-1}\right)^{\vee} \left(\frac{\partial \mathbf{g}_{ab}}{\partial \theta_2}\mathbf{g}_{ab}^{-1}\right)^{\vee} \cdots \left(\frac{\partial \mathbf{g}_{ab}}{\partial \theta_n}\mathbf{g}_{ab}^{-1}\right)^{\vee}\right],$$

$$\dot{\theta} = \left[\dot{\theta}_1 \ \dot{\theta}_2 \ \cdots \ \dot{\theta}_n\right]^{T}.$$

The physical meanings of each column of the Jacobian matrix J_{ab}^{a} are the twist coordinates of the corresponding joint in the absolute coordinate frame. Therefore, the absolute twist coordinates of the point attached to the rigid end effector are given with equation (11.10) or (11.11). Given a fixed absolute coordinate frame, we can directly write out the Jacobian matrix J_{ab}^{a} whose columns are the twists of the joint in the chain. For the sake of length, the exponential production theory of screw is not fully developed here. Please refer to [11] for the details.

The reciprocal screw theory [11–17] indicates that the constraints exerted on the point attached to the rigid end effector are reciprocal to the twist of the point. Therefore,

$$\boldsymbol{\tau}^{T}\Delta\boldsymbol{v}_{ab}^{a} = \left(\boldsymbol{\tau}^{T}\Delta\mathbf{J}_{ab}^{a}\right)\dot{\theta} = 0 \tag{11.12}$$

where $\boldsymbol{\tau}$ indicates the wrench coordinates in the absolute coordinate system, the superscript T indicates the transposal of a vector or matrix, and $\Delta = \begin{bmatrix} 0 & \mathbf{I}_3 \\ \mathbf{I}_3 & 0 \end{bmatrix}$ and $\mathbf{I}_3 = \begin{bmatrix} 1 & 0 & 0 \\ 0 & 1 & 0 \\ 0 & 0 & 1 \end{bmatrix}$.

Because of the independency and arbitrariness of $\dot{\theta}$, the necessary and sufficient criteria for equation (11.12) can be simplified as:

$$\left(\mathbf{J}_{ab}^{a}\right)^{T} \Delta\boldsymbol{\tau} = 0 \tag{11.13}$$

The above analysis is only based on a rigid model, therefore, it is not difficult to find that the requirements of zero stresses at the two extreme configurations [10] and throughout the deployment process [7] for a deployable structure are naturally satisfied.

11.3 MECHANISM THEORY OF A DEPLOYABLE STRUCTURE UNIT

Any deployable structure is made up of a number of primary structure units. Therefore, the mobility analysis results of a structure unit are also able to reveal the mechanism properties of the whole structure. As a result, the mobility of an SLE and its combinations will be investigated thoroughly with the reciprocal screw theory based on exponential production in this section.

11.3.1 Mobility Analysis of an SLE

The SLE is a fundamental component in the deployable structures shown in Figure 11.1 and Figure 11.2. The mobility properties of an SLE are investigated from the viewpoint of mechanism in this section. Figure 11.3 shows two SLEs connected by two bi-joints, one of which is fixed with the base while the other is prescribed as an end effector.

For simplification, one might suppose that the plane determined by the SLE of A_1C_1 and B_1C_1 is denoted by π_1, and the plane determined by the SLE of A_2C_2 and B_2C_2 is denoted by π_2. For the sake of convenience, we can also assume that the subtended angle by the two planes π_1 and π_2 is represented by α, the x-axis is superimposed with the intersection of planes π_1 and π_2, and the y-axis passes through point A_1 and is restricted within plane π_1, the z-axis is determined in accordance with the right-hand rule.

Now, we can investigate the twists and the constraints of point B_1 attached to the end effector B_1B_2 in the coordinate system shown in Figure 11.3 with the above analysis process. Firstly, we can directly obtain the twists in the fixed absolute coordinate frame:

$$\mathbf{J}_{ab}^{a_1} = \begin{bmatrix} \$_{A_1} & \$_{C_1} & \$_{B_1} \end{bmatrix} \tag{11.14}$$

where $\$_{A_1} = (0 \ \ 0 \ \ 1 \ \ y_{A_1} \ \ -x_{A_1} \ \ 0)^T$, $\$_{C_1} = (0 \ \ 0 \ \ 1 \ \ y_{C_1} \ \ -x_{C_1} \ \ 0)^T$, $\$_{B_1} = (0 \ \ 0 \ \ 1 \ \ y_{B_1} \ \ -x_{B_1} \ \ 0)^T$.

It is not difficult to find that the rank deficiency criterion for equation (11.14) is that $\det\left(\begin{bmatrix} y_{C_1} - y_{A_1} & y_{B_1} - y_{A_1} \\ x_{C_1} - x_{A_1} & x_{B_1} - x_{A_1} \end{bmatrix}\right) = 0$, which can be equivalently expressed as $\frac{y_{C_1}-y_{A_1}}{x_{C_1}-x_{A_1}} = \frac{y_{B_1}-y_{A_1}}{x_{B_1}-x_{A_1}}$ for the case shown in Figure 11.3. That is to say, the rank deficiency occurs only when points B_1, C_1, and A_1 stay in a line, which can be avoided by considering the dimensions of the linkages.

It is not difficult to find that the wrenches exerted on the point B_1 of the end effector B_1B_2 by the kinematic chain with equation (11.13) are:

$$\boldsymbol{\tau}_{(B_1B_2)1} = \begin{bmatrix} \$^r_{11} & \$^r_{12} & \$^r_{13} \end{bmatrix} \begin{bmatrix} F_{11} \\ F_{12} \\ F_{13} \end{bmatrix} \tag{11.15}$$

where F_{1i} $(i = 1,2,3)$ represent the tensities of the wrenches at the point B_1 exerted by the kinematic chain $A_1C_1B_1$, and $\$^r_{11} = \begin{pmatrix} 0 & 0 & 1 & 0 & 0 & 0 \end{pmatrix}^T$ indicates a unit force constraint along the z-axis; $\$^r_{12} = \begin{pmatrix} 0 & 0 & 0 & 1 & 0 & 0 \end{pmatrix}^T$, $\$^r_{13} = \begin{pmatrix} 0 & 0 & 0 & 0 & 1 & 0 \end{pmatrix}^T$, and $\$^r_{12}$ and $\$^r_{13}$ denote torque units about the x- and y-axes, individually. Therefore, we can draw the following conclusion from the above analysis.

Conclusion 1. The end effector B_1B_2 can only move within the plane π_1, which is shown in Figure 11.3.

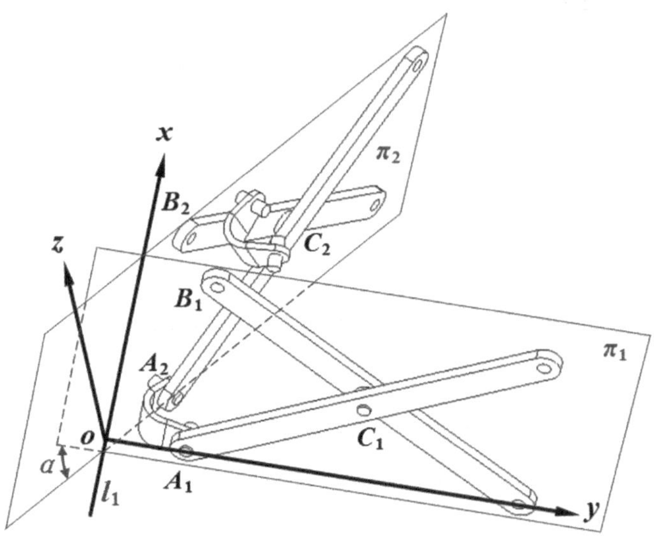

FIGURE 11.3 A structure consists of two SLEs.

11.3.2 Mobility Analysis of SLE Combinations

Next, we can also investigate the twist and constraint of the point B_2 attached to the end effector B_1B_2 from another SLE within plane π_2 in the coordinate system shown in Figure 11.3.

$$\mathbf{J}_{ab}^{a_2} = \begin{bmatrix} \$_{A_2} & \$_{C_2} & \$_{B_2} \end{bmatrix} \tag{11.16}$$

where

$$\$_{A_2} = \begin{pmatrix} 0 & \sin\alpha & -\cos\alpha & -y_{A_2}\cos\alpha - z_{A_2}\sin\alpha & x_{A_2}\cos\alpha & x_{A_2}\sin\alpha \end{pmatrix}^T,$$

$$\$_{C_2} = \begin{pmatrix} 0 & \sin\alpha & -\cos\alpha & -y_{C_2}\cos\alpha - z_{C_2}\sin\alpha & x_{C_2}\cos\alpha & x_{C_2}\sin\alpha \end{pmatrix}^T,$$

$$\$_{B_2} = \begin{pmatrix} 0 & \sin\alpha & -\cos\alpha & -y_{B_2}\cos\alpha - z_{B_2}\sin\alpha & x_{B_2}\cos\alpha & x_{B_2}\sin\alpha \end{pmatrix}^T.$$

With equation (11.13), it is not difficult to find that the wrenches exerted on point B_2 of the end effector B_1B_2 are:

$$\tau_{(B_1B_2)2} = \begin{bmatrix} \$_{21}^r & \$_{22}^r & \$_{23}^r \end{bmatrix} \begin{bmatrix} F_{21} \\ F_{22} \\ F_{23} \end{bmatrix} \tag{11.17}$$

where F_{2i} $(i = 1,2,3)$ indicate the tensities of the wrenches at the point B_1 exerted by the kinematic chain $A_2C_2B_2$, $\$_{21}^r = \begin{pmatrix} 0 & -\sin\alpha & \cos\alpha & 0 & 0 & 0 \end{pmatrix}^T$, $\$_{22}^r = \begin{pmatrix} 0 & 0 & 0 & 1 & 0 & 0 \end{pmatrix}^T$, $\$_{23}^r = \begin{pmatrix} 0 & 0 & 0 & 0 & \cos\alpha & \sin\alpha \end{pmatrix}^T$; and $\$_{21}^r$ represents a unit force along the normal of the plane π_2, $\$_{22}^r$ and $\$_{23}^r$ represent torques about the x-axis and the direction $\begin{pmatrix} 0 & \cos\alpha & \sin\alpha \end{pmatrix}^T$, individually. Therefore, we can draw another conclusion:

Conclusion 2. The end effector B_1B_2 can only move within the plane restricted by the SLE of A_2C_2 and B_2C_2 which is denoted by π_2.

Suppose

$$\boldsymbol{\tau}_{B_1B_2} = \boldsymbol{\tau}_{(B_1B_2)1} + \boldsymbol{\tau}_{(B_1B_2)2} = \$_{B_1B_2}^r F \tag{11.18}$$

where $\$_{B_1B_2}^r = \begin{bmatrix} \$_{11}^r & \$_{12}^r & \$_{13}^r & \$_{21}^r & \$_{22}^r & \$_{23}^r \end{bmatrix}$, $F = \begin{pmatrix} F_{11} & F_{12} & F_{13} & F_{21} & F_{22} & F_{23} \end{pmatrix}^T$. Also suppose that the twist of link B_1B_2 under the constraints depicted with equation (11.18) is denoted by $\$_{B_1B_2}$. According to reciprocal screw theory, we can find that

$$\$_{B_1B_2}^T \Delta \boldsymbol{\tau}_{B_1B_2} = \left(\$_{B_1B_2}^T \Delta \$_{B_1B_2}^r \right) F = 0. \tag{11.19}$$

Considering the arbitrariness of F, one immediately obtains

$$\left(\$^{r}_{B_1B_2}\right)^T \Delta\$_{B_1B_2} = 0. \tag{11.20}$$

Solving equation (11.20) yields

$$\$_{B_1B_2} = k\begin{pmatrix}0 & 0 & 0 & 1 & 0 & 0\end{pmatrix}^T \tag{11.21}$$

where k indicates the magnitude of the twist. $\$_{B_1B_2}$ denotes a translation along the x-axis.

From Conclusions 1 and 2, one also immediately finds that the bi-link (end effector) B_1B_2 can only move within the intersection of the plane π_1 and π_2, which interprets equation (11.21) in geometry. Consequently, we obtain that in a deployable structure, a pair of multi-joints hinged with two or more non-coplanar SLEs should have and only have one visible relative translation along the common intersection of these SLE planes.

Now, we might turn to analyze the case when the rank deficiency of $\$^{r}_{B_1B_2}$ occurs. It is not difficult to find that this only arises at $\sin\alpha = 0$, the solutions of which are $\alpha = 0$ or $\alpha = \pi$. For the case $\alpha = 0$ or $\alpha = \pi$, the two SLEs are coplanar.

For this case, the twist solutions of equation (11.20) will be:

$$\$_{B_1B_2} = \begin{bmatrix}0 & 0 & 1 & 0 & 0 & 0\\ 0 & 0 & 0 & 1 & 0 & 0\\ 0 & 0 & 0 & 0 & 1 & 0\end{bmatrix}^T \begin{bmatrix}k_1\\ k_2\\ k_3\end{bmatrix} \tag{11.22}$$

where k_1, k_2, and k_3 indicate the magnitudes of the twists; $\$_{B_1B_2}$ denotes that link B_1B_2 has three DOFs, one is a rotation about the z-axis and the other two are two translations along the x- and y-directions, individually. Of course, the structure in such a case can not be used as a deployable structure because it does not have a determinate configuration even after full deployment.

However, when the size of multi-joint B_1B_2 reduces zero, the two SLEs are directly hinged at one superimposed point $B_1(B_2)$, which is shown by the left subfigure of Figure 11.4. And as a result, the rotational component, denoted by screw $k_1(0\ \ 0\ \ 1\ \ 0\ \ 0\ \ 0)^T$, in equation (11.22) becomes a self rotation about the contracted point $B_1(B_2)$. Therefore, besides the self rotation, the visible free motions of the superimposed point $B_1(B_2)$ with respect to the fixed frame can be expressed as

$$\$_{B_1B_2} = \begin{bmatrix}0 & 0 & 0 & 1 & 0 & 0\\ 0 & 0 & 0 & 0 & 1 & 0\end{bmatrix}^T \begin{bmatrix}k_2\\ k_3\end{bmatrix}. \tag{11.23}$$

Equation (11.23) indicates that the superimposed point $B_1(B_2)$ shown on the left in Figure 11.4 has two translational degrees of freedom along the x- and

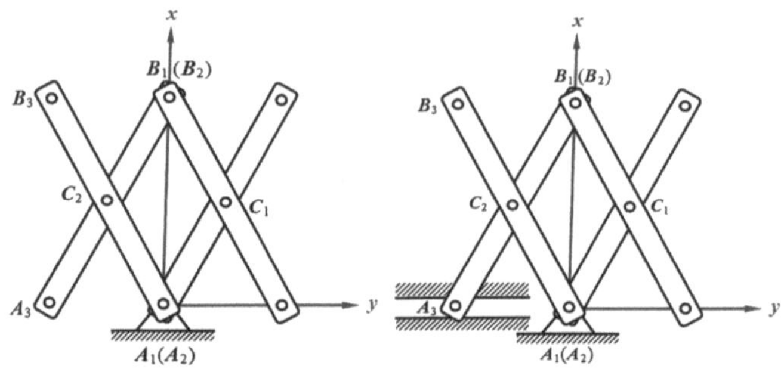

FIGURE 11.4 A planar structure consists of two SLEs.

y-axes, respectively. Now, a further problem arises—what will the free motion of the point $B_1(B_2)$ be if the slide shown on the right in Figure 11.4 is added to any point of the structure, A_3 for example?

For this case, one might find that the kinematic chain $A_2C_2B_2$ now becomes a hybrid one which consists of two sub kinematic chains, C_2A_2 and C_2A_3. If a slide with a direction of $(a \quad b \quad 0)$ (where $\sqrt{a^2+b^2}=1$) is introduced at point A_3. The twists of the link A_3C_2 jointed by the hybrid chain $B_2 <^{C_2A_2}_{C_2A_3}$ can be equivalently obtained with reciprocal screw theory.

Firstly, the twists of the PR chain at the superimposed point A_3 are:

$$\mathbf{J}_{A_S} = \begin{bmatrix} \$_{A_{3S}} & \$_{A_{3R}} \end{bmatrix} \tag{11.24}$$

where $\$_{A_{3S}} = \begin{pmatrix} 0 & 0 & 0 & a & b & 0 \end{pmatrix}^T, \$_{A_{3R}} = \begin{pmatrix} 0 & 0 & 1 & y_{A_3} & -x_{A_3} & 0 \end{pmatrix}^T$.

The terminal constraints exerted on A_3C_2 by the sub-chain are

$$\tau_1 = \begin{bmatrix} \$^{r1}_{A_3} & \$^{r2}_{A_3} & \$^{r3}_{A_3} & \$^{r4}_{A_3} \end{bmatrix} \begin{bmatrix} f_1 \\ f_2 \\ f_3 \\ f_4 \end{bmatrix}. \tag{11.25}$$

where $\$^{r1}_{A_3} = \begin{pmatrix} b & -a & 0 & 0 & 0 & -ax_{A_3}-by_{A_3} \end{pmatrix}^T, \$^{r2}_{A_3} = \begin{pmatrix} 0 & 0 & 1 & 0 & 0 & 0 \end{pmatrix}^T, \$^{r3}_{A_3} = \begin{pmatrix} 0 & 0 & 0 & 1 & 0 & 0 \end{pmatrix}^T, \$^{r4}_{A_3} = \begin{pmatrix} 0 & 0 & 0 & 0 & 1 & 0 \end{pmatrix}^T$, f_i $(i = 1,2,3,4)$ represent the tensities of the wrenches.

The twists of the RR chain at the superimposed point C_2 are

$$\mathbf{J}_{A_2C_2} = \begin{bmatrix} \$_{A_2} & \$_{C_2} \end{bmatrix} \tag{11.26}$$

where
$\$_{A_2} = \begin{pmatrix} 0 & 0 & 1 & y_{A_2} & -x_{A_2} & 0 \end{pmatrix}^T, \$_{C_2} = \begin{pmatrix} 0 & 0 & 1 & y_{C_2} & -x_{C_2} & 0 \end{pmatrix}^T$.

The terminal constraints exerted on A_3C_2 at the point C_2 are

$$\tau_2 = \begin{bmatrix} \$^{r1}_{C_2} & \$^{r2}_{C_2} & \$^{r3}_{C_2} & \$^{r4}_{C_2} \end{bmatrix} \begin{bmatrix} f_5 \\ f_6 \\ f_7 \\ f_8 \end{bmatrix} \tag{11.27}$$

where $\$^{r1}_{C_2} = \begin{pmatrix} x_{A_2} - x_{C_2} & y_{A_2} - y_{C_2} & 0 & 0 & 0 & y_{A_2}x_{C_2} - x_{A_2}y_{C_2} \end{pmatrix}^T$, $\$^{r2}_{C_2} = \begin{pmatrix} 0 & 0 & 1 & 0 & 0 & 0 \end{pmatrix}^T$, $\$^{r3}_{C_2} = \begin{pmatrix} 0 & 0 & 0 & 1 & 0 & 0 \end{pmatrix}^T$, $\$^{r4}_{C_2} = \begin{pmatrix} 0 & 0 & 0 & 0 & 1 & 0 \end{pmatrix}^T$, f_i $(i = 5,6,7,8)$ represent the tensities of the screws.

From Eqs. (11.25) and (11.27), we can find that the terminal constraints exerted on link A_3C_2 are

$$\tau_{A_3C_2} = \tau_1 + \tau_2 = \begin{bmatrix} \$^{r1}_{A_3} & \$^{r1}_{C_2} & \$^{r2}_{A_3} & \$^{r3}_{A_3} & \$^{r4}_{A_3} \end{bmatrix} \begin{bmatrix} f_1 \\ f_5 \\ f_2 + f_6 \\ f_3 + f_7 \\ f_4 + f_8 \end{bmatrix}. \tag{11.28}$$

Consequently, the twist of the link A_3C_2 can be solved with reciprocal screw theory

$$(\$_{A_3C_2})^T \boldsymbol{\Delta} \tau_{A_3C_2} = 0 \tag{11.29}$$

which can also be expressed by considering the arbitrariness of $f_i (i = 1,2,\cdots,8)$ in the following form

$$(\$^r_{A_3C_2})^T \boldsymbol{\Delta} \$_{A_3C_2} = 0 \tag{11.30}$$

where $\$^r_{A_3C_2} = \begin{bmatrix} \$^{r1}_{A_3} & \$^{r1}_{C_2} & \$^{r2}_{A_3} & \$^{r3}_{A_3} & \$^{r4}_{A_3} \end{bmatrix}$.

The twist of the link A_3C_2, denoted by $\$_{A_3C_2}$, is

$$\$_{A_3C_2} = k_1 \begin{pmatrix} 0 \\ 0 \\ 1 \\ \dfrac{\frac{a}{b}\left(x_{A_2}y_{C_2} - y_{A_2}x_{C_2} - x_{A_2}y_{A_3} + x_{C_2}y_{A_3}\right) - \left(\frac{a}{b}\right)^2 x_{A_3}\left(x_{A_2} - x_{C_2}\right)}{\frac{a}{b}\left(x_{A_2} - x_{C_2}\right) + \left(y_{A_2} - y_{C_2}\right)} + \frac{a}{b}x_{A_3} + y_{A_3} \\ \dfrac{\left(x_{A_2}y_{C_2} - y_{A_2}x_{C_2} - x_{A_2}y_{A_3} + x_{C_2}y_{A_3}\right) - \frac{a}{b}x_{A_3}\left(x_{A_2} - x_{C_2}\right)}{\frac{a}{b}\left(x_{A_2} - x_{C_2}\right) + \left(y_{A_2} - y_{C_2}\right)} \\ 0 \end{pmatrix}. \tag{11.31}$$

When $a = 0$ and $b = 1$, equation (11.31) can be simplified as

$$\$_{A_3C_2} = k_1 \begin{pmatrix} 0 \\ 0 \\ 1 \\ y_{A_3} \\ -\dfrac{x_{C_2} - x_{A_2}}{y_{C_2} - y_{A_2}}\left(y_{A_3} - y_{A_2}\right) - x_{A_2} \\ 0 \end{pmatrix}. \tag{11.32}$$

Equation (11.32) represents an instantaneous rotation about point B_3. Therefore, the twists of the superimposed point $B_1(B_2)$ jointed with the hybrid kinematic chain $B_2 <^{C_2A_2}_{C_2A_3}$ now become

$$\mathbf{J}_{A_3C_2B_2} = \begin{bmatrix} \$_{A_3C_2} & \$_{B_2} \end{bmatrix} \begin{bmatrix} k_1 \\ k_2 \end{bmatrix} \tag{11.33}$$

where $\$_{B_2} = \begin{pmatrix} 0 & 0 & 1 & y_{B_2} & -x_{B_2} & 0 \end{pmatrix}^T$, and k_1, k_2 denote the magnitudes of the twists.

The terminal constraints of the hybrid chain to the superimposed point $B_1(B_2)$ can be similarly obtained with reciprocal screw theory:

$$\tau_{(B_1B_2)2} = \begin{bmatrix} \$^{r1}_{B_2} & \$^{r2}_{B_2} & \$^{r3}_{B_2} & \$^{r4}_{B_2} \end{bmatrix} \begin{bmatrix} c_1 \\ c_2 \\ c_3 \\ c_4 \end{bmatrix} \tag{11.34}$$

where $\$^{r1}_{B_2} = \begin{pmatrix} x_{B_3} - x_{B_2} & y_{B_3} - y_{B_2} & 0 & 0 & 0 & y_{B_3}x_{B_2} - x_{B_3}y_{B_2} \end{pmatrix}^T$, $\$^{r2}_{B_2} = \begin{pmatrix} 0 & 0 & 1 & 0 & 0 & 0 \end{pmatrix}^T$, $\$^{r3}_{B_2} = \begin{pmatrix} 0 & 0 & 0 & 1 & 0 & 0 \end{pmatrix}^T$, $\$^{r4}_{B_2} = \begin{pmatrix} 0 & 0 & 0 & 0 & 1 & 0 \end{pmatrix}^T$, and $c_i (i = 1,2,3,4)$ represent the tensities of the wrenches.

The terminal constraints of the kinematic chain $A_1C_1B_1$ have been expressed by equation (11.15). Therefore, the free motions that the superimposed point $B_1(B_2)$ has are just those expressed with equation (11.33). From the above analysis, we have known that $\$_{B_2}$ represents a self-rotation about the superimposed point $B_1(B_2)$. Therefore, the visible motion of the point is depicted by equation (11.31) or (11.32). This is the case for the deployable gate shown in Figure 11.1.

From the above analysis, we can find that coplanar hinged SLEs have two visible translations, and, therefore, one guide should be added to the structure during deployment, which might be presented by Figure 11.1.

11.4 APPLICATIONS OF DEPLOYABLE STRUCTURE UNITS

Any two adjacent SLE planes of a deployable structure are either coplanar or subtended by a specified angle. The following sections will present a number of typical deployable structures utilized in engineering.

11.4.1 Flat Deployable Structures

From the above analysis, we can find that the two bi-joints connected by two adjacent SLEs can only make a relative translation along the common intersection of the two adjacent planes determined by the hinged SLEs of the bi-joints. Therefore, if we add a third SLE, and suppose that all links of the SLE are identical and intermediately jointed, and also let the subtended angles of every two planes be $\alpha = \frac{\pi}{3}$, the three links B_1B_2, B_3B_4, and B_5B_6 can make three translations along the three intersections l_1, l_2, and l_3, which can be obtained with a similar analysis process. If the three SLEs are identical and their joints are at the intermediate points of the links, the three intersecting lines l_1, l_2, and l_3 will be parallel, the three links B_1B_2, B_3B_4, and B_5B_6 will make one same translation with respect to A_1A_2, A_3A_4, and A_5A_6 synchronously. Therefore, the structure shown on the left in Figure 11.5 can be a unit of flat deployable structure. The top view of such a unit is shown on the right in Figure 11.5, which is an equilateral triangle.

To make a flat deployable structure, the multi-joints A_iA_{i+1} and B_iB_{i+1} ($i = 1,2,3$) should be manufactured as Figure 11.6 shows. The subtended angles of each adjacent two joint planes in the multi-joint shown in Figure 11.6 are identical and equal to $\frac{\pi}{3}$. The deployable structure hinged with two hexad joints is shown in Figure 11.7.

A similar procedure can be utilized to analyze a structure unit, when the subtended angle of two adjacent SLE planes is $\alpha = \frac{\pi}{2}$. The unit of such a structure hinged by two quaternion joints is shown in Figure 11.8.

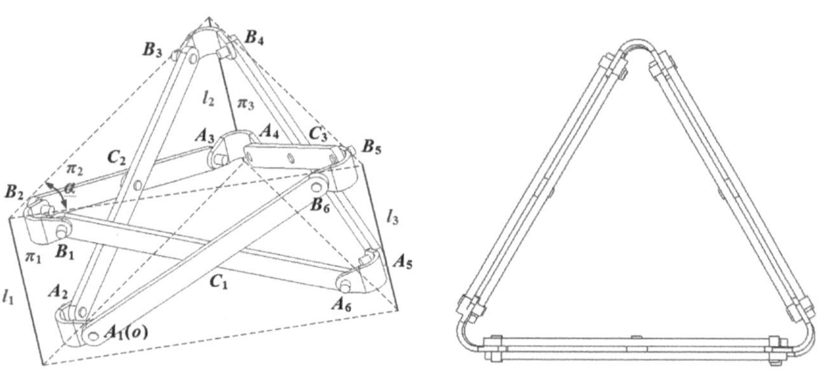

FIGURE 11.5 An equilateral triangle deployable structure unit.

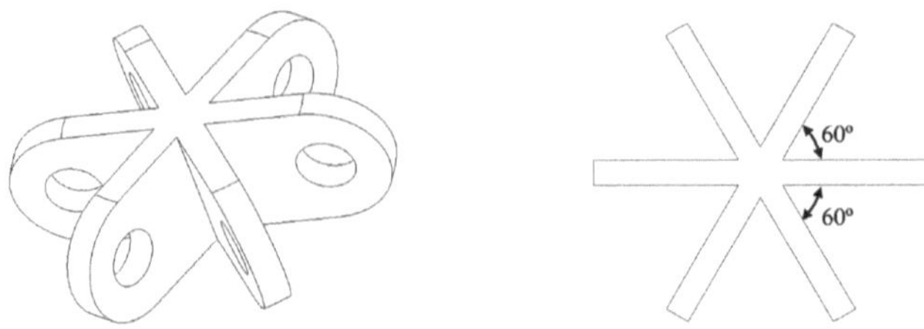

Left: the isotropic view of the joint Right: the top view of the joint

FIGURE 11.6 A hexad joint for the flat deployable structure of equilateral triangle.

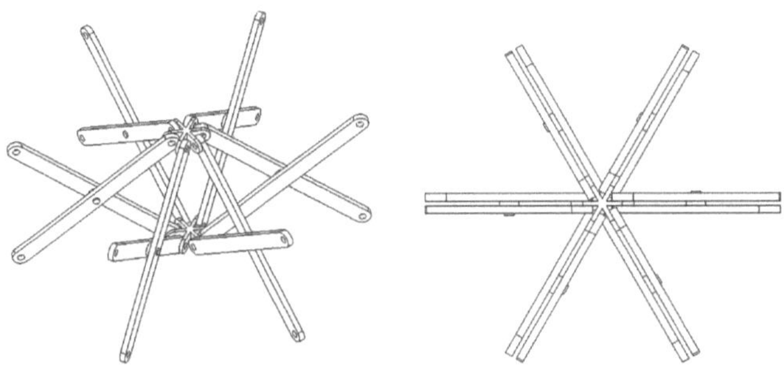

Left: the isotropic view of the structure unit Right: the top view of the structure unit

FIGURE 11.7 A hexad flat deployable structure unit.

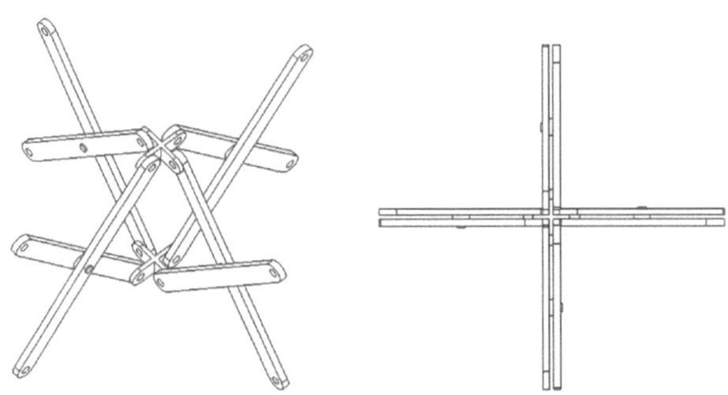

Left: the isotropic view of the structure unit Right: the top view of the structure unit

FIGURE 11.8 A quaternion flat deployable structure unit.

A deployable structure made up of quaternion structure units is shown in the left of Figure 11.2, the front, top, and left views of which are shown in Figure 11.9. When the subtended angle of two adjacent SLE planes is $\alpha = \frac{2\pi}{3}$, the unit of such a structure hinged by two trinary joints is shown in Figure 11.10.

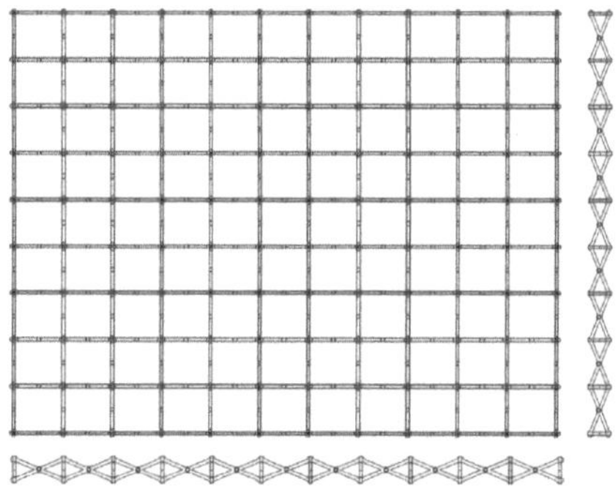

FIGURE 11.9 A square flat deployable structure.

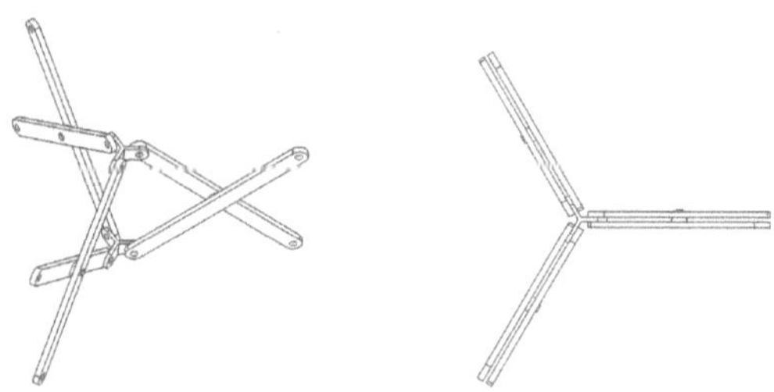

Left: the isotropic view of the structure unit Right: the top view of the structure unit

FIGURE 11.10 A trinary flat deployable structure unit.

11.4.2 Cylindrical and Spherical Deployable Structures

As is discussed in Section 11.3, the multi-joint should only have a free translation along the intersection of its directly hinged SLE planes. When the SLE planes remain coplanar and the multi-joint degenerates to be a point, the degenerate point has one relative translation with respect to the fixed joint if a guide is introduced to any other point of the structure. A cylindrical deployable structure consists of identical SLEs whose joints are at the middle of the links in the longitudinal direction while in the transactional direction, the SLEs are identical but not jointed at the middle point of the links, which is shown in Figure 11.11. Besides the general mechanism properties discussed above, the most obvious characteristics of the cylindrical deployable structure are that the projections of the transactional deployable structure are in radial directions from a same point, o, which is just the center of the dome arc.

The basic geometry constraint for a space deployable structure is that the configuration is able to be stored in a compact shape [9]. Therefore, we immediately have

$$l_i + k_i = l_{i+1} + k_{i+1} (i = 1,2,3, \cdots). \tag{11.35}$$

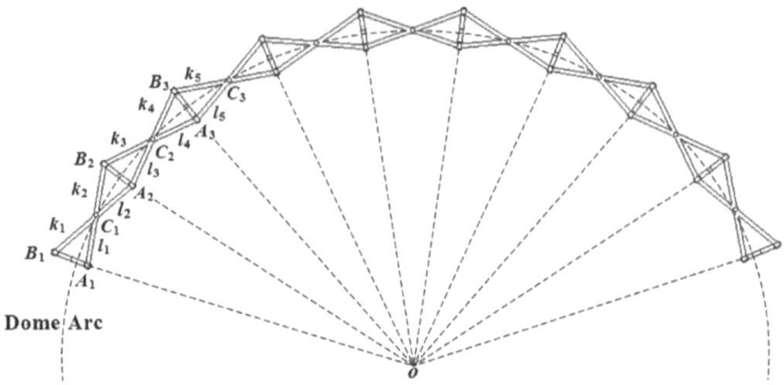

FIGURE 11.11 A cylindrical deployable structure.

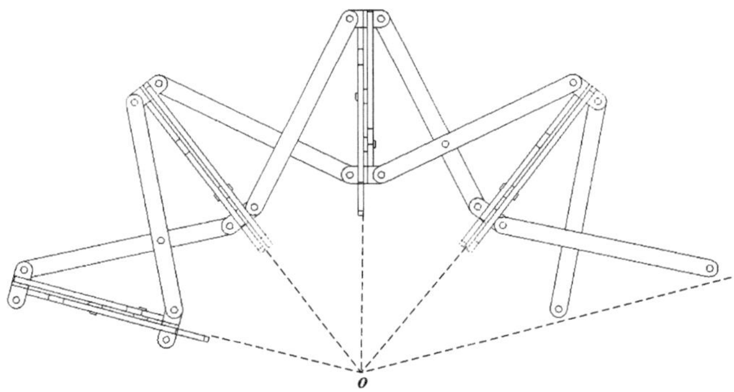

FIGURE 11.12 A big circle section of a spherical deployable structure.

For a spherical deployable structure, any big circle section is a radial deployable structure whose center is the sphere center which is shown in Figure 11.12. An isotropic view of a part of a spherical deployable structure is shown in Figure 11.13.

From the above analysis, we can find that in order to form a flat, cylindrical, or spherical deployable structure with SLEs, the pair of multi-joints connected by two or more adjacent SLEs should only make one relative translation along the common intersection of the adjacent planes determined by the hinged SLEs of the pair of multi-joints. If the SLEs remain in one plane, another guide should be added to make sure that the pair of pivots of any two SLEs only have one relative translation, which might be interpreted by the deployable gate structure shown in Figure 11.1.

11.5 CONCLUSIONS

This chapter addresses the mechanism theory and application of deployable structures based on SLEs. It starts from the mobility analysis of simple planar

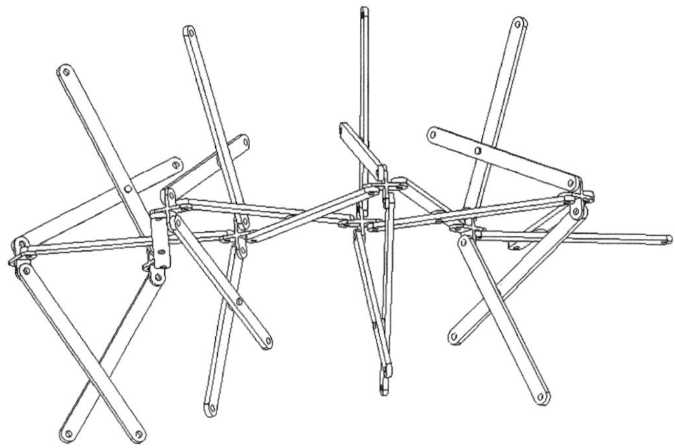

FIGURE 11.13 Part view of a spherical deployable structure.

linkages, and then investigates the mechanism theory of forming the spatial deployable units utilized in flat, cylindrical, and spherical deployable structures. For a deployable structure unit, the pair of multi-joints should have one relative translation along the common intersecting line of all planes determined by their hinged SLEs if these planes do not remain coplanar; however, the common hinges of coplanar adjacent SLEs should also have one relative translation when an additional guide is given. The theoretical results obtained in this chapter have the general adaptation for all kinds of deployable structures.

REFERENCES

[1] C. Hachem, E. Karni, A. Hanaor, Evaluation of biological deployable systems, International Journal of Space Structures 20 (2005) 189–200.

[2] F. Escrig, P. Valcarcel, Geometry of expandable space structures, International Journal of Space Structures 11 (1996) 257–274.

[3] W. Chen, G. Fu, et al., A new design conception for large span deployable flat grid structures, International Journal of Space Structures 17 (2002) 293–299.

[4] W.J. Chen, L.Y. Luo, et al., A study on space masts basedon octahedral truss family, International Journal of Space Structures 16 (2001) 75–82.

[5] A. Hanaor, R. Levy, Evaluation of deployable structures for space enclosures, International Journal of Space Structures 16 (2001) 211–229.

[6] Y. Chen, Z. You, T. Tarnai, Threefold-symmetric bricard linkages for deployable structures, International Journal of Solids and Structures 42 (2005) 2287–2301.

[7] T. Langbecker, Kinematic analysis of deployable scissor structures, International Journal of Space Structures 14 (1999) 1–15.

[8] Y. Rosenfeld, R.D. Logcher, New concepts for deployable collapsible structures, International Journal of Space Structures 3 (1988) 20–32.

[9] F. Escrig, J.P. Valcarcel, J. Sanchez, Deployable cover on a swimming pool in seville, Journal of the International Association for Shell and Spatial Structures 37 (1996) 39–70.

[10] C.J. Gantes, E. Konitopoulou, Geometric design of arbitrarily curved bi-stable deployable arches with discrete joint size, International Journal of Solids and Structures 41 (2004) 5517–5540.

[11] R.M. Murray, Z. Li, S.S. Sastry, A Mathematical Introduction to Robotic Manipulation, CRC Press, Boca Raton, 1994.
[12] K.H. Hunt, Kinematic Geometry of Mechanisms, Oxford University Press, Oxford, 1978.
[13] J. Phillips, Freedom in Machinery, Introducing Screw Theory, vol. 1, Cambridge University Press, Cambridge, 1984.
[14] J. Phillips, Freedom in machinery, Screw Theory Exemplified, vol. 2, Cambridge University Press, Cambridge, 1990.
[15] R.S. Ball, A Treatise on The Theory of Screws, Cambridge University Press, Cambridge, 1998.
[16] J.R. Phillips, K.H. Hunt, On the theorem of three axes in the spatial motion of three bodies, Australian Journal of Exerted Science 15 (1964) 267–287.
[17] K.J. Waldron, The constraint analysis of mechanisms, Journal of Mechanisms 1 (1966) 101–114, Pergamon Press.

Chapter 12

Structure Synthesis of Spatial Mechanisms

In mechanical engineering, a practical concept design often starts from the synthesis of a mechanism with the prescribed degrees of freedom (DOFs). Therefore, mechanism synthesis plays a vital role in real engineering. Alizade [1,2] and Gogu [3,4], respectively, addressed the structural synthesis of spatial parallel and hybrid manipulators with a revised Grübler–Kutzbach formula of DOF based on the algebraic summations of links, joints and the number of constraints induced by them. However, just as has been pointed out in previous chapters and [5–8], the restrictions of such methods are still an objective problem. This chapter will focus on the synthesis problem of a mechanism and its kinematic chains with the analytical process of DOFs of a mechanism.

In the history of mechanism synthesis, a significant example is that the creation of linkages to produce exact straight-line motion was an important engineering as well as a mathematical problem of the 19th century [9, 10]. While many engineers and mathematicians were searching for a 4-, 5-, or 6-bar straight-line linkage, all suffered from the fact that they could not attain such a motion. In the middle of 19th century, Peaucellier investigated an 8-bar linkage shown in Figure 12.1 and discovered he could generate an exact straight-line motion from a rotary input. This invention was recognized by several mathematicians as being very important to the design of general mathematical calculators [9, 10]. This eight-link linkage was the one of the first to produce exact straight-line motion and was independently invented by a French engineer named Peaucellier and by a Russian mathematician named Lipkin [9, 10], and is often called the Peaucellier–Lipkin eight-link linkage. From then on, many scholars probed the problem of attaining an exact straight-line motion only with links and revolute joints, and according to the Grübler–Kutzbach DOF criterion, they draw a conclusion that it is important to construct such a mechanism unless at least six links and seven revolute joints are used [11].

However, the above conclusion is just based on the traditional Grübler–Kutzbach formula of DOF. And therefore, it is partial. The famous Sarrus linkage is such an exception. The so-called Sarrus linkage, which is shown in Figure 12.2, is a linkage that converts circular to linear motion using a hinged square [12]. The whole square end effector C_1C_2 can make exact straight-line

Advanced Theory of Constraint and Motion Analysis for Robot Mechanisms.
http://dx.doi.org/10.1016/B978-0-12-420162-0.00012-6

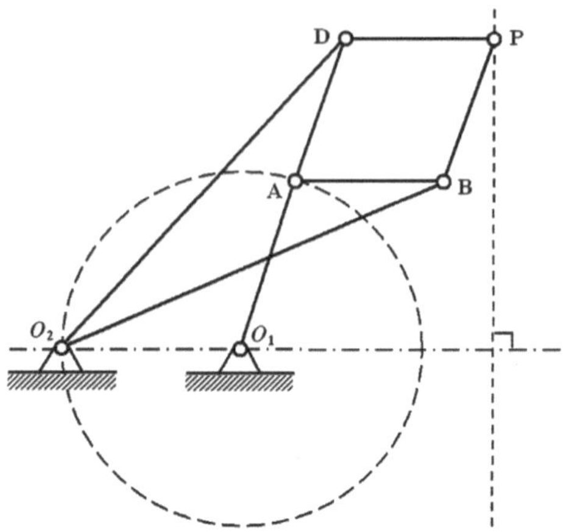

FIGURE 12.1 Structure of Peaucellier–Lipkin eight-link linkage.

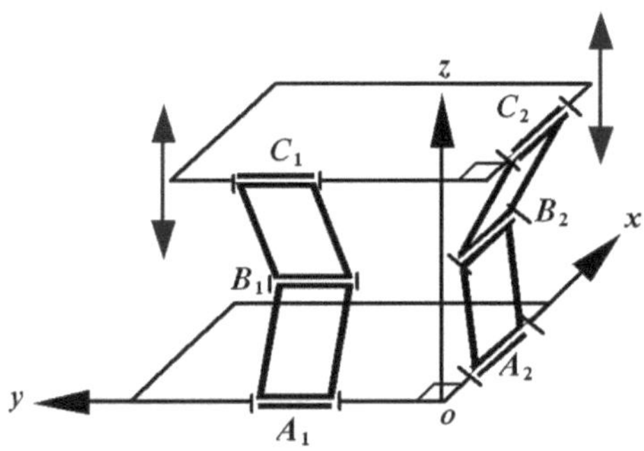

FIGURE 12.2 Structure of the Sarrus linkage.

translation along the z-axis which shows better properties both in mechanical structure and in kinematics than those of the Peaucellier–Lipkin eight-link linkage. Therefore, the restrictions of the traditional formulae of DOF are explicit. In the following, the general process of mechanism synthesis based on analytical methodology will be thoroughly proposed.

12.1 GENERAL PROCESS OF MECHANISM SYNTHESIS

As mentioned above, the Sarrus linkage just represents one way to achieve limited but exact straight-line motion from a combination of revolute joints. What is ridiculous is that Sarrus linkage has better structural and kinematic properties than those of Peaucellier–Lipkin linkage, while the former does not obey the traditional Grübler–Kutzbach formula of DOF and some of its reform forms. That is, the traditional formulae of DOF can by no means validly and robustly guide the mechanism synthesis. Hence, it is of primary importance to find out a valid way to correctly synthesize all kinds of mechanisms required in real engineering applications. This chapter explores this problem and finds that, similar to mechanism analysis, only analytical methods can fulfill such tasks for mechanism synthesis. In fact, the mechanism synthesis is an inverse process of the analytical approaches proposed in [5–8]. The main steps to analytically synthesize a mechanism with the specified motions are listed below.

Step 1: Express the free motions required for the prescribed end effector in a Cartesian coordinate system.

Although any analytical approach might be able to accomplish the work, this chapter strongly suggests here that the reciprocal screw theory should be applied for its succinct representation. Therefore, the Plücker coordinates of the specified motions should be expressed in a Cartesian coordinate system. This chapter supposes that the screws of the free motions of the end effector are denoted by $\mathbf{\$}^{\omega}_{End}$.

Step 2: Solve the constraint(s) applied to the end effector.

According to reciprocities between free motions and constraints of an end effector, the constraints applied to the end effector can be solved with the following equation if the free motions are expressed in Plücker coordinates:

$$(\mathbf{\$}^{\omega}_{End})^T \Delta \mathbf{\$}^{r}_{End} = 0 \tag{12.1}$$

where $\mathbf{\Delta} = \begin{bmatrix} \mathbf{0}_3 & \mathbf{I}_3 \\ \mathbf{I}_3 & \mathbf{0}_3 \end{bmatrix}$, $\mathbf{0}_3 = \begin{bmatrix} 0 & 0 & 0 \\ 0 & 0 & 0 \\ 0 & 0 & 0 \end{bmatrix}$, $\mathbf{I}_3 = \begin{bmatrix} 1 & 0 & 0 \\ 0 & 1 & 0 \\ 0 & 0 & 1 \end{bmatrix}$ and $\mathbf{\$}^{r}_{End}$ denotes any constraint applied to the end effector.

Step 3: Decide the number of kinematic chains, $m(m \geq 1)$, that will be used to connect the end effector with the fixed base.

The mechanism will be a single open chain manipulator if only one kinematic chain is introduced to the end effector; and the mechanism will be a closed chain manipulator provided that more than one kinematic chains were used to connect the end effector with the fixed base.

Step 4: Synthesize the terminal constraint(s) of each kinematic chain.

Suppose

$$\mathbf{\$}^{r}_{End} = \begin{bmatrix} \mathbf{\$}^{r1}_{End} & \mathbf{\$}^{r2}_{End} & \cdots & \mathbf{\$}^{rn}_{End} \end{bmatrix} \tag{12.2}$$

where n indicates the dimension of the terminal constraint basis of the end effector.

Suppose that the terminal constraint(s) of the $i^{\text{th}}(i = 1,2,\ldots,m)$ kinematic chain is denoted by $\mathbf{\$}^{r}_{i}$, the terminal constraint(s) of the chain can be synthesized with:

$$\mathbf{\$}^{r}_{i} = \mathbf{\$}^{r}_{End}\mathbf{K}_i \tag{12.3}$$

where $\mathbf{K}_i = [\boldsymbol{K}_{i1} \quad \boldsymbol{K}_{i2} \quad \cdots \quad \boldsymbol{K}_{in_i}]$, and $\boldsymbol{K}_{ij} = (k_{i1} \quad k_{i2} \quad \cdots \quad k_{in})^T$ and $j = 1,2,\ldots,n_i$.

For a feasible mechanism, if the end effector only has the required free motions, the result terminal constraints of all these m kinematic chains, $\bigcup_{i=1}^{m} \$_i^r$ *should be equivalent to* $\$_{End}^r$. **This is called the construction criterion 1 of the feasible kinematic chains.**

The necessity and sufficiency of this criterion can be directly deduced from equation (12.1) with linear algebra theory.

Step 5: Solve the twist basis of the ith kinematic chain with the terminal constraints, $\$_i^r$, synthesized in step 4.

With reciprocal screw theory, a basis of the twist(s) of the ith kinematic chain, denoted by $\$_{B_i}^{\omega}$, can be obtained by solving the following equation:

$$(\$_i^r)^T \boldsymbol{\Delta} \$_{B_i}^{\omega} = 0 \tag{12.4}$$

where $\$_i^r$ represents the terminal constraint(s) of the ith kinematic chain synthesized in step 4.

Step 6: Synthesize the twist(s) of the ith kinematic chain with the twist basis of the ith chain, $\$_{B_i}^{\omega}$, obtained in step 5.

Suppose

$$\$_{B_i}^{\omega} = \left[\$_{B_i}^{\omega 1} \quad \$_{B_i}^{\omega 2} \quad \cdots \quad \$_{B_i}^{\omega n_i} \right] \tag{12.5}$$

where n_i indicates the dimension of the twist basis of the ith $(i = 1,2,\ldots,m)$ kinematic chain.

According to linear algebra, any twist of the ith kinematic chain can be expressed as the linear combinations of the twist basis of the chain:

$$\$_i^{\omega a} = \$_{B_i}^{\omega} \boldsymbol{c}_i \tag{12.6}$$

where $\boldsymbol{c}_i = (c_1 \quad c_2 \quad \cdots \quad c_{n_i})^T$.

Consequently, the twists of each kinematic chain can be synthesized through equation (12.6). *However, in order to keep the twists of the chain equivalent to the twist basis of the* ith *chain, the rank of the total twists synthesized through equation* (12.6) *should equal the dimension of the twist basis of the chain.* **This is called the construction criterion 2 of the feasible kinematic chains.**

The necessity and sufficiency of this criterion can be immediately obtained from equation (12.4). According to construction criteria 1 and 2, the required synthesis target of a mechanism can be gradually accomplished with the above six steps. Obviously, with these six steps, a different person might synthesize different kinematic chains and different mechanisms. However, all the end effectors of the mechanisms synthesized with the same criteria will surely have the same specified free motion(s).

12.2 ANALYSIS OF STRUCTURE SYNTHESIS

This section will further discuss the procedure of structure synthesis through some representative examples.

12.2.1 Synthesis of Lower Limbs of a Humanoid Robot

According to the biologic kinematics, the lower limb of human being can be roughly simplified to the architecture shown in Figure 12.3. The kinematic chain of the lower limb of a human being is made up of one revolute joint and two spherical joints in series. To investigate the kinematic chains with screw theory, we should set a coordinate system. Because the analysis results are independent with the selection of coordinate system, we can choose a coordinate system such that most of the screw coordinates can have the simplest expressions.

Suppose that the origin of the coordinate system is superimposed with the center of the hip joint (spherical joint A), the x-axis is parallel to the axis of the knee joint (revolute joint B), the z-axis is along line AB, and the y-axis can be determined naturally. In this coordinate system, we can assume the coordinates of joint A, B, and C to be $A\begin{pmatrix}0 & 0 & 0\end{pmatrix}, B\begin{pmatrix}0 & 0 & z_B\end{pmatrix}$, and $C\begin{pmatrix}0 & y_C & z_C\end{pmatrix}$, respectively.

The screw matrix, $\$_{ABC}$, includes:

(1) The twists of the spherical join A are

$$\$_1 = [1 \quad 0 \quad 0 \quad 0 \quad 0 \quad 0]^T, \$_2 = [0 \quad 1 \quad 0 \quad 0 \quad 0 \quad 0]^T,$$
$$\$_3 = [0 \quad 0 \quad 1 \quad 0 \quad 0 \quad 0]^T;$$

(2) The twist of the revolute joint B is

$$\$_4 = [1 \quad 0 \quad 0 \quad 0 \quad z_B \quad 0]^T;$$

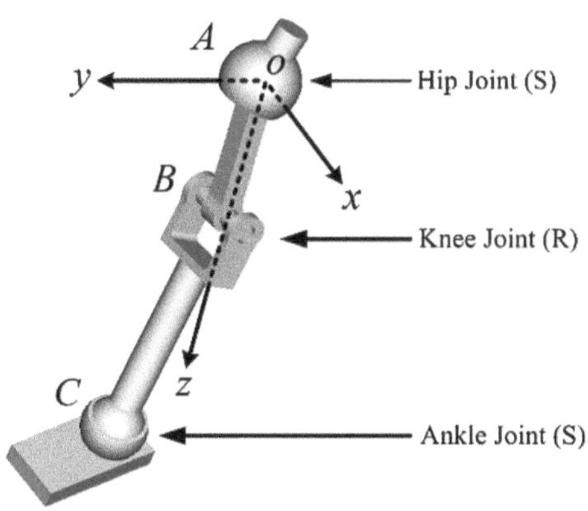

FIGURE 12.3 Rough architecture (SRS) for the lower limb of a human being.

(3) The twists of the spherical joint C are

$$\$_5 = [1 \ \ 0 \ \ 0 \ \ 0 \ \ z_C \ \ -y_C]^T, \$_6 = [0 \ \ 1 \ \ 0 \ \ -z_C \ \ 0 \ \ 0]^T,$$
$$\$_7 = [0 \ \ 0 \ \ 1 \ \ y_C \ \ 0 \ \ 0]^T.$$

Therefore,

$$\$_{ABC} = \begin{bmatrix} 1 & 0 & 0 & 1 & 1 & 0 & 0 \\ 0 & 1 & 0 & 0 & 0 & 1 & 0 \\ 0 & 0 & 1 & 0 & 0 & 0 & 1 \\ 0 & 0 & 0 & 0 & 0 & -z_C & y_C \\ 0 & 0 & 0 & z_B & z_C & 0 & 0 \\ 0 & 0 & 0 & 0 & -y_C & 0 & 0 \end{bmatrix}. \tag{12.7}$$

In general cases, we have:

$$\text{Rank}(\$_{ABC}) = 6. \tag{12.8}$$

Theoretically, we can select any twists whose maximum linearly independent set covers that of $\$_{ABC}$, but the rank is not larger than Rank($\$_{ABC}$). To simplify the manufacturing process and save processing costs, we need to select or combine the maximum linearly independent set of $\$_{ABC}$. In what follows, we will discuss and compare the possible simplifications and the architecture schemes of the lower limb of a humanoid robot.

1. When the maximal linearly independent set are selected as

$$\$_{ABC}^{M_1} = [\$_1 \ \ \$_2 \ \ \$_3 \ \ \$_4 \ \ \$_5 \ \ \$_6] \tag{12.9}$$

the kinematic chain of the lower limb of a human being can be designed as Figure 12.4 or Figure 12.5 shows.

In the case of Figure 12.4, $\$_{ABC}^{M_1}$ can be divided as:

(1) The twists of the hip joint (spherical join A) are

$$\$_1 = [1 \ \ 0 \ \ 0 \ \ 0 \ \ 0 \ \ 0]^T, \$_2 = [0 \ \ 1 \ \ 0 \ \ 0 \ \ 0 \ \ 0]^T,$$
$$\$_3 = [0 \ \ 0 \ \ 1 \ \ 0 \ \ 0 \ \ 0]^T;$$

(2) The twists of the knee joint (revolute joint B) are

$$\$_4 = [1 \ \ 0 \ \ 0 \ \ 0 \ \ z_B \ \ 0]^T;$$

(3) The twists of the ankle joint (universal joint C) are

$$\$_5 = [1 \ \ 0 \ \ 0 \ \ 0 \ \ z_C \ \ -y_C]^T, \$_6 = [0 \ \ 1 \ \ 0 \ \ -z_C \ \ 0 \ \ 0]^T.$$

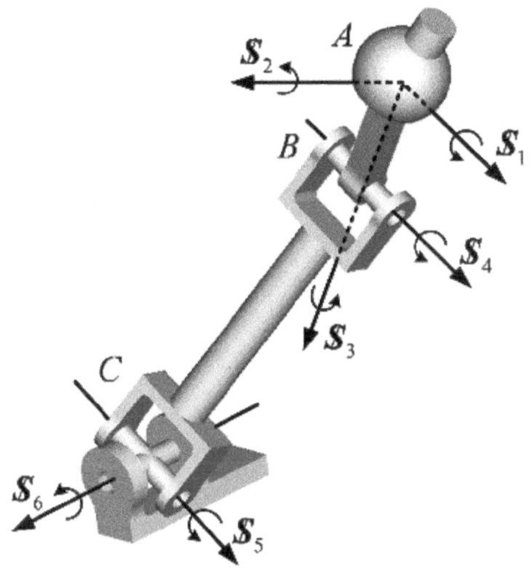

FIGURE 12.4 First SRU architecture for the lower limb of a humanoid robot.

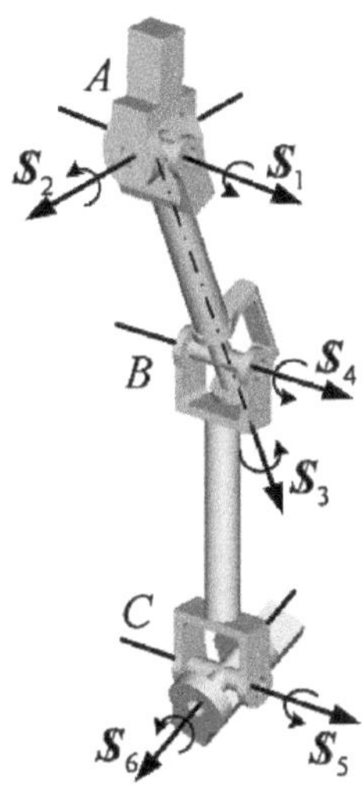

FIGURE 12.5 First 3U architecture for the lower limb of a humanoid robot.

In the case of Figure 12.5, $\$_{ABC}^{M_1}$ can be divided as:

(1) The twists of the hip joint (universal join A) are

$$\$_1 = [1 \quad 0 \quad 0 \quad 0 \quad 0 \quad 0]^T, \$_2 = [0 \quad 1 \quad 0 \quad 0 \quad 0 \quad 0]^T;$$

(2) The twists of the knee joint (universal joint B) are

$$\$_3 = [0\ \ 0\ \ 1\ \ 0\ \ 0\ \ 0]^T, \$_4 = [1\ \ 0\ \ 0\ \ 0\ \ z_B\ \ 0]^T;$$

(3) The twists of the ankle joint (universal joint C) are

$$\$_5 = [1\ \ 0\ \ 0\ \ 0\ \ z_C\ \ -y_C]^T, \$_6 = [0\ \ 1\ \ 0\ \ -z_C\ \ 0\ \ 0]^T.$$

2. When the maximal linearly independent set are selected as:

$$\$_{ABC}^{M_2} = [\$_1\ \ \$_2\ \ \$_3\ \ \$_4\ \ \$_5\ \ \$_7] \tag{12.10}$$

the kinematic chain can be designed as Figure 12.4 or Figure 12.5 shows.
In the case of Figure 12.6, $\$_{ABC}^{M_2}$ can be divided as:

(1) The twists of the hip joint (spherical join A) are

$$\$_1 = [1\ \ 0\ \ 0\ \ 0\ \ 0\ \ 0]^T, \$_2 = [0\ \ 1\ \ 0\ \ 0\ \ 0\ \ 0]^T,$$
$$\$_3 = [0\ \ 0\ \ 1\ \ 0\ \ 0\ \ 0]^T;$$

(2) The twists of the knee joint (revolute joint B) are

$$\$_4 = [1\ \ 0\ \ 0\ \ 0\ \ z_B\ \ 0]^T;$$

(3) The twists of the ankle joint (universal joint C) are

$$\$_5 = [1\ \ 0\ \ 0\ \ 0\ \ z_C\ \ -y_C]^T, \$_7 = [0\ \ 0\ \ 1\ \ y_C\ \ 0\ \ 0]^T.$$

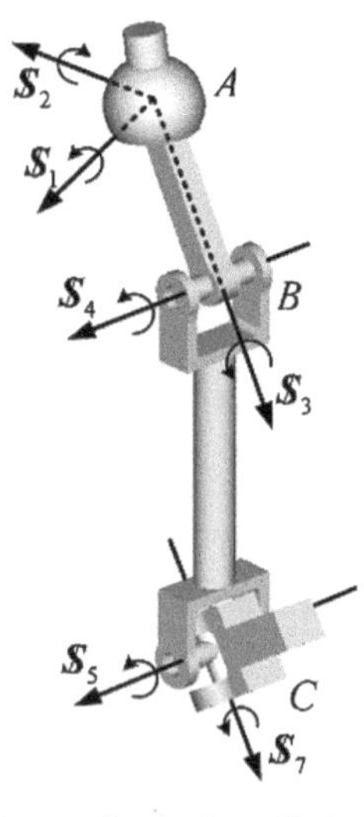

FIGURE 12.6 Second SRU architecture for the lower limb of a humanoid robot.

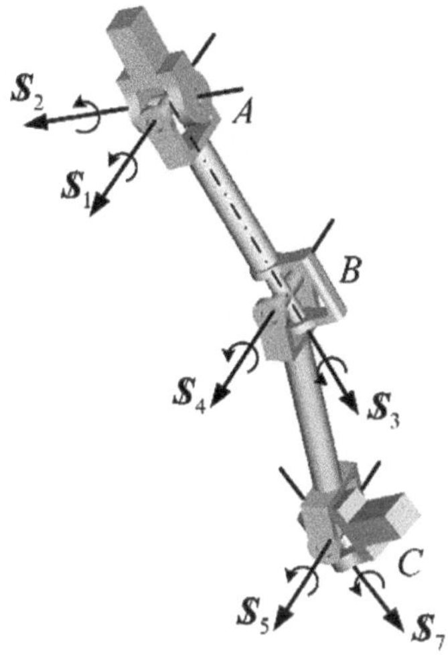

FIGURE 12.7 Second 3U architecture for the lower limb of a humanoid robot.

In the case of Figure 12.7, $\$^{M_2}_{ABC}$ can be divided as:

(1) The twists of the hip joint (universal join A) are

$$\$_1 = [1 \quad 0 \quad 0 \quad 0 \quad 0 \quad 0]^T, \$_2 = [0 \quad 1 \quad 0 \quad 0 \quad 0 \quad 0]^T;$$

(2) The twists of the knee joint (universal joint B) are

$$\$_3 = [0 \quad 0 \quad 1 \quad 0 \quad 0 \quad 0]^T, \$_4 = [1 \quad 0 \quad 0 \quad 0 \quad z_B \quad 0]^T;$$

(3) The twists of the ankle joint (universal joint C) are

$$\$_5 = [1 \quad 0 \quad 0 \quad 0 \quad z_C \quad -y_C]^T, \$_7 = [0 \quad 0 \quad 1 \quad y_C \quad 0 \quad 0]^T.$$

3. When the maximal linearly independent set are selected as:

$$\$^{M_3}_{ABC} = [\$_1 \quad \$_2 \quad \$_3 \quad \$_4 \quad \$_5 \quad \$_{6,7}] \qquad (12.11)$$

where $\$_{6,7}$ are the linear combinations of $\$_6$ and $\$_7$, in the case of Figure 12.8, we have:

$$\$_{6,7} = \$_6 \sin\alpha + \$_7 \cos\alpha$$

that is,

$$\$_{6,7} = [0 \quad \sin\alpha \quad \cos\alpha \quad y_C\cos\alpha - z_C\sin\alpha \quad 0 \quad 0]^T$$

where α is the angle from the thigh to the leg, the kinematic chain can be designed as Figure 12.8 shows.

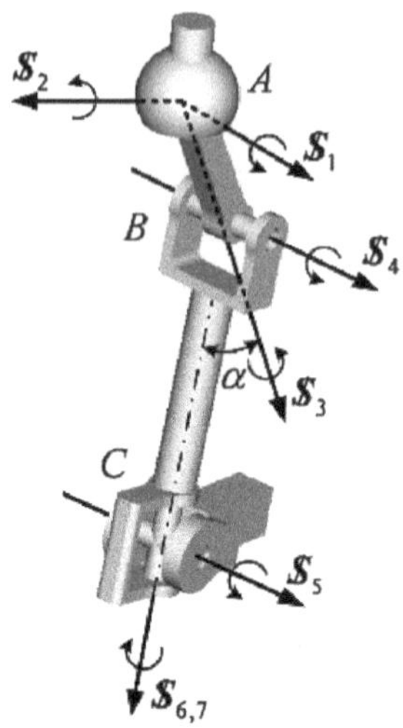

FIGURE 12.8 Third SRU architecture for the lower limb of a humanoid robot.

In the case of Figure 12.8, $\$_{ABC}^{M_3}$ can be divided as:

(1) The twists of the hip joint (spherical join A) are

$$\$_1 = [1 \quad 0 \quad 0 \quad 0 \quad 0 \quad 0]^T, \$_2 = [0 \quad 1 \quad 0 \quad 0 \quad 0 \quad 0]^T,$$
$$\$_3 = [0 \quad 0 \quad 1 \quad 0 \quad 0 \quad 0]^T;$$

(2) The twists of the knee joint (revolute joint B) are

$$\$_4 = [1 \quad 0 \quad 0 \quad 0 \quad z_B \quad 0]^T;$$

(3) The twists of the ankle joint (universal joint C) are

$$\$_5 = [1 \quad 0 \quad 0 \quad 0 \quad z_C \quad -y_C]^T,$$
$$\$_{6,7} = [0 \quad \sin\alpha \quad \cos\alpha \quad y_C\cos\alpha - z_C\sin\alpha \quad 0 \quad 0]^T.$$

4. When the maximal linearly independent set are selected as:

$$\$_{ABC}^{M_4} = [\$_1 \quad \$_2 \quad \$_{2,3,7} \quad \$_4 \quad \$_5 \quad \$_6] \tag{12.12}$$

where $\$_{2,3,7}$ are the linear combinations of $\$_2$, $\$_3$, and $\$_7$, in the case of Figure 12.9, we have:

$$\$_{2,3,7} = -\$_2\sin\gamma - \$_3\cos\gamma + \frac{z_B}{y_C}(\$_7 - \$_3)\sin\gamma,$$

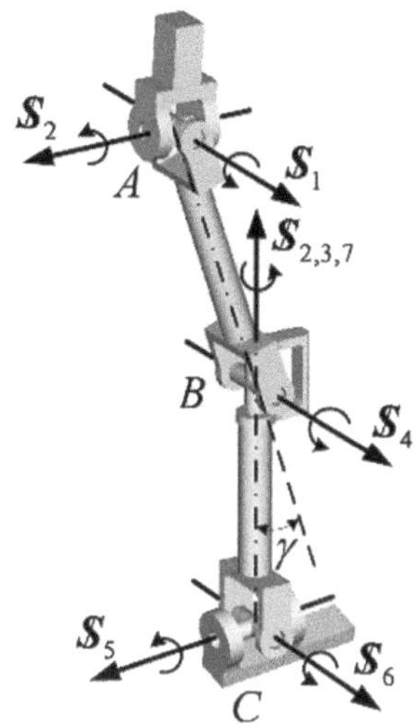

FIGURE 12.9 Third 3U architecture for lower limb of humanoid robot.

that is,

$$\$_{2,3,7} = [0 \quad -\sin\gamma \quad -\cos\gamma \quad z_B \sin\gamma \quad 0 \quad 0]^T,$$

where γ is the angle from the thigh to the leg, the kinematic chain can be designed as Figure 12.9 shows.

In the case of Figure 12.9, $\$^{M_4}_{ABC}$ can be divided as:

(1) The twists of the hip joint (universal join A) are

$$\$_1 = [1 \quad 0 \quad 0 \quad 0 \quad 0 \quad 0]^T, \$_2 = [0 \quad 1 \quad 0 \quad 0 \quad 0 \quad 0]^T;$$

(2) The twists of the knee joint (universal joint B) are

$$\$_{2,3,7} = [0 \quad -\sin\gamma \quad -\cos\gamma \quad z_B \sin\gamma \quad 0 \quad 0]^T,$$
$$\$_4 = [1 \quad 0 \quad 0 \quad 0 \quad z_B \quad 0]^T;$$

(3) The twists of the ankle joint (universal joint C) are

$$\$_5 = [1 \quad 0 \quad 0 \quad 0 \quad z_C \quad -y_C]^T, \$_6 = [0 \quad 1 \quad 0 \quad -z_C \quad 0 \quad 0]^T.$$

5. When the maximal linearly independent set are selected as:

$$\$^{M_5}_{ABC} = [\$_1 \quad \$_2 \quad \$_{2,3,6} \quad \$_4 \quad \$_5 \quad \$_{6,7}] \qquad (12.13)$$

where, $\$_{2,3,6}$ are the linear combinations of $\$_2$,$\$_3$ and $\$_6$, in the case of Figure 12.10, there are:

$$\$_{2,3,6} = \$_2 \sin\psi - \$_3 \cos\psi + \frac{z_B}{z_C}(\$_6 - \$_2)\sin\psi,$$

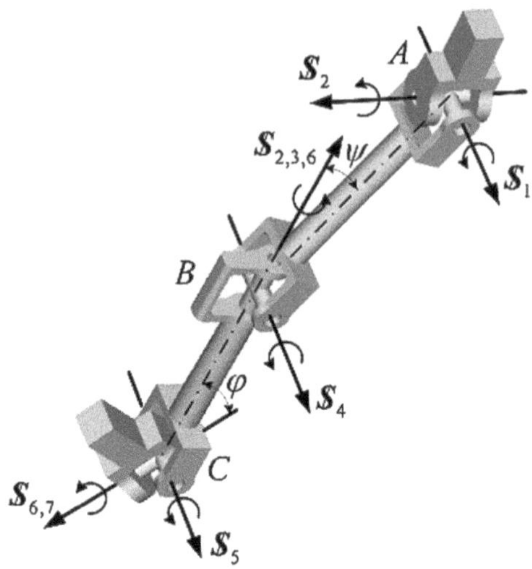

FIGURE 12.10 Fourth 3U architecture for the lower limb of a humanoid robot.

that is,

$$\$_{2,3,6} = [0 \quad \sin\psi \quad -\cos\psi \quad -z_B \sin\psi \quad 0 \quad 0]^T,$$

where ψ is the angle from the thigh to the leg; $\$_{6,7}$ are the linear combinations of $\$_6$ and $\$_7$, in the case of Figure 12.10, we have:

$$\$_{6,7} = \$_7 \cos(\varphi - \psi) + \$_6 \sin(\varphi - \psi),$$

that is,

$$\$_{6,7} = [0 \quad \sin(\varphi - \psi) \quad \cos(\varphi - \psi) \quad y_C \cos(\varphi - \psi) - z_C \sin(\varphi - \psi) \quad 0 \quad 0]^T$$

where ϕ is the angle from the side axis of foot to the leg, the kinematic chain can be designed as Figure 12.10 shows.

In the case of Figure 12.10, $\$_{ABC}^{M_5}$ can be divided as:

(1) The twists of the hip joint (spherical join A) are

$$\$_1 = [1 \quad 0 \quad 0 \quad 0 \quad 0 \quad 0]^T, \$_2 = [0 \quad 1 \quad 0 \quad 0 \quad 0 \quad 0]^T;$$

(2) The twists of the knee joint (revolute joint B) are

$$\$_{2,3,6} = [0 \quad \sin\psi \quad -\cos\psi \quad -z_B \sin\psi \quad 0 \quad 0]^T,$$
$$\$_4 = [1 \quad 0 \quad 0 \quad 0 \quad z_B \quad 0]^T;$$

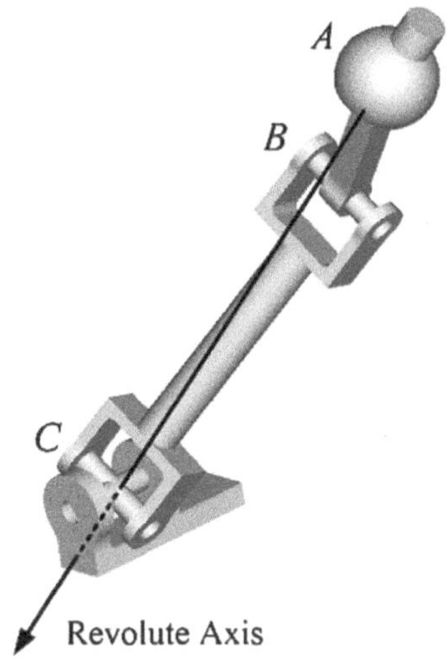

FIGURE 12.11 First SRU architecture for lower limb of a humanoid robot when turning (Case 1).

(3) The twists of the ankle joint (spherical joint *C*) are

$$\$_5 = [1 \quad 0 \quad 0 \quad 0 \quad z_C \quad -y_C]^T,$$
$$\$_{6,7} = [0 \quad \sin(\varphi - \psi) \quad \cos(\varphi - \psi) \quad y_C \cos(\varphi - \psi) - z_C \sin(\varphi - \psi) \quad 0 \quad 0]^T.$$

Of course, when we select or combine other kinds of maximum linearly independent screw sets, we can create the new corresponding kinematic chains. For the sake of brevity, we do not list all of the possible combinations. In the following section, we will discuss the best ones within the above architecture schemes.

A well-modeled humanoid robot should have both the same mobility as the human being and have the merits of feasibility of control and manufacturing, and low processing costs. Considering these requirements, we can find that the architecture scheme of Figure 12.4, Figure 12.5, and Figure 12.9 might be acceptable, because the ankle joints of these schemes can directly execute three orthogonal revolute motions shown in Figure 12.11, Figure 12.12, and Figure 12.13, respectively, while the other cases can not. Besides, the joints of these three schemes are all low pairs which are easily manufactured.

In addition, we can find that when the foot is rotating around the crus, the drawbacks of the architecture scheme in Figure 12.12 are obvious—the locus of the foot is a circle. Although the foot loci of the cases in Figure 12.11 and Figure 12.13 are all point and the case of Figure 12.11 is also widely adopted by many robot engineers, the feasibility of control and manufacturing for spherical joint A in Figure 12.11 or its equivalent substitutions (three revolute joints) are not as good as those of the universal joint A in Figure 12.13. The joints in the case of Figure 12.13 are all universal joints, which are easily controlled and manufactured with adequate precision and low cost.

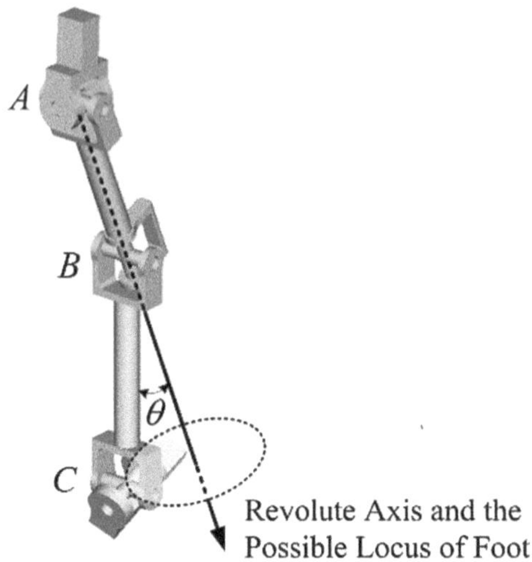

FIGURE 12.12 Third 3U architecture for lower limb of a humanoid robot when turning (Case 2).

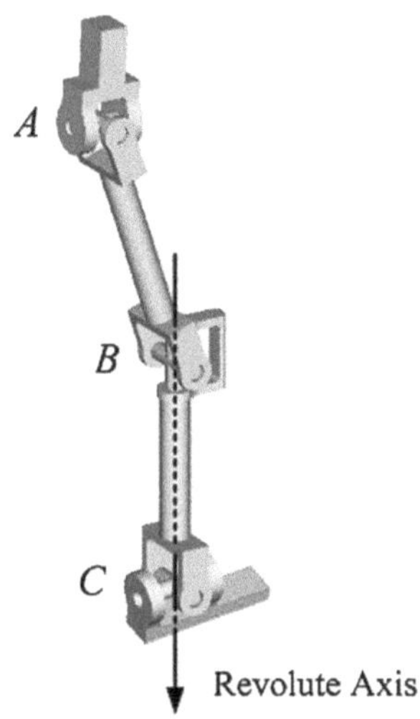

FIGURE 12.13 Third 3U architecture for lower limb of a humanoid robot when turning (Case 3).

12.2.2 Synthesis of an Independent Suspension of Invariable Alignment Parameter

A suspension is a group of equipments that connect the wheel and the vehicle body. Its major functions are to determine the track of the wheel during jumping up and down and to reduce the shock to the vehicle body resulting from the forces and moments induced by uneven roads. Therefore, it plays important roles in handling ability, stability, and comfort.

As the motions of the end effector of a mechanism are guided by its kinematic chains, we will first focus on the synthesis of kinematic chains of a mechanism and then analyze an independent suspension with flexural links of anisotropic stiffness.

12.2.2.1 The Simple Kinematic Chains for Full Planar Motions

The full planar motions are defined as the possibility of translations in any two directions of a plane and one rotation about the normal of the plane. A rigid end effector shown in Figure 12.14 is subjected to a serial kinematic chain, denoted by AB, which consists of a number of single DOF revolute joints $A, J_1, J_2, \ldots J_n, B$. Suppose the end effector only has full planar motions within the yoz-plane. So the kinematic chain only provides planar terminal constraints which can be expressed in screws below

$$\$_C = \left[\$_{C_1} \quad \$_{C_2} \quad \$_{C_3}\right] \tag{12.14}$$

where $\$_C$ represents the constraint screw matrix of the chain exerted on the end effector, $\$_{C_1} = [1 \ \ 0 \ \ 0 \ \ 0 \ \ 0 \ \ 0]^T$ which represents a force constraint perpendicular to the plane yoz, $\$_{C_2} = [0 \ \ 0 \ \ 0 \ \ 0 \ \ 1 \ \ 0]^T$ which denotes a moment constraint about the y-axis and $\$_{C_3} = [0 \ \ 0 \ \ 0 \ \ 0 \ \ 0 \ \ 1]^T$ which denotes a moment constraint about the z-axis.

Consequently, the kinematic screws for such terminal constraints represented in equation (12.8) should satisfy the following equation [5,6]:

$$\$_C^T \mathbf{\Delta} \$_F = 0 \tag{12.15}$$

where $\mathbf{\Delta} = \begin{bmatrix} 0 & \mathbf{I} \\ \mathbf{I} & 0 \end{bmatrix}$ and $\mathbf{I} = \begin{bmatrix} 1 & 0 & 0 \\ 0 & 1 & 0 \\ 0 & 0 & 1 \end{bmatrix}$, $\$_F$ represents the screw matrix of free motions allowed by the constraints denoted expressed by $\$_C$.

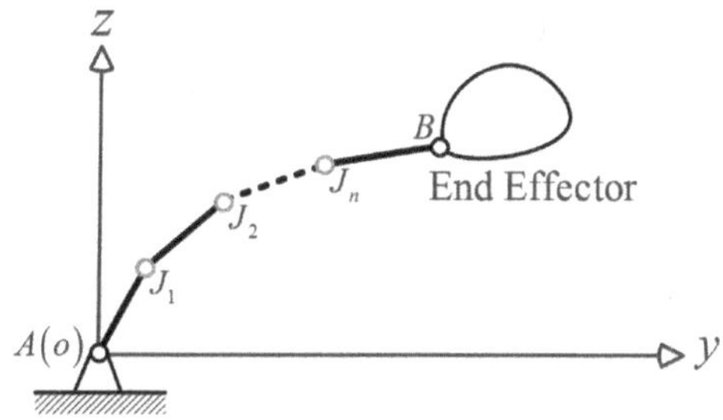

FIGURE 12.14 End effector via a planar kinematic chain.

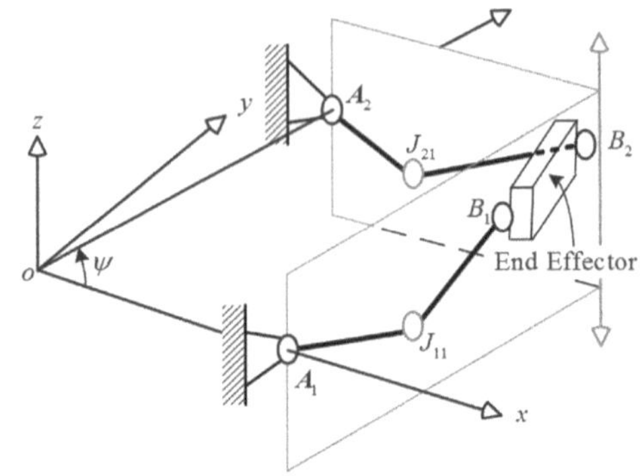

FIGURE 12.15 Coordinate frame and the rigid mechanism.

Substituting equation (12.14) into equation (12.15), we obtain a set of base screws of the free motions:

$$\$_F = [\$_{F_1} \quad \$_{F_2} \quad \$_{F_3}] \tag{12.16}$$

where $\$_F$ denotes the screw matrix of free motions of the end effector, $\$_{F_1} = [1 \quad 0 \quad 0 \quad 0 \quad 0 \quad 0]^T$ which represents a base screw of pure rotation about the x-axis, $\$_{F_2} = [0 \quad 0 \quad 0 \quad 0 \quad 1 \quad 0]^T$ which represents a base screw of pure translation along the y-axis, and $\$_{F_3} = [0 \quad 0 \quad 0 \quad 0 \quad 0 \quad 1]^T$ which represents a base screw of pure translation along the z-axis.

Equation (12.16) indicates that at least three single DOF joints are needed to generate the full planar motions for the end effector. However, the prismatic joint is not fit for passively actuated occasions because of its possible violation of the translation requirements under loads in applications. Therefore, the following discussion will focus on the kinematic chains that only consist of single DOF revolute joints.

Suppose that an end effector shown in Figure 12.15 is connected with the fixed base through two RRR kinematic chains. The axes of revolute joints A_1 and A_2 on the vehicle body are intersected at point o. For simplicity, o is supposed to be the origin of the coordinate system, x-axis is assumed to be the axis of joints A_1 and the xoy-plane is determined by the axes of joints A_1 and A_2, z-axis is determined by the right-hand rule. Suppose the subtended angle of axes A_1 and A_2 is ψ $(0° < \psi < 180°)$, the distance from joint A_1 to the origin, o, is a and the distance from joint A_2 to the origin is b. So the coordinates of joints A_1 and A_2 are $(a \quad 0 \quad 0)$ and $(b\cos\psi \quad b\sin\psi \quad 0)$, individually. Further, we can suppose the coordinates of the other joints in the two kinematic chains $A_1J_{11}B_1$ and $A_2J_{21}B_2$ are $J_{11}(a \quad y_{J_{11}} \quad z_{J_{11}})$, $B_1(a \quad y_{B_1} \quad z_{B_1})$, $J_{21}(x_{J_{21}} \quad y_{J_{21}} \quad z_{J_{21}})$, and $B_2(x_{B_2} \quad y_{B_2} \quad z_{B_2})$, respectively.

In the coordinate frame, *oxyz*, the directional vectors of joints A_1, J_{11}, and B_1 can be expressed by

$$s_1 = (1 \quad 0 \quad 0)^{\mathrm{T}}. \tag{12.17}$$

The directional vectors of joints A_2, J_{21}, and B_2 can be expressed by

$$s_2 = (\cos\psi \quad \sin\psi \quad 0)^{\mathrm{T}}. \tag{12.18}$$

Therefore, the screws of the kinematic chain, A_1B_1, are

$$\$_{A_1B_1} = [\$_{A_1} \quad \$_{J_{11}} \quad \$_{B_1}] \tag{12.19}$$

where $\$_{A_1B_1}$ indicates the screw matrix of the kinematic chain A_1B_1, and $\$_{A_1} = [1 \quad 0 \quad 0 \quad 0 \quad 0 \quad 0]^{\mathrm{T}}$ which represents a pure rotation about the x-axis, $\$_{J_{11}} = [1 \quad 0 \quad 0 \quad 0 \quad z_{J_{11}} \quad -y_{J_{11}}]^{\mathrm{T}}$ which represents a rotation about the axis of joint J_{11}, and $\$_{B_1} = [1 \quad 0 \quad 0 \quad 0 \quad z_{B_1} \quad -y_{B_1}]^{\mathrm{T}}$ which represents a rotation about the axis of joint B_1.

Consequently, the constraints exerted on the end effector can be found by substituting equation (12.19) into equation (12.15)

$$\$^r_{A_1B_1} = \left[\$^{r1}_{A_1B_1} \quad \$^{r2}_{A_1B_1} \quad \$^{r3}_{A_1B_1}\right] \tag{12.20}$$

where $\$^r_{A_1B_1}$ represents the inverse screw matrix of the kinematic chain, A_1B_1, $\$^{r1}_{A_1B_1} = [0 \quad 0 \quad 0 \quad 0 \quad 1 \quad 0]^T$ which represents a moment constraint about the y-axis, $\$^{r2}_{A_1B_1} = [0 \quad 0 \quad 0 \quad 0 \quad 0 \quad 1]^T$ which represents a moment constraint about the z-axis, and $\$^{r3}_{A_1B_1} = [1 \quad 0 \quad 0 \quad 0 \quad 0 \quad 0]^T$ which represents a force constraint in the x-direction.

The screw matrix of the kinematic chain A_2B_2 is

$$\$_{A_2B_2} = [\$_{A_2} \quad \$_{J_{21}} \quad \$_{B_2}] \tag{12.21}$$

where

$$\$_{A_2} = [\cos\psi \quad \sin\psi \quad 0 \quad 0 \quad 0 \quad 0]^{\mathrm{T}},$$

$$\$_{J_{21}} = \left[\cos\psi \quad \sin\psi \quad 0 \quad -z_{J_{21}}\sin\psi \quad z_{J_{21}}\cos\psi \quad x_{J_{21}}\sin\psi - y_{J_{21}}\cos\psi\right]^{\mathrm{T}},$$

$$\$_{B_2} = \left[\cos\psi \quad \sin\psi \quad 0 \quad -z_{B_2}\sin\psi \quad z_{B_2}\cos\psi \quad x_{B_2}\sin\psi - y_{B_2}\cos\psi\right]^{\mathrm{T}}.$$

The constraints exerted on the end effector can be found by substituting equation (12.21) into equation (12.15)

$$\$^r_{A_2B_2} = \left[\$^{r1}_{A_2B_2} \quad \$^{r2}_{A_2B_2} \quad \$^{r3}_{A_2B_2}\right] \tag{12.22}$$

where $\$^{r1}_{A_2B_2} = [0 \quad 0 \quad 0 \quad -\sin\psi \quad \cos\psi \quad 0]^T$, $\$^{r2}_{A_2B_2} = [0 \quad 0 \quad 0 \quad 0 \quad 0 \quad 1]^T$, $\$^{r3}_{A_2B_2} = [\cos\psi \quad \sin\psi \quad 0 \quad 0 \quad 0 \quad 0]^T$.

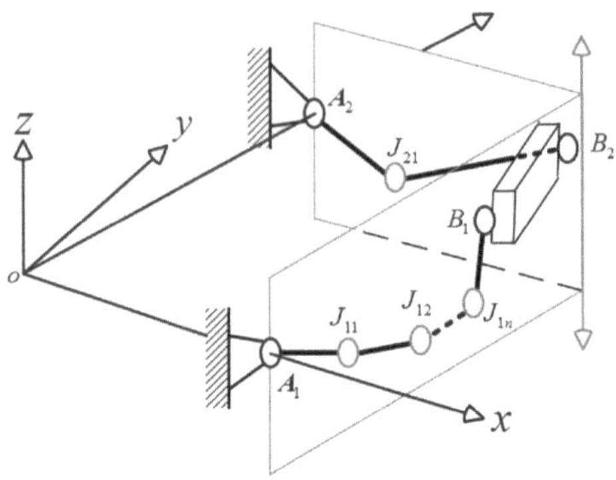

FIGURE 12.16 Effect of the increasing number of the joints on the motion of the end effector.

Associating equations (12.20) and (12.22), we can obtain the terminal constraints exerted on the end effector which can be expressed by

$$\$_C = \left[\$^r_{A_1B_1} \ \$^r_{A_2B_2} \right]. \tag{12.23}$$

Therefore, the free motion of the knuckle allowed by the two kinematic chains can be solved by substituting equation (12.23) into equation (12.15):

$$\$_F = [0 \ \ 0 \ \ 0 \ \ 0 \ \ 0 \ \ 1]^T. \tag{12.24}$$

Equation (12.24) will always hold so long as the two planes specified by the kinematic chains A_1B_1 and A_2B_2 are neither parallel nor coplanar.

What follows will discuss the free motion of the end effector when one or more identical revolute joints are added to its planar kinematic chains. For the sake of simplicity, we can suppose that there are some identical joints, $J_1, \ldots, J_n \left(n \geq 1\right)$, added in the kinematic chain A_1B_1 which are shown in Figure 12.16. The kinematic screws of this chain will become

$$\$_{A_1B_1} = [\$_{A_1} \ \ \$_{J_1} \ \ \$_{J_2} \ \ \cdots \ \ \$_{J_n} \ \ \$_{B_1}] \tag{12.25}$$

where $\$_{A_1} = [1 \ \ 0 \ \ 0 \ \ 0 \ \ 0 \ \ 0]^{\mathrm{T}}$, $\$_{J_1} = [1 \ \ 0 \ \ 0 \ \ 0 \ \ z_{J_1} \ \ -y_{J_1}]^{\mathrm{T}}$, $\$_{J_2} = [1 \ \ 0 \ \ 0 \ \ 0 \ \ z_{J_2} \ \ -y_{J_2}]^{\mathrm{T}}$, ... , $\$_{J_n} = [1 \ \ 0 \ \ 0 \ \ 0 \ \ z_{J_2} \ \ -y_{J_2}]^{\mathrm{T}}$, $\$_{B_1} = [1 \ \ 0 \ \ 0 \ \ 0 \ \ z_{B2} \ \ -y_{B2}]^{\mathrm{T}}$.

As a result, the constraints exerted on the end effector can also be found from equation (12.15)

$$\$^r_{A_1B_1} = \left[\$^{r1}_{A_1B_1} \ \ \$^{r2}_{A_1B_1} \ \ \$^{r3}_{A_1B_1} \right] \tag{12.26}$$

which is identical to equation (12.25). This indicates that there is no effect on the terminal constraints of the kinematic chain when one or more identical revolute joints are added to the chain A_1B_1. The same is true for the chain A_2B_2.

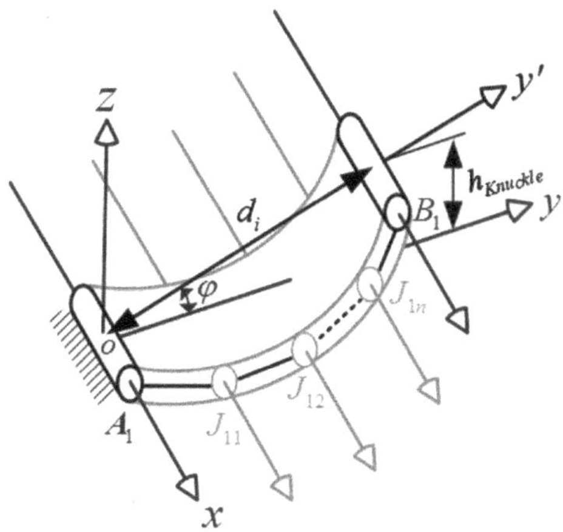

FIGURE 12.17 Infinite number of revolute joints in each chain of the end effector.

Therefore, the free motion of the end effector which will be used as the knuckle of the suspension is not affected by the increasing number of revolute joints in each chain. In addition, the free motion of the knuckle does not contain explicitly the joints' coordinates. So the knuckle will still keep the same translation represented by equation (12.24) even if the number of revolute joints increases to infinity in every kinematic chain.

Suppose that the kinematic chain, A_1B_1, consists of infinite revolute joints when the rigid links in Figure 12.16 are replaced by one flexural link shown in Figure 12.17, which can be bent very easily about the x-direction but has very high stiffness in other directions.

Suppose that the link A_1B_1 is a continuous plate which is made up of anisotropic elastic material. According to the jumping span for a knuckle, we can establish the maximum deflection model of the link. The deployed length of the link is supposed to be l, so

$$l = \sqrt{h^2 + d^2} \tag{12.27}$$

where h denotes the half jumping distance for the wheel and d indicates the distance of the link A_1B_1 when it moves to the horizontal plane. The relationship of these parameters is shown in Figure 12.18.

Suppose that the flexural stiffness about x-direction is E_xI_x. When the knuckle moves to the horizontal plane, the horizontal component of joint force $F\cos\varphi$ will be the maximum, where φ shown in Figure 12.17 denotes the angle from horizontal plane to line A_1B_1. Generally, the bending equation for the link should be

$$\frac{\partial^2 z}{\partial y^2} + \frac{F\cos\varphi}{E_xI_x}z = 0. \tag{12.28}$$

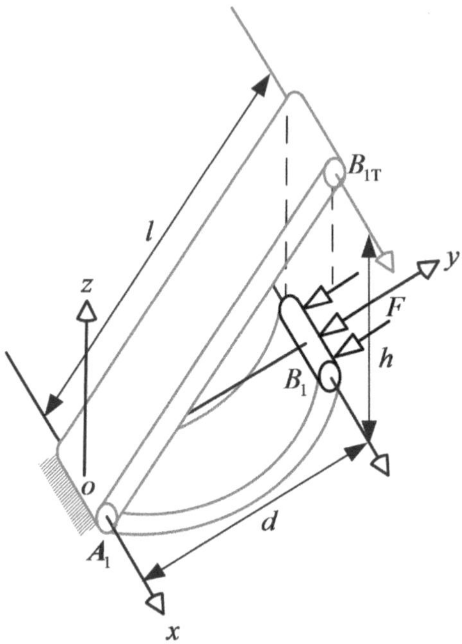

FIGURE 12.18 Anisotropic elastic link of the knuckle.

The solution of equation (12.28) can be expressed as

$$z = c_1 \cos \sqrt{\frac{F \cos \varphi}{E_x I_x}} y + c_2 \sin \sqrt{\frac{F \cos \varphi}{E_x I_x}} y \tag{12.29}$$

where c_1 and c_2 are coefficients prescribed by the initial conditions which can be obtained immediately from Figure 12.17 and Figure 12.18:

$$\begin{cases} z(0) = c_1 = 0 \\ z(d_i) = c_1 \cos \sqrt{\frac{F \cos \varphi}{E_x I_x}} d_i + c_2 \sin \sqrt{\frac{F \cos \varphi}{E_x I_x}} d_i = 0 \end{cases} \tag{12.30}$$

where $d_i = \frac{d}{\cos \varphi}$ and d represents the shortest distance between joints A_1 and B_1 in the horizontal case.

Equation (12.30) can be further simplified as

$$\begin{cases} c_1 = 0 \\ F = \dfrac{E_x I_x n^2 \pi^2 \cos \varphi}{d^2} \end{cases} \tag{12.31}$$

where $n = 1,2,\ldots$ is the natural number which represents the order of bending mode, F indicates the force exerted on the free end of the link by the knuckle and its value is determined by the physical parameters E_x, I_x, and d which provide the optimization bases for the anisotropic flexural link.

Substituting equation (12.31) into equation (12.30) presents

$$z = c_2 \sin \frac{n\pi y \cos\varphi}{d}. \tag{12.32}$$

The whole length of the center line of the link is assumed to be constant which has the following relation:

$$l = \int_0^{d_i} \sqrt{1 + \left(\frac{\partial z}{\partial y}\right)^2} dy. \tag{12.33}$$

Substituting equation (12.32) into equation (12.33) yields

$$\int_0^{d_i} \sqrt{1 + \frac{n^2\pi^2 c_2^2}{d^2} \cos^2 \frac{n\pi y}{d}} dy = l. \tag{12.34}$$

Because there is no analytical integral for the left side of equation (12.34), we can take the Taylor series for the integrand

$$\int_0^{d_i} \left[1 + \frac{n^2\pi^2 c_2^2}{d^2} + \frac{n^4\pi^4 c_2^2}{2d^4} \left(1 + \frac{n^2\pi^2 c_2^2}{d^2}\right)^{-\frac{1}{2}} y^2 + \cdots\right] dy = l. \tag{12.35}$$

Equation (12.35) can be approximately simplified by neglecting the third and higher items

$$\frac{6n^2\pi^2 + n^4\pi^4}{6d} c_2^2 - \frac{n^6\pi^6}{12d^3} c_2^4 + \frac{n^8\pi^8}{16d^5} c_2^6 = l - d_i. \tag{12.36}$$

Rearrange equation (12.36) to form a standard equation:

$$c_2^6 - \frac{4d^2}{3n^2\pi^2} c_2^4 + \frac{8d^4\left(6 + n^2\pi^2\right)}{3n^6\pi^6} c_2^2 - \frac{16d^5}{n^8\pi^8}\left(l - d_i\right) = 0. \tag{12.37}$$

Suppose that $k = -\frac{4d_i^2}{3n^2\pi^2}$, $m = \frac{8d^4(6+n^2\pi^2)}{3n^6\pi^6}$, $n = -\frac{16d_i^5}{n^8\pi^8}(l - d_i)$, and $p = m - k^{\frac{2}{3}}$, $q = \frac{2k^3}{27} - \frac{1}{3}km + n$, we can now solve c_2^2 from equation (12.37)

$$c_2^2 = \sqrt[3]{\sqrt{\frac{q^2}{4} + \frac{p^3}{27}} - \frac{q}{2}} - \sqrt[3]{\sqrt{\frac{q^2}{4} + \frac{p^3}{27}} + \frac{q}{2}}. \tag{12.38}$$

From Figure 12.18, we find that

$$c_2 = -\sqrt{\sqrt[3]{\sqrt{\frac{q^2}{4}+\frac{p^3}{27}}-\frac{q}{2}} - \sqrt[3]{\sqrt{\frac{q^2}{4}+\frac{p^3}{27}}+\frac{q}{2}}}. \tag{12.39}$$

Substituting equation (12.39) into equation (12.32) yields

$$z = -\sqrt{\sqrt[3]{\sqrt{\frac{q^2}{4}+\frac{p^3}{27}}-\frac{q}{2}} - \sqrt[3]{\sqrt{\frac{q^2}{4}+\frac{p^3}{27}}+\frac{q}{2}}}\sin\frac{n\pi y\cos\varphi}{d}. \tag{12.40}$$

The above analysis shows that the flat flexural link does be able to act as the infinite-R kinematic chain in engineering.

12.2.2.2 Theoretical Analysis

The bending moment about the x-axis, $M_x(y)$, resulting from the deflection of the link is

$$M_x(y) = -Fz = \frac{E_x J_x n^2\pi^2\cos\varphi}{d^2}\sqrt{\sqrt[3]{\sqrt{\frac{q^2}{4}+\frac{p^3}{27}}-\frac{q}{2}} - \sqrt[3]{\sqrt{\frac{q^2}{4}+\frac{p^3}{27}}+\frac{q}{2}}} \times \sin\frac{n\pi y\cos\varphi}{d} \tag{12.41}$$

where $\cos\varphi = \frac{d}{\sqrt{d^2+h_{Knuckle}^2}}$ and $h_{Knuckle}$ denotes the jumped height of the knuckle from the horizontal plane at this instant and $h_{Knuckle} \leq h$ which is shown in Figure 12.15.

Equation (12.41) indicates that the maximum bending moment about the x-axis for the anisotropic flexural link is $\frac{E_x I_x n^2\pi^2}{d\sqrt{d^2+h_{Knuckle}^2}}\sqrt{\sqrt[3]{\sqrt{\frac{q^2}{4}+\frac{p^3}{27}}-\frac{q}{2}} - \sqrt[3]{\sqrt{\frac{q^2}{4}+\frac{p^3}{27}}+\frac{q}{2}}}$ which provides the optimization relationship between the parameters E_x, I_x, d, and h for the anisotropic flexural link.

For the other anisotropic flexural link, we can obtain similar results. Therefore, the knuckle will be subject to a pair of pushing forces F expressed by equation (12.31) besides the three pairs of constraints denoted by equation (12.23). The statics relationship is shown in Figure 12.19.

Considering the symmetry of the two anisotropic flexural links, the pushing forces at the ends of the two links are identical, which are denoted by F in Figure 12.19. Under the action of the two Fs, the additional lateral forces at the ends of the two links should also be identical, which is represented by F'. The

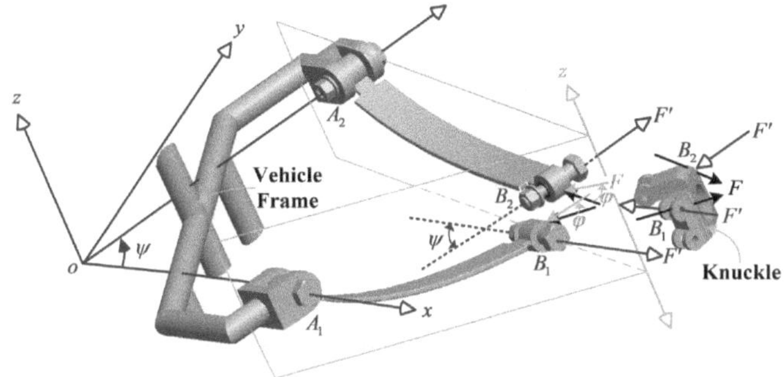

FIGURE 12.19 Statics relationship of the knuckle and the anisotropic links.

equilibrium equations for the knuckle should be

$$2F\cos\varphi\cos\frac{180-\psi}{2}-2F'\cos\frac{\psi}{2}=0 \tag{12.42}$$

where ψ is the subtended angle of the axes of joints A_1.

Consequently, we obtain

$$F'=F\cos\varphi\tan\frac{\psi}{2}. \tag{12.43}$$

Substituting equation (12.31) into equation (12.43) produces

$$F'=\frac{E_x I_x n^2\pi^2}{d^2}\cos^2\varphi\tan\frac{\psi}{2}. \tag{12.44}$$

To investigate the effect of F' on the deformation of the anisotropic flexural link, we can analyze the equivalent cantilever shown in Figure 12.20. Suppose that the absolute coordinates of B_1 are $(0 \quad d_i \quad h_1)$ where $d_i=\frac{d}{\cos\varphi}$ and d is the distance from the joint A_1 to the vertical guide which is shown in

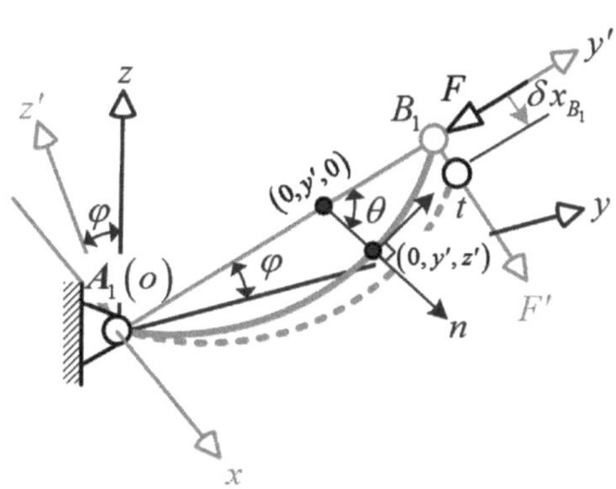

FIGURE 12.20 Deformation of the anisotropic flexural link.

Figure 12.18. So the local coordinates in the $oxy'z'$-frame are $(0 \quad d_i \quad 0)$. In the local coordinate frame $oxy'z'$, the normal direction of the bending curve at point $(0 \quad y' \quad z')$ is

$$\tan\theta = -\frac{dy'}{dz'} = -\frac{d_i}{n\pi c_2}\sec\frac{n\pi y'}{d_i}. \tag{12.45}$$

So the normal line equation passing through point $(0 \quad y' \quad z')$ in the $y'oz'$-$y'oz'$-plane is

$$Z - z' = -\frac{d_i}{n\pi c_2}\sec\frac{n\pi y'}{d_i}\left(Y - y'\right) \tag{12.46}$$

where $(Y \quad Z)$ represent the coordinates of any point on the normal line.

The cross point of the normal line and the y'-axis can be obtained by letting $Z = 0$ in equation (12.45)

$$\begin{cases} Y = \dfrac{n\pi c_2 z'}{d_i}\cos\dfrac{n\pi y'}{d_i} + y' \\ Z = 0. \end{cases} \tag{12.47}$$

So the distance between points $(y' \quad z')$ and $(Y \quad Z)$ is

$$s = \sqrt{\left(\frac{n\pi c_2 z'}{d_i}\cos\frac{n\pi y'}{d_i}\right)^2 + z'^2}. \tag{12.48}$$

Now, the bending moment and the torque about the point $(y' \quad z')$ under the action of F' can be expressed as

$$\begin{cases} M_n\left(y'\right) = F'\left(d_i - \dfrac{n\pi c_2 z'}{d_i}\cos\dfrac{n\pi y'}{d_i} - y'\right)\sin\theta \\ T_t\left(y'\right) = F'\left[s - \left(d_i - \dfrac{n\pi c_2 z'}{d_i}\cos\dfrac{n\pi y'}{d_i} - y'\right)\cos\theta\right]. \end{cases} \tag{12.49}$$

Associating equations (12.45), (12.48), and (12.49), we can obtain the following explicit expression for the bending moment and torque:

$$\begin{cases} M_n\left(y'\right) = F'\left(d_i - \dfrac{n\pi c_2 z'}{d_i}\cos\dfrac{n\pi y'}{d_i} - y'\right)\dfrac{d_i\sec\dfrac{n\pi y'}{d_i}}{\sqrt{\left(n\pi c_2\right)^2 + \left(d_i\sec\dfrac{n\pi y'}{d_i}\right)^2}} \\ T_t\left(y'\right) = F'\left[\sqrt{\left(\dfrac{n\pi c_2 z'}{d_i}\cos\dfrac{n\pi y'}{d_i}\right)^2 + z'^2} - \dfrac{d_i - \dfrac{n\pi c_2 z'}{d_i}\cos\dfrac{n\pi y'}{d_i} - y'}{\sqrt{1 + \left(\dfrac{d_i}{n\pi c_2}\sec\dfrac{n\pi y'}{d_i}\right)^2}}\right]. \end{cases} \tag{12.50}$$

The flexion, δx_{B_1}, can be obtained in line with the Mohr integral theory:

$$\delta x_{B_1} = \int_0^{d_i} \frac{M_n(y')}{E_n I_n} \frac{\partial M_n(y')}{\partial F'} dy' + \int_0^{d_i} \frac{T_t(y')}{G_t I_t} \frac{\partial T_t(y')}{\partial F'} dy' \tag{12.51}$$

where $M_n(y')$ denotes the moment about the normal direction at point y', E_n is the Young's elastic module about the normal direction at point y', I_n is the inertia moment of the transverse section of the link about the normal direction at point y', $T_t(y')$ represents the torque about the tangent direction at point y', G_t is the torsional modulus of elasticity of the material about the tangent direction at point y', I_t is the torsional inertia moment of the transverse section of the link about the tangent direction at point y'.

Substituting equation (12.50) into equation (12.51) presents

$$\begin{aligned}\delta x_{B_1} = {} & \frac{F' d_i^2}{E_n I_n} \int_0^{d_i} \frac{\left(d_i - \frac{n\pi c_2^2}{2d_i} \sin \frac{2n\pi y'}{d_i} - y'\right)^2}{\frac{1}{2}(n\pi c_2)^2 \left(1 + \cos \frac{2n\pi y'}{d_i}\right) + d_i^2} dy' \\ & + \frac{F'}{G_t I_t} \int_0^{d_i} \frac{\left[\begin{array}{l} c_2 \sin \frac{n\pi y'}{d_i} \sqrt{1 + \cos \frac{2n\pi y'}{d_i} + \left(\frac{n\pi c_2}{d_i}\right)^2 \left(1 + \cos \frac{2n\pi y'}{d_i}\right)^2 + \left(\frac{d_i}{n\pi c_2}\right)^2} \\ - \cos \frac{n\pi y'}{d_i} \left(d_i - \frac{n\pi c_2^2}{2d_i} \sin \frac{2n\pi y'}{d_i} - y'\right) \end{array}\right]^2}{\cos \frac{n\pi y'}{d_i} \sqrt{\frac{1}{2}\left(1 + \cos \frac{2n\pi y'}{d_i}\right) + \left(\frac{d_i}{n\pi c_2}\right)^2}} dy'.\end{aligned} \tag{12.52}$$

To simplify the engineering calculation, equation (12.50) can be approximately rewritten as

$$\begin{cases} M_n(y') = F'(d_i - y') \\ T_t(y') = -F' z'. \end{cases} \tag{12.53}$$

Associating equations (12.40), (12.51) and (12.53), we immediately obtain

$$\begin{aligned}\delta x_{B_1} &= \int_0^d \frac{F'(d_i - y)^2}{E_n I_n} dy + \int_0^d \frac{F'\left(\sqrt[3]{\sqrt{\frac{q^2}{4} + \frac{p^3}{27}} - \frac{q}{2}} - \sqrt[3]{\sqrt{\frac{q^2}{4} + \frac{p^3}{27}} + \frac{q}{2}}\right) \sin^2 \frac{n\pi y}{d_i}}{G_t I_t} dy \\ &= F' \left[\frac{d_i^3}{3E_n I_n} + \frac{d_i \left(\sqrt[3]{\sqrt{\frac{q^2}{4} + \frac{p^3}{27}} - \frac{q}{2}} - \sqrt[3]{\sqrt{\frac{q^2}{4} + \frac{p^3}{27}} + \frac{q}{2}}\right)}{2G_t I_t}\right].\end{aligned} \tag{12.54}$$

Considering the symmetry of the two anisotropic flexural links, A_1B_1 and A_2B_2, the knuckle will have a displacement in the direction of the bisection line of the angle ψ. The displacement of the knuckle, which is assumed to be a rigid body, and the deflections of the anisotropic flexural links are shown

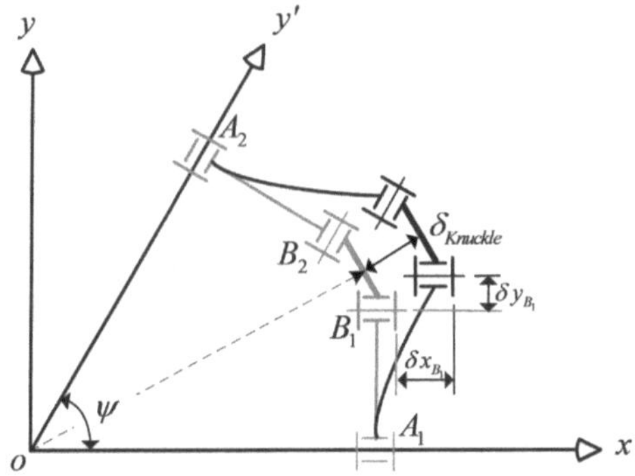

FIGURE 12.21 Displacement resulting from the deflections of the links.

in Figure 12.21, in which δx_{B_1} is expressed by equation (12.54) and δy_{B_1} represents the vertical displacement accompanying δx_{B_1}.

From Figure 12.21, we immediately find that

$$\delta_{Knuckle} = \frac{\delta x_{B_1}}{\cos\frac{\psi}{2}} = \frac{F'}{\cos\frac{\psi}{2}} \times \left[\frac{\left(d^2 + h_{Knuckle}^2\right)^{\frac{3}{2}}}{3E_n I_n} + \frac{\sqrt{d^2 + h_{Knuckle}^2}\left(\sqrt[3]{\sqrt{\frac{q^2}{4} + \frac{p^3}{27}} - \frac{q}{2}} - \sqrt[3]{\sqrt{\frac{q^2}{4} + \frac{p^3}{27}} + \frac{q}{2}}\right)}{2G_t I_t} \right]. \tag{12.55}$$

Equation (12.55) indicates that $\delta_{Knuckle}$ is always larger than δx_{B_1} for $0 < \frac{\psi}{2} < \frac{\pi}{2}$. The optimizing task is to reduce the value of δx_{B_1}. Substituting equation (12.44) into equation (12.55) yields

$$\delta_{Knuckle} = \frac{E_x I_x n^2 \pi^2 \cos^2\varphi \sin\frac{\psi}{2}}{d^2 \cos^2\frac{\psi}{2}} \times \left[\frac{\left(d^2 + h_{Knuckle}^2\right)^{\frac{3}{2}}}{3E_n I_n} + \frac{\sqrt{d^2 + h_{Knuckle}^2}\left(\sqrt[3]{\sqrt{\frac{q^2}{4} + \frac{p^3}{27}} - \frac{q}{2}} - \sqrt[3]{\sqrt{\frac{q^2}{4} + \frac{p^3}{27}} + \frac{q}{2}}\right)}{2G_t I_t} \right]. \tag{12.56}$$

Therefore, the displacement of the knuckle resulting from the elasticity of the flexural links can be optimized with equation (12.56).

12.2.2.3 Numerical Simulations and Model Tests

Suppose that the undeformed links are those at the horizontal plane which are shown in Figure 12.19. The primary parameters are listed in Table 12.1.

Numerical simulations are conducted based on equation (12.56). The displacement of the knuckle resulting from the flexion of the anisotropic elastic links with respect to the jumped height is shown in Figure 12.22. The minus sign in the lateral error indicates that the direction of displacement of the knuckle is reverse to the dashed arrow shown in Figure 12.21.

The largest lateral alignment errors in the horizontal plane do not exceed 0.1 mm which occurs at the upper limit and lower limit of the journey of the knuckle. Therefore, the largest error of the linear alignment is within the interval between 0 and 0.1 mm when the knuckle is jumping 100 mm up and down. Comparing with the rigid guidance mechanism, the flexural suspension provides a very good precision that the vehicle could allow. However, this is obtained under the hypothesis that the undeformed links are those located in the xoy-plane shown in Figure 12.19. If we suppose that the joint $B_i (i = 1,2)$ is located in a plane that is h_1 away from the xoy plane, where h_1 can be optimized, the largest alignment error can be minimized. The different alignment errors of the knuckle because of the flexion of the anisotropic elastic links with respect to the jumped height are shown in Figure 12.23. The largest alignment error of the suspension is restricted to 0.04 mm when $h_1 = 50$ mm which is shown by the left graph of Figure 12.23. In vehicle engineering, the generally used interval

TABLE 12.1 Parameters of the Flat Flexural Links of the Knuckle

E_x	I_x	E_n	I_n	G_t	I_t	d (mm)	ψ	h (mm)
2.03×10^5	4.17	2.03×10^5	1.04×10^4	6.09×10^4	16.7	150	$\frac{\pi}{2}$	100

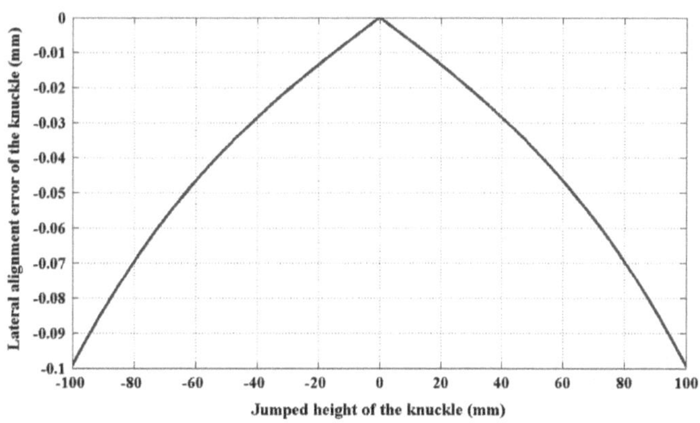

FIGURE 12.22 Lateral alignment error of the knuckle in jumping up and down.

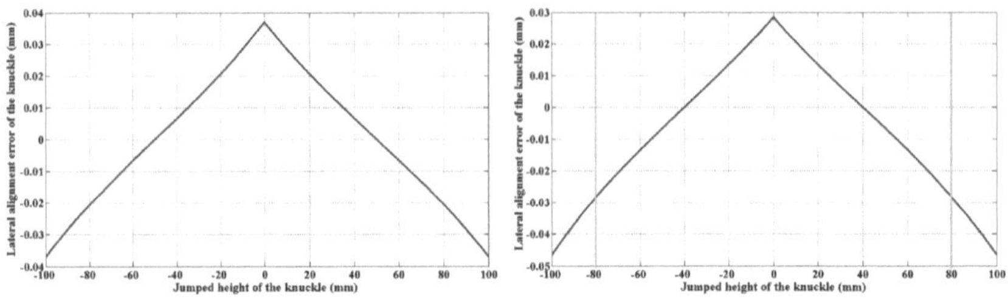

FIGURE 12.23 Different horizontal displacements of the knuckle subjected to h_1.

of journey of the knuckle in statistics is [−40 mm, 40 mm] and consequently the largest alignment error of the suspension is restricted to 0.03 mm within this interval when $h_1 = 40$ mm which is shown on the right of Figure 12.23.

Therefore, equation (12.56) can be utilized to optimize the design of the flexural links of the suspension proposed in the chapter. Theoretical analysis and numerical simulations indicate that the rotational stiffness about the lateral direction should be as low as possible while the normal and torsional stiffness should be as high as possible. To reduce the alignment error of the suspension during jumping up and down shown in Figures 12.22 and 12.23, we can increase the stiffness about the normal direction and the tangent direction on the one hand, and increase the number of the identical anisotropic flexural links to connect the knuckle and the vehicle frame on the other. Figure 12.24, for example, shows the case when two identical anisotropic flexural links are used on each side of the knuckle.

To further increase the increase the stiffness about the normal direction and the tangent direction, we can use some parallel ribs to correspondingly connect the upper flexural links and the lower links with revolute joints. The strengthened ribs are jointed for the two flexural links which is shown in Figure 12.25. This provides much higher bending and torsional stiffness for the suspension than the left one in Figure 12.22.

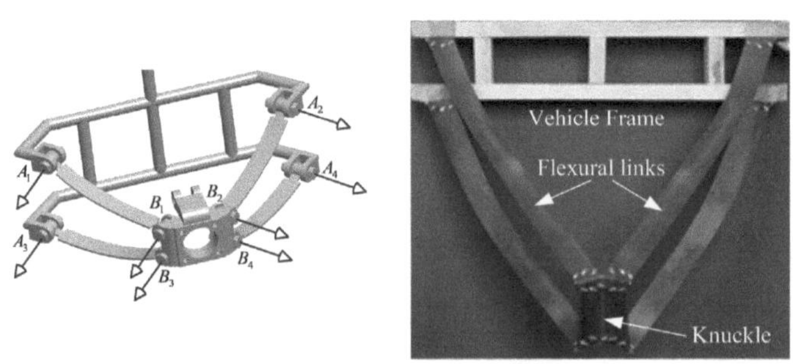

(a) CAD model **(b)** Test model

FIGURE 12.24 Double flexural links on each side of the knuckle and the model.

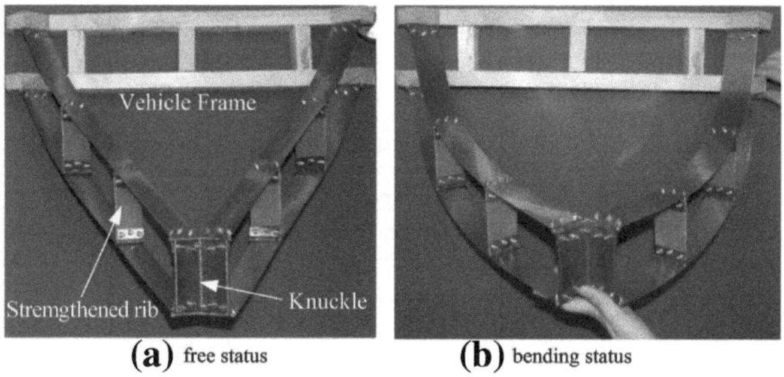

FIGURE 12.25 Rib-jointed two flexural links and the qualitative test.

Model tests show that the rib jointed flexural links are much better at keeping the alignment parameters of the knuckle.

12.3 CONCLUSIONS

The application of DOF analysis theory in the structure synthesis and the design innovation of mechanisms have been studied. Through presenting general steps in mechanism synthesis, the primary method for the innovative design of mechanism with the universal significance is also studied. The procedure of structure synthesis is discussed in detail through a lower limb of human being and an independent suspension of invariable alignment parameter. A well-modeled humanoid robot has been designed, which has both the same mobility as a human being and has the merits of feasibility of control and manufacturing, and also low processing costs. The synthesis of kinematic chains of a mechanism for the suspension is investigated and then the independent suspension with flexural links of anisotropic stiffness is analyzed.

REFERENCES

[1] R. Alizade, Ç. Bayram, Structural synthesis of parallel manipulators, Mechanism and Machine Theory 39 (2004) 857–870.
[2] R. Alizade, C. Bayram, E. Gezgin, Structural synthesis of serial platform manipulators, Mechanism and Machine Theory 39 (2004) 857–870.
[3] G. Gogu, Mobility of mechanisms: a critical review, Mechanism and Machine Theory 40 (2005) 1068–1097.
[4] G. Gogu, Mobility and spatiality of parallel robots revisited via theory of linear transformations, European Journal of Mechanics A/Solids 24 (2005) 690–711.
[5] J.S. Zhao, K. Zhou, Z.J. Feng, A theory of degrees of freedom for mechanisms, Mechanism and Machine Theory 39 (2004) 621–643.
[6] J.S. Zhao, K. Zhou, D.Z. Mao, Y.F. Gao, Y.F. Fang, A new method to study the degree of freedom of spatial parallel mechanisms, The International Journal of Advanced Manufacturing Technology 23 (2004) 288–294.

[7] J.S. Zhao, Z.J. Feng, K. Zhou, D.W. Jin, Re-analysis on the degree-of-freedom configuration of the platforms in spatial parallel mechanisms with constraints spaces, The International Journal of Advanced Manufacturing Technology 28 (2006) 190–196.

[8] J.S. Zhao, Z.J. Feng, J.X. Dong, Computation of the configuration degree of freedom of a spatial parallel mechanism by using reciprocal screw theory, Mechanism and Machine Theory 41 (2006) 1486–1504.

[9] A.B. Kempe, How to Draw a Straight Line, Macmillan, London, 1877, <http://kmoddl.library.cornell.edu/tutorials/11/.html>, 2013 (accessed 27.08.13).

[10] <http://kmoddl.library.cornell.edu/model.php?m=244.html>, 2013, (accessed 27.08.13).

[11] R.L. Norton, Design of Machinery—An Introduction to the Synthesis and Analysis of Mechanisms and Machines, McGraw–Hill Companies Inc., 2001.

[12] Eric W. Weisstein, "Sarrus Linkage," From MathWorld–A Wolfram Web Resource. <http://mathworld.wolfram.com/SarrusLinkage.html>, (accessed 27.08.13).

Chapter 13

Workspace Synthesis of Spatial Mechanisms

The major contributions of former works are mainly in finding the reachable workspace of a manipulator [1–4], and we have also proposed a systematic methodology to investigate the reachable workspace of a manipulator in Chapter 5. Compared to the reachable workspace, the dexterous workspace is seldom addressed in former works. Lately, Dai and Shah [5] investigated the orientation capability of a serial manipulator by examining the rotatability of its equivalent mechanism and by proposing the workspace region decomposition, based on the polynomial discriminant derived from the virtual angle analysis of the end effector link of the equivalent mechanism. Wang and Tang [6] proposed a dimensional design approach to satisfy the requirement of motion platform realizing orientation capability. Bonev and Ryu [7] presented a method for the computation of the orientation workspace of 6-degrees-of-freedom (DOFs) parallel manipulators, defined the set of all attainable orientations of the mobile platform about a fixed point.

Nevertheless, we still find that it is urgently necessary to find an efficient method to optimize the design of a spatial parallel manipulator with any desired dexterous workspace. In this chapter, we mainly discuss the dexterous workspace of a spatial parallel manipulator while not necessarily considering the shape of the whole reachable workspace, which has been repeatedly addressed.

13.1 THEOREM OF DEXTEROUS WORKSPACE OF SPATIAL PARALLEL MANIPULATOR

As mentioned in Chapter 5, the reachable workspace of a manipulator is defined as the volume within which the reference point, the geometry center of the manipulator, can be reached [8]. The dexterous workspace of a manipulator can be defined as the volume within which the reference point can be reached with a desired orientation [8].

If we assume that the geometric center point on the end effector connected with $n(n \geq 2)n(n \geq 2)$ kinematic chains is C, the coordinate systems' relationships can be shown in Figure 13.1 where the origin of the local coordinate system is superimposed with point C. The end vertexes of each kinematic chain

Advanced Theory of Constraint and Motion Analysis for Robot Mechanisms.
http://dx.doi.org/10.1016/B978-0-12-420162-0.00013-8

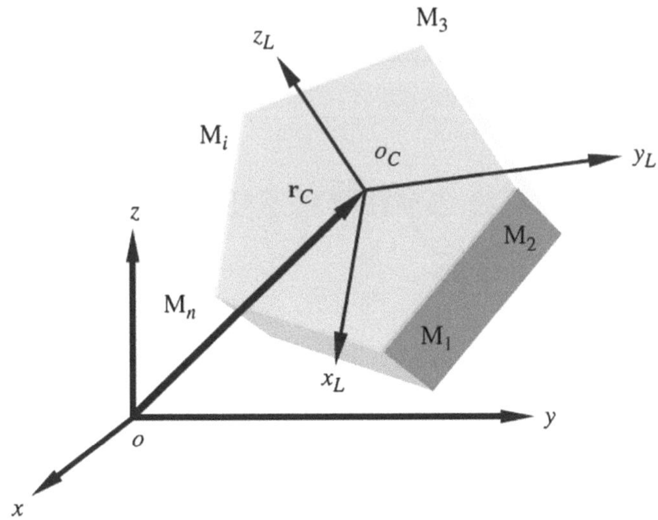

FIGURE 13.1 Coordinate system of a manipulator.

connected with the manipulator are denoted by $M_i(i = 1,2,\ldots,n)$, the end vertexes of each kinematic chain connected with the fixed base are denoted by $B_i(i = 1,2,\ldots,n)$. After knowing the DOF of the manipulator from Chapter 4 [9], we can select the least number of variables to depict the kinematic relationships of the manipulator and the actuations. In the absolute coordinate system, the vectors of the terminal vertexes of each kinematic chain connected with the fixed base are $\boldsymbol{r}_1,\boldsymbol{r}_2,\ldots,\boldsymbol{r}_n$, the coordinates of C are (x,y,z) and the vectors of the vertexes B_i and point C are denoted by $\boldsymbol{r}_{B_i}$ and $\boldsymbol{r}_C$. In the local coordinate system, the vectors of the vertexes M_i is denoted by $\boldsymbol{r}_{M_i}^L$.

With Euler pose angles, we can easily get the transformation matrix $\mathbf{R}$ from the end effector's local coordinate system to the absolute one shown in Figure 13.1. So, we obtain:

$$\boldsymbol{r}_{M_i} = \boldsymbol{r}_C + \boldsymbol{r}_{CM_i} = \boldsymbol{r}_C + \mathbf{R}\boldsymbol{r}_{M_i}^L \tag{13.1}$$

where $i = 1,2,\ldots,n, r_{CM_i}$ denotes the vector of CM_i in the absolute coordinate system.

Because $\boldsymbol{r}_{M_i}^L$ usually has very simple expressions in the local coordinate system, we can therefore easily obtain $\boldsymbol{r}_{M_i}$ in the absolute coordinate system. Considering the geometric and kinematic constraints of the system, we can get a set of equations as follows:

$$\begin{cases} \left\|\boldsymbol{r}_{M_1} - \boldsymbol{r}_{B_1}\right\| = l_1 \\ \left\|\boldsymbol{r}_{M_2} - \boldsymbol{r}_{B_2}\right\| = l_2 \\ \quad\vdots \\ \left\|\boldsymbol{r}_{Mn} - \boldsymbol{r}_{Bn}\right\| = l_n \end{cases} \tag{13.2}$$

where l_i $\left(i = 1,2,\ldots,n\right)$ denotes the length of the ith kinematic chain.

With equation (13.2), we can construct the following equations:

$$G_i\left(x,y,z,\alpha,\beta,\gamma\right) = \left\|\boldsymbol{r}_{M_i} - \boldsymbol{r}_{B_i}\right\|^2 - l_i^2 \tag{13.3}$$

where α,β, and γ denote the Euler pose angles.

Therefore, the workspace of the manipulator, denoted by Ω_t can be denoted as:

$$\Omega_t = \left\{\left(x,y,z,\alpha,\beta,\gamma\right) \middle| \mathbf{G}\left(x,y,z,\alpha,\beta,\gamma\right) = \mathbf{0}\right\} \tag{13.4}$$

where

$$\mathbf{G} = \begin{bmatrix} G_1\left(x,y,z,\alpha,\beta,\gamma\right) \\ G_2\left(x,y,z,\alpha,\beta,\gamma\right) \\ \vdots \\ G_n\left(x,y,z,\alpha,\beta,\gamma\right) \end{bmatrix}.$$

That is, the total workspace of the manipulator is a set of points in the three-dimensional space where we can find at least one set of parameters (α,β,γ) such that the three Cartesian coordinates (x, y, z) at this point satisfy that $\mathbf{G}\left(x,y,z,\alpha,\beta,\gamma\right) = 0$.

Therefore, the dexterous workspace, Ω_d, can be expressed as:

$$\Omega_d = \left\{\left(x,y,z\right) \middle| \mathbf{G}\left(x,y,z,\alpha,\beta,\gamma\right) = \mathbf{0},\quad \alpha \in \left[\alpha_{\min},\alpha_{\max}\right], \right.$$
$$\left. \beta \in \left[\beta_{\min},\beta_{\max}\right],\quad \gamma \in \left[\gamma_{\min},\gamma_{\max}\right]\right\}. \tag{13.5}$$

Naturally, we have the following relationship:

$$\Omega_d \subseteq \Omega_t. \tag{13.6}$$

Define functions $L_i\left(x,y,z,\alpha,\beta,\gamma\right)$:

$$L_i\left(x,y,z,\alpha,\beta,\gamma\right) = \left(x_{M_i} - x_{B_i}\right)^2 + \left(y_{M_i} - y_{B_i}\right)^2 + \left(z_{M_i} - z_{B_i}\right)^2 \tag{13.7}$$

with equation (13.2), we know that

$$l_i = \sqrt{L_i}. \tag{13.8}$$

For any point $\left(x,y,z\right) \in \Omega_d$, equation (13.4) will always hold. If the space Ω_d is two-connected (as a result, there exists no inner singularity), the extreme values of function (13.7) must be located either on the boundary surfaces, denoted by S_d, of Ω_d or on the points, denoted by $\mathrm{P}(x,y,z,\alpha,\beta,\gamma)$, where $\left.\frac{\partial L_i\left(x,y,z,\alpha,\beta,\gamma\right)}{\partial\left(x,y,z,\alpha,\beta,\gamma\right)}\right|_P = \mathbf{0}$.

Obviously, the solutions of $\frac{\partial L_i\left(x,y,z,\alpha,\beta,\gamma\right)}{\partial\left(x,y,z,\alpha,\beta,\gamma\right)} = \mathbf{0}$ are identical to those of $\frac{\partial G_i\left(x,y,z,\alpha,\beta,\gamma\right)}{\partial\left(x,y,z,\alpha,\beta,\gamma\right)} = \mathbf{0}$. Therefore, we can obtain the solutions of $\frac{\partial L_i\left(x,y,z,\alpha,\beta,\gamma\right)}{\partial\left(x,y,z,\alpha,\beta,\gamma\right)} = \mathbf{0}$ by solving $\frac{\partial G_i\left(x,y,z,\alpha,\beta,\gamma\right)}{\partial\left(x,y,z,\alpha,\beta,\gamma\right)} = \mathbf{0}$.

Consequently, the algorithm to optimize the lengths of the legs of a spatial parallel manipulator can be given here:

Step 1: search the point set $\Omega_s = \left\{\left(x,y,z,\alpha,\beta,\gamma\right) \middle| \frac{\partial L_i\left(x,y,z,\alpha,\beta,\gamma\right)}{\partial\left(x,y,z,\alpha,\beta,\gamma\right)} = \mathbf{0}\right\}$;

Step 2: find the cross set of Ω_s and the specified dexterous workspace Ω_d, denoted by

$$\Omega_c = \Omega_d \cap \Omega_s;$$

Step 3: find the point set of the possible extreme values, denoted by Ω_e,

$$\Omega_e = \{(x,y,z,\alpha,\beta,\gamma) \,|(x,y,z,\alpha,\beta,\gamma) \in S_d, (x,y,z,\alpha,\beta,\gamma) \in \Omega_c\};$$

Step 4: solve the function values set at Ω_e,

$$L_i = \{L_i(x,y,z,\alpha,\beta,\gamma)|\,(x,y,z,\alpha,\beta,\gamma) \in \Omega_e\};$$

Step 5: obtain the extreme values of the length of the ith$(i = 1,2,\ldots,n)leg$,

$$L_i^{\min} = \min(L_i) \text{ and} L_i^{\max} = \max(L_i);$$

Step 6: if the dexterous workspace has the symmetrical characteristics as those of the manipulator structure, we can find that the optimum length ranges are identical and equal $[\sqrt{L_i^{\min}}, \sqrt{L_i^{\max}}]$ [10]; if the dexterous workspace does not have the symmetrical characteristics as those of the manipulator, we can repeat the above five steps to find the extreme values of each leg.

13.2 OPTIMUM DESIGN OF SPATIAL MANIPULATORS AIMING AT A DESIRED DEXTEROUS WORKSPACE

13.2.1 Optimum Design of a Schoenflies-Type Parallel Manipulator

The architecture of a Schoenflies-type parallel manipulator [11] is shown in Figure 13.2. The thorough analysis of Schoenflies-type parallel manipulators with 3-DOF translations and 1-DOF rotation can be found in [12]. The manipulator proposed herein is actuated by four identical UPU (universal joint + prismatic joint + universal joint) kinematic chains. The manipulator has four DOFs (one rotational DOF and three orthogonal translational DOFs), which can be analyzed according to [13]. Correspondingly, we can only select four independent variables, one set of which are the absolute coordinates of the geometry center of the manipulator, denoted by (x,y,z); and the other is the rotational angle about tghe z-axis, denoted by α. As a result, the transformation matrix from the local coordinate system to the absolute we can be expressed as:

$$\mathbf{R} = \begin{bmatrix} \cos\alpha & -\sin\alpha & 0 \\ \sin\alpha & \cos\alpha & 0 \\ 0 & 0 & 1 \end{bmatrix}.$$

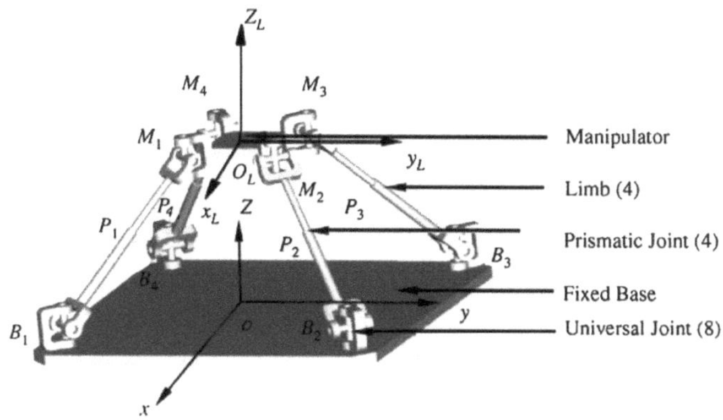

FIGURE 13.2 Architecture of a Schoenflies-type parallel manipulator.

Suppose the coordinates of the ith$(i = 1,2,\ldots,n)$ vertex of the end effector in the local coordinate system are denoted by $M_i\left(x_{M_i}^L, y_{M_i}^L, z_{M_i}^L\right)$, the absolute coordinates $M_i(x_{M_i}, y_{M_i}, z_{M_i})$ can be obtained:

$$\begin{bmatrix} x_{M_i} \\ y_{M_i} \\ z_{M_i} \end{bmatrix} = \begin{bmatrix} x \\ y \\ z \end{bmatrix} + \mathbf{R} \begin{bmatrix} x_{M_i}^L \\ y_{M_i}^L \\ z_{M_i}^L \end{bmatrix} = \begin{bmatrix} x \\ y \\ z \end{bmatrix} + \begin{bmatrix} \cos\alpha & -\sin\alpha & 0 \\ \sin\alpha & \cos\alpha & 0 \\ 0 & 0 & 1 \end{bmatrix} \begin{bmatrix} x_{Pi}^L \\ y_{Pi}^L \\ z_{Pi}^L \end{bmatrix}$$

$$= \begin{bmatrix} x + x_{M_i}^L \cos\alpha - y_{M_i}^L \sin\alpha \\ y + x_{M_i}^L \sin\alpha + y_{M_i}^L \cos\alpha \\ z + z_{M_i}^L \end{bmatrix}.$$

The actuation condition of the ith kinematic chain will be:

$$\|M_i B_i\|^2 - L_i^2 = 0. \tag{13.9}$$

Expand the above equation to yield:

$$\left(x + x_{M_i}^L \cos\alpha - y_{M_i}^L \sin\alpha - x_{B_i}\right)^2 + \left(y + x_{M_i}^L \sin\alpha + y_{M_i}^L \cos\alpha - y_{B_i}\right)^2 + \left(z + z_{M_i}^L - z_{B_i}\right)^2 - L_i^2 = 0.$$

According to equation (13.3), we can suppose:

$$\begin{aligned} \boldsymbol{G}_i\left(x,y,z,\alpha\right) = {} & \left(x + x_{M_i}^L \cos\alpha - y_{M_i}^L \sin\alpha - x_{B_i}\right)^2 \\ & + \left(y + x_{M_i}^L \sin\alpha + y_{M_i}^L \cos\alpha - y_{B_i}\right)^2 \\ & + \left(z + z_{M_i}^L - z_{B_i}\right)^2 - L_i^2. \end{aligned}$$

There will be:

$$G(x,y,z,\alpha)=\begin{bmatrix}\left(x+x_{M_1}^L\cos\alpha-y_{M_1}^L\sin\alpha-x_{B_1}\right)^2\\ \quad+\left(y+x_{M_1}^L\sin\alpha+y_{M_1}^L\cos\alpha-y_{B_1}\right)^2\\ \quad+\left(z+z_{M_1}^L-z_{B_1}\right)^2-L_1^2\\ \left(x+x_{M_2}^L\cos\alpha-y_{M_2}^L\sin\alpha-x_{B_2}\right)^2\\ \quad+\left(y+x_{M_2}^L\sin\alpha+y_{M_2}^L\cos\alpha-y_{B_2}\right)^2\\ \quad+\left(z+z_{M_2}^L-z_{B_2}\right)^2-L_2^2\\ \left(x+x_{M_3}^L\cos\alpha-y_{M_3}^L\sin\alpha-x_{B_3}\right)^2\\ \quad+\left(y+x_{M_3}^L\sin\alpha+y_{M_3}^L\cos\alpha-y_{B_3}\right)^2\\ \quad+\left(z+z_{M_3}^L-z_{B_3}\right)^2-L_3^2\\ \left(x+x_{M_4}^L\cos\alpha-y_{M_4}^L\sin\alpha-x_{B_4}\right)^2\\ \quad+\left(y+x_{M_4}^L\sin\alpha+y_{M_4}^L\cos\alpha-y_{B_4}\right)^2\\ \quad+\left(z+z_{M_4}^L-z_{B_4}\right)^2-L_4^2\end{bmatrix}. \tag{13.10}$$

Suppose the desired dexterous workspace is a cylinder and the rotational angle range at any point within the cylinder is $\alpha\in\left[-\frac{\pi}{3},\frac{\pi}{3}\right]$, the cylinder radius is R and its z-coordinate range is that $z\in\left[h_1,h_2\right]$. As a result, the required dexterous workspace shown in Figure 13.3 can be expressed in the absolute

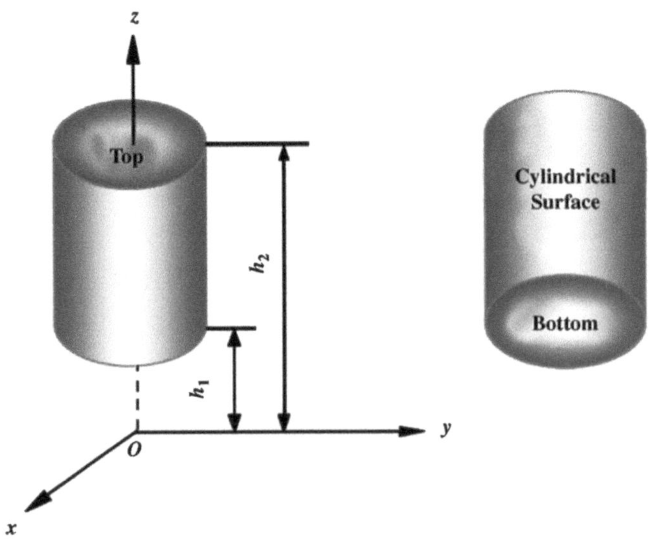

FIGURE 13.3 The required dexterous workspace.

coordinate system as:

$$\Omega_d = \left\{ (x,y,z,\alpha) \left| \begin{array}{l} x = \rho\cos\theta \\ y = \rho\sin\theta \end{array} \right., \quad \rho \in [0,R], \quad \theta \in [0,2\pi]; \right.$$
$$\left. h_1 \le z \le h_2; \quad \alpha \in \left[-\frac{\pi}{3}, +\frac{\pi}{3} \right] \right\}. \tag{13.11}$$

According to [11], the dexterous workspace expressed with equation (13.11) is doubly connected domain and does not consist of any singular point inside the workspace provided that $\frac{\|M_1M_2\|}{\|B_1B_2\|} \neq \frac{\|M_1M_4\|}{\|B_1B_4\|}$. The required dexterous workspace compared to the total workspace of the manipulator is shown in Figure 13.4. With equation (13.11), equation (13.10) can also be transformed into:

$$G(x,y,z,\alpha) = \begin{bmatrix} \left(\rho\cos\theta + x_{M_1}^L\cos\alpha - y_{M_1}^L\sin\alpha - x_{B_1}\right)^2 \\ \quad + \left(\rho\sin\theta + x_{M_1}^L\sin\alpha + y_{M_1}^L\cos\alpha - y_{B_1}\right)^2 \\ \quad + \left(z + z_{M_1}^L - z_{B_1}\right)^2 - L_1^2 \\ \left(\rho\cos\theta + x_{M_2}^L\cos\alpha - y_{M_2}^L\sin\alpha - x_{B_2}\right)^2 \\ \quad + \left(\rho\sin\theta + x_{M_2}^L\sin\alpha + y_{M_2}^L\cos\alpha - y_{B_2}\right)^2 \\ \quad + \left(z + z_{M_2}^L - z_{B_2}\right)^2 - L_2^2 \\ \left(\rho\cos\theta + x_{M_3}^L\cos\alpha - y_{M_3}^L\sin\alpha - x_{B_3}\right)^2 \\ \quad + \left(\rho\sin\theta + x_{M_3}^L\sin\alpha + y_{M_3}^L\cos\alpha - y_{B_3}\right)^2 \\ \quad + \left(z + z_{M_3}^L - z_{B_3}\right)^2 - L_3^2 \\ \left(\rho\cos\theta + x_{M_4}^L\cos\alpha - y_{M_4}^L\sin\alpha - x_{B_4}\right)^2 \\ \quad + \left(\rho\sin\theta + x_{M_4}^L\sin\alpha + y_{M_4}^L\cos\alpha - y_{B_4}\right)^2 \\ \quad + \left(z + z_{M_4}^L - z_{B_4}\right)^2 - L_4^2 \end{bmatrix}. \tag{13.12}$$

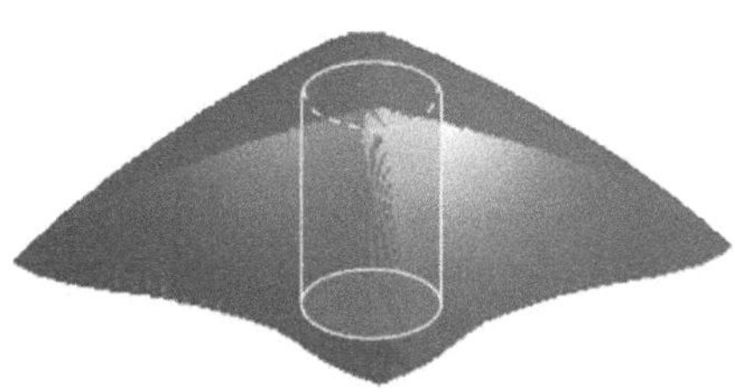

FIGURE 13.4 The required dexterous workspace and the reachable workspace.

The workspace of the manipulator can be naturally expressed as:

$$\boldsymbol{G}\left(x,y,z,\alpha\right) = 0. \tag{13.13}$$

To find the optimum design parameters under the desired dexterous workspace, we can firstly analyze the partial derivatives of equation (13.12):

$$\frac{\partial G_i\left(x,y,z,\alpha\right)}{\partial\left(\rho,\theta,z,\alpha\right)} = 2\begin{bmatrix} I_{i1} & I_{i2} & I_{i3} & I_{i4} \end{bmatrix}^T \tag{13.14}$$

where $i = 1,2,3,4$;

$$\begin{aligned}
I_{11} &= \cos\theta\left(\rho\cos\theta + x_{M_1}^L\cos\alpha - y_{M_1}^L\sin\alpha - x_{B_1}\right)\\
&\quad + \sin\theta\left(\rho\sin\theta + x_{M_1}^L\sin\alpha + y_{M_1}^L\cos\alpha - y_{B_1}\right);\\
I_{12} &= -\rho\sin\theta\left(\rho\cos\theta + x_{M_1}^L\cos\alpha - y_{M_1}^L\sin\alpha - x_{B_1}\right)\\
&\quad + \rho\cos\theta\left(\rho\sin\theta + x_{M_1}^L\sin\alpha + y_{M_1}^L\cos\alpha - y_{B_1}\right);\\
I_{13} &= z + z_{M_1}^L - z_{B_1};\\
I_{14} &= \left(-x_{M_1}^L\sin\alpha - y_{M_1}^L\cos\alpha\right)\left(\rho\cos\theta + x_{M_1}^L\cos\alpha - y_{M_1}^L\sin\alpha - x_{B_1}\right)\\
&\quad + \left(x_{M_1}^L\cos\alpha - y_{M_1}^L\sin\alpha\right)\left(\rho\sin\theta + x_{M_1}^L\sin\alpha + y_{M_1}^L\cos\alpha - y_{B_1}\right);\\
I_{21} &= \cos\theta\left(\rho\cos\theta + x_{M_2}^L\cos\alpha - y_{M_2}^L\sin\alpha - x_{B_2}\right)\\
&\quad + \sin\theta\left(\rho\sin\theta + x_{M_2}^L\sin\alpha + y_{M_2}^L\cos\alpha - y_{B_2}\right);\\
I_{22} &= -\rho\sin\theta\left(\rho\cos\theta + x_{M_2}^L\cos\alpha - y_{M_2}^L\sin\alpha - x_{B_2}\right)\\
&\quad + \rho\cos\theta\left(\rho\sin\theta + x_{M_2}^L\sin\alpha + y_{M_2}^L\cos\alpha - y_{B_2}\right);\\
I_{23} &= z + z_{M_2}^L - z_{B_2};\\
I_{24} &= \left(-x_{M_2}^L\sin\alpha - y_{M_2}^L\cos\alpha\right)\left(\rho\cos\theta + x_{M_2}^L\cos\alpha - y_{M_2}^L\sin\alpha - x_{B_2}\right)\\
&\quad + \left(x_{M_2}^L\cos\alpha - y_{M_2}^L\sin\alpha\right)\left(\rho\sin\theta + x_{M_2}^L\sin\alpha + y_{M_2}^L\cos\alpha - y_{B_2}\right)\\
I_{31} &= \cos\theta\left(\rho\cos\theta + x_{M_3}^L\cos\alpha - y_{M_3}^L\sin\alpha - x_{B_3}\right)\\
&\quad + \sin\theta\left(\rho\sin\theta + x_{M_3}^L\sin\alpha + y_{M_3}^L\cos\alpha - y_{B_3}\right);\\
I_{32} &= -\rho\sin\theta\left(\rho\cos\theta + x_{M_3}^L\cos\alpha - y_{M_3}^L\sin\alpha - x_{B_3}\right)\\
&\quad + \rho\cos\theta\left(\rho\sin\theta + x_{M_3}^L\sin\alpha + y_{M_3}^L\cos\alpha - y_{B_3}\right);\\
I_{33} &= z + z_{M_3}^L - z_{B_3};
\end{aligned}$$

$$
\begin{aligned}
I_{34} &= \left(-x_{M_3}^L \sin\alpha - y_{M_3}^L \cos\alpha\right)\left(\rho\cos\theta + x_{M_3}^L \cos\alpha - y_{M_3}^L \sin\alpha - x_{B_3}\right) \\
&\quad + \left(x_{M_3}^L \cos\alpha - y_{M_3}^L \sin\alpha\right)\left(\rho\sin\theta + x_{M_3}^L \sin\alpha + y_{M_3}^L \cos\alpha - y_{B_3}\right); \\
I_{41} &= \cos\theta\left(\rho\cos\theta + x_{M_4}^L \cos\alpha - y_{M_4}^L \sin\alpha - x_{B_4}\right) \\
&\quad + \sin\theta\left(\rho\sin\theta + x_{M_4}^L \sin\alpha + y_{M_4}^L \cos\alpha - y_{B_4}\right); \\
I_{42} &= -\rho\sin\theta\left(\rho\cos\theta + x_{M_4}^L \cos\alpha - y_{M_4}^L \sin\alpha - x_{B_4}\right) \\
&\quad + \rho\cos\theta\left(\rho\sin\theta + x_{M_4}^L \sin\alpha + y_{M_4}^L \cos\alpha - y_{B_4}\right); \\
I_{43} &= z + z_{M_4}^L - z_{B_4}; \\
I_{44} &= \left(-x_{M_4}^L \sin\alpha - y_{M_4}^L \cos\alpha\right)\left(\rho\cos\theta + x_{M_4}^L \cos\alpha - y_{M_4}^L \sin\alpha - x_{B_4}\right) \\
&\quad + \left(x_{M_4}^L \cos\alpha - y_{M_4}^L \sin\alpha\right)\left(\rho\sin\theta + x_{M_4}^L \sin\alpha + y_{M_4}^L \cos\alpha - y_{B_4}\right).
\end{aligned}
$$

Given the parameters of the fixed base, the moving manipulator and the desired dexterous workspace, we can minimize the length of each kinematic chain to obtain the optimum objective function. If we herein presume $\|M_1M_2\| = 2b$ and $\|M_1M_4\| = 2a$, the coordinates of the manipulator's four vertexes in the local coordinate system can be denoted as:

$$M_1(a, -b, 0), \quad M_2(a, b, 0), \quad M_3(-a, b, 0), \quad M_4(-a, -b, 0).$$

Similarly, if we presume $\|B_1B_2\| = 2q$ and $\|B_1B_4\| = 2p$, the coordinates of the manipulator's four vertexes in the absolute coordinate system can be denoted as:

$$B_1(p, -q, 0), \quad B_2(p, q, 0), \quad B_3(-p, q, 0), \quad B_4(-p, -q, 0).$$

If we herein also suppose that $h_1 = 600$ mm, $h_2 = 1200$ mm, $R = 200$ mm, $a = 200$ mm, $b = 300$ mm, $p = 1000$ mm, and $q = 1200$ mm, and suppose that $\frac{a}{b} \neq \frac{p}{q}$ (to avoid the *similar singularity*[11]), Therefore, searching the desired dexterous workspace will be naturally transformed to be the optimization problem below.

Because $\frac{\partial L_i}{\partial(\rho,\theta,z,\alpha)} = 0$ are the necessary conditions for this problem, if we let $\frac{\partial L_i}{\partial(\rho,\theta,z,\alpha)} = 0$, there will be $\frac{\partial G_i(x,y,z,\alpha)}{\partial(\rho,\theta,z,\alpha)} = 0$, we can therefore find that $z = z_{B_i} - z_{M_i}^L = 0$ which does not belong to the desired dexterous workspace expressed in equation (13.11). That is, according to step 2, $\Omega_c = \{\Phi\}$, where Φ denotes empty set.

Consequently, the extrema of the length function must locate on the boundaries of the volume. That is, the extrema of the function should locate on the top, bottom circular surfaces or the cylindrical surface shown in Figure 13.4.

Firstly, we search the solutions at the bottom circular surface at $z = h_1$, we can use the following set of equations for the $i^{\text{th}}(i = 1,2,3,4)$ kinematic chain to find the extreme point(s):

$$\begin{cases} I_{i1} = 0 \\ I_{i2} = 0 \\ I_{i4} = 0\,. \end{cases}$$

Regarding the first kinematic chain, there will be:

$$\begin{cases} \cos\theta\left(\rho\cos\theta + x^L_{p_1}\cos\alpha - y^L_{p_1}\sin\alpha - x_{B_1}\right) \\ \quad + \sin\theta\left(\rho\sin\theta + x^L_{p_1}\sin\alpha + y^L_{p_1}\cos\alpha - y_{B_1}\right) = 0 \\ -\rho\sin\theta\left(\rho\cos\theta + x^L_{p_1}\cos\alpha - y^L_{p_1}\sin\alpha - x_{B_1}\right) \\ \quad + \rho\cos\theta\left(\rho\sin\theta + x^L_{p_1}\sin\alpha + y^L_{p_1}\cos\alpha - y_{B_1}\right) = 0 \\ \left(-x^L_{p_1}\sin\alpha - y^L_{p_1}\cos\alpha\right)\left(\rho\cos\theta + x^L_{p_1}\cos\alpha - y^L_{p_1}\sin\alpha - x_{B_1}\right) \\ \quad + \left(x^L_{p_1}\cos\alpha - y^L_{p_1}\sin\alpha\right)\left(\rho\sin\theta + x^L_{p_1}\sin\alpha + y^L_{p_1}\cos\alpha - y_{B_1}\right) = 0. \end{cases} \tag{13.15}$$

Substituting equation (13.11) for the corresponding parameters in (13.15) yields:

$$\begin{cases} \cos\theta\left(\rho\cos\theta + a\cos\alpha + b\sin\alpha - p\right) \\ \quad + \sin\theta\left(\rho\sin\theta + a\sin\alpha - b\cos\alpha + q\right) = 0 \\ -\rho\sin\theta\left(\rho\cos\theta + a\cos\alpha + b\sin\alpha - p\right) \\ \quad + \rho\cos\theta\left(\rho\sin\theta + a\sin\alpha - b\cos\alpha + q\right) = 0 \\ \left(-a\sin\alpha + b\cos\alpha\right)\left(\rho\cos\theta + a\cos\alpha + b\sin\alpha - p\right) \\ \quad + \left(a\cos\alpha + b\sin\alpha\right)\left(\rho\sin\theta + a\sin\alpha - b\cos\alpha + q\right) = 0. \end{cases} \tag{13.16}$$

Considering $a = 200$ mm, $b = 300$ mm, $p = 1000$ mm, and $q = 1200$ mm, we can compute the solution of equation (13.16):

$$\begin{cases} \rho = 0 \\ \theta = 39.8056^\circ \\ \alpha = 6.1155^\circ\,. \end{cases}$$

Therefore, the length of the kinematic chain at this point is an extremum:

$$l_1^1 = 1342.9779.$$

Now, we also need to analyze the boundary points of the circular surface at $z = h_1$. According to equation (13.11), we need to analyze the length values of the kinematic chain at $\begin{cases} \rho = R \\ \theta = 0 \\ \alpha = -\frac{\pi}{3} \end{cases}$ and $\begin{cases} \rho = R \\ \theta = 0 \\ \alpha = \frac{\pi}{3} \end{cases}$, which are computed individually as:

$$l_1^2 = 1431.7821 \text{ and } l_1^3 = 1431.7821.$$

According to equation (13.11), we also need to analyze the length values of the kinematic chain at $\begin{cases} \rho = R \\ \theta = \pi \\ \alpha = -\frac{\pi}{3} \end{cases}$ and $\begin{cases} \rho = R \\ \theta = \pi \\ \alpha = \frac{\pi}{3} \end{cases}$, which are computed individually as:

$$l_1^4 = 1725.6437 \text{ and } l_1^5 = 1600.6730.$$

Therefore

$$\begin{cases} l_1^{\min} = 1342.9779 \\ l_1^{\max} = 1725.6437. \end{cases}$$

For the second kinematic chain at $z = h_1$, there will be:

$$\begin{cases} \cos\theta\left(\rho\cos\theta + a\cos\alpha - b\sin\alpha - p\right) \\ \quad + \sin\theta\left(\rho\sin\theta + a\sin\alpha + b\cos\alpha - q\right) = 0 \\ -\rho\sin\theta\left(\rho\cos\theta + a\cos\alpha - b\sin\alpha - p\right) \\ \quad + \rho\cos\theta\left(\rho\sin\theta + a\sin\alpha + b\cos\alpha - q\right) = 0 \\ \left(-a\sin\alpha - b\cos\alpha\right)\left(\rho\cos\theta + a\cos\alpha - b\sin\alpha - p\right) \\ \quad + \left(a\cos\alpha - b\sin\alpha\right)\left(\rho\sin\theta + a\sin\alpha + b\cos\alpha - q\right) = 0. \end{cases} \tag{13.17}$$

For $a = 200$ mm, $b = 300$ mm, $p = 1000$ mm, and $q = 1200$ mm, the solution of equation (13.17) will be:

$$\begin{cases} \rho = 0 \\ \theta = -39.8056^\circ \\ \alpha = -6.1155^\circ. \end{cases}$$

Therefore, the length of the kinematic chain at this point is:

$$l_2^1 = 1342.9779.$$

Now, we need to analyze the boundary points of the circular surface at $z = h_1$, too. The length values of the kinematic chain at $\begin{cases} \rho = R \\ \theta = 0 \\ \alpha = -\frac{\pi}{3} \end{cases}$ and $\begin{cases} \rho = R \\ \theta = 0 \\ \alpha = \frac{\pi}{3} \end{cases}$ can be computed individually:

$$l_2^2 = 1431.7821, \quad l_2^3 = 1431.7821.$$

Similarly, the length values of the kinematic chain at $\begin{cases} \rho = R \\ \theta = \pi \\ \alpha = -\frac{\pi}{3} \end{cases}$ and $\begin{cases} \rho = R \\ \theta = \pi \\ \alpha = \frac{\pi}{3} \end{cases}$ can be computed individually as:

$$l_2^4 = 1600.6730, \quad l_2^5 = 1725.6437.$$

Therefore

$$\begin{cases} l_2^{\min} = 1342.9779 \\ l_2^{\max} = 1725.6437. \end{cases}$$

For the third kinematic chain at $z = h_1$, there will be:

$$\begin{cases} \cos\theta\left(\rho\cos\theta - a\cos\alpha - b\sin\alpha + p\right) \\ \quad + \sin\theta\left(\rho\sin\theta - a\sin\alpha + b\cos\alpha - q\right) = 0 \\ -\rho\sin\theta\left(\rho\cos\theta - a\cos\alpha - b\sin\alpha + p\right) \\ \quad + \rho\cos\theta\left(\rho\sin\theta - a\sin\alpha + b\cos\alpha - q\right) = 0 \\ \left(a\sin\alpha - b\cos\alpha\right)\left(\rho\cos\theta - a\cos\alpha - b\sin\alpha + p\right) \\ \quad - \left(a\cos\alpha + b\sin\alpha\right)\left(\rho\sin\theta - a\sin\alpha + b\cos\alpha - q\right) = 0. \end{cases} \tag{13.18}$$

For $a = 200$ mm, $b = 300$ mm, $p = 1000$ mm, and $q = 1200$ mm, the solution of equation (13.18) will be:

$$\begin{cases} \rho = 0 \\ \theta = -39.8056° \\ \alpha = -6.1155°. \end{cases}$$

Therefore, the length of the kinematic chain at this point is:

$$l_3^1 = 1342.9779.$$

According to equation (13.11), we need to analyze the length values of the kinematic chain at $\begin{cases} \rho = R \\ \theta = 0 \\ \alpha = -\frac{\pi}{3} \end{cases}$ and $\begin{cases} \rho = R \\ \theta = 0 \\ \alpha = \frac{\pi}{3} \end{cases}$, which are computed individually as:

$$l_3^2 = 1600.6730, \quad l_3^3 = 1725.6437.$$

Besides, we also need to analyze the length values of the kinematic chain at $\begin{cases} \rho = R \\ \theta = \pi \\ \alpha = -\frac{\pi}{3} \end{cases}$ and $\begin{cases} \rho = R \\ \theta = \pi \\ \alpha = \frac{\pi}{3} \end{cases}$, which are computed individually as:

$$l_3^4 = 1431.7821, \quad l_3^5 = 1431.7821.$$

Therefore

$$\begin{cases} l_3^{\min} = 1342.9779 \\ l_3^{\max} = 1725.6437. \end{cases}$$

For the fourth kinematic chain at $z = h_1$, there will be:

$$\begin{cases} \cos\theta\left(\rho\cos\theta - a\cos\alpha + b\sin\alpha + p\right) \\ \quad + \sin\theta\left(\rho\sin\theta - a\sin\alpha - b\cos\alpha + q\right) = 0 \\ -\rho\sin\theta\left(\rho\cos\theta - a\cos\alpha + b\sin\alpha + p\right) \\ \quad + \rho\cos\theta\left(\rho\sin\theta - a\sin\alpha - b\cos\alpha + q\right) = 0 \\ \left(a\sin\alpha + b\cos\alpha\right)\left(\rho\cos\theta - a\cos\alpha + b\sin\alpha + p\right) \\ \quad + \left(-a\cos\alpha + b\sin\alpha\right)\left(\rho\sin\theta - a\sin\alpha - b\cos\alpha + q\right) = 0. \end{cases} \tag{13.19}$$

For $a = 200$ mm, $b = 300$ mm, $p = 1000$ mm, and $q = 1200$ mm, the solution of equation (13.19) will be:

$$\begin{cases} \rho = 0 \\ \theta = -39.8056^\circ \\ \alpha = -6.1155^\circ. \end{cases}$$

Therefore, the length of the kinematic chain at this point is:

$$l_4^1 = 1342.9779.$$

The length values of the kinematic chain at $\begin{cases} \rho = R \\ \theta = 0 \\ \alpha = -\frac{\pi}{3} \end{cases}$ and $\begin{cases} \rho = R \\ \theta = 0 \\ \alpha = \frac{\pi}{3} \end{cases}$ are computed individually as:

$$l_4^2 = 1600.6730, \quad l_4^3 = 1725.6437.$$

The length values of the kinematic chain at $\begin{cases} \rho = R \\ \theta = \pi \\ \alpha = -\frac{\pi}{3} \end{cases}$ and $\begin{cases} \rho = R \\ \theta = \pi \\ \alpha = \frac{\pi}{3} \end{cases}$ are computed individually as:

$$l_4^4 = 1431.7821, \quad l_4^5 = 1431.7821.$$

Therefore

$$\begin{cases} l_4^{\min} = 1342.9779 \\ l_4^{\max} = 1725.6437. \end{cases}$$

Therefore, in the lower circular plane, the maximum and the minimum lengths of all of the kinematic chains are identical. So, we can set the length variable ranges of each kinematic chain for the lower circular plane case to be:

$$\begin{cases} l_{lower}^{\min} = 1342.9779 \\ l_{lower}^{\max} = 1725.6437. \end{cases} \tag{13.20}$$

Similarly, we can search the extrema of the function (13.2) in the upper circular plane at $z = 1200$ mm, the solutions are:

$$\begin{cases} l_{upper}^{\min} = 1769.1806 \\ l_{upper}^{\max} = 2014.4096. \end{cases} \tag{13.21}$$

With the like procedures, we can analyze the extrema at the cylindrical surface shown in Figure 13.4. However, it is not difficult to find that the extrema of the objective function must locate either on the circle of the lower circular plane $z = h_1$ or on the upper circular plane $z = h_2$. Consequently, they must between the values of equations (13.20) and (13.21).

Therefore, to satisfy the desired dexterous workspace of equation (13.11), the length range of the identical kinematic chain should meet:

$$\begin{cases} l_{\min} \leq 1342.9779 \\ l_{\max} \geq 2014.4096. \end{cases} \tag{13.22}$$

For example, if we design each kinematic chain of the manipulator with $\begin{cases} l_{\min} = 1300 \\ l_{\max} = 2100 \end{cases}$, the dexterous workspace of the manipulator not only meets the initial needs but also has a certain surplus. The relationship between the workspace of the manipulator, the desired dexterous workspace, and the parallel manipulator is shown in Figure 13.5. When all of the kinematic chains shrink to $l_{\min} = 1300$, the smallest angle from the chain to the fixed base, denoted by

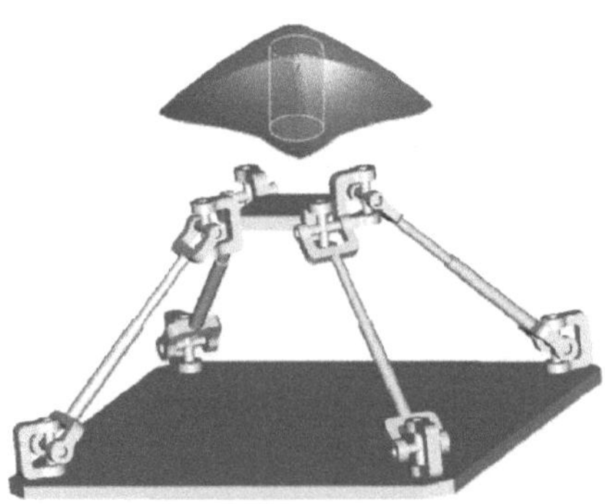

FIGURE 13.5 The required dexterous workspace and the mechanism.

β shown in Figure 13.2, can also be calculated:

$$\beta = \arccos\left(\frac{\|B_1B_3\| - \|M_1M_3\|}{2L_{\min}}\right) = \arccos\left(\frac{\sqrt{1600^2+1800^2}}{2\times 1300}\right)$$
$$= 22.1384^\circ > 0. \tag{13.23}$$

Equation (13.23) expects that the manipulator will not reduce to the *boundary singularity* [11] even when all of its legs shrink to their least lengths, which also demonstrates the feasibility of manufacturing if we select $\begin{cases} l_{\min} = 1300 \\ l_{\max} = 2100 \end{cases}$ to design the kinematic chains of the manipulator. From the above analysis, we can find that, in applications, we only need to analyze one of the legs of the manipulator because the desired dexterous workspace has the symmetrical characteristics of the manipulator.

13.2.2 Optimum Design of a Kind of Gough–Stewart Spatial Parallel Manipulator

The architecture of a Gough–Stewart spatial parallel manipulator is shown in Figure 13.6. The manipulator is actuated by six identical SPS (spherical joint, prismatic joint, and spherical joint) kinematic chains. The manipulator has six DOFs (three rotational DOFs and three orthogonal translational DOFs). Therefore, if we presume the absolute coordinates of the geometry center of the manipulator to be (x, y, z) and the rotational order is 3-1-3, the rotational angles are denoted by ψ,θ, and ϕ individually. Therefore, the transformation matrix

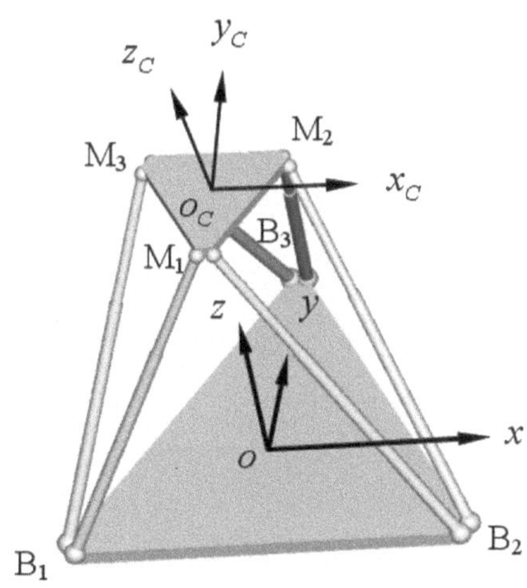

FIGURE 13.6 Architecture of Gough–Stewart platform.

from the end pose to the initial one can be denoted as:

$$\mathbf{R}=\begin{bmatrix}\cos\psi & -\sin\psi & 0\\ \sin\psi & \cos\psi & 0\\ 0 & 0 & 1\end{bmatrix}\begin{bmatrix}1 & 0 & 0\\ 0 & \cos\theta & -\sin\theta\\ 0 & \sin\theta & \cos\theta\end{bmatrix}\begin{bmatrix}\cos\phi & -\sin\phi & 0\\ \sin\phi & \cos\phi & 0\\ 0 & 0 & 1\end{bmatrix}.$$

Assume that

$$\begin{cases} q_0=\cos\frac{\theta}{2}\cos\frac{\psi+\phi}{2}\\ q_1=\sin\frac{\theta}{2}\cos\frac{\psi-\phi}{2}\\ q_2=\sin\frac{\theta}{2}\sin\frac{\psi-\phi}{2}\\ q_3=\cos\frac{\theta}{2}\sin\frac{\psi+\phi}{2}\end{cases}$$

where $q_i\,(i=0,1,2,3)$ is a set of quaternions.

Naturally, we have:

$$q_0^2+q_1^2+q_2^2+q_3^2=1. \tag{13.24}$$

The rotational matrix can be expressed in the form of quaternions:

$$\mathbf{R}=\begin{bmatrix}2(q_0^2+q_1^2)-1 & 2(q_1q_2-q_0q_3) & 2(q_1q_3+q_0q_2)\\ 2(q_1q_2+q_0q_3) & 2(q_0^2+q_2^2)-1 & 2(q_2q_3-q_0q_1)\\ 2(q_1q_3-q_0q_2) & 2(q_2q_3+q_0q_1) & 2(q_0^2+q_3^2)-1\end{bmatrix}. \tag{13.25}$$

Suppose the coordinates of the ith$(i=1,2,\ldots,6)$ vertex of the end effector in the local coordinate system are denoted by $P_i\left(x_{M_i}^C\ y_{M_i}^C\ z_{M_i}^C\right)$, the absolute coordinates $P_i(x_{M_i}\ y_{M_i}\ z_{M_i})$ can be obtained:

$$\begin{bmatrix}x_{M_i}\\ y_{M_i}\\ z_{M_i}\end{bmatrix}=\begin{bmatrix}x\\ y\\ z\end{bmatrix}+\mathbf{R}\begin{bmatrix}x_{M_i}^C\\ y_{M_i}^C\\ z_{M_i}^C\end{bmatrix}. \tag{13.26}$$

Substitute equation (13.25) in equation (13.26), there will be:

$$\begin{bmatrix}x_{M_i}\\ y_{M_i}\\ z_{M_i}\end{bmatrix}=\begin{bmatrix}x+x_{M_i}^C[2(q_0^2+q_1^2)-1]+2y_{M_i}^C(q_1q_2-q_0q_3)+2z_{M_i}^C(q_1q_3+q_0q_2)\\ y+2x_{M_i}^C(q_1q_2+q_0q_3)+y_{M_i}^C[2(q_0^2+q_2^2)-1]+2z_{M_i}^C(q_2q_3-q_0q_1)\\ z+2x_{M_i}^C(q_1q_3-q_0q_2)+2y_{M_i}^C(q_2q_3+q_0q_1)+z_{M_i}^C[2(q_0^2+q_3^2)-1]\end{bmatrix}.$$

Assume

$$\boldsymbol{F}_i\left(x,y,z,q_0,q_1,q_2,q_3\right)=\|M_iB_i\|^2-L_i^2. \tag{13.27}$$

Expanding equation (13.27) yields:

$$\begin{aligned}\boldsymbol{F}_i=&\{x+x_{M_i}^C[2(q_0^2+q_1^2)-1]+2y_{M_i}^C(q_1q_2-q_0q_3)\\ &+2z_{M_i}^C(q_1q_3+q_0q_2)-x_{B_i}\}^2\end{aligned}$$

$$
\begin{aligned}
&+\{y+2x_{M_i}^C(q_1q_2+q_0q_3)+y_{M_i}^C[2(q_0^2+q_2^2)-1]\\
&\quad+2z_{M_i}^C(q_2q_3-q_0q_1)-y_{B_i}\}^2\\
&+\{z+2x_{M_i}^C(q_1q_3-q_0q_2)+2y_{M_i}^C(q_2q_3+q_0q_1)\\
&\quad+z_{M_i}^C[2(q_0^2+q_3^2)-1]-z_{B_i}\}^2-L_i^2
\end{aligned} \tag{13.28}
$$

The workspace of the manipulator can be denoted as:

$$\boldsymbol{F}_i = 0, (i = 1,2,\ldots,6).$$

Suppose the desired dexterous workspace is shown in Figure 13.7. If we presume the dexterous rotational angle range at any point within the adroit workspace are:

$$\psi,\theta,\phi \in \left[-\frac{\pi}{6}, +\frac{\pi}{6}\right] \tag{13.29}$$

We now can find out the variable ranges of $q_i (i = 0,1,2,3)$ from equation (13.29):

Because

$$\frac{\partial q_1}{\partial(\psi,\theta,\phi)} = \begin{bmatrix} \frac{\partial q_1}{\partial \psi} \\ \frac{\partial q_1}{\partial \theta} \\ \frac{\partial q_1}{\partial \phi} \end{bmatrix} = \frac{1}{2}\begin{bmatrix} -\sin\frac{\theta}{2}\sin\frac{\psi-\phi}{2} \\ \cos\frac{\theta}{2}\cos\frac{\psi-\phi}{2} \\ \sin\frac{\theta}{2}\sin\frac{\psi-\phi}{2} \end{bmatrix}.$$

Let $\frac{\partial q_1}{\partial(\psi,\theta,\phi)} = 0$, we can find that the solutions do not belong to the variable ranges of (13.29). Therefore, the extreme values for $q_1(\psi,\theta,\phi)$ must locate at

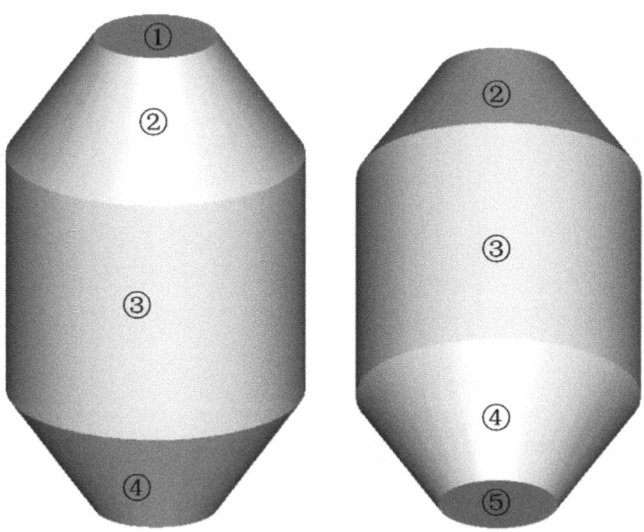

FIGURE 13.7 The desired dexterous workspace and the boundaries.

the vertex of variable ranges (13.29). It is not hard to find that:

$$q_1\left(\psi,\theta,\phi\right) \in \left[-\sin\frac{\pi}{12}, \sin\frac{\pi}{12}\right]. \tag{13.30}$$

Similarly, we can obtain that

$$\begin{cases} q_2\left(\psi,\theta,\phi\right) \in \left[-\frac{1}{2}\sin\frac{\pi}{12}, \frac{1}{2}\sin\frac{\pi}{12}\right] \\ q_3\left(\psi,\theta,\phi\right) \in \left[-\frac{1}{2}\cos\frac{\pi}{12}, \frac{1}{2}\cos\frac{\pi}{12}\right]. \end{cases} \tag{13.31}$$

From equation (13.24), we can find that q_0 is not independent when $q_i(i = 0,1,2,3)$ are given. Therefore, equations (13.24) can be naturally transformed into:

$$\begin{cases} -\sin\frac{\pi}{12} \le q_1 \le \sin\frac{\pi}{12} \\ -\frac{1}{2}\sin\frac{\pi}{12} \le q_2 \le \frac{1}{2}\sin\frac{\pi}{12} \\ -\frac{1}{2}\cos\frac{\pi}{12} \le q_3 \le \frac{1}{2}\cos\frac{\pi}{12} \\ q_0^2 + q_1^2 + q_2^2 + q_3^2 = 1. \end{cases} \tag{13.32}$$

We now consider the parametric equations for the boundaries of the desired dexterous workspace. Suppose the distance from the bottom circular surface of the dexterous workspace to the *xoy*-plane is denoted by *h*, the workspace whose dimensions are given in Figure 13.8, can be depicted as equations (13.33) to (13.37):

$$\begin{cases} x^2 + y^2 = 10000 \\ z = h \end{cases} \tag{13.33}$$

$$\begin{cases} x^2 + y^2 = \left[100 + \frac{7(z-h)}{10}\right]^2 \\ h \le z \le h + 200 \end{cases} \tag{13.34}$$

$$\begin{cases} x^2 + y^2 = 240^2 \\ h + 200 \le z \le h + 600 \end{cases} \tag{13.35}$$

$$\begin{cases} x^2 + y^2 = \left[660 - \frac{7(z+h)}{10}\right]^2 \\ h + 600 \le z \le h + 800 \end{cases} \tag{13.36}$$

$$\begin{cases} x^2 + y^2 \le 10000 \\ z = h + 800. \end{cases} \tag{13.37}$$

Now, the problem left is to search the extreme values of $L_i\left(i = 1,2,\ldots,6\right)$ on the boundaries of the desired dexterous workspace according to the theorem. Because the lengths of all of the kinematic chains of the manipulator are identical, the problem, in fact, reduces to search the extreme values of the

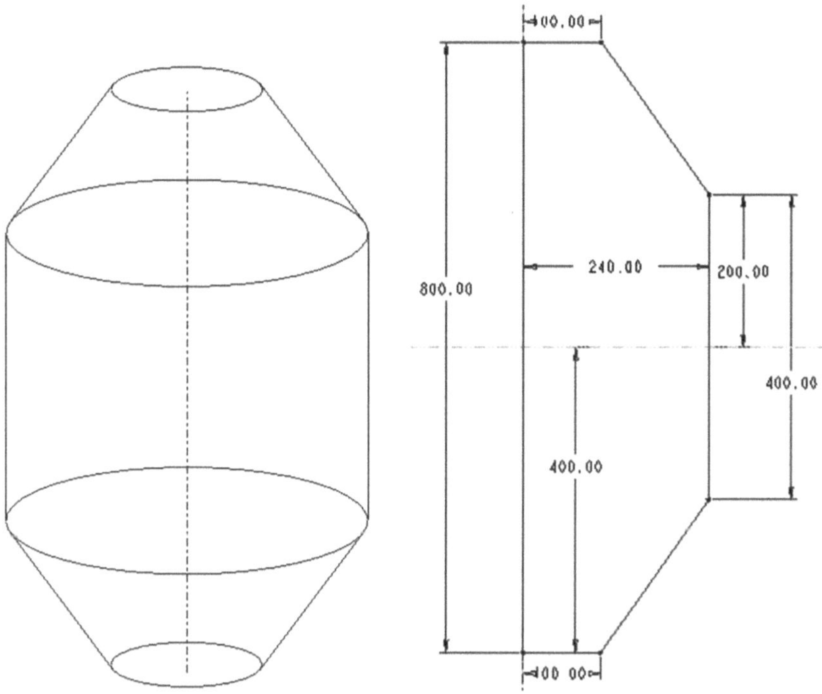

FIGURE 13.8 Desired dexterous workspace and the dimensions.

length of any one kinematic chain on the boundaries of the adroit workspace, say, the first one.

Regarding the first kinematic chain, we can obtain from equation (13.28):

$$\begin{aligned} \boldsymbol{F}_1 = &\Big\{x + x^C_{M_1}[2(q_0^2 + q_1^2) - 1] + 2y^C_{M_1}(q_1q_2 - q_0q_3) \\ &\quad + 2z^C_{M_1}(q_1q_3 + q_0q_2) - x_{B_1}\Big\}^2 \\ &+ \Big\{y + 2x^C_{M_1}(q_1q_2 + q_0q_3) + y^C_{M_1}[2(q_0^2 + q_2^2) - 1] \\ &\quad + 2z^C_{M_1}(q_2q_3 - q_0q_1) - y_{B_1}\Big\}^2 \\ &+ \Big\{z + 2x^C_{M_1}(q_1q_3 - q_0q_2) + 2y^C_{M_1}(q_2q_3 + q_0q_1) \\ &\quad + z^C_{M_1}[2(q_0^2 + q_3^2) - 1] - z_{B_1}\Big\}^2 - L_1^2 \end{aligned} \tag{13.38}$$

$$\frac{\partial \boldsymbol{F}_1}{\partial\left(x, y, z, q_1, q_2, q_3\right)} = 2\begin{bmatrix} I_1 & I_2 & I_3 & I_4 & I_5 & I_6 \end{bmatrix}^T \tag{13.39}$$

where

$$
\begin{aligned}
I_1 &= x + x_{M_1}^C\left[2(q_0^2+q_1^2)-1\right] + 2y_{M_1}^C(q_1q_2-q_0q_3) \\
&\quad +2z_{M_1}^C(q_1q_3+q_0q_2) - x_{B_1}; \\
I_2 &= y + 2x_{M_1}^C(q_1q_2+q_0q_3) + y_{M_1}^C\left[2(q_0^2+q_2^2)-1\right] \\
&\quad +2z_{M_1}^C(q_2q_3-q_0q_1) - y_{B_1}; \\
I_3 &= z + 2x_{M_1}^C(q_1q_3-q_0q_2) + 2y_{M_1}^C(q_2q_3+q_0q_1) \\
&\quad +z_{M_1}^C\left[2(q_0^2+q_3^2)-1\right] - z_{B_1}; \\
I_4 &= 2\left(2q_1x_{M_1}^C + q_2y_{M_1}^C + q_3z_{M_1}^C\right)\left\{x + x_{M_1}^C\left[2(q_0^2+q_1^2)-1\right]\right. \\
&\quad \left. +2y_{M_1}^C(q_1q_2-q_0q_3) + 2z_{M_1}^C(q_1q_3+q_0q_2) - x_{B_1}\right\} \\
&\quad +2\left(q_2x_{M_1}^C - q_0z_{M_1}^C\right)\left\{y + 2x_{M_1}^C(q_1q_2+q_0q_3) + y_{M_1}^C\left[2(q_0^2+q_2^2)-1\right]\right. \\
&\quad \left. +2z_{M_1}^C(q_2q_3-q_0q_1) - y_{B_1}\right\} \\
&\quad +2\left(q_3x_{M_1}^C + q_0y_{M_1}^C\right)\left\{z + 2x_{M_1}^C(q_1q_3-q_0q_2) + 2y_{M_1}^C(q_2q_3+q_0q_1)\right. \\
&\quad \left. +z_{M_1}^C\left[2(q_0^2+q_3^2)-1\right] - z_{B_1}\right\}; \\
I_6 &= -2\left(q_0y_{M_1}^C - q_1z_{M_1}^C\right)\left\{x + x_{M_1}^C\left[2(q_0^2+q_1^2)-1\right] + 2y_{M_1}^C(q_1q_2-q_0q_3)\right. \\
&\quad \left. +2z_{M_1}^C(q_1q_3+q_0q_2) - x_{B_1}\right\} \\
&\quad +\left(q_0x_{M_1}^C + q_2z_{M_1}^C\right)\left\{y + 2x_{M_1}^C(q_1q_2+q_0q_3) + y_{M_1}^C\left[2(q_0^2+q_2^2)-1\right]\right. \\
&\quad \left. +2z_{M_1}^C(q_2q_3-q_0q_1) - y_{B_1}\right\} \\
&\quad +2\left(q_1x_{M_1}^C + q_2y_{M_1}^C + 2q_3z_{M_1}^C\right)\left\{z + 2x_{M_1}^C(q_1q_3-q_0q_2)\right. \\
&\quad \left. +2y_{M_1}^C(q_2q_3+q_0q_1) + z_{M_1}^C\left[2(q_0^2+q_3^2)-1\right] - z_{B_1}\right\}.
\end{aligned}
$$

Let

$$
\begin{cases}
\dfrac{\partial \mathbf{F}_1}{\partial\left(x,y,z,q_1,q_2,q_3\right)} = \mathbf{0} \\
q_0^2+q_1^2+q_2^2+q_3^2 = 1
\end{cases}
\tag{13.40}
$$

According to Figure 13.9, in the absolute coordinate system, the coordinates of spherical joints on the fixed base can be individually expressed in the form of circumcircle radius, R, as:

$$
B_1\left(-\frac{\sqrt{3}}{2}R,\ -\frac{1}{2}R, 0\right),\quad B_2\left(\frac{\sqrt{3}}{2}R,\ -\frac{1}{2}R, 0\right), B_3\left(0, R, 0\right).
$$

In the local coordinate system shown in Figure 13.8, the coordinates of spherical joints on the moving manipulator can be individually expressed in the

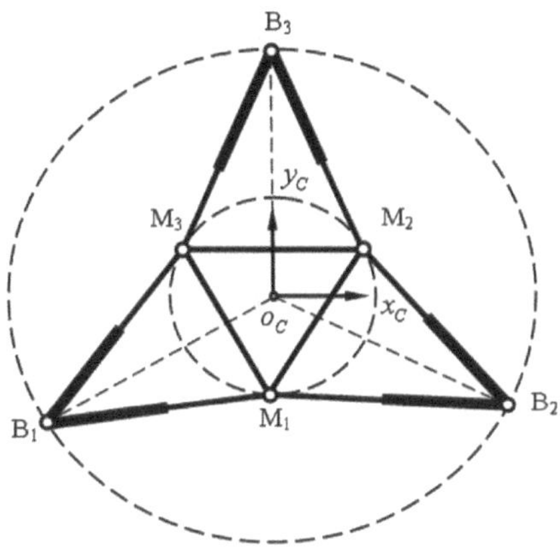

FIGURE 13.9 Top view of Gough–Stewart manipulator.

form of circumcircle radius, r, as:

$$M_3\left(0,-r,0\right),M_2\left(\frac{\sqrt{3}}{2}r,\frac{1}{2}r,0\right),\quad M_3\left(-\frac{\sqrt{3}}{2}r,\frac{1}{2}r,0\right).$$

Assume $r = 500$ mm and $R = 1500$ mm, the length extrema points within the boundaries of the dexterous workspace of each kinematic chain of the manipulator can be calculated by comparing the solutions of equation (13.40) with the vertexes of equations (13.32) to (13.37). First, we can solve the solutions of equation (13.40):

$$\begin{cases} x = -1299.71242046722 \\ y = -1249.98133139307 \\ z = -4.26840083861197. \end{cases}$$

However, the solutions do not belong to the desired dexterous workspace, therefore, the length extrema must locate at the boundary points. As a result, if the values of the $L_i (i = 1,2,\ldots,6)$ from equation (13.28) at every boundary point are calculated, the minimum one is the least length and the maximum one is the largest length.

For the sake of simplicity, we herein only enumerate the detail process of searching the length extrema points within the boundary of the first circular surface of the desired dexterous workspace shown in Figure 13.7.

The parametric expressions of the circular surface are prescribed by equation (13.37), assume $h = 1500$ mm, equation (13.37) can be denoted as:

$$\begin{cases} x^2 + y^2 \leq 10000 \\ z = 2300. \end{cases} \tag{13.41}$$

Substitute z in equation (13.38) with $z = 2300$, we can find that

$$L_1 = \sqrt{\begin{array}{l}\{x + x^C_{M_1}[2(q_0^2 + q_1^2) - 1] + 2y^C_{M_1}(q_1q_2 - q_0q_3) \\ \quad + 2z^C_{M_1}(q_1q_3 + q_0q_2) - x_{B_1}\}^2 \\ \quad + \{y + 2x^C_{M_1}(q_1q_2 + q_0q_3) + y^C_{M_1}[2(q_0^2 + q_2^2) - 1] \\ \quad + 2z^C_{M_1}(q_2q_3 - q_0q_1) - y_{B_1}\}^2 \\ \quad + \{2300 + 2x^C_{M_1}(q_1q_3 - q_0q_2) + 2y^C_{M_1}(q_2q_3 + q_0q_1) \\ \quad + z^C_{M_1}[2(q_0^2 + q_3^2) - 1] - z_{B_1}\}^2\end{array}}. \tag{13.42}$$

Following, we obtain the extrema of L_1 under the conditions of (13.41):

$$\begin{cases} L_1^{\min} = 2256.60346756914\text{mm} \\ L_1^{\max} = 3170.91813590044\text{mm}. \end{cases}$$

Similar procedures can be utilized to solve the extrema of (13.42) at the other boundary points of equations (13.32) and (13.41).

Through comparing the extrema of the length function on the five boundary surfaces shown in Figure 13.7, we can find that the best length range of each kinematic chain can be set as:

$$\begin{cases} L^{\min} \le 1579.39832870369\text{mm} \\ L^{\max} \ge 3170.91813590044\text{mm}. \end{cases}$$

However, the least length of each kinematic chain should also satisfy that the distance from the manipulator to the fixed base is greater than zero. That is, for the first kinematic chain B_1M_1, its least length should be larger than its projection line on the fixed-base plane shown in Figure 13.9. Because the projection line can be calculated in the triangle $o_C B_1 M_1$, least length of kinematic chain B_1M_1 can be denoted as:

$$L^{least}_{B_1M_1} > \sqrt{\|o_C B_1\|^2 + \|o_C M_1\|^2 - 2\|o_C B_1\|\,\|o_C M_1\| \cos\angle B_1 o_C M_1}.$$

Therefore, the length range of each kinematic chain will be:

$$\begin{cases} L^{\max} \ge 3170.91813590044\text{mm} \\ L^{\min} > 1322.87565553230\text{mm} \\ L^{\min} \le 1579.39832870369\text{mm}. \end{cases}$$

For an example, if we select $\begin{cases} L^{\min} = 1500\text{mm} \\ L^{\max} = 3200\text{mm} \end{cases}$ to design the manipulator shown in Figure 13.7, the dexterous workspace will be a little larger than the desired one shown in Figure 13.9.

In this example, we assume that the length and length range of every kinematic chain are identical individually. In applications, it does not matter if the length and length range of every kinematic chain are not identical. The only thing that remains to be done is repeating the above process for each kinematic chain.

13.2.3 Interval Analysis on the Perturbation Space of a Suspension Mechanism

In applications, considering the load of the end effector, the dexterous workspace of the manipulator will change within a certain interval. For example, the suspension mechanism proposed in this section, shown in Figure 13.10, has only one translational DOF along the z-axis in theory. That is to say, the alignment parameters are invariable during jounce and rebound. However, considering the accelerating and braking forces, the inertial forces resulting from turning and so on, the capacity of the suspension mechanism to trace a straight line relies on the stiffness of the mechanism.

The structure of a rectilinear suspension is shown in Figure 13.10. The knuckle is connected with the vehicle body via four identical branches, $A_iB_i (i = 1,2,3,4)$, whose axial length can be changed automatically within a required distance range. According to Chapter 4, we can find that the free motion of the knuckle is a straight line along the z-axis. And each branch only subject to one shearing force along the joint axes of both ends, one torque about its axis and one bending moment about the direction perpendicular to branch axis and the end joint axes simultaneously. To investigate the perturbation space of the suspension, the statics should be first analyzed.

Generally, there are driving and braking forces along the x-direction, the lateral forces along the y-direction, the vertical forces from the ground and the moments and torques about the x-, y- and z-axes. Since the knuckle just tracks along the z-axis, the external loads transferred from the road to the knuckle are withstood by the suspension member bars except the force along the

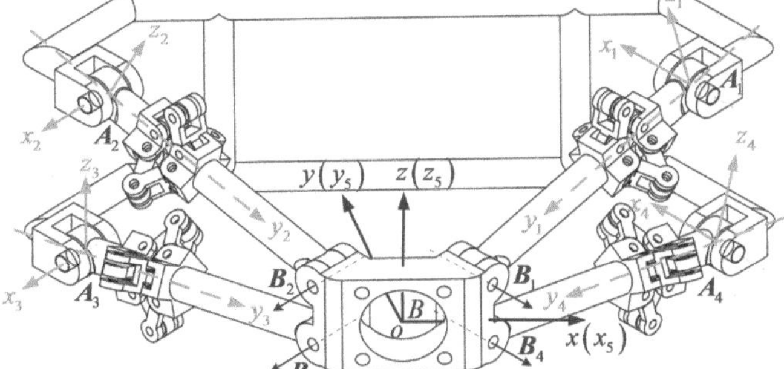

FIGURE 13.10 Structure of the suspension mechanism.

z-axis, which is withstood by the suspension spring. As a result, the external loads on the knuckle can be decomposed as two forces and three moments in the coordinate frame $oxyz$, namely $\boldsymbol{F}_x, \boldsymbol{F}_y, \boldsymbol{M}_x, \boldsymbol{M}_y$, and $\boldsymbol{M}_z$, correspondingly. Suppose the initial absolute coordinates of the center of the knuckle are denoted by $\boldsymbol{r}_B = (0\ 0\ 0)^T$ and the rotated angle about x-, y-, and z-axes are represented by α, β, and γ, respectively.

To simplify the analysis, the knuckle is treated as a rigid body within the tolerance of engineering applications. The coordinates of the four joints B_1, B_2, B_3, and B_4 are $\boldsymbol{r}_{B_1}^L = (l\ 0\ m), \boldsymbol{r}_{B_2}^L = (-l\ 0\ m), \boldsymbol{r}_{B_3}^L = (-l\ 0\ -m)$, and $\boldsymbol{r}_{B_4}^L = (l\ 0\ -m)$ in the knuckle coordinate frame, individually. So the absolute coordinates of these joints are expressed by

$$\boldsymbol{r}_{B_i} = \mathbf{R}\boldsymbol{r}_{B_i}^L + \delta\boldsymbol{r}_B \tag{13.43}$$

where $i = 1,2,3,4$, $\boldsymbol{r}_{B_i}^L$ denotes the coordinates of joint B_i in the local frame $ox_5y_5z_5$, and $\boldsymbol{r}_{B_i}$ represents the absolute coordinates B_i in the frame $oxyz$, and

$$\mathbf{R} = \begin{bmatrix} \cos\beta\cos\gamma & -\cos\beta\sin\gamma & \sin\beta \\ \sin\alpha\sin\beta\cos\gamma + \cos\alpha\sin\gamma & -\sin\alpha\sin\beta\sin\gamma + \cos\alpha\cos\gamma & -\sin\alpha\cos\beta \\ -\cos\alpha\sin\beta\cos\gamma + \sin\alpha\sin\gamma & \cos\alpha\sin\beta\sin\gamma + \sin\alpha\cos\gamma & \cos\alpha\cos\beta \end{bmatrix},$$

$\delta\boldsymbol{r}_B = (\delta x\quad \delta y\quad 0)^T$.

Considering the tiny values of the rotational angles α, β, and γ, the transformation matrix can be approximately simplified as

$$\mathbf{R} = \begin{bmatrix} 1 & -\gamma & \beta \\ \gamma & 1 & -\alpha \\ -\beta & \alpha & 1 \end{bmatrix}.$$

So the displacements of joint $B_i (i = 1,2,3,4)$ under the external forces $\boldsymbol{F}_x, \boldsymbol{F}_y, \boldsymbol{M}_x, \boldsymbol{M}_y$, and $\boldsymbol{M}_z$ can be expressed as

$$\boldsymbol{s}_{B_i} = \boldsymbol{r}_{B_i} - \boldsymbol{r}_{B_i}^L = (\mathbf{R} - \mathbf{I})\,\boldsymbol{r}_{B_i}^L + \delta\boldsymbol{r}_B \tag{13.44}$$

where $\mathbf{I} = \begin{bmatrix} 1 & 0 & 0 \\ 0 & 1 & 0 \\ 0 & 0 & 1 \end{bmatrix}$.

The initial directional vector of the joint axis of each branch, represented by $\boldsymbol{x}_i$, and the local coordinate axes in the absolute coordinate frame are expressed by

$$\begin{cases} \boldsymbol{x}_i = (\cos\varphi_i\quad \sin\varphi_i\quad 0)^T \\ \boldsymbol{y}_i = (-\cos\phi_i\sin\varphi_i\quad \cos\phi_i\cos\varphi_i\quad -\sin\phi_i)^T \\ \boldsymbol{z}_i = (-\sin\phi_i\sin\varphi_i\quad \sin\phi_i\cos\varphi_i\quad \cos\phi_i)^T \end{cases}$$

where $\phi_i = \arcsin\frac{h_i - z}{L_i}, L_i = \sqrt{d_i^2 + (h_i - z)^2}, (i = 1,2,3,4), \phi_i$ represents the angle from the absolute x-axis to the x_i-axis, y_i-axis is along the branch axis of A_iB_i and d_i represents the shortest length of the branch A_iB_i,h_i is the difference of the z-coordinates of joint A_i and B_i, and z denotes the travel of the knuckle from the horizontal plane xoy.

Therefore, the rotational matrix from the branch coordinate frame to the absolute coordinate frame is

$$\mathbf{R}_i^B = \begin{bmatrix} \cos\varphi_i & -\cos\phi_i \sin\varphi_i & -\sin\phi_i \sin\varphi_i \\ \sin\varphi_i & \cos\phi_i \cos\varphi_i & \sin\phi_i \cos\varphi_i \\ 0 & -\sin\phi_i & \cos\phi_i \end{bmatrix}$$

And the joint axial direction vectors after deformations are

$$\boldsymbol{x}_i' = \mathbf{R}\boldsymbol{x}_i = \begin{bmatrix} 1 & -\gamma & \beta \\ \gamma & 1 & -\alpha \\ -\beta & \alpha & 1 \end{bmatrix} \begin{bmatrix} \cos\varphi_i \\ \sin\varphi_i \\ 0 \end{bmatrix} = \begin{bmatrix} \cos\varphi_i - \gamma\sin\varphi_i \\ \sin\varphi_i + \gamma\cos\varphi_i \\ \alpha\sin\varphi_i - \beta\cos\varphi_i \end{bmatrix}. \quad (13.45)$$

Expanding equation (13.44) generates the displacements of the loaded end of each branch:

$$\begin{cases} \boldsymbol{s}_{B_1} = \begin{bmatrix} 0 & -\gamma & \beta \\ \gamma & 0 & -\alpha \\ -\beta & \alpha & 0 \end{bmatrix} \begin{bmatrix} l \\ 0 \\ m \end{bmatrix} + \begin{bmatrix} \delta x \\ \delta y \\ 0 \end{bmatrix} = \begin{bmatrix} \delta x + m\beta \\ \delta y - m\alpha + l\gamma \\ -l\beta \end{bmatrix} \\ \boldsymbol{s}_{B_2} = \begin{bmatrix} 0 & -\gamma & \beta \\ \gamma & 0 & -\alpha \\ -\beta & \alpha & 0 \end{bmatrix} \begin{bmatrix} -l \\ 0 \\ m \end{bmatrix} + \begin{bmatrix} \delta x \\ \delta y \\ 0 \end{bmatrix} = \begin{bmatrix} \delta x + m\beta \\ \delta y - m\alpha - l\gamma \\ l\beta \end{bmatrix} \\ \boldsymbol{s}_{B_3} = \begin{bmatrix} 0 & -\gamma & \beta \\ \gamma & 0 & -\alpha \\ -\beta & \alpha & 0 \end{bmatrix} \begin{bmatrix} -l \\ 0 \\ -m \end{bmatrix} + \begin{bmatrix} \delta x \\ \delta y \\ 0 \end{bmatrix} = \begin{bmatrix} \delta x - m\beta \\ \delta y + m\alpha - l\gamma \\ l\beta \end{bmatrix} \\ \boldsymbol{s}_{B_4} = \begin{bmatrix} 0 & -\gamma & \beta \\ \gamma & 0 & -\alpha \\ -\beta & \alpha & 0 \end{bmatrix} \begin{bmatrix} l \\ 0 \\ -m \end{bmatrix} + \begin{bmatrix} \delta x \\ \delta y \\ 0 \end{bmatrix} = \begin{bmatrix} \delta x - m\beta \\ \delta y + m\alpha + l\gamma \\ -l\beta \end{bmatrix}. \end{cases} \quad (13.46)$$

The displacement of joint $B_i (i = 1,2,3,4)$ in its axial direction, $\boldsymbol{x}_i$, can be obtained by neglecting the second and higher infinite items:

$$\begin{cases} \delta_1 = \boldsymbol{s}_{B_1} \cdot \boldsymbol{x}_1 = (\delta x + m\beta)\cos\varphi_1 + (\delta y - m\alpha + l\gamma)\sin\varphi_1 \\ \delta_2 = \boldsymbol{s}_{B_2} \cdot \boldsymbol{x}_2 = (\delta x + m\beta)\cos\varphi_2 + (\delta y - m\alpha - l\gamma)\sin\varphi_2 \\ \delta_3 = \boldsymbol{s}_{B_3} \cdot \boldsymbol{x}_3 = (\delta x - m\beta)\cos\varphi_3 + (\delta y + m\alpha - l\gamma)\sin\varphi_3 \\ \delta_4 = \boldsymbol{s}_{B_4} \cdot \boldsymbol{x}_4 = (\delta x - m\beta)\cos\varphi_4 + (\delta y + m\alpha + l\gamma)\sin\varphi_4. \end{cases} \quad (13.47)$$

The torsional angle about the axis of each branch is

$$\psi_i = \frac{T_i L_i}{G I_p} \tag{13.48}$$

where T_i denotes the torque about the ith branch axis, G stands for the shearing modulus of the branch, and I_p represents the polar moment of inertia of the branch.

The deflection angle about the x_i-axis is

$$\theta_i = \frac{F_i L_i^2}{2EI} + \frac{M_i L_i}{EI} \tag{13.49}$$

where F_i denotes the shearing force exerted on the ith branch, M_i denotes the bending moment exerted on the ith branch by the knuckle, and E represents the elastic modulus of the branch and Iindicates the moment of inertia of the cross section about the y_i-axis.

The deflections of the loaded end of each branch in x_i-direction is

$$\delta_i = \frac{F_i L_i^3}{3EI} + \frac{M_i L_i^2}{2EI}. \tag{13.50}$$

Figure 13.11 shows the relationships of the torsional angle and deflection angle. The vector which is perpendicular to $\boldsymbol{x}_i'$ and z_i can be expressed as

$$\boldsymbol{p}_i = \boldsymbol{x}_i' \times \boldsymbol{z}_i = \begin{bmatrix} \cos\phi_i \left(\sin\varphi_i + \gamma\cos\varphi_i\right) - \sin\phi_i \cos\varphi_i \left(\alpha \sin\varphi_i - \beta\cos\varphi_i\right) \\ -\cos\phi_i \left(\cos\varphi_i - \gamma\sin\varphi_i\right) - \sin\phi_i \sin\varphi_i \left(\alpha\sin\varphi_i - \beta\cos\varphi_i\right) \\ \sin\phi_i \end{bmatrix}.$$

Therefore, the subtended angle between vectors $\boldsymbol{p}_i$ and $\boldsymbol{x}_i$ is

$$\cos\left(\frac{\pi}{2} - \theta_i\right) = \frac{\boldsymbol{p}_i \cdot \boldsymbol{x}_i}{|\boldsymbol{p}_i|} \approx \gamma\cos\phi_i - \sin\phi_i \left(\alpha\sin\varphi_i - \beta\cos\varphi_i\right).$$

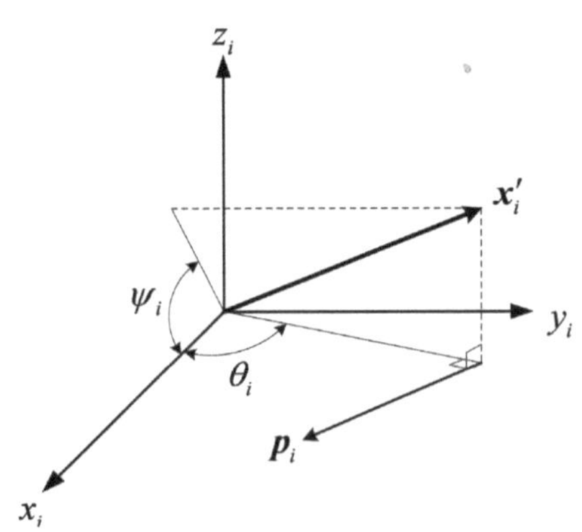

FIGURE 13.11 Angular relationships in the deflection of a branch.

As a result, the deflection angle is obtained

$$\theta_i = \arcsin\left[\gamma \cos\phi_i - \sin\phi_i \left(\alpha \sin\varphi_i - \beta \cos\varphi_i\right)\right]. \tag{13.51}$$

Similarly, the torsional angles also can be expressed as

$$\psi_i = -\arcsin\left[\gamma \sin\phi_i - \cos\phi_i \left(\alpha \sin\varphi_i - \beta \cos\varphi_i\right)\right]. \tag{13.52}$$

Associating equations (13.47)–(13.52), we find that

$$\begin{cases} \frac{F_i L_i^3}{3EI} + \frac{M_i L_i^2}{2EI} = \delta_i \\ \frac{F_i L_i^2}{2EI} + \frac{M_i L_i}{EI} = \theta_i \\ \frac{T_i L_i}{GI_p} = \psi_i . \end{cases} \tag{13.53}$$

Therefore, F_i, M_i, and T_i can be expressed as

$$\begin{cases} F_i = \frac{6EI(2\delta_i - \theta_i L_i)}{L_i^3} \\ M_i = \frac{2EI(2\theta_i L_i - 3\delta_i)}{L_i^2} \\ T_i = \frac{GI_p \psi_i}{L_i} . \end{cases} \tag{13.54}$$

These forces represented in equation (13.54) can be expressed as the following form in the absolute coordinate frame:

$$\boldsymbol{F}_i = F_i \mathbf{R}_i^B \begin{bmatrix} 1 \\ 0 \\ 0 \end{bmatrix} = F_i \begin{bmatrix} \cos\varphi_i & -\cos\phi_i \sin\varphi_i & -\sin\phi_i \sin\varphi_i \\ \sin\varphi_i & \cos\phi_i \cos\varphi_i & \sin\phi_i \cos\varphi_i \\ 0 & -\sin\phi_i & \cos\phi_i \end{bmatrix} \begin{bmatrix} 1 \\ 0 \\ 0 \end{bmatrix}$$

$$= F_i \begin{bmatrix} \cos\varphi_i \\ \sin\varphi_i \\ 0 \end{bmatrix} \tag{13.55a}$$

$$\boldsymbol{M}_i = M_i \mathbf{R}_i^B \begin{bmatrix} 0 \\ 0 \\ 1 \end{bmatrix} = M_i \begin{bmatrix} \cos\varphi_i & -\cos\phi_i \sin\varphi_i & -\sin\phi_i \sin\varphi_i \\ \sin\varphi_i & \cos\phi_i \cos\varphi_i & \sin\phi_i \cos\varphi_i \\ 0 & -\sin\phi_i & \cos\phi_i \end{bmatrix} \begin{bmatrix} 0 \\ 0 \\ 1 \end{bmatrix}$$

$$= M_i \begin{bmatrix} -\sin\phi_i \sin\varphi_i \\ \sin\phi_i \cos\varphi_i \\ \cos\phi_i \end{bmatrix} \tag{13.55b}$$

$$\boldsymbol{T}_i = T_i \mathbf{R}_i^B \begin{bmatrix} 0 \\ 1 \\ 0 \end{bmatrix} = T_i \begin{bmatrix} \cos\varphi_i & -\cos\phi_i \sin\varphi_i & -\sin\phi_i \sin\varphi_i \\ \sin\varphi_i & \cos\phi_i \cos\varphi_i & \sin\phi_i \cos\varphi_i \\ 0 & -\sin\phi_i & \cos\phi_i \end{bmatrix} \begin{bmatrix} 0 \\ 1 \\ 0 \end{bmatrix}$$

$$= T_i \begin{bmatrix} -\cos\phi_i \sin\varphi_i \\ \cos\phi_i \cos\varphi_i \\ -\sin\phi_i \end{bmatrix}. \tag{13.55c}$$

The additional moment of $\boldsymbol{F}_i (i = 1,2,3,4)$ about the center of the knuckle can be expressed as

$$\begin{cases} \boldsymbol{M}_{\mathbf{F}_1} = \begin{bmatrix} l \\ 0 \\ m \end{bmatrix} \times \boldsymbol{F}_1 = F_1 \begin{bmatrix} 0 & -m & 0 \\ m & 0 & -l \\ 0 & l & 0 \end{bmatrix} \begin{bmatrix} \cos\varphi_1 \\ \sin\varphi_1 \\ 0 \end{bmatrix} = F_1 \begin{bmatrix} -m\sin\varphi_1 \\ m\cos\varphi_1 \\ l\sin\varphi_1 \end{bmatrix} \\ \boldsymbol{M}_{\mathbf{F}_2} = \begin{bmatrix} -l \\ 0 \\ m \end{bmatrix} \times \boldsymbol{F}_2 = F_2 \begin{bmatrix} 0 & -m & 0 \\ m & 0 & l \\ 0 & -l & 0 \end{bmatrix} \begin{bmatrix} \cos\varphi_2 \\ \sin\varphi_2 \\ 0 \end{bmatrix} = F_2 \begin{bmatrix} -m\sin\varphi_2 \\ m\cos\varphi_2 \\ -l\sin\varphi_2 \end{bmatrix} \\ \boldsymbol{M}_{\mathbf{F}_3} = \begin{bmatrix} -l \\ 0 \\ -m \end{bmatrix} \times \boldsymbol{F}_3 = F_3 \begin{bmatrix} 0 & m & 0 \\ -m & 0 & l \\ 0 & -l & 0 \end{bmatrix} \begin{bmatrix} \cos\varphi_3 \\ \sin\varphi_3 \\ 0 \end{bmatrix} = F_3 \begin{bmatrix} m\sin\varphi_3 \\ -m\cos\varphi_3 \\ -l\sin\varphi_3 \end{bmatrix} \\ \boldsymbol{M}_{\mathbf{F}_4} = \begin{bmatrix} l \\ 0 \\ -m \end{bmatrix} \times \boldsymbol{F}_4 = F_4 \begin{bmatrix} 0 & m & 0 \\ -m & 0 & -l \\ 0 & l & 0 \end{bmatrix} \begin{bmatrix} \cos\varphi_4 \\ \sin\varphi_4 \\ 0 \end{bmatrix} = F_4 \begin{bmatrix} m\sin\varphi_4 \\ -m\cos\varphi_4 \\ l\sin\varphi_4 \end{bmatrix}. \end{cases} \tag{13.55d}$$

Based upon the discussion above, the static equilibrium equations of the knuckle can be expressed as:

$$\begin{cases} \sum_{i=1}^{4} F_i \cos\varphi_i - F_x = 0 \\ \sum_{i=1}^{4} F_i \sin\varphi_i - F_y = 0 \\ -\sum_{i=1}^{4} \left(M_i \sin\phi_i \sin\varphi_i + T_i \cos\phi_i \sin\varphi_i\right) - mF_1 \sin\varphi_1 - mF_2 \sin\varphi_2 \\ \qquad + mF_3 \sin\varphi_3 + mF_4 \sin\varphi_4 - M_x = 0 \\ \sum_{i=1}^{4} \left(M_i \sin\phi_i \cos\varphi_i + T_i \cos\phi_i \cos\varphi_i\right) + mF_1 \cos\varphi_1 + mF_2 \cos\varphi_2 \\ \qquad - mF_3 \cos\varphi_3 - mF_4 \cos\varphi_4 - M_y = 0 \\ \sum_{i=1}^{4} \left(M_i \cos\phi_i - T_i \sin\phi_i\right) + lF_1 \sin\varphi_1 - lF_2 \sin\varphi_2 - lF_3 \sin\varphi_3 \\ \qquad + lF_4 \sin\varphi_4 - M_z = 0. \end{cases} \tag{13.56}$$

There are five equations and five unknowns, $\delta x, \delta y, \alpha, \beta$, and γ, in equation set (13.56), so the perturbation of the knuckle can be known when the external forces F_x, F_y, M_x, M_y, and M_z are specified.

Solving equation (13.56), we can find that the position and pose parameters of the knuckle are expressed as the functions of the external loads, material, and geometric parameters of the suspension. However, all of the above variables are either very difficult to gain or changed within certain intervals. Therefore, the mapping of these variables into the perturbation space of the knuckle should be investigated with the interval analysis method [14].

Suppose that the structure parameters of the suspension mechanism are shown in Table 13.1 and the external loads are shown in Table 13.2. Accordingly, the variable intervals of the center of the knuckle and the rotated angles, responses to sequence 1, are shown in Figures 13.12–13.15.

TABLE 13.1 Parameters of the Suspension Mechanism

G (GPa)	E (GPa)	I_p(mm^4)	I(mm^4)	m (mm)
206	80	381	547	75
l (mm)	h_i(mm)	z (mm)	d_i (mm)	ϕ_i(°)
50	0	[−100,100]	250	135;225;135;225

TABLE 13.2 The External Loads

F_x(N)	F_y(N)	$M_x(N \cdot m)$	$M_y(N \cdot m)$	$M_z(N \cdot m)$
Sequence 1				
[−1000,1000]	[−1000,1000]	0	0	0
Sequence 2				
0	0	[−100,100]	[−100,100]	0

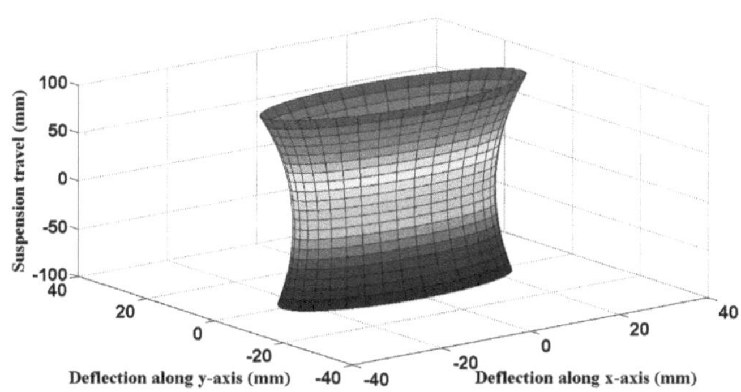

FIGURE 13.12 Deflections of the knuckle center of sequence 1.

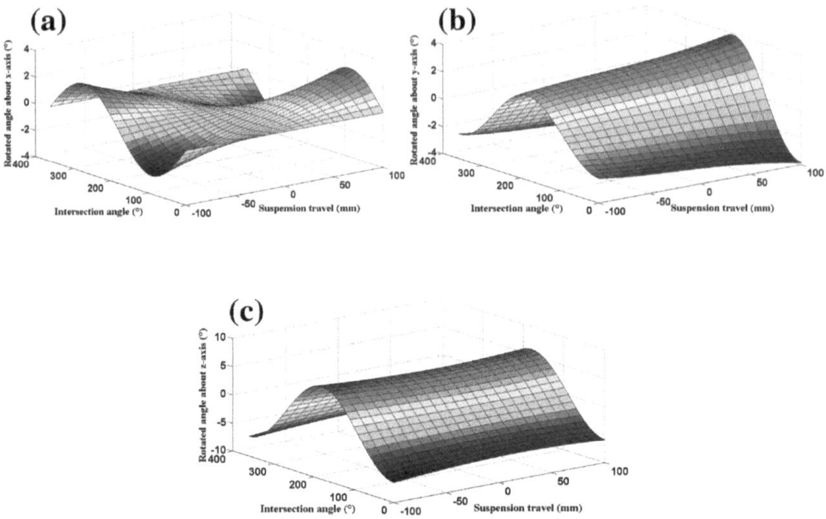

FIGURE 13.13 Rotated angle about x-, y-, z-axes of sequence 1.

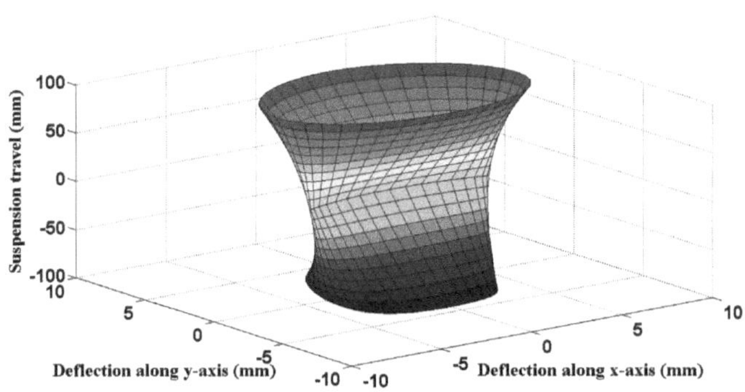

FIGURE 13.14 Deflections of the knuckle center of sequence 2.

In Figures 13.13–13.15, the coordinate axis, intersection angle, indicates the subtended angle between the external component force (or moment) and the x-axis which is within $[0,360°]$. From the above simulation results, we can find that the position and orientation of the knuckle are changing within an interval under the external loads. Figure 13.12 indicates that the stiffness of the mechanism along x-axis and y-axis are different and the stiffness along y-axis is much bigger. From Figure 13.13 we know that the external force could not only cause deflections but also rotations. Similarly conclusions can be drawn from Figures 13.14 and 13.15.

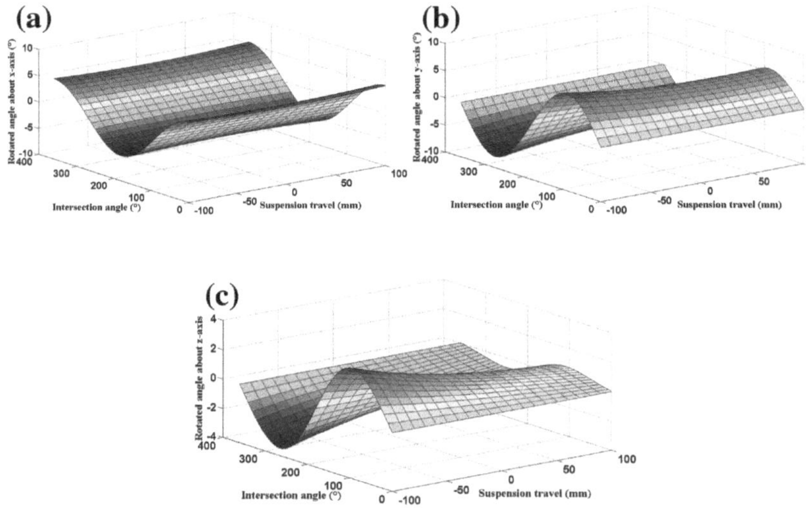

FIGURE 13.15 Rotated angle about x-, y-, z-axes of sequence 2.

13.3 CONCLUSIONS

This chapter exploits a new algorithm to optimize the design of a spatial parallel manipulator for the purpose of obtaining a desired dexterous workspace rather than the whole reachable workspace. With the analysis of the DOF of a manipulator, we can select the least number of variables to depict the kinematic constraints of each leg of a manipulator. The optimum parameters can be obtained by searching the extreme values of the objective functions with the prescribed adroit workspace. An example is utilized to demonstrate the significant advantages of this method in the dexterous workspace synthesis. In applications, this method can be widely used to synthesize, optimize, and create all kinds of new spatial parallel manipulators with a desired dexterous workspace.

REFERENCES

[1] D. Sen, T.S. Mruthyunjaya, A centro-based characterization of singularities in the workspace of planar closed-loop manipulators, Mechanism and Machine Theory 33 (1998) 1091–1104.

[2] D. Sen, T.S. Mruthyunjaya, A computational geometry approach for determination of boundary of workspaces of planar manipulators with arbitrary topology, Mechanism and Machine Theory 34 (1999) 149–169.

[3] M.Z.A. Majid, Z. Huang, Y.L. Yao, Workspace analysis of a six-degrees of freedom, three-prismatic-prismatic-spheric-revolute parallel manipulator, The International Journal of Advanced Manufacturing Technology 16 (2000) 441–449.

[4] I.A. Bonev, J. Ryu, A geometrical method for computing the constant-orientation workspace of 6-PRRS parallel manipulators, Mechanism and Machine Theory 36 (2001) 1–13.

[5] J.S. Dai, P. Shah, Orientation capability of planar manipulators using virtual joint angle analysis, Mechanism and Machine Theory 38 (2003) 241–252.
[6] J. Wang, X. Tang, Analysis and dimensional design of a novel hybrid machine tool, International Journal of Machine Tools & Manufacture 43 (2003) 647–655.
[7] I.A. Bonev, J. Ryu, A new approach to orientation workspace analysis of 6-DOF parallel manipulators, Mechanism and Machine Theory 36 (2001) 15–28.
[8] A. Kumar, K.J. Waldron, The workspaces of a mechanical manipulator, Journal of Mechanical Design 103 (1981) 665–672.
[9] J.S. Zhao, K. Zhou, D.Z. Mao, Y.F. Gao, Y.F. Fang, A new method to study the degree of freedom of spatial parallel mechanisms, The International Journal of Advanced Manufacturing Technology 23 (2004) 288–294.
[10] J.S. Zhao, M. Chen, K. Zhou, J.X. Dong, Z.J. Feng, Workspace of parallel manipulators with symmetric identical kinematic chains, Mechanism and Machine Theory 41 (2006) 632–645.
[11] J.S. Zhao, Y.Z. Fu, K. Zhou, Z.J. Feng, Mobility properties of a schoenflies type parallel manipulator, Robotics and Computer-Integrated Manufacturing 22 (2006) 124–133.
[12] X. Kong, C.M. Gosselin, Type synthesis of 3T1R 4-DOF parallel manipulators based on screw theory, IEEE Transactions on Robotics and Automation 20 (2004) 181–190.
[13] J.S. Zhao, K. Zhou, Z.J. Feng, A theory of degrees of freedom for mechanisms, Mechanism and Machine Theory 39 (2004) 621–643.
[14] R.E. Moore, R.B. Kearfott, M.J. Cloud, Introduction to Interval Analysis, Society for Industrial and Applied Mathematics, Philadelphia, 2009.

Chapter 14

Kinematic Synthesis of Spatial Mechanisms

In this chapter, we adopt the natural coordinates to deal with the kinematics model for different spatial mechanisms. After obtaining the constraint equations in natural coordinates, we can conduct a kinematics analysis of the mechanism and then build a dynamics model with Newton–Euler's formulation. Regarding kinematics, many works have addressed this problem in the past. Szkodny [1] presented formulas connecting driving forces of actuators with natural coordinates. Homogeneous transformations were used to describe the positions and orientations of the systems. These transformations afford the possibility for a joint description of the position and orientation, which is an invaluable advantage in describing kinematics, and especially, dynamics of parallel manipulators. The detailed description and application of the advantages of natural coordinates can be found in [2–4]. In what follows, we will discuss the kinematic synthesis of spatial robot mechanisms.

14.1 KINEMATIC SYNTHESIS OF A SPATIAL PARALLEL MANIPULATOR WITH THREE PURE ROTATIONAL DEGREES OF FREEDOM (DOFS)

Figure 14.1 shows a parallel manipulator, and it is not difficult to find that the three UPS kinematic chains are all zero terminal constraints.

Assume that the radius of the base triangular $B_1B_2B_3$ is R and the height of the central column is h, then the coordinates of the vertexes B_1, B_2, and B_3 can be expressed in the absolute coordinate system shown in Figures 14.1 and 14.2 whose origin is superimposed with the spherical joint on the top of the column CO, and x-axis and y-axis parallel to CB_2 and CB_3, respectively.

Assume the coordinates of the three vertexes of the end effector $M_1M_2M_3$ to be $M_1(x_{M_1}, y_{M_1}, z_{M_1})$, $M_2(x_{M_2}, y_{M_2}, z_{M_2})$ and $M_3(x_{M_3}, y_{M_3}, z_{M_3})$.

Therefore, the kinematic screws of chain $B_1P_1M_1$ can be denoted as:

$$\$_{B_1M_1} = \begin{bmatrix} \$_1 & \$_2 & \$_3 & \$_4 & \$_5 & \$_6 \end{bmatrix} \tag{14.1}$$

Advanced Theory of Constraint and Motion Analysis for Robot Mechanisms.
http://dx.doi.org/10.1016/B978-0-12-420162-0.00014-X

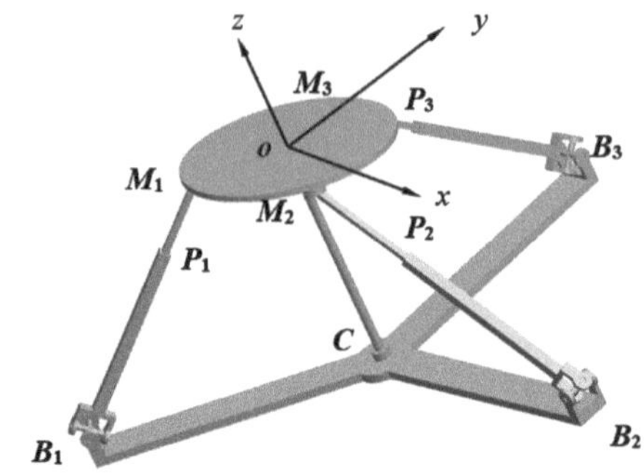

FIGURE 14.1 A parallel manipulator with 1S and 3UPS kinematic chains.

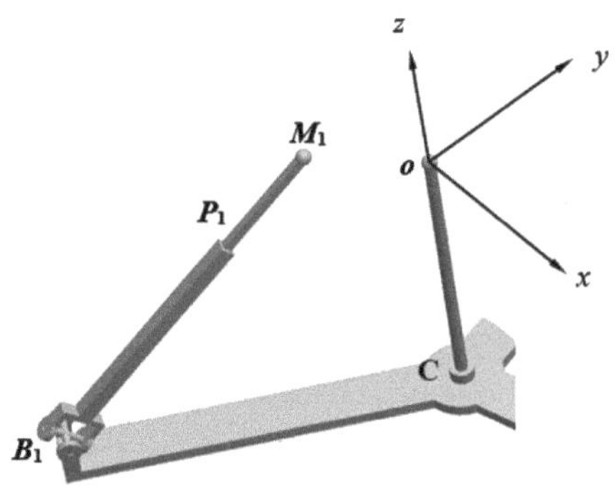

FIGURE 14.2 A UPS kinematic chain and the fixed base.

where $\$_1 = \begin{bmatrix} -\frac{\sqrt{2}}{2} & \frac{\sqrt{2}}{2} & 0 & -\frac{\sqrt{2}}{2}h & \frac{\sqrt{2}}{2}h & R \end{bmatrix}^T$.

The direction vector of limb B_1M_1 can be expressed as:

$$e_1 = \frac{\begin{bmatrix} x_{M_1} + \frac{\sqrt{2}}{2}R & y_{M_1} + \frac{\sqrt{2}}{2}R & z_{M_1} + h \end{bmatrix}^T}{\sqrt{\left(x_{M_1} + \frac{\sqrt{2}}{2}R\right)^2 + \left(y_{M_1} + \frac{\sqrt{2}}{2}R\right)^2 + \left(z_{M_1} + h\right)^2}}. \tag{14.2}$$

The direction vector of $\boldsymbol{u}_B^2$ can be expressed as the cross production of its orthogonal revolute axis $\boldsymbol{u}_B^1 = \begin{bmatrix} -\frac{\sqrt{2}}{2} & \frac{\sqrt{2}}{2} & 0 \end{bmatrix}$ and $\boldsymbol{e}_1$:

$$\boldsymbol{u}_B^2 = \frac{\boldsymbol{u}_B^1 \times \boldsymbol{e}_1}{\|\boldsymbol{u}_B^1 \times \boldsymbol{e}_1\|} = \frac{\begin{bmatrix} z_{M_1} + h & z_{M_1} + h & -(x_{M_1} + y_{M_1}) - \sqrt{2}R \end{bmatrix}^T}{\sqrt{(x_{M_1} + y_{M_1} + \sqrt{2}R)^2 + 2(z_{M_1} + h)^2}} \tag{14.3}$$

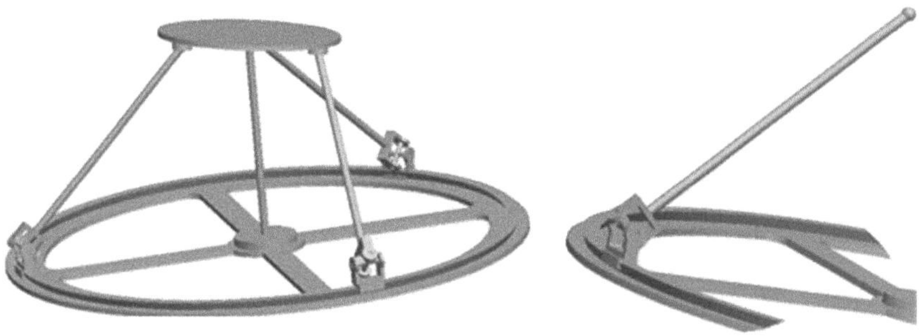

FIGURE 14.3 A parallel manipulator with a circular guide and its PUS kinematic chain.

$$\$_2 = \frac{\begin{bmatrix} z_{M_1}+h & z_{M_1}+h & -(x_{M_1}+y_{M_1})-\sqrt{2}R & \begin{matrix}\frac{\sqrt{2}}{2}R(x_{M_1}+y_{M_1}) \\ +h^2+R^2+hz_{M_1}\end{matrix} & \begin{matrix}-\frac{\sqrt{2}}{2}R(x_{M_1}+y_{M_1}) \\ -(h^2+R^2+hz_{M_1})\end{matrix} & 0 \end{bmatrix}^T}{\sqrt{(x_{M_1}+y_{M_1}+\sqrt{2}R)^2+2(z_{M_1}+h)^2}},$$

$$\$_3 = \frac{\begin{bmatrix} 0 & 0 & 0 & x_{M_1}+\frac{\sqrt{2}}{2}R & y_{M_1}-\frac{\sqrt{2}}{2}R & z_{M_1}+h \end{bmatrix}^T}{\sqrt{\left(x_{M_1}+\frac{\sqrt{2}}{2}R\right)^2+\left(y_{M_1}-\frac{\sqrt{2}}{2}R\right)^2+(z_{M_1}+h)^2}},$$

$$\$_4 = \begin{bmatrix} 1 & 0 & 0 & 0 & z_{M_1} & -y_{M_1} \end{bmatrix}^T,$$

$$\$_5 = \begin{bmatrix} 0 & 1 & 0 & -z_{M_1} & 0 & x_{M_1} \end{bmatrix}^T, \quad \$_6 = \begin{bmatrix} 0 & 0 & 1 & y_{M_1} & -x_{M_1} & 0 \end{bmatrix}^T.$$

Therefore, if we assume K to have the following form,

$$\mathbf{K} = \$_{B_1M_1} = \begin{bmatrix} \$_1 & \$_2 & \$_3 & \$_4 & \$_5 & \$_6 \end{bmatrix}. \quad (14.4)$$

We can synthesize the kinematic chain $B_1P_1M_1$ shown in Figure 14.2, whose terminal constraint is zero.

With a similar process, we can also synthesize a number of PUS kinematic chains if we properly set up the absolute coordinate system. For example, the PUS kinematic chains in the parallel manipulators shown in Figures 14.3–14.6 are also zero terminal constraints.

The structure schemes in Figures 14.1 and 14.3–14.6 can all meet the needs of the DOF requirements. Following, we will analyze the kinematics of such a mechanism. For the sake of simplicity, we only investigate the kinematics model of the structure form shown in Figure 14.1.

Firstly, we investigate the controllability of the actuation scheme for the manipulator shown in Figure 14.1. When the actuations are given, the three prismatic joints can be considered as fixed ones at this configuration. Now, let us consider the terminal constraints exerted on the end effector.

FIGURE 14.4 A parallel manipulator with equilateral triangle guide and its PUS kinematic chain.

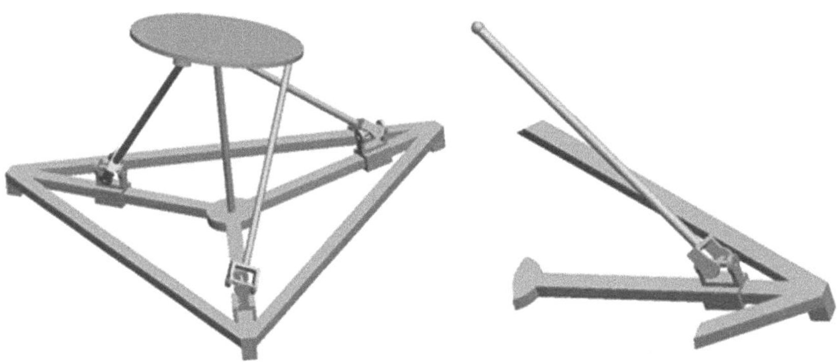

FIGURE 14.5 A parallel manipulator with radial guide and its PUS kinematic chain.

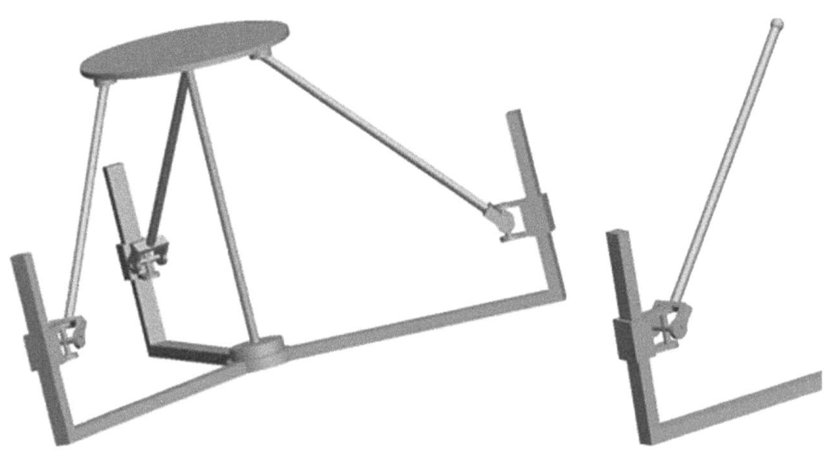

FIGURE 14.6 A parallel manipulator with vertical guide and its PUS kinematic chain.

When the actuation is assigned to the prismatic joint P_1, the kinematic screws of chain B_1M_1 can be denoted as:

$$\$_{B_1M_1} = \begin{bmatrix} \$_1 & \$_2 & \$_3 & \$_4 & \$_5 & \$_6 \end{bmatrix}. \tag{14.5}$$

The terminal constraint of the chain $\$_{B_1M_1}$ can be obtained:

$$\$^r_{B_1M_1} = \frac{\left[x_{M_1} + \frac{\sqrt{2}}{2}R \; y_{M_1} + \frac{\sqrt{2}}{2}R \; z_{M1} + h \; hy_{M_1} - \frac{\sqrt{2}}{2}Rz_{M_1} \; \frac{\sqrt{2}}{2}Rz_{M1} - hx_{M_1} \; \frac{\sqrt{2}}{2}Rx_{M_1} - \frac{\sqrt{2}}{2}Ry_{M_1}\right]^T}{\sqrt{\left(x_{M_1} + \frac{\sqrt{2}}{2}R\right)^2 + \left(y_{M_1} - \frac{\sqrt{2}}{2}R\right)^2 + (z_{M_1} + h)^2}} \tag{14.6}$$

Similarly, when the actuations are exerted on each kinematic chain, their terminal constraints can also be gained:

$$\$^r_{B_2M_2} = \frac{\left[x_{M_2} - R \; y_{M_2} \; z_{M2} + h \; hy_{M_2} \; - Rz_{M_2} - hx_{M_2} \; Ry_{M_2}\right]^T}{\sqrt{\left(x_{M_2} - R\right)^2 + y^2_{M_1} + (z_{M_2} + h)^2}} \tag{14.7}$$

$$\$^r_{B_3M_3} = \frac{\left[x_{M_3} \; y_{M_3} - R \; z_{M3} + h \; Rz_{M_3} + hy_{M_3} - hx_{M_3} \; - Rx_{M_3}\right]^T}{\sqrt{x^2_{M_3} + \left(y_{M_3} - R\right)^2 + \left(z_{M_3} + h\right)^2}} \tag{14.8}$$

The vertical column applies three orthogonal forces to the end effector through its top spherical joint o, which can be expressed in screw matrix as:

$$\$_{Co} = \begin{bmatrix} 1 & 0 & 0 & 0 & 0 & 0 \\ 0 & 1 & 0 & 0 & 0 & 0 \\ 0 & 0 & 1 & 0 & 0 & 0 \end{bmatrix}. \tag{14.9}$$

Therefore, if we ignore the magnitudes of the reciprocal screws of equations (14.7), (14.8), and (14.9), the terminal constraints exerted on the end effector can be expressed as:

$$\mathbf{S}_C = \begin{bmatrix} 1 & 0 & 0 & 0 & 0 & 0 \\ 0 & 1 & 0 & 0 & 0 & 0 \\ 0 & 0 & 1 & 0 & 0 & 0 \\ x_{M_1} + \frac{\sqrt{2}}{2}R & y_{M_1} + \frac{\sqrt{2}}{2}R & z_{M_1} + h & hy_{M_1} - \frac{\sqrt{2}}{2}Rz_{M_1} & \frac{\sqrt{2}}{2}Rz_{M_1} - hx_{M_1} & \frac{\sqrt{2}}{2}Rx_{M_1} - \frac{\sqrt{2}}{2}Ry_{M_1} \\ x_{M_2} - R & y_{M_2} & z_{M_2} - h & hy_{M_2} & -Rz_{M_2} - hx_{M_2} & Ry_{M_2} \\ x_{M_3} & y_{M_3} - R & z_{M_3} - h & Rz_{M_3} + hy_{M_3} & -hx_{M_3} & -Rx_{M_3} \end{bmatrix}. \tag{14.10}$$

The controllability problem of the end effector can be further simplified to find whether the determinant of the sub matrix of equation (14.10) is zero:

$$\mathbf{S}_{Csub} = \begin{bmatrix} hy_{M_1} - \frac{\sqrt{2}}{2}Rz_{M_1} & \frac{\sqrt{2}}{2}Rz_{M_1} - hx_{M_1} & \frac{\sqrt{2}}{2}Rx_{M_1} - \frac{\sqrt{2}}{2}Ry_{M_1} \\ hy_{M_2} & -Rz_{M_2} - hx_{M_2} & Ry_{M_2} \\ Rz_{M_3} + hy_{M_3} & -hx_{M_3} & -Rx_{M_3} \end{bmatrix}. \tag{14.11}$$

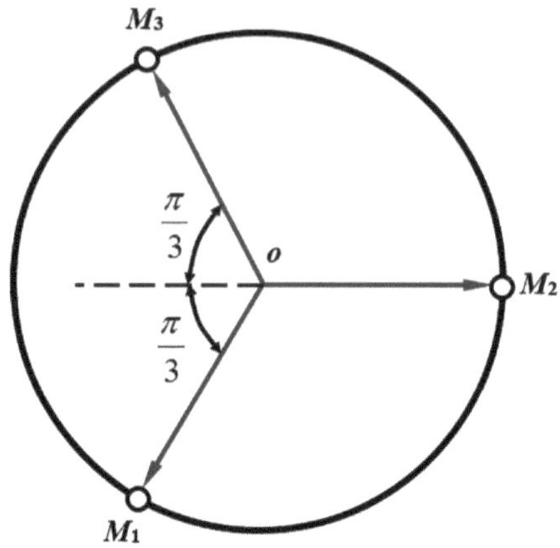

FIGURE 14.7 Top view of the end effector.

If $|\mathbf{S}_{Csub}| \neq 0$, then the manipulator is controllable; otherwise, the manipulator is singular at this configuration.

Prior to compute $|\mathbf{S}_{Csub}|$, we should establish the kinematic equations of the system with natural coordinates:

(1) the actuation equations are

$$\begin{cases} (x_{M_1} - x_{B_1})^2 + (y_{M_1} - y_{B_1})^2 + (z_{M_1} - z_{B_1})^2 = l_1^2 \\ (x_{M_2} - x_{B_2})^2 + (y_{M_2} - y_{B_2})^2 + (z_{M_2} - z_{B_2})^2 = l_2^2 \\ (x_{M_3} - x_{B_3})^2 + (y_{M_3} - y_{B_3})^2 + (z_{M_3} - z_{B_3})^2 = l_3^2 \end{cases} \tag{14.12}$$

where l_1,l_2 and l_3 denote the three actuations.

(2) the fixed distance constraint equations

Assume the three spherical joints are equiangular, distributed in the top view of the end effector shown in Figure 14.7. The vectors of the three spherical joints will always have the following relationship:

$$\overrightarrow{oM_2} = k_1 \overrightarrow{oM_1} + k_2 \overrightarrow{oM_3} \tag{14.13}$$

where $k_1 = k_2 = -1$ here.

If the radius of the manipulator is assumed to be r, we can write the following fixed relationship equations:

$$\begin{cases} x_{M_1}^2 + y_{M_1}^2 + z_{M_1}^2 = r^2 \\ x_{M_2}^2 + y_{M_2}^2 + z_{M_2}^2 = r^2 \\ (x_{M_1} - x_{M_2})^2 + (y_{M_1} - y_{M_2})^2 + (z_{M_1} - z_{M_2})^2 = 3r^2 \\ x_{M_1} + x_{M_3} = -x_{M_2} \\ y_{M_1} + y_{M_3} = -y_{M_2} \\ z_{M_1} + z_{M_3} = -z_{M_2} \end{cases} \tag{14.14}$$

Therefore, the kinematics equations of the system can be expressed as:

$$\boldsymbol{P}(\boldsymbol{x}) = 0 \tag{14.15}$$

where

$$\boldsymbol{x} = \begin{bmatrix} x_{M_1} & y_{M_1} & z_{M_1} & x_{M_2} & y_{M_2} & z_{M_2} & x_{M_3} & y_{M_3} & z_{M_3} \end{bmatrix}^T,$$

$$\boldsymbol{P}(\boldsymbol{x}) = \begin{bmatrix} (x_{M_1} - x_{B_1})^2 + (y_{M_1} - y_{B_1})^2 + (z_{M_1} - z_{B_1})^2 - l_1^2 \\ (x_{M_2} - x_{B_2})^2 + (y_{M_2} - y_{B_2})^2 + (z_{M_2} - z_{B_2})^2 - l_2^2 \\ (x_{M_3} - x_{B_3})^2 + (y_{M_3} - y_{B_3})^2 + (z_{M_3} - z_{B_3})^2 - l_3^2 \\ x_{M_1}^2 + y_{M_1}^2 + z_{M_1}^2 - r^2 \\ x_{M_2}^2 + y_{M_2}^2 + z_{M_2}^2 - r^2 \\ (x_{M_1} - x_{M_2})^2 + (y_{M_1} - y_{M_2})^2 + (z_{M_1} - z_{M_2})^2 - 3r^2 \\ x_{M_1} + x_{M_2} + x_{M_3} \\ y_{M_1} + y_{M_2} + y_{M_3} \\ z_{M_1} + z_{M_2} + z_{M_3} \end{bmatrix},$$

$$\frac{\partial \boldsymbol{P}(\boldsymbol{x})}{\partial \boldsymbol{x}} = 2\begin{bmatrix} x_{M_1} - x_{B_1} & y_{M_1} - y_{B_1} & z_{M_1} - z_{B_1} & 0 & 0 & 0 & 0 & 0 & 0 \\ 0 & 0 & 0 & x_{M_2} - x_{B_2} & y_{M_2} - y_{B_2} & z_{M_2} - z_{B_2} & 0 & 0 & 0 \\ 0 & 0 & 0 & 0 & 0 & 0 & x_{M_3} - x_{B_3} & y_{M_3} - y_{B_3} & z_{M_3} - z_{B_3} \\ x_{M_1} & y_{M_1} & z_{M_1} & 0 & 0 & 0 & 0 & 0 & 0 \\ 0 & 0 & 0 & x_{M_2} & y_{M_2} & z_{M_2} & 0 & 0 & 0 \\ x_{M_1} - x_{M_2} & y_{M_1} - y_{M_2} & z_{M_1} - z_{M_2} & x_{M_2} - x_{M_1} & y_{M_2} - y_{M_1} & z_{M_2} - z_{M_1} & 0 & 0 & 0 \\ \frac{1}{2} & 0 & 0 & \frac{1}{2} & 0 & 0 & \frac{1}{2} & 0 & 0 \\ 0 & \frac{1}{2} & 0 & 0 & \frac{1}{2} & 0 & 0 & \frac{1}{2} & 0 \\ 0 & 0 & \frac{1}{2} & 0 & 0 & \frac{1}{2} & 0 & 0 & \frac{1}{2} \end{bmatrix}.$$

Therefore, the singularity criteria of the manipulator can also be expressed as:

$$\left| \frac{\partial \boldsymbol{P}(\boldsymbol{x})}{\partial \boldsymbol{x}} \right| = 0. \tag{14.16}$$

However, it will be more economical to find the singularities of the manipulator by computing the determinant of equation (14.11) than by calculating equation (14.16), though they have the same solutions. So long as $\left| \frac{\partial \boldsymbol{P}(\boldsymbol{x})}{\partial \boldsymbol{x}} \right| \neq 0$, the forward displacement of the manipulator can be iterated with numerical method, such as the Newton–Raphson method.

$$\boldsymbol{x}_{i+1} = \boldsymbol{x}_i - \left[\frac{\partial \boldsymbol{P}(\boldsymbol{x}_i)}{\partial \boldsymbol{x}_i} \right]^{-1} \boldsymbol{P}(\boldsymbol{x}_i) \tag{14.17}$$

Therefore, $\boldsymbol{x}$ can be obtained if the actuations $l_i (i = 1,2,3)$ are given; and as a result, the forward displacement of the end effector can be solved. Suppose the initial normal direction vector of the end effector plane is:

$$\boldsymbol{n}_0 = [0 \quad 0 \quad 1]^T. \tag{14.18}$$

The normal direction vector of the end effector plane, denoted by $\boldsymbol{n}(\boldsymbol{x})$, is the function of its configuration. When $\boldsymbol{x}$ are obtained, $\boldsymbol{n}(\boldsymbol{x})$ can be expressed as:

$$\boldsymbol{n}(\boldsymbol{x}) = \begin{bmatrix} n_x(x) \\ n_y(x) \\ n_z(x) \end{bmatrix} = \frac{\overrightarrow{M_1M_2} \times \overrightarrow{M_1M_3}}{\|\overrightarrow{M_1M_2} \times \overrightarrow{M_1M_3}\|} \tag{14.19}$$

where $n_x(\boldsymbol{x}), n_y(\boldsymbol{x}), n_z(\boldsymbol{x})$ are the direction cosine of the normal direction vector.

As a result, the angular displacement relative to the absolute coordinate system can be solved:

$$\begin{cases} \alpha = \arccos(n_x(x)) \\ \beta = \arccos(n_y(x)) - \frac{\pi}{2} \\ \gamma = \arccos(n_z(x)) - \frac{\pi}{2}. \end{cases} \tag{14.20}$$

The derivative of equation (14.15) with respect to time can be expressed as:

$$\begin{cases} (x_{M_1} - x_{B_1})\dot{x}_{M_1} + (y_{M_1} - y_{B_1})\dot{y}_{M_1} + (z_{M_1} - z_{B_1})\dot{z}_{M_1} - l_1\dot{i}_1 = 0 \\ (x_{M_2} - x_{B_2})\dot{x}_{M_2} + (y_{M_2} - y_{B_2})\dot{y}_{M_2} + (z_{M_2} - z_{B_2})\dot{z}_{M_2} - l_2\dot{i}_2 = 0 \\ (x_{M_3} - x_{B_3})\dot{x}_{M_3} + (y_{M_3} - y_{B_3})\dot{y}_{M_3} + (z_{M_3} - z_{B_3})\dot{z}_{M_3} - l_3\dot{i}_3 = 0 \\ x_{M_1}\dot{x}_{M_1} + y_{M_1}\dot{y}_{M_1} + z_{M_1}\dot{z}_{M_1} = 0 \\ x_{M_2}\dot{x}_{M_2} + y_{M_2}\dot{y}_{M_2} + z_{M_2}\dot{z}_{M_2} = 0 \\ (x_{M_1} - x_{M_2})(\dot{x}_{M_1} - \dot{x}_{M_2}) + (y_{M_1} - y_{M_2})(\dot{y}_{M_1} - \dot{y}_{M_2}) + (z_{M_1} - z_{M_2})(\dot{z}_{M_1} - \dot{z}_{M_2}) = 0 \\ \dot{x}_{M_1} + \dot{x}_{M_2} + \dot{x}_{M_3} = 0 \\ \dot{y}_{M_1} + \dot{y}_{M_2} + \dot{y}_{M_3} = 0 \\ \dot{y}_{M_1} + \dot{z}_{M_2} + \dot{z}_{M_3} = 0. \end{cases} \tag{14.21}$$

In matrix form, equation (14.21) can be rewritten as:

$$\mathbf{A}\dot{\boldsymbol{x}} = \boldsymbol{c} \tag{14.22}$$

where

$$\mathbf{A} = \begin{bmatrix} x_{M_1} - x_{B_1} & y_{M_1} - y_{B_1} & z_{M_1} - z_{B_1} & 0 & 0 & 0 & 0 & 0 & 0 \\ 0 & 0 & 0 & x_{M_2} - x_{B_2} & y_{M_2} - y_{B_2} & z_{M_2} - z_{B_2} & 0 & 0 & 0 \\ 0 & 0 & 0 & 0 & 0 & 0 & x_{M_3} - x_{B_3} & y_{M_3} - y_{B_3} & z_{M_3} - z_{B_3} \\ x_{M_1} & y_{M_1} & z_{M_1} & 0 & 0 & 0 & 0 & 0 & 0 \\ 0 & 0 & 0 & x_{M_2} & y_{M_2} & z_{M_2} & 0 & 0 & 0 \\ x_{M_1} - x_{M_2} & y_{M_1} - y_{M_2} & z_{M_1} - z_{M_2} & x_{M_2} - x_{M_1} & y_{M_2} - y_{M_1} & z_{M_2} - z_{M_1} & 0 & 0 & 0 \\ 1 & 0 & 0 & 1 & 0 & 0 & 1 & 0 & 0 \\ 0 & 1 & 0 & 0 & 1 & 0 & 0 & 1 & 0 \\ 0 & 0 & 1 & 0 & 0 & 1 & 0 & 0 & 1 \end{bmatrix},$$

$$\dot{\boldsymbol{x}} = \begin{bmatrix} \dot{x}_{M_1} & \dot{y}_{M_1} & \dot{z}_{M_1} & \dot{x}_{M_2} & \dot{y}_{M_2} & \dot{z}_{M_2} & \dot{x}_{M_3} & \dot{y}_{M_3} & \dot{z}_{M_3} \end{bmatrix}^T,$$

$$\boldsymbol{c} = \begin{bmatrix} l_1\dot{l}_1 & l_2\dot{l}_2 & l_3\dot{l}_3 & 0 & 0 & 0 & 0 & 0 & 0 \end{bmatrix}^T.$$

The general velocity of the natural coordinates can be solved by left multiplying $\mathbf{A}^{-1}$ at both sides of equation (14.22) so long as $|\mathbf{A}| \neq 0$. Assume

$$\begin{cases} \boldsymbol{v}_{M_1} = \begin{bmatrix} \dot{x}_{M_1} & \dot{y}_{M_1} & \dot{z}_{M_1} \end{bmatrix}^T \\ \boldsymbol{v}_{M_2} = \begin{bmatrix} \dot{x}_{M_2} & \dot{y}_{M_2} & \dot{z}_{M_2} \end{bmatrix}^T \end{cases} \tag{14.23}$$

$$\boldsymbol{v}_{M_1} - \boldsymbol{v}_{M_2} = \boldsymbol{\omega} \times \boldsymbol{r}_{M_1M_2}. \tag{14.24}$$

$\boldsymbol{r}_{M_1M_1}$ cross multiplies both sides of equation (14.23) yields:

$$\begin{aligned} \boldsymbol{r}_{M_1M_2} \times (\boldsymbol{v}_{M_1} - \boldsymbol{v}_{M_2}) &= \boldsymbol{r}_{M_1M_2} \times (\boldsymbol{\omega} \times \boldsymbol{r}_{M_1M_2}) \\ &= \left(\|\boldsymbol{r}_{M_1M_2}\|^2 \mathbf{I} - \boldsymbol{r}_{M_1M_2}\boldsymbol{r}_{M_1M_2}^T \right) \boldsymbol{\omega}. \end{aligned} \tag{14.25}$$

According to Theorem 7.2, we assume that

$$\begin{cases} \mathbf{B}_{12} = \|\boldsymbol{r}_{M_1M_2}\|^2 \ \ \mathbf{I} - \boldsymbol{r}_{M_1M_2}\boldsymbol{r}_{M_1M_2}^T \\ \mathbf{B}_{13} = \|\boldsymbol{r}_{M_1M_3}\|^2 \ \ \mathbf{I} - \boldsymbol{r}_{M_1M_3}\boldsymbol{r}_{M_1M_3}^T. \end{cases} \tag{14.26}$$

The angular velocity of the end effector can be therefore expressed with

$$\boldsymbol{\omega} = \mathbf{B}^{-1}\boldsymbol{q} \tag{14.27}$$

where $\mathbf{B} = \mathbf{B}_{12} + \mathbf{B}_{13}$, $\boldsymbol{q} = \boldsymbol{r}_{M_1M_2} \times (\boldsymbol{v}_{M_1} - \boldsymbol{v}_{M_2}) + \boldsymbol{r}_{M_1M_3} \times (\boldsymbol{v}_{M_3} - \boldsymbol{v}_{M_1})$.

The derivative of equation set (14.22) with respect to time can be expressed as:

$$\mathbf{A}\ddot{\boldsymbol{x}} = \boldsymbol{d} \tag{14.28}$$

where

$$\ddot{\boldsymbol{x}} = [\, \ddot{x}_1 \quad \ddot{y}_1 \quad \ddot{z}_1 \quad \ddot{x}_2 \quad \ddot{y}_2 \quad \ddot{z}_2 \quad \ddot{x}_3 \quad \ddot{y}_3 \quad \ddot{z}_3 \,]^T \tag{14.29}$$

$$\boldsymbol{d} = \begin{bmatrix} l_1\ddot{l}_1 + \dot{l}_1^2 - (\dot{x}_{M_1})^2 - (\dot{y}_{M_1})^2 - (\dot{z}_{M_1}) \\ l_2\ddot{l}_2 + \dot{l}_2^2 - (\dot{x}_{M_2})^2 - (\dot{y}_{M_2})^2 - (\dot{z}_{M_2})^2 \\ l_3\ddot{l}_3 + \dot{l}_3^2 - (\dot{x}_{M_3})^2 - (\dot{y}_{M_3})^2 - (\dot{z}_{M_3})^2 \\ -(\dot{x}_{M_1})^2 - (\dot{y}_{M_1})^2 - (\dot{z}_{M_1})^2 \\ -(\dot{x}_{M_2})^2 - (\dot{y}_{M_2})^2 - (\dot{z}_{M_2})^2 \\ -(\dot{x}_{M_1} - \dot{x}_{M_2})^2 - (\dot{y}_{M_1} - \dot{y}_{M_2})^2 - (\dot{z}_{M_1} - \dot{z}_{M_2})^2 \\ 0 \\ 0 \\ 0 \end{bmatrix}. \tag{14.30}$$

The derivative of equation set (14.24) with respect to time can be expressed as:

$$\dot{\boldsymbol{v}}_{M_1} - \dot{\boldsymbol{v}}_{M_2} = \dot{\boldsymbol{\omega}} \times \boldsymbol{r}_{M_1M_2} + \boldsymbol{\omega} \times (\boldsymbol{\omega} \times \boldsymbol{r}_{M_1M_2}). \tag{14.31}$$

$\boldsymbol{r}_{M_1M_2}$ cross multiplies both sides of equation (14.31) yields:

$$\boldsymbol{r}_{M_1M_2} \times (\dot{\boldsymbol{v}}_{M_1} - \dot{\boldsymbol{v}}_{M_2}) = \boldsymbol{r}_{M_1M_2} \times (\dot{\boldsymbol{\omega}} \times \boldsymbol{r}_{M_1M_2}) + \boldsymbol{r}_{M_1M_2} \times \left[\boldsymbol{\omega} \times (\boldsymbol{\omega} \times \boldsymbol{r}_{M_1M_2})\right].$$

Therefore, with Theorem 7.2, the angular acceleration of the end effector, $\boldsymbol{\varepsilon}$, can be obtained:

$$\boldsymbol{\varepsilon} = \dot{\boldsymbol{\omega}} = (\mathbf{B}_{12} + \mathbf{B}_{13})^{-1} \boldsymbol{w} \tag{14.32}$$

where $\boldsymbol{w} = \boldsymbol{r}_{M_1M_2} \times (\dot{\boldsymbol{v}}_{M_1} - \dot{\boldsymbol{v}}_{M_2}) + (\boldsymbol{r}_{M_1M_2} \cdot \boldsymbol{\omega})\boldsymbol{\omega} \times \boldsymbol{r}_{M_1M_2} + \boldsymbol{r}_{M_1M_3} \times (\dot{\boldsymbol{v}}_{M_1} - \dot{\boldsymbol{v}}_{M_3}) + (\boldsymbol{r}_{M_1M_3} \cdot \boldsymbol{\omega})\boldsymbol{\omega} \times \boldsymbol{r}_{M_1M_3}$.

In the following, we will investigate the variable range of each limb of the manipulator. According to the requirements of the manipulator's specifications, we can find the extremes of each limb. The transformation matrix from the extreme configuration to the initial configuration can be expressed as:

$$\begin{aligned}\mathbf{T} &= \begin{bmatrix} 1 & 0 & 0 \\ 0 & \cos\alpha & -\sin\alpha \\ 0 & \sin\alpha & \cos\alpha \end{bmatrix} \begin{bmatrix} \cos\beta & 0 & \sin\beta \\ 0 & 1 & 0 \\ -\sin\beta & 0 & \cos\beta \end{bmatrix} \begin{bmatrix} \cos\gamma & \sin\gamma & 0 \\ \sin\gamma & \cos\gamma & 0 \\ 0 & 0 & 1 \end{bmatrix} \\ &= \begin{bmatrix} \cos\beta\cos\gamma & -\cos\beta\cos\gamma & \sin\beta \\ \sin\alpha\sin\beta\cos\gamma + \cos\alpha\sin\gamma & -\sin\alpha\sin\beta\sin\gamma + \cos\alpha\cos\gamma & -\sin\alpha\cos\beta \\ -\cos\alpha\sin\beta\cos\gamma + \sin\alpha\sin\gamma & \cos\alpha\sin\beta\sin\gamma + \sin\alpha\cos\gamma & \cos\alpha\cos\beta \end{bmatrix}.\end{aligned} \tag{14.33}$$

According to the above analysis, we conduct numerical simulation with the following parameters:

The length range of each leg: $l_i \in [1800,3800]$, $i = 1,2,3$;

The radius of the manipulator circumcircle: $r = 400$;

The radius of the base circumcircle: $R = 1200$.

If we presume that the velocity of each leg of the Gough–Stewart platform changes with respect to time as Figure 14.2 shows, and the maximum velocity of each leg, the velocity, angular velocity, acceleration, and angular acceleration with respect to time will be shown in Figures 14.3–14.6, respectively. The total time of simulation is 20 s, where $t_1 = 10$ s, $t_2 = 16$ s and $t_3 = 20$ s. Numerical simulations show that this method with natural coordinates can effectively and efficiently accomplish real-time calculations.

14.2 KINEMATICS OF A SPATIAL PARALLEL MANIPULATOR WITH TWO ROTATIONAL AND ONE TRANSLATIONAL DOF

In this section, we propose a fast kinematics model for the dynamics of a 2R1T spatial parallel manipulator with natural coordinates. This parallel manipulator is designed particularly to test certain sensors of an automobile for two rotational and one vertical vibration. The fact that the constraint equations in this model are either quadratic or linear and the coordinates are fully Cartesian facilitates the processing of displacements, velocities, and accelerations. This is because the derivative matrix of the constraint equations only consists of linear and even constant elements, which has remarkable advantages in building kinematics and dynamics models than those through rotation matrix, the elements of which often contain quadratic or transcendental functions. The fast kinematics modeling of this method will surely result in quick analysis of dynamics. The actuators' driving forces and the constraints of the joints concerned are built up with Newton–Euler's formulation. The theoretical analysis and simulations are verified through real experiments in engineering.

The specifications of the manipulator are that the end effector of the parallel manipulator should have two rotational DOFs about the x-direction and the y-direction, and one translational DOF along the z-axis.

Therefore, if we create a proper absolute coordinate system $oxyz$, whose origin is superimposed with the initial rotational center of the end effector of the manipulator, the free mobility space might be immediately expressed with a base of the space:

$$\$_F = \begin{bmatrix} \$_1^F & \$_2^F & \$_3^F \end{bmatrix} \tag{14.34}$$

where

$\$_1^F = [1 \;\; 0 \;\; 0 \;\; 0 \;\; 0 \;\; 0]^T$, which denotes a free rotation about the x-axis;
$\$_2^F = [0 \;\; 1 \;\; 0 \;\; 0 \;\; 0 \;\; 0]^T$, which denotes a free rotation about the y-axis;
$\$_3^F = [0 \;\; 0 \;\; 0 \;\; 0 \;\; 0 \;\; 1]^T$, which denotes a free translation along the z-axis.

According to reciprocal screw theory, the reciprocal screw(s) of a screw matrix $\$$ can be obtained by solving

$$\$^T \mathbf{\Delta} \$^r = 0 \tag{14.35}$$

where $\$^r$ denote the reciprocal screw of $\$$, and $\mathbf{\Delta} = \begin{bmatrix} 0 & \mathbf{I} \\ \mathbf{I} & 0 \end{bmatrix}$ and $\mathbf{I}$ is a 3×3 identity matrix.

Consequently, the constraint space spanned by the terminal constraints of all kinematic chains, denoted by $\$_C$, can be immediately obtained with reciprocal screw theory:

$$\$_C = \begin{bmatrix} \$_1^C & \$_2^C & \$_3^C \end{bmatrix} \tag{14.36}$$

where

$\$_1^C = \begin{bmatrix} 1 & 0 & 0 & 0 & 0 & 0 \end{bmatrix}^T$, which denotes a force along the x-axis;
$\$_2^C = \begin{bmatrix} 0 & 1 & 0 & 0 & 0 & 0 \end{bmatrix}^T$, which denotes a force along the y-axis;
$\$_3^C = \begin{bmatrix} 0 & 0 & 0 & 0 & 0 & 1 \end{bmatrix}^T$, which denotes a moment about the z-axis.

Consequently, the terminal constraint of every kinematic chain must be expressed as the linear combinations of the basis (14.36). As a result, the rank of the constraints of each kinematic chain should satisfy

$$\text{Rank}(\$_i^r) \leq \dim(\$_C) = 3. \tag{14.37}$$

Suppose the terminal constraints of such a chain is denoted by $\$_i^r$, there will be:

$$-m_E \boldsymbol{a}_o - m_E \boldsymbol{g} + \sum_{i=1}^{3} \boldsymbol{F}_{SM_i} + \boldsymbol{F}_o + \boldsymbol{F}_{ext} = 0 \tag{14.38}$$

where

$$\mathbf{K}_i = \begin{bmatrix} k_{11} & k_{12} & k_{13} \\ k_{21} & k_{22} & k_{23} \\ & \vdots & \\ k_{m1} & k_{m2} & k_{m3} \end{bmatrix}_{m\times 3} \tag{14.39}$$

and $i, m = 1,2,3, \ldots$

To safeguard the relationship of (14.37), there should be,

$$\text{Rank}(\mathbf{K}_i) \leq 3. \tag{14.40}$$

If we simply let

$$\mathbf{K}_1 = \begin{bmatrix} 1 & 0 & 0 \\ 0 & 1 & 0 \\ 0 & 0 & 1 \end{bmatrix} \tag{14.41}$$

$\$_1^r$ can be obtained

$$\$_1^r = \begin{bmatrix} 1 & 0 & 0 & 0 & 0 & 0 \\ 0 & 1 & 0 & 0 & 0 & 0 \\ 0 & 0 & 0 & 0 & 0 & 1 \end{bmatrix}^T. \tag{14.42}$$

FIGURE 14.8 Coordinate system for the parallel manipulator.

Therefore, the twists of the chain can be obtained through reciprocal screw theory:

$$\$_1 = \begin{bmatrix} 1 & 0 & 0 & 0 & 0 & 0 \\ 0 & 1 & 0 & 0 & 0 & 0 \\ 0 & 0 & 0 & 0 & 0 & 1 \end{bmatrix}^T. \tag{14.43}$$

Naturally, equation (14.43) denotes the twists of a UP kinematic chain. That is, if we use a UP kinematic chain to connect the end effector and the fixed base, the end effector can only execute two rotations about the x- and the y-directions, and one translation along the z-axis. If we exert three actuations to these U and P kinematic pairs, the end effector must be controllable. However, the strength of such a structure is very weak for it is a pure series. In order to strengthen the manipulator, we have to utilize parallel structure.

As is analyzed above, the UP kinematic chain can maintain the three motions of the end effector. The other kinematic chains, therefore, should be assigned to zero terminal constraints. So, we can synthesize three identical zero terminal constraint kinematic chains to provide actuations for the end effector by prescribing $\mathbf{K}_1 = \mathbf{K}_2 = \mathbf{K}_3 = 0$.

Reciprocally, the rank of the twists of such a kinematic chain with zero terminal constraint is six. Therefore, any chain that consists of six independent twists at least is applicable. It is not difficult to find that a PSS kinematic chain has zero terminal constraint, the analysis of which is omitted here for the sake of simplicity. So we can design the structure of the manipulator as Figure 14.8 shows and the three actuators are assigned to the three sliders.

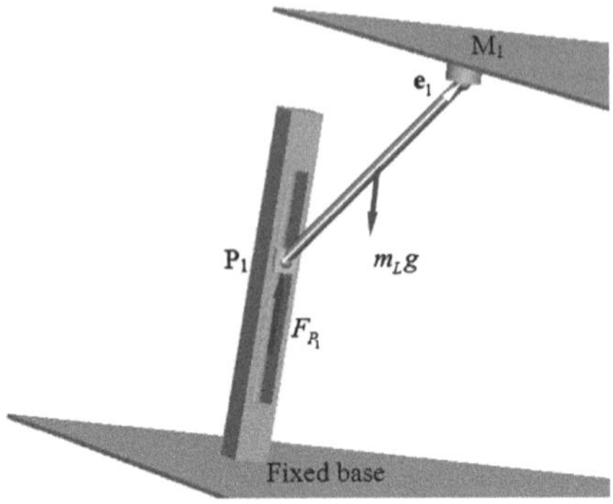

FIGURE 14.9 Kinematic relationship of limb P_1M_1.

Next, we investigate the controllability of the actuation scheme for the manipulator shown in Figure 14.8. When the actuations are given to the three sliders, P_1, P_2, and P_3, the three prismatic joints can be considered as fixed ones at this configuration. Next, let us consider the terminal constraints exerted on the end effector under these actuations.

For the kinematic chain P_1M_1 shown in Figure 14.9, when the actuation is assigned to the prismatic joint P_1, the twists of the chain can be denoted as:

$$\$_{P_1M_1} = \begin{bmatrix} \$_{11} & \$_{12} & \$_{13} & \$_{14} & \$_{15} & \$_{16} \end{bmatrix} \tag{14.44}$$

where

$$\$_{11} = \begin{bmatrix} 1 & 0 & 0 & 0 & z_{P_1} & -y_{P_1} \end{bmatrix}^T, \quad \$_{12} = \begin{bmatrix} 0 & 1 & 0 & -z_{P_1} & 0 & x_{P_1} \end{bmatrix}^T,$$
$$\$_{13} = \begin{bmatrix} 0 & 0 & 1 & y_{P_1} & -x_{P_1} & 0 \end{bmatrix}^T, \quad \$_{14} = \begin{bmatrix} 1 & 0 & 0 & 0 & z_1 & -y_1 \end{bmatrix}^T,$$
$$\$_{15} = \begin{bmatrix} 0 & 1 & 0 & -z_1 & 0 & x_1 \end{bmatrix}^T, \quad \$_{16} = \begin{bmatrix} 0 & 0 & 1 & y_1 & -x_1 & 0 \end{bmatrix}^T.$$

The terminal constraint of the chain $\$_{P_1M_1}$ can be obtained with equation (14.35):

$$\$^r_{P_1M_1} = \begin{bmatrix} x_1 - x_{P_1} & y_1 - y_{P_1} & z_1 - z_{P_1} & y_{P_1}z_1 - z_{P_1}y_1 & z_{P_1}x_1 - x_{P_1}z_1 & x_{P_1}y_1 - y_{P_1}x_1 \end{bmatrix}^T \tag{14.45}$$

Similarly, when the actuations are exerted on the other two prismatic joints, the terminal constraints of their kinematic chains can be gained in the same Cartesian coordinate system:

$$\$^r_{P_2M_2} = \begin{bmatrix} x_2 - x_{P_2} & y_2 - y_{P_2} & z_2 - z_{P_2} & y_{P_2}z_2 - z_{P_2}y_2 & z_{P_2}x_2 - x_{P_2}z_2 & x_{P_2}y_2 - y_{P_2}x_2 \end{bmatrix}^T \tag{14.46}$$

$$\$^r_{P_3M_3} = \begin{bmatrix} x_3 - x_{P_3} & y_3 - y_{P_3} & z_3 - z_{P_3} & y_{P_3}z_3 - z_{P_3}y_3 & z_{P_3}x_3 - x_{P_3}z_3 & x_{P_3}y_3 - y_{P_3}x_3 \end{bmatrix}^T \tag{14.47}$$

Therefore, when the three actuations are assigned to the three prismatic joints, the terminal constraints exerted on the end effector can be obtained from equations (14.42), (14.45), (14.46), and (14.47):

$$\$^T_C = \begin{bmatrix} 1 & 0 & 0 & 0 & 0 & 0 \\ 0 & 1 & 0 & 0 & 0 & 0 \\ 0 & 0 & 0 & 0 & 0 & 1 \\ x_1 - x_{P_1} & y_1 - y_{P_1} & z_1 - z_{P_1} & y_{P_1}z_1 - z_{P_1}y_1 & z_{P_1}x_1 - x_{P_1}z_1 & x_{P_1}y_1 - y_{P_1}x_1 \\ x_2 - x_{P_2} & y_2 - y_{P_2} & z_2 - z_{P_2} & y_{P_2}z_2 - z_{P_2}y_2 & z_{P_2}x_2 - x_{P_2}z_2 & x_{P_2}y_2 - y_{P_2}x_2 \\ x_3 - x_{P_3} & y_3 - y_{P_3} & z_3 - z_{P_3} & y_{P_3}z_3 - z_{P_3}y_3 & z_{P_3}x_3 - x_{P_3}z_3 & x_{P_3}y_3 - y_{P_3}x_3 \end{bmatrix} \tag{14.48}$$

When the determinant of matrix (14.48) is zero, the end effector is uncontrollable. This criterion can be further simplified—the determinant of a sub matrix $\$_{Csub}$ is zero where

$$\$_{Csub} = \begin{bmatrix} y_{P_1}z_1 - z_{P_1}y_1 & z_{P_1}x_1 - x_{P_1}z_1 & z_1 - z_{P_1} \\ y_{P_2}z_2 - z_{P_2}y_2 & z_{P_2}x_2 - x_{P_2}z_2 & z_2 - z_{P_2} \\ y_{P_3}z_3 - z_{P_3}y_3 & z_{P_3}x_3 - x_{P_3}z_3 & z_3 - z_{P_3} \end{bmatrix}. \tag{14.49}$$

That is, the end effector is controllable unless the following equation holds:

$$|\$_{Csub}| = 0. \tag{14.50}$$

If $|\$_{Csub}| \neq 0$, the manipulator is controllable; otherwise, the manipulator is singular at this configuration. Prior to computing $|\$_{Csub}|$, we should establish the kinematic equations of the system with natural coordinates :

(1) The actuating constraint equations

$$\begin{cases} (x_1 - x_{P_1})^2 + (y_1 - y_{P_1})^2 + (z_1 - z_{P_1})^2 = l_1^2 \\ (x_2 - x_{P_2})^2 + (y_2 - y_{P_2})^2 + (z_2 - z_{P_2})^2 = l_2^2 \\ (x_3 - x_{P_3})^2 + (y_3 - y_{P_3})^2 + (z_3 - z_{P_3})^2 = l_3^2 \end{cases} \tag{14.51}$$

where l_1, l_2, and l_3 denote the lengths of the three limbs.

(2) The rigid distance constraint equations

$$\begin{cases} (x_2 - x_1)^2 + (y_2 - y_1)^2 + (z_2 - z_1)^2 = l^2_{M_1M_2} \\ (x_3 - x_2)^2 + (y_3 - y_2)^2 + (z_3 - z_2)^2 = l^2_{M_2M_3} \\ (x_1 - x_3)^2 + (y_1 - y_3)^2 + (z_1 - z_3)^2 = l^2_{M_3M_1} \end{cases} \tag{14.52}$$

where $l_{M_1M_2}$, $l_{M_2M_3}$, and $l_{M_3M_1}$ denote the rigid distances between joints M_1, M_2, and M_3.

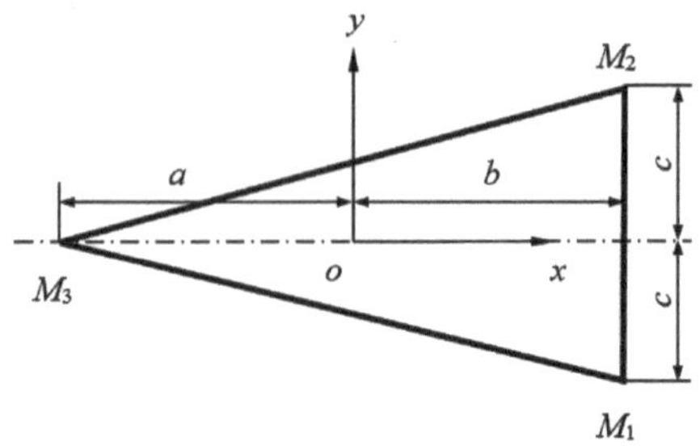

FIGURE 14.10 Top view of the end effector.

(3) The directional constraint equations

Because the end effector cannot rotate about the z-axis, which can be found from Figure 14.8, we know that the projecting line of M_1M_2 in the xoy-plane will always be parallel to the y-axis. Therefore, we have

$$x_2 - x_1 = 0. \tag{14.53}$$

From Figure 14.10, we can obtain the following vector equation:

$$2b(\overrightarrow{oM_3}) + a(\overrightarrow{oM_1} + \overrightarrow{oM_2}) = 0. \tag{14.54}$$

Expanding equation (14.54) yields

$$2b\begin{bmatrix} x_3 - x_o \\ y_3 - y_o \\ z_3 - z_o \end{bmatrix} + a\begin{bmatrix} x_1 + x_2 - 2x_o \\ y_1 + y_2 - 2y_o \\ z_1 + z_2 - 2z_o \end{bmatrix} = 0$$

which can be further simplified as

$$\begin{bmatrix} x_o \\ y_o \\ z_o \end{bmatrix} = \frac{1}{2(a+b)}\begin{bmatrix} a(x_1 + x_2) + 2bx_3 \\ a(y_1 + y_2) + 2by_3 \\ a(z_1 + z_2) + 2bz_3 \end{bmatrix}. \tag{14.55}$$

According to the establishment of the coordinate system, we know that

$$\begin{cases} x_o = 0 \\ y_o = 0 \end{cases}. \tag{14.56}$$

Therefore, we obtain the following constraint equations

$$\begin{cases} a(x_1 + x_2) + 2bx_3 = 0 \\ a(y_1 + y_2) + 2by_3 = 0 \end{cases}. \tag{14.57}$$

Therefore, the kinematics equations of the system can be obtained from equations (14.51), (14.52), (14.53), and (14.57):

$$\boldsymbol{P}(\boldsymbol{x}) = 0 \tag{14.58}$$

where

$$\boldsymbol{x} = \begin{bmatrix} x_1 & y_1 & z_1 & x_2 & y_2 & z_2 & x_3 & y_3 & z_3 \end{bmatrix}^T \tag{14.59}$$

$$\boldsymbol{P}(\boldsymbol{x}) = \begin{bmatrix} (x_1 - x_{P_1})^2 + (y_1 - y_{P_1})^2 + (z_1 - z_{P_1})^2 = l_1^2 \\ (x_2 - x_{P_2})^2 + (y_2 - y_{P_2})^2 + (z_2 - z_{P_2})^2 = l_2^2 \\ (x_3 - x_{P_3})^2 + (y_3 - y_{P_3})^2 + (z_3 - z_{P_3})^2 = l_3^2 \\ (x_2 - x_1)^2 + (y_2 - y_1)^2 + (z_2 - z_1)^2 = l_{M_1M_2}^2 \\ (x_3 - x_2)^2 + (y_3 - y_2)^2 + (z_3 - z_2)^2 = l_{M_2M_3}^2 \\ (x_1 - x_3)^2 + (y_1 - y_3)^2 + (z_1 - z_3)^2 = l_{M_3M_1}^2 \\ x_2 - x_1 \\ a(x_1 + x_2) + 2bx_3 \\ a(y_1 + y_2) + 2by_3 \end{bmatrix} \tag{14.60}$$

$$\frac{\partial \boldsymbol{P}(\boldsymbol{x})}{\partial \boldsymbol{x}} = 2\begin{bmatrix} x_1 - x_{P_1} & y_1 - y_{P_1} & z_1 - z_{P_1} & 0 & 0 & 0 & 0 & 0 & 0 \\ 0 & 0 & 0 & x_2 - x_{P_2} & y_2 - y_{P_2} & z_2 - z_{P_2} & 0 & 0 & 0 \\ 0 & 0 & 0 & 0 & 0 & 0 & x_3 - x_{P_3} & y_3 - y_{P_3} & z_3 - z_{P_3} \\ x_1 - x_2 & y_1 - y_2 & z_1 - z_2 & x_2 - x_1 & y_2 - y_1 & z_2 - z_1 & 0 & 0 & 0 \\ 0 & 0 & 0 & x_2 - x_3 & y_2 - y_3 & z_2 - z_3 & x_3 - x_2 & y_3 - y_2 & z_3 - z_2 \\ x_1 - x_3 & y_1 - y_3 & z_1 - z_3 & 0 & 0 & 0 & x_3 - x_1 & y_3 - y_1 & z_3 - z_1 \\ -\frac{1}{2} & 0 & 0 & \frac{1}{2} & 0 & 0 & 0 & 0 & 0 \\ \frac{1}{2}a & 0 & 0 & \frac{1}{2}a & 0 & 0 & b & 0 & 0 \\ 0 & \frac{1}{2}a & 0 & 0 & \frac{1}{2}a & 0 & 0 & b & 0 \end{bmatrix}. \tag{14.61}$$

Therefore, the singularity criteria of the manipulator can also be expressed as:

$$\left| \frac{\partial \boldsymbol{P}(\boldsymbol{x})}{\partial \boldsymbol{x}} \right| = 0. \tag{14.62}$$

However, it will be more efficient to find the singularities of the manipulator by computing the determinant of equation (14.50) than by calculating equation (14.62) though they have the same solutions. So long as $\left| \frac{\partial \boldsymbol{P}(\boldsymbol{x})}{\partial \boldsymbol{x}} \right| \neq 0$, the forward displacement of the manipulator can be iterated with numerical method, such as Newton–Raphson method, which we do not intend to discuss here.

$$\boldsymbol{x}_{i+1} = \boldsymbol{x}_i - \left[\frac{\partial \boldsymbol{P}(\boldsymbol{x}_i)}{\partial \boldsymbol{x}} \right]^{1} \boldsymbol{P}(\boldsymbol{x}_i) \tag{14.63}$$

Therefore, $\boldsymbol{x}$ can be obtained if the actuations z_i, $i = 1,2,3$ are given; and as a result, the forward displacement of the end effector can be solved. Suppose the initial normal direction vector of the end effector plane is:

$$\boldsymbol{n}_0 = [0 \ \ 0 \ \ 1]^T. \tag{14.64}$$

The normal direction vector of the end effector plane, denoted by $\boldsymbol{n}(\boldsymbol{x})$, is the function of its configuration. When X are obtained, $\boldsymbol{n}(\boldsymbol{x})$ can be expressed as:

$$\boldsymbol{n}(\boldsymbol{x}) = [\, n_x(\boldsymbol{x}) \ \ n_y(\boldsymbol{x}) \ \ n_z(\boldsymbol{x})\,]^T = \frac{\overrightarrow{M_1M_2} \times \overrightarrow{M_1M_3}}{\| \overrightarrow{M_1M_2} \times \overrightarrow{M_1M_3} \|} \tag{14.65}$$

where $(\boldsymbol{n}_x(\boldsymbol{x}), \boldsymbol{n}_y(\boldsymbol{x}), \boldsymbol{n}_z(\boldsymbol{x})$ are the direction cosine of the normal direction vector.

As a result, the angular displacements of the end effector relative to the absolute coordinate system are defined as the acute angles between the corresponding axes and the end effector plane, which can be solved:

$$\begin{cases} \alpha_x = \arcsin\,(n_x(\boldsymbol{x})) \\ \alpha_y = \arcsin\,(n_y(\boldsymbol{x})) \\ \alpha_z = \arcsin\,(n_z(\boldsymbol{x})) \end{cases} \tag{14.66}$$

where α_x, α_y, and α_z denote the angles from the x-, y-, and z-axes of the absolute coordinate system to the end effector plane.

The derivative of equation (14.60) with respect to time can be expressed as:

$$\mathbf{A}\dot{\boldsymbol{x}} = \boldsymbol{c} \tag{14.67}$$

where

$$\mathbf{A} = \begin{bmatrix} x_1 - x_{P_1} & y_1 - y_{P_1} & z_1 - z_{P_1} & 0 & 0 & 0 & 0 & 0 & 0 \\ 0 & 0 & 0 & x_2 - x_{P_2} & y_2 - y_{P_2} & z_2 - z_{P_2} & 0 & 0 & 0 \\ 0 & 0 & 0 & 0 & 0 & 0 & x_3 - x_{P_3} & y_3 - y_{P_3} & z_3 - z_{P_3} \\ x_1 - x_2 & y_1 - y_2 & z_1 - z_2 & x_2 - x_1 & y_2 - y_1 & z_2 - z_1 & 0 & 0 & 0 \\ 0 & 0 & 0 & x_2 - x_3 & y_2 - y_3 & z_2 - z_3 & x_3 - x_2 & y_3 - y_2 & z_3 - z_2 \\ x_1 - x_3 & y_1 - y_3 & z_1 - z_3 & 0 & 0 & 0 & x_3 - x_1 & y_3 - y_1 & z_3 - z_1 \\ -1 & 0 & 0 & 1 & 0 & 0 & 0 & 0 & 0 \\ a & 0 & 0 & a & 0 & 0 & 2b & 0 & 0 \\ 0 & a & 0 & 0 & a & 0 & 0 & 2b & 0 \end{bmatrix} \tag{14.68}$$

$$\dot{\boldsymbol{x}} = \left[\dot{x}_{M_1} \ \ \dot{y}_{M_1} \ \ \dot{z}_{M_1} \ \ \dot{x}_{M_2} \ \ \dot{y}_{M_2} \ \ \dot{z}_{M_2} \ \ \dot{x}_{M_3} \ \ \dot{y}_{M_3} \ \ \dot{z}_{M_3} \right]^T \tag{14.69}$$

$$c = [\,(z_1 - z_{P_1})\dot{z}_{P_1} \quad (z_2 - z_{P_2})\dot{z}_{P_2} \quad (z_3 - z_{P_3})\dot{z}_{P_3} \quad 0 \quad 0 \quad 0 \quad 0 \quad 0 \quad 0\,]^T. \tag{14.70}$$

The general velocity of the natural coordinates can be solved by left multiplying $\mathbf{A}^{-1}$ at both sides of equation (14.67) so long as $|\mathbf{A}| \neq 0$. Assume

$$\begin{cases} v_1 = [\,\dot{x}_1 \quad \dot{y}_1 \quad \dot{z}_1\,]^T \\ v_2 = [\,\dot{x}_2 \quad \dot{y}_2 \quad \dot{z}_2\,]^T \\ v_3 = [\,\dot{x}_3 \quad \dot{y}_3 \quad \dot{z}_3\,]^T. \end{cases} \tag{14.71}$$

From equation (14.71), we have

$$v_2 - v_1 = \omega \times r_{M_1M_2} \tag{14.72}$$

where $r_{M_1M_2}$ denotes a vector from point M_1 to point M_2.

Left cross multiplying both sides of equation (14.72) with $r_{M_1M_2}$ yields:

$$r_{M_1M_2} \times (v_2 - v_1) = r_{M_1M_2} \times (\omega \times r_{M_1M_2}) = \left(\|r_{M_1M_2}\|^2 I - r_{M_1M_2} r_{M_1M_2}^T\right)\omega. \tag{14.73}$$

Similarly, we have

$$r_{M_1M_3} \times (v_3 - v_1) = r_{M_1M_3} \times (\omega \times r_{M_1M_3}) = \left(\|r_{M_1M_3}\|^2 I - r_{M_1M_3} r_{M_1M_3}^T\right)\omega. \tag{14.74}$$

Assume

$$\begin{cases} \mathbf{B} = \left(\|r_{M_1M_2}\|^2 + \|r_{M_1M_3}\|^2\right) I - \left(\|r_{M_1M_2} r_{M_1M_2}^T + r_{M_1M_3} r_{M_1M_3}^T\right) \\ q = r_{M_1M_2} \times (v_2 - v_1) + r_{M_1M_3} \times (v_3 - v_1) \end{cases} \tag{14.75}$$

It is not difficult to prove that **B** will not be singular unless the three points $M_1, M_2,$ andM_3 just remain collinear. Therefore, the angular velocity of the end effector can be obtained:

$$\omega = \mathbf{B}^{-1} q. \tag{14.76}$$

The derivative of equation set (14.67) with respect to time can be expressed as:

$$\mathbf{A}\ddot{x} = d \tag{14.77}$$

where

$$\ddot{x} = \left[\ddot{x}_1 \quad \ddot{y}_1 \quad \ddot{z}_1 \quad \ddot{x}_2 \quad \ddot{y}_2 \quad \ddot{z}_2 \quad \ddot{x}_3 \quad \ddot{y}_3 \quad \ddot{z}_3\right]^T \tag{14.78}$$

$$
\boldsymbol{d} = \begin{bmatrix} (z_1 - z_{P_1})\ddot{z}_{P_1} - (\dot{x}_1)^2 - (\dot{y}_1)^2 - (\dot{z}_1 - \dot{z}_{P_1})^2 \\ (z_2 - z_{P_2})\ddot{z}_{P_2} - (\dot{x}_2)^2 - (\dot{y}_2)^2 - (\dot{z}_2 - \dot{z}_{P_2})^2 \\ (z_3 - z_{P_3})\ddot{z}_{P_3} - (\dot{x}_3)^2 - (\dot{y}_3)^2 - (\dot{z}_3 - \dot{z}_{P_3})^2 \\ -(\dot{x}_2 - \dot{x}_1)^2 - (\dot{y}_2 - \dot{y}_1)^2 - (\dot{z}_2 - \dot{z}_1)^2 \\ -(\dot{x}_3 - \dot{x}_2)^2 - (\dot{y}_3 - \dot{y}_2)^2 - (\dot{z}_3 - \dot{z}_2)^2 \\ -(\dot{x}_1 - \dot{x}_3)^2 - (\dot{y}_1 - \dot{y}_3)^2 - (\dot{z}_1 - \dot{z}_3)^2 \\ 0 \\ 0 \\ 0 \end{bmatrix}. \tag{14.79}
$$

The derivative of equation set (14.72) with respect to time can be expressed as:

$$
\dot{\boldsymbol{v}}_2 - \dot{\boldsymbol{v}}_1 = \dot{\boldsymbol{\omega}} \times \boldsymbol{r}_{M_1M_2} + \boldsymbol{\omega} \times \left(\boldsymbol{\omega} \times \boldsymbol{r}_{M_1M_2}\right). \tag{14.80}
$$

Left cross multiplying both sides of equation (14.80) with $\boldsymbol{r}_{M_1M_2}$ yields:

$$
\boldsymbol{r}_{M_1M_2} \times \left(\dot{\boldsymbol{v}}_2 - \dot{\boldsymbol{v}}_1\right) = \boldsymbol{r}_{M_1M_2} \times \left(\dot{\boldsymbol{\omega}} \times \boldsymbol{r}_{M_1M_2}\right) + \boldsymbol{r}_{M_1M_2} \times \left[\boldsymbol{\omega} \times \left(\boldsymbol{\omega} \times \boldsymbol{r}_{M_1M_2}\right)\right]. \tag{14.81}
$$

Similarly, we have

$$
\boldsymbol{r}_{M_1M_3} \times \left(\dot{\boldsymbol{v}}_3 - \dot{\boldsymbol{v}}_1\right) = \boldsymbol{r}_{M_1M_3} \times \left(\dot{\boldsymbol{\omega}} \times \boldsymbol{r}_{M_1M_3}\right) + \boldsymbol{r}_{M_1M_3} \times \left[\boldsymbol{\omega} \times \left(\boldsymbol{\omega} \times \boldsymbol{r}_{M_1M_3}\right)\right]. \tag{14.82}
$$

Assume

$$
\begin{aligned} \boldsymbol{w} = {} & \boldsymbol{r}_{M_1M_2} \times \left(\dot{\boldsymbol{v}}_2 - \dot{\boldsymbol{v}}_1\right) + (\boldsymbol{r}_{M_1M_2} \cdot \boldsymbol{\omega})\boldsymbol{\omega} \times \boldsymbol{r}_{M_1M_2} + \left[\boldsymbol{r}_{M_1M_3} \times \left(\dot{\boldsymbol{v}}_3 - \dot{\boldsymbol{v}}_1\right) \right. \\ & \left. + (\boldsymbol{r}_{M_1M_3} \cdot \boldsymbol{\omega})\boldsymbol{\omega} \times \boldsymbol{r}_{M_1M_3}\right] \end{aligned} \tag{14.83}
$$

Therefore,

$$
\mathbf{B}\dot{\boldsymbol{\omega}} = \boldsymbol{w}. \tag{14.84}
$$

Consequently, the angular acceleration of the end effector, $\boldsymbol{\varepsilon}$, can be obtained:

$$
\boldsymbol{\varepsilon} = \dot{\boldsymbol{\omega}} = \mathbf{B}^{-1}\boldsymbol{w}. \tag{14.85}
$$

In the following we will investigate the dynamics of the parallel manipulator. To completely model the driving forces of the actuators, the reactions of every joint and the gravity effects as well as the viscous friction at each joint, we here establish the dynamics formulation of the manipulator by the Newton–Euler approach.

14.2.1 Dynamics of Each Limb

A limb of the parallel manipulator is shown in Figure 14.9. As we have known the positions, velocities and accelerations of points M_i, $i = 1,2,3$, we can easily obtain the kinematic properties of the limbs.

As is shown in Figure 14.9, the unit direction of this limb is

$$e_i = \frac{1}{L_i}[x_i - x_{P_i} \quad y_i - y_{P_i} \quad z_i - z_{P_i}]^T. \tag{14.86}$$

Because the velocity of point M_i can be expressed as

$$v_{M_i} = v_{P_i} + \omega_i \times L_i e_i \tag{14.87}$$

where $v_{P_i} = [0\ 0\ \dot{z}_{P_i}]$ and ω_i denotes the angular velocity of the limb about the spherical joint P_i.

Substituting equation (14.71) into equation (14.87) yields

$$\omega_i \times e_i = \frac{1}{L_i}\begin{bmatrix} \dot{x}_i \\ \dot{y}_i \\ \dot{z}_i - \dot{z}_{P_i} \end{bmatrix}. \tag{14.88}$$

Before making a further conclusion, we first prove a proposition:

A two-force rigid straight link jointed its connecting neighbors by two sphere pairs at both sides will always make instantaneous rotation whose axis is perpendicular to the link axis.

Proof. Without loss of generality, we suppose a Cartesian coordinate system whose origin is superimposed with one sphere pair. The kinetic energy of the rigid link can be expressed as:

$$T = \frac{1}{2}\omega^T J\omega$$

where ω denotes the instantaneous angular velocity about the origin of the coordinate system, and J denotes the inertia matrix of the rigid link about the origin of the coordinate system.

While the work contributed by the forces exerted on the two-force rigid link through the other sphere pair is:

$$W = (r \times F) \cdot \theta$$

where r denotes the vector of the action point of the force F, θ denotes the vector of angular displacement.

Obviously, such a SS rigid body system is a holonomic system. Therefore, we have the following variation formula:

$$\delta T - \delta W = 0$$

where δ denotes the variation operation.

$$(J\omega) \cdot \delta\omega - (r \times F) \cdot \delta\theta = 0$$

Considering $\boldsymbol{\theta} = \boldsymbol{\omega} t$, we immediately have:

$$[\boldsymbol{J\omega} - t(\boldsymbol{r} \times \boldsymbol{F})] \cdot \delta\boldsymbol{\omega} = 0.$$

Because $\delta\boldsymbol{\omega}$ is an independent variation vector, we have:

$$\boldsymbol{J\omega} - t(\boldsymbol{r} \times \boldsymbol{F}) = 0.$$

Consequently,

$$\boldsymbol{\omega} = t\boldsymbol{J}^{-1}(\boldsymbol{r} \times \boldsymbol{F}).$$

Obviously, $\boldsymbol{\omega} \perp \boldsymbol{r}$.

As a matter of fact, if there are a number of forces exerted on the rigid link simultaneously at the point $k_i\boldsymbol{r}(i = 1,2,\ldots)$, we can find that

$$W = \left(\sum_i k_i\boldsymbol{r} \times \boldsymbol{F}\right) \cdot \boldsymbol{\theta} = \left(\sum_i k_i\right)(\boldsymbol{r} \times \boldsymbol{F}) \cdot \boldsymbol{\theta}.$$

Similarly, we at last have

$$\boldsymbol{\omega} = t\left(\sum_i k_i\right)\boldsymbol{J}^{-1}(\boldsymbol{r} \times \boldsymbol{F}).$$

Obviously, there are $\boldsymbol{\omega} \perp \boldsymbol{r}$, too.

Therefore, the proposition is proved.

Based on the above proposition, it is not difficult to find that taking cross product with $\boldsymbol{e}_i$ at both sides of equation (14.88) yields

$$\boldsymbol{\omega}_i = \frac{1}{L_i}\boldsymbol{e}_i \times \begin{bmatrix} \dot{x}_i \\ \dot{y}_i \\ \dot{z}_i - \dot{z}_{P_i} \end{bmatrix}. \tag{14.89}$$

Similarly, we can obtain the tangential angular acceleration of the limb, $\boldsymbol{\varepsilon}_i$:

$$\boldsymbol{\varepsilon}_i = \frac{1}{L_i}\boldsymbol{e}_i \times \begin{bmatrix} \ddot{x}_i \\ \ddot{y}_i \\ \ddot{z}_i - \ddot{z}_{P_i} \end{bmatrix}. \tag{14.90}$$

For simplicity, the actions and reactions of spherical joints P_i and M_i might be properly assumed as along the axis of the limb. Therefore, we can immediately get the Newton's equation for slider P_i

$$\boldsymbol{F}_{P_i} - m_{P_i}(\boldsymbol{a}_{P_i} + \boldsymbol{g}) - \boldsymbol{F}_{S_{P_i}} - f\boldsymbol{v}_{P_i} = 0 \tag{14.91}$$

where

$$\boldsymbol{F}_{P_i} = [0 \quad 0 \quad F_{P_i}]^T \tag{14.92}$$

And $\boldsymbol{F}_{P_i}$ denotes the strength of the vertical force of the actuator $\boldsymbol{F}_{P_i}$.

$$\boldsymbol{a}_{P_i} = [0 \quad 0 \quad \ddot{z}_{P_i}]^T \tag{14.93}$$

And $\boldsymbol{F}_{SP_i}$ denotes the force vector at the spherical joint P_i exerted by the limb P_iM_i on the slider, $\boldsymbol{g}$ denotes the acceleration due to gravity and f denotes the coefficient of viscous friction at the prismatic joint P_i.

Similarly, the Newton's equation for the limb P_iM_i can be expressed below:

$$\boldsymbol{F}_{SP_i} - m_i(\boldsymbol{a}_i + \boldsymbol{g}) - \boldsymbol{F}_{SM_i} = 0 \tag{14.94}$$

where m_i is the mass of the limb P_iM_i, and $\boldsymbol{a}_i$ is the acceleration of the center of the limb and we have the following relationship:

$$\boldsymbol{a}_i = \boldsymbol{a}_{P_i} + \frac{L_i}{2}\boldsymbol{\omega}_i \times (\boldsymbol{\omega}_i \times \boldsymbol{e}_i) + \frac{L_i}{2}\boldsymbol{\varepsilon}_i \times \boldsymbol{e}_i. \tag{14.95}$$

Associating equations (14.91) and (14.94), we obtain the Newton's equation for the limb:

$$\boldsymbol{F}_{P_i} - m_{P_i}(\boldsymbol{a}_{P_i} + \boldsymbol{g}) - f\boldsymbol{v}_{P_i} - m_i(\boldsymbol{a}_i + \boldsymbol{g}) - \boldsymbol{F}_{SM_i} = 0. \tag{14.96}$$

Therefore,

$$\boldsymbol{F}_{SM_i} = \boldsymbol{F}_{P_i} - \boldsymbol{D}_i \tag{14.97}$$

where $\boldsymbol{D}_i = m_{p_i}(\boldsymbol{a}_{P_i} + \boldsymbol{g}) - f\boldsymbol{v}_{P_i} + m_i(\boldsymbol{a}_i + \boldsymbol{g})$.

14.2.2 Dynamics of the End Effector

From equation (14.55), we know that the acceleration of the mass center of the end effector is

$$\boldsymbol{a}_o = \begin{bmatrix} \ddot{x}_o \\ \ddot{y}_o \\ \ddot{z}_o \end{bmatrix} = \frac{1}{2(a+b)} \begin{bmatrix} a(\ddot{x}_1 + \ddot{x}_2) + 2b\ddot{x}_3 \\ a(\ddot{y}_1 + \ddot{y}_2) + 2b\ddot{y}_3 \\ a(\ddot{z}_1 + \ddot{z}_2) + 2b\ddot{z}_3 \end{bmatrix}. \tag{14.98}$$

Therefore, we obtain the Newton's equation for the end effector of the parallel manipulator:

$$- m_E\boldsymbol{a}_o - m_E\boldsymbol{g} + \sum_{i=1}^{3} \boldsymbol{F}_{SM_i} + \boldsymbol{F}_o + \boldsymbol{F}_{ext} = 0 \tag{14.99}$$

where m_E denotes the mass of the end effector, and

$$\boldsymbol{F}_o = [F_x \; F_y \; - m_U g - f z_o]^T \tag{14.100}$$

m_U denotes the total mass of the column below the universal joint and f denotes the coefficient of viscous friction at the prismatic joint of the column, and F_x and F_y are two unknown components of F_o along x-axis and y-axis. F_{ext} denotes the force load of the end effector.

Associating equations (14.96) and (14.99), we obtain the Newton's equation for the system

$$\sum_{i=1}^{3} \boldsymbol{F}_{P_i} = \boldsymbol{F} \tag{14.101}$$

where

$$\boldsymbol{F} = \begin{bmatrix} F_1 \\ F_2 \\ F_3 \end{bmatrix} = \sum_{i=1}^{3} [m_{P_i}(\boldsymbol{a}_{P_i} + \boldsymbol{g}) + f_i \mathbf{v}_{P_i} + m_i(\boldsymbol{a}_i + \boldsymbol{g})] + m_E(\boldsymbol{a}_o + \boldsymbol{g}) - \boldsymbol{F}_o - \boldsymbol{F}_{ext}. \tag{14.102}$$

Suppose the moment of inertia of the end effector is denoted by $\mathbf{I}_o$ in its local coordinate system, the moment of inertia in the fixed Cartesian coordinate system can be obtained:

$$\mathbf{I} = \mathbf{R}\mathbf{I}_o\mathbf{R}^T \tag{14.103}$$

where $\mathbf{R}$ is the transformation matrix from the local coordinate system to the fixed one specified by

$$\mathbf{R} = [\boldsymbol{n}_1 \; \boldsymbol{n}_2 \; \boldsymbol{n}_3] \tag{14.104}$$

And where

$$\boldsymbol{n}_2 = \frac{\overrightarrow{M_1 M_2}}{\| \overrightarrow{M_1 M_2} \|} \tag{14.105}$$

$\boldsymbol{n}_3 = \mathbf{n}(\boldsymbol{x})$ which is depicted by equation (14.65).

$$\boldsymbol{n}_1 = \boldsymbol{n}_2 \times \boldsymbol{n}_3 \tag{14.106}$$

The Euler's equation for the end effector can be obtained

$$-\mathbf{I}\boldsymbol{\varepsilon} - \boldsymbol{\omega} \times \mathbf{I}\boldsymbol{\omega} + \sum_{i=1}^{3} (\boldsymbol{r}_{M_i} \times \boldsymbol{F}_{MS_i}) + \boldsymbol{M}_{ext} + \boldsymbol{M}_o - \sum_{i=1}^{3} [c(\boldsymbol{\omega} - \boldsymbol{\omega}_i)] = 0 \tag{14.107}$$

where $\boldsymbol{M}_{ext}$ denotes the moment load of the end effector, $\boldsymbol{c}$ denotes the moment coefficient of the viscous frictions at the spherical joints, $\boldsymbol{M}_o$ denotes the moment applied by the universal joint to the end effector and

$$\boldsymbol{M}_o = [0 \quad 0 \quad M_z]^T \tag{14.108}$$

and where M_z denotes the torque from the column kinematic chain.

Substituting equation (14.97) into equation (14.107) yields

$$\sum_{i=1}^{3} (\boldsymbol{r}_{M_i} \times \boldsymbol{F}_{P_i}) = \mathbf{G} \tag{14.109}$$

Where

$$\mathbf{G} = \sum_{i=1}^{3} \boldsymbol{r}_{M_1} \times \boldsymbol{D}_i + \sum_{i=1}^{3} [c(\boldsymbol{\omega} - \boldsymbol{\omega}_i)] + \mathbf{I}\boldsymbol{\varepsilon} + \boldsymbol{\omega} \times \mathbf{I}\boldsymbol{\omega} - \boldsymbol{M}_{ext} - \boldsymbol{M}_o. \tag{14.110}$$

Substituting equation (14.92) into equation (14.109) and expanding the left of equation (14.109) yields

$$\sum_{i=1}^{3}(\boldsymbol{r}_{M_i} \times \boldsymbol{F}_{P_i}) = \sum_{i=1}^{3}\left(\begin{bmatrix} 0 & -z_{M_i} & y_{M_i} \\ z_{M_i} & 0 & -x_{M_i} \\ -y_{M_i} & x_{M_i} & 0 \end{bmatrix}\begin{bmatrix} 0 \\ 0 \\ F_{P_i} \end{bmatrix}\right)$$

$$= \begin{bmatrix} \sum_{i=1}^{3} y_{M_i} \quad F_{P_i} \\ -\sum_{i=1}^{3} x_{M_i} \quad F_{P_i} \\ 0 \end{bmatrix}. \tag{14.111}$$

Correspondingly, if equation (14.110) is expressed as $\mathbf{G} = [G_1\ G_2\ G_3]^T$, then equation (14.109) can be expressed as:

$$\begin{bmatrix} \sum_{i=1}^{3} y_{M_i} \quad F_{P_i} \\ -\sum_{i=1}^{3} x_{M_i} \quad F_{P_i} \\ 0 \end{bmatrix} = \begin{bmatrix} G_1 \\ G_2 \\ G_3 \end{bmatrix}. \tag{14.112}$$

Utilizing the z-direction component of equation (14.101) and the x-direction and y-direction components of equation (14.112), we obtain

$$\begin{cases} \sum_{i=1}^{3} F_{P_i} = F_3 \\ \sum_{i=1}^{3} y_{M_i} F_{P_i} = G_1 \\ -\sum_{i=1}^{3} x_{M_i} F_{P_i} = G_2. \end{cases} \tag{14.113}$$

Equation (14.113) might be rewritten in matrix multiplication form:

$$\mathbf{A}\boldsymbol{F}_p = \boldsymbol{c} \tag{14.114}$$

where

$$\mathbf{A} = \begin{bmatrix} 1 & 1 & 1 \\ y_{M_1} & y_{M_2} & y_{M_3} \\ -x_{M_1} & -x_{M_2} & -x_{M_3} \end{bmatrix}$$

$$\boldsymbol{F}_p = [F_{P1} \quad F_{P2} \quad F_{P3}]^T$$

$$\boldsymbol{c} = [F_3 \quad G_1 \quad G_2]^T.$$

Supposing $|\mathbf{A}| = 0$, we can find that there should be:

$$(y_{M_2} - y_{M_1})(x_{M_3} - x_{M_1}) - (y_{M_3} - y_{M_1})(x_{M_2} - x_{M_1}) = 0. \qquad (14.115)$$

If equation (14.115) holds, the three points M_1, M_2, and M_3 should remain collinear, which is in conflict with the reality. And therefore, $|\mathbf{A}| \neq 0$. So, the forces of the three actuators can be solved:

$$\boldsymbol{F}_P = \mathbf{A}^{-1}\boldsymbol{c}. \qquad (14.116)$$

After obtaining $\boldsymbol{F}_P$, the forces of every joint can be easily solved.

14.3 KINEMATIC OF THE SUSPENSION MECHANISM WITH INVARIABLE ORIENTATION PARAMETERS

Figure 14.11 shows a new vehicle suspension mechanism that can guide the wheel to jounce and rebound with invariable orientation parameters. The end effector (knuckle), $C_1C_2C_3C_4$, is connected with the vehicle body through four identical planar RRR kinematic chains, $A_iB_iC_i (i = 1,2,3,4)$. The two planes that the kinematic chains $A_1B_1C_1$ and $A_2B_2C_2$ locate are either parallel or superimposed to each other. Similarly, the two planes that the kinematic chains $A_3B_3C_3$ and $A_4B_4C_4$ locate are either parallel or superimposed to each other.

For simplicity, we can first investigate the free motion(s) of the knuckle, $C_1C_2C_3C_4$, under the constraints of the upper two planar RRR-chains, $A_1B_1C_1$

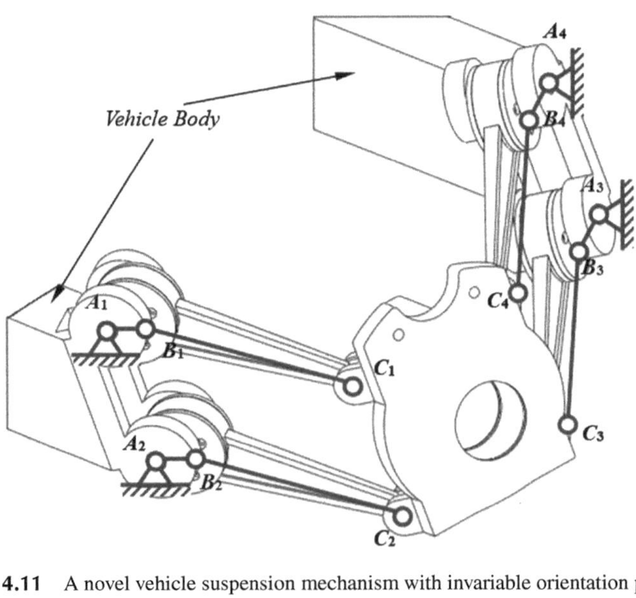

FIGURE 14.11 A novel vehicle suspension mechanism with invariable orientation parameters.

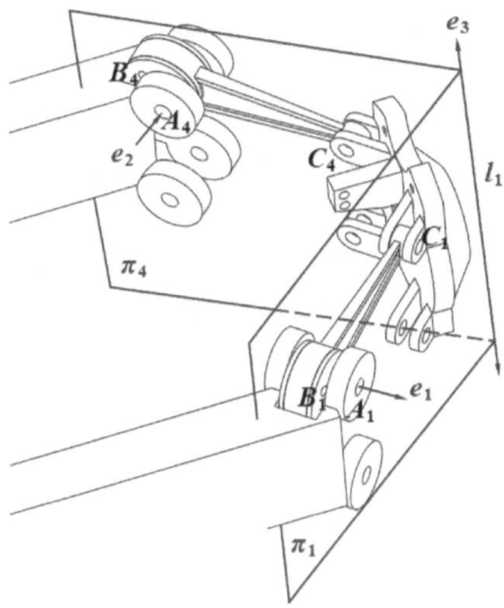

FIGURE 14.12 Geometric interpretation for the mechanism to generate straight line motion.

and $A_4B_4C_4$ shown in Figure 14.4. π_1-plane indicates the plane that $A_1B_1C_1$ kinematic chain determines, and π_4-plane indicates the plane that $A_4B_4C_4$ kinematic chain locates. The intersection of these two planes is denoted by l_{14} in Figure 14.12.

Geometrically, we can find that the planar RRR-chain, $A_1B_1C_1$, can only allow the connected knuckle, $C_1C_2C_3C_4$, to have motions parallel to π_1-plane. Similarly, the planar RRR-chain, $A_4B_4C_4$, can also allow the connected knuckle, $C_1C_2C_3C_4$, to only have motions parallel to π_4-plane. As a result, $C_1C_2C_3C_4$ only has a free motion parallel to the intersection of these two planes, l_{14}. The next section will present a theoretical analysis for this conclusion.

For the sake of expression, we can establish a fixed coordinate frame with the vehicle body. Suppose the unit normals of these two planes, $\boldsymbol{\pi}_1$ and $\boldsymbol{\pi}_4$, are $\boldsymbol{e}_1$ and $\boldsymbol{e}_2$, respectively. For convenience in the following discussion, we can also assume that $\boldsymbol{e}_3 = \frac{\boldsymbol{e}_1\times\boldsymbol{e}_2}{\|\boldsymbol{e}_1\times\boldsymbol{e}_2\|}$. At the fixed coordinate frame, suppose the absolute coordinates of the six revolute joints are $\boldsymbol{r}_{A_1}$,$\boldsymbol{r}_{B_1}$,$\boldsymbol{r}_{C_1}$,$\boldsymbol{r}_{A_4}$,$\boldsymbol{r}_{B_4}$ and $\boldsymbol{r}_{C_4}$.

The screws of the kinematic chain, $A_1B_1C_1$, are immediately obtained with the coordinates of the revolute joints, $\boldsymbol{r}_{A_1}$,$\boldsymbol{r}_{B_1}$ and $\boldsymbol{r}_{C_1}$:

$$\boldsymbol{\$}_{A_1B_1C_1} = [\boldsymbol{\$}_{A_1}\ \boldsymbol{\$}_{B_1}\ \boldsymbol{\$}_{C_1}]$$

where $\boldsymbol{\$}_{A_1} = \begin{bmatrix} \boldsymbol{e}_1 \\ \boldsymbol{r}_{A_1}\times\boldsymbol{e}_1 \end{bmatrix}$, $\boldsymbol{\$}_{B_1} = \begin{bmatrix} \boldsymbol{e}_1 \\ \boldsymbol{r}_{B_1}\times\boldsymbol{e}_1 \end{bmatrix}$, $\boldsymbol{\$}_{C_1} = \begin{bmatrix} \boldsymbol{e}_1 \\ \boldsymbol{r}_{C_1}\times\boldsymbol{e}_1 \end{bmatrix}$.

The terminal constraint of the kinematic chain, $A_1B_1C_1$, can be obtained with reciprocal screw theory:

$$\boldsymbol{\$}_{A_1B_1C_1}\boldsymbol{\Delta\$}^r_{A_1B_1C_1} = 0 \tag{14.117}$$

where $\boldsymbol{\Delta} = \begin{bmatrix} \mathbf{0}_{3\times3} & \mathbf{I}_{3\times3} \\ \mathbf{I}_{3\times3} & \mathbf{0}_{3\times3} \end{bmatrix}$, $\mathbf{0}_{3\times3} = \begin{bmatrix} 0 & 0 & 0 \\ 0 & 0 & 0 \\ 0 & 0 & 0 \end{bmatrix}$, and $\mathbf{I}_{3\times3} = \begin{bmatrix} 1 & 0 & 0 \\ 0 & 1 & 0 \\ 0 & 0 & 1 \end{bmatrix}$, and $\$^r_{A_1B_1C_1}$ denotes the reciprocal screws of $\$_{A_1B_1C_1}$.

So, the terminal constraints of kinematic chain, $A_1B_1C_1$, can be immediately obtained from equation (14.117):

$$\$^r_{A_1B_1C_1} = \left[\$^{r1}_{A_1B_1C_1} \ \ \$^{r2}_{A_1B_1C_1} \ \ \$^{r3}_{A_1B_1C_1}\right] \tag{14.118}$$

where $\$^{r1}_{A_1B_1C_1} = \begin{bmatrix} \boldsymbol{e}_1 \\ \boldsymbol{r}_{C_1} \times \boldsymbol{e}_1 \end{bmatrix}$ which indicates a force that passes through point C_1 and is perpendicular to the π_1-plane, $\$^{r2}_{A_1B_1C_1} = \begin{bmatrix} 0 \\ \boldsymbol{e}_3 \end{bmatrix}$ which indicates a torque about the $\boldsymbol{e}_3$-axis, $\$^{r3}_{A_1B_1C_1} = \begin{bmatrix} 0 \\ \boldsymbol{e}_4 \end{bmatrix}$ and $\boldsymbol{e}_4 = \boldsymbol{e}_3 \times \boldsymbol{e}_1$, $\$^{r3}_{A_1B_1C_1}$ indicates a torque about the $\boldsymbol{e}_4$-axis. $\boldsymbol{e}_4 = \boldsymbol{e}_2$ when $\pi_1 \perp \pi_2$.

Similarly, the screws of kinematic chain, $A_4B_4C_4$, can be expressed in the fixed coordinate frame:

$$\$_{A_4B_4C_4} = [\$_{A_4} \ \ \$_{B_4} \ \ \$_{C_4}]$$

where $\$_{A_4} = \begin{bmatrix} \boldsymbol{e}_2 \\ \boldsymbol{r}_{A_4} \times \boldsymbol{e}_2 \end{bmatrix}$, $\$_{B_4} = \begin{bmatrix} \boldsymbol{e}_2 \\ \boldsymbol{r}_{B_4} \times \boldsymbol{e}_2 \end{bmatrix}$, $\$_{C_4} = \begin{bmatrix} \boldsymbol{e}_2 \\ \boldsymbol{r}_{C_4} \times \boldsymbol{e}_2 \end{bmatrix}$.

The terminal constraints of the kinematic chain, $A_4B_4C_4$, can be obtained similarly to equation (14.117):

$$\$^r_{A_4B_4C_4} = \left[\$^{r1}_{A_4B_4C_4} \ \ \$^{r2}_{A_4B_4C_4} \ \ \$^{r3}_{A_4B_4C_4}\right] \tag{14.119}$$

where $\$^{r1}_{A_4B_4C_4} = \begin{bmatrix} \boldsymbol{e}_2 \\ \boldsymbol{r}_{C_4} \times \boldsymbol{e}_2 \end{bmatrix}$ which indicates a force passing through point C_4 and perpendicular to the plane π_4, $\$^{r2}_{A_4B_4C_4} = \begin{bmatrix} 0 \\ \boldsymbol{e}_3 \end{bmatrix}$ which indicates a torque about the $\boldsymbol{e}_3$-axis, $\$^{r3}_{A_4B_4C_4} = \begin{bmatrix} 0 \\ \boldsymbol{e}_5 \end{bmatrix}$ and $\boldsymbol{e}_5 = \boldsymbol{e}_2 \times \boldsymbol{e}_3$,$\$^{r3}_{A_4B_4C_4}$ indicates a torque about the $\boldsymbol{e}_5$-axis. $\boldsymbol{e}_5 = \boldsymbol{e}_1$ when $\pi_1 \perp \pi_2$.

Now, we can investigate the geometric constraints of the kinematic chains of the knuckle shown in Figure 14.13.

As is assumed above, the two planes, π_1 and π_2, that the kinematic chains $A_1B_1C_1$ and $A_2B_2C_2$ locate are either parallel or superimposed on each other. The two planes, π_3 and π_4, that the kinematic chains $A_3B_3C_3$ and $A_4B_4C_4$ locate are either parallel or superimposed on each other. Therefore,

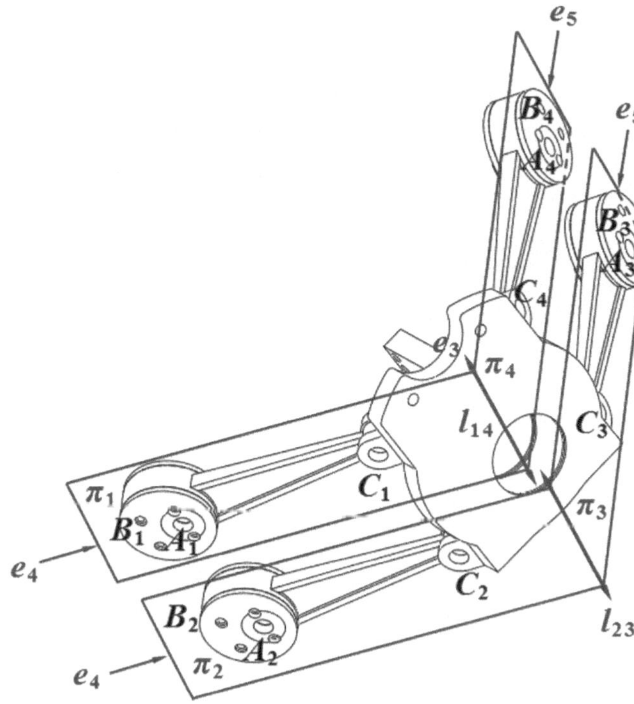

FIGURE 14.13 Kinematic chains and the planes they locate.

the intersections of π_1- and π_4-planes, π_2- and π_3-planes are either parallel or superimposed to be one line. That is, lines l_{14} and l_{23} are either parallel to or overlapped with each other. Similarly, we can obtain the terminal constraints of the other two kinematic chains, $A_2B_2C_2$ and $A_3B_3C_3$, individually:

$$\$^r_{A_2B_2C_2} = \begin{bmatrix} \$^{r1}_{A_2B_2C_2} & \$^{r2}_{A_2B_2C_2} & \$^{r3}_{A_2B_2C_2} \end{bmatrix} \tag{14.120}$$

$$\$^r_{A_3B_3C_3} = \begin{bmatrix} \$^{r1}_{A_3B_3C_3} & \$^{r2}_{A_3B_3C_3} & \$^{r3}_{A_3B_3C_3} \end{bmatrix} \tag{14.121}$$

where $\$^{r1}_{A_2B_2C_2} = \begin{bmatrix} \boldsymbol{e}_1 \\ \boldsymbol{r}_{C_2} \times \boldsymbol{e}_1 \end{bmatrix}$ which indicates a force passing through point C_2 and perpendicular to the plane π_2 shown in Figure 14.13, $\$^{r2}_{A_2B_2C_2} = \begin{bmatrix} 0 \\ \boldsymbol{e}_3 \end{bmatrix}$ which indicates a torque about the $\boldsymbol{e}_3$-axis, $\$^{r3}_{A_2B_2C_2} = \begin{bmatrix} 0 \\ \boldsymbol{e}_4 \end{bmatrix}$ and $\boldsymbol{e}_4 = \boldsymbol{e}_4 \times \boldsymbol{e}_1$, $\$^{r3}_{A_2B_2C_2}$ indicates a torque about the $\boldsymbol{e}_4$-axis; $\$^{r1}_{A_3B_3C_3} = \begin{bmatrix} \boldsymbol{e}_2 \\ \boldsymbol{r}_{C_3} \times \boldsymbol{e}_2 \end{bmatrix}$ which indicates a force passing through point C_3 and perpendicular to the

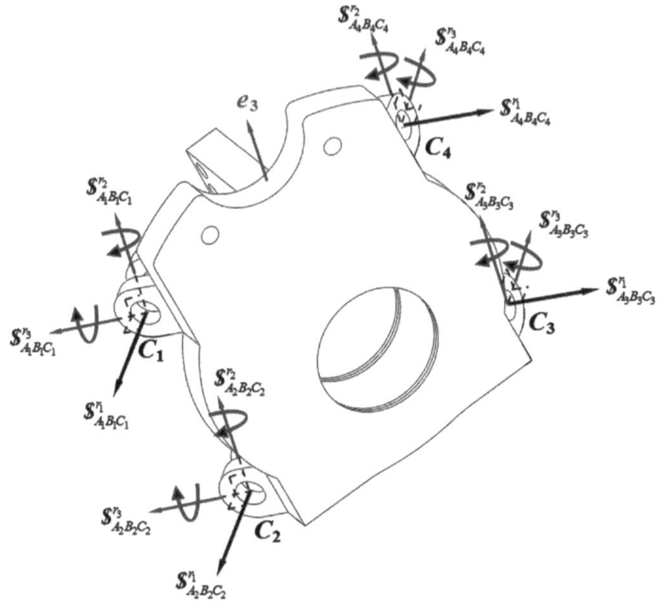

FIGURE 14.14 Constraints exerted on the knuckle by the four kinematic chains.

plane π_3, $\$^{r2}_{A_3B_3C_3} = \begin{bmatrix} 0 \\ e_3 \end{bmatrix}$ which indicates a torque about the e_3-axis, $\$^{r3}_{A_3B_3C_3} = \begin{bmatrix} 0 \\ e_5 \end{bmatrix}$ and $e_5 = e_2 \times e_3$, $\$^{r3}_{A_3B_3C_3}$ indicates a torque about the e_5-axis.

It is not difficult to find that $\$^{r2}_{A_1B_1C_1}$, $\$^{r2}_{A_2B_2C_2}$, $\$^{r2}_{A_3B_3C_3}$ and $\$^{r2}_{A_4B_4C_4}$ are over constraints of a torque about the e_3-axis shown in Figures 14.12 and 14.14; $\$^{r3}_{A_1B_1C_1}$ and $\$^{r3}_{A_2B_2C_2}$ are over constraints of a torque about the e_4-axis shown in Figures 14.13 and 14.14; $\$^{r3}_{A_3B_3C_3}$ and $\$^{r3}_{A_4B_4C_4}$ are over constraints of a torque about the e_5-axis shown in Figures 14.13 and 14.14; $\$^{r1}_{A_1B_1C_1}$ and $\$^{r1}_{A_2B_2C_2}$ are over constraints of force along the direction of e_1 shown in Figure 14.14; $\$^{r1}_{A_3B_3C_3}$ and $\$^{r1}_{A_4B_4C_4}$ are over constraints of force along the direction of e_2 shown in Figure 14.14. All these overconstraints are helpful in enhancing the stiffness and the stability of the suspension.

From equations (14.118) to (14.121), we can find that the constraints exerted on the knuckle are shown in Figure 14.14. Suppose that $\pi_1 \perp \pi_4$ and $\pi_2 \perp \pi_3$ and the knuckle sustains a set of orthogonal torsional loads $\boldsymbol{M}_{e_1}$, $\boldsymbol{M}_{e_2}$, and $\boldsymbol{M}_{e_3}$; and a set of force loads $\boldsymbol{F}_{e_1}$, $\boldsymbol{F}_{e_2}$ and $\boldsymbol{F}_{e_3}$. Then, we can therefore assume that the four torsional overconstraints $\$^{r2}_{A_1B_1C_1}$, $\$^{r2}_{A_2B_2C_2}$, $\$^{r2}_{A_3B_3C_3}$, and $\$^{r2}_{A_4B_4C_4}$ are identical and all equal $\frac{1}{4}\boldsymbol{M}_{e_3}$; the two torsional overconstraints $\$^{r3}_{A_1B_1C_1}$ and $\$^{r3}_{A_2B_2C_2}$ are identical and all equal $\frac{1}{2}\boldsymbol{M}_{e_2}$; the two torsional overconstraints $\$^{r3}_{A_3B_3C_3}$ and $\$^{r3}_{A_4B_4C_4}$ are identical and all equal $\frac{1}{2}\boldsymbol{M}_{e_1}$; the two force overconstraints $\$^{r1}_{A_1B_1C_1}$ and $\$^{r1}_{A_2B_2C_2}$ are identical and all equal $\frac{1}{2}\boldsymbol{F}_{e_1}$; the two force overconstraints $\$^{r1}_{A_3B_3C_3}$ and $\$^{r1}_{A_4B_4C_4}$ are identical and all equal $\frac{1}{2}\boldsymbol{F}_{e_2}$. Consequently, the constraints (overconstraints are excluded) exerted on the knuckle by the four kinematic chains can be expressed as:

$$\$^{r}_{C_1C_2C_3C_4} = [\$^{r1}_{A_1B_1C_1} \quad \$^{r2}_{A_1B_1C_1} \quad \$^{r3}_{A_1B_1C_1} \quad \$^{r1}_{A_4B_4C_4} \quad \$^{r3}_{A_4B_4C_4}]. \quad (14.122)$$

The free motion of the end effector, P_3Q_3, can be obtained by solving :

$$\$^{r}_{C_1C_2C_3C_4} \Delta \$^{F}_{C_1C_2C_3C_4} = 0$$

where $\$^{F}_{P_3Q_3}$ denotes the screw of the free motion of the end effector.

$$\$^{F}_{C_1C_2C_3C_4} = \begin{bmatrix} 0 \\ \boldsymbol{e}_3 \end{bmatrix} \quad (14.123)$$

Equation (14.123) indicates that the end effector, $C_1C_2C_3C_4$, strictly has one translational DOF along the $\boldsymbol{e}_3$-axis shown in Figure 14.12. This again demonstrates the correctness of the geometry deduction obtained above. Because both the right and the left suspensions of a vehicle can track a straight line during the jumping of the wheels in theory, their distances and the orientation parameters will always remain invariable. This not only eliminates the shambling shocks resulting from wheel jumping but also reduces the abrasion of the wheels. Therefore, this kind of independent suspension can greatly improve the ride and handling of advanced vehicles. For the same mechanical specifications, the whole weight of the suspension mechanism shown in Figure 14.11 can be made much lower.

14.3.1 Jumping Distance and Motion Constraint Equations of the Suspension

From the above analysis, we can draw a conclusion that each kinematic chain sustains two torques and one force perpendicular to the plane that the kinematic chain determines, and the end effector that will be used as the knuckle of the suspension in application always tracks an exact straight line. That is, in any kinematic chain plane, π_1-plane for an example, the RRR-kinematic chain can be equivalently replaced by a slider-crank linkage shown in Figure 14.15.

To be utilized as a suspension mechanism, the workspace of the slider-crank should be firstly analyzed.

14.3.2 Space of Jounce and Rebound of the Suspension Mechanism

Suppose the link A_1B_1 shown in Figure 14.15 is a crank that can make a circular rotation about the fixed axis passing through revolute joint A_1. Also suppose that the four RRR-kinematic chains, $A_iB_iC_i (i = 1,2,3,4)$, are identical which is not necessary in general and which can be proved with the previous process;

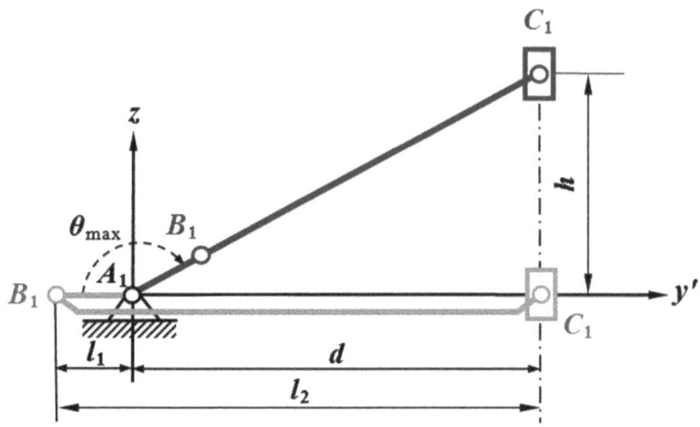

FIGURE 14.15 An equivalent mechanism in motion.

and the lengths of links A_iB_i and B_iC_i are l_1 and l_2, respectively. Also suppose that the subtended angle of the two planes, π_1 and π_2, is $\frac{\pi}{2}$, the distance from joint C_1 to joint C_4 is l_3, and the distance from joint A_1 to joint A_4 is l_4. From Figure 14.15, we can find that the distance the end effector moved from the horizontal plane to the upper limitation is:

$$h = \sqrt{(l_1 + l_2)^2 - d^2}. \tag{14.124}$$

The span of jounce and rebound of the end effector is $2h$ which can be found from Figure 14.15. Therefore, h is the primary specification parameter of a suspension, and the displacement of the end effector, z, should satisfy:

$$-\sqrt{(l_1 + l_2)^2 - d^2} < z < \sqrt{(l_1 + l_2)^2 - d^2}. \tag{14.125}$$

14.3.3 Displacement Constrain Equations and the Kinematics Analysis of the Suspension Mechanism

Without loss of generality, the equivalent slider-crank mechanism is assumed within the π_1-plane determined by links A_1B_1 and B_1C_1. The slider C_1 shown in Figures 14.15 and 14.16 denotes the knuckle, $C_1C_2C_3C_4$, and the dot dashed line denotes the prescribed straight line. Obviously, $d = \frac{\sqrt{2}}{2}(l_4 - l_3)$ for $\pi_1 \perp \pi_4$ here.

Assume that the angle from minus $\boldsymbol{e}_2$-axis to link A_1B_1 is denoted by θ and the angle from minus $\boldsymbol{e}_2$-axis to link C_1B_1 is denoted by φ which are shown in Figure 14.16. We get

$$\begin{cases} -l_1\cos\theta + l_2\cos\varphi = d \\ l_1\sin\theta + l_2\sin\varphi = z. \end{cases} \tag{14.126}$$

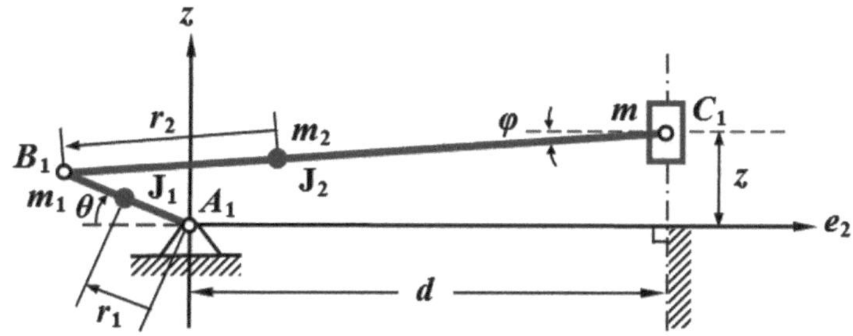

FIGURE 14.16 Displacement representation of the spatial mechanism shown in Figure 14.15.

Substituting $d = \frac{\sqrt{2}}{2}(l_4 - l_3)$ into equation (14.126) and eliminating variable φ yields:

$$z = l_1 \sin\theta \pm \sqrt{l_2^2 - \left[\frac{\sqrt{2}}{2}(l_4 - l_3) + l_1 \cos\theta\right]^2}. \tag{14.127}$$

The two possible solutions of equation (14.127) correspond to two feasible configurations of the mechanism. Under the configuration shown in Figure 14.16, we have

$$z = l_1 \sin\theta + \sqrt{l_2^2 - \left[\frac{\sqrt{2}}{2}(l_4 - l_3) + l_1 \cos\theta\right]^2}. \tag{14.128}$$

Equation (14.128) indicates that the displacement of the end effector CD is a unit function of the rotational input, θ. Similarly, given a value of the displacement of the wheel, z, we can obtain the variables φ and θ. From equation (14.126), we can also obtain the following equation by eliminating θ:

$$2dl_2 \cos\varphi + 2zl_2 \sin\varphi - (d^2 + z^2 + l_2^2 - l_1^2) = 0. \tag{14.129}$$

Substituting

$$\cos\varphi = \frac{1 - \tan^2\frac{\varphi}{2}}{1 + \tan^2\frac{\varphi}{2}}$$

and

$$\sin\varphi = \frac{2\tan\frac{\varphi}{2}}{1 + \tan^2\frac{\varphi}{2}}$$

into equation (14.129) yields:

$$(d^2 + z^2 + l_2^2 - l_1^2 + 2dl_2)\tan^2\frac{\varphi}{2} - 4zl_2\tan\frac{\varphi}{2} + (d^2 + z^2 + l_2^2 - l_1^2 + 2dl_2) = 0. \tag{14.130}$$

Therefore, we can obtain φ by solving equation (14.130):

$$\varphi = 2\arctan\frac{2zl_2 \pm \sqrt{4l_2^2(z^2+d^2)-(d^2+z^2+l_2^2-l_1^2)^2}}{d^2+z^2+l_2^2-l_1^2+2dl_2}. \qquad (14.131)$$

When the knuckle moves above the horizontal line shown in Figure 14.16, we should select the following solution:

$$\varphi = 2\arctan\frac{2zl_2 - \sqrt{4l_2^2(z^2+d^2)-(d^2+z^2+l_2^2-l_1^2)^2}}{d^2+z^2+l_2^2-l_1^2+2dl_2}. \qquad (14.132)$$

When the knuckle moves below the horizontal line, the solution should be selected as:

$$\varphi = 2\arctan\frac{2zl_2 + \sqrt{4l_2^2(z^2+d^2)-(d^2+z^2+l_2^2-l_1^2)^2}}{d^2+z^2+l_2^2-l_1^2+2dl_2}. \qquad (14.133)$$

Substituting equation (14.131), into equation (14.126), we obtain the following solutions corresponding to equations (14.132) and (14.133), respectively:

$$\theta = \arccos\left(\frac{l_2}{l_1}\cos\varphi - \frac{d}{l_1}\right) \qquad (14.134)$$

$$\theta = -\arccos\left(\frac{l_2}{l_1}\cos\varphi - \frac{d}{l_1}\right). \qquad (14.135)$$

Differentiating equation (14.126) with respect to time yields:

$$\begin{cases} \dot{\theta}l_1\sin\theta - \dot{\varphi}l_2\sin\varphi = 0 \\ \dot{\theta}l_1\cos\theta + \dot{\varphi}l_2\cos\varphi = \dot{z}. \end{cases} \qquad (14.136)$$

We can immediately obtain the following relationship:

$$\begin{cases} \dot{\theta} = \dot{z}\dfrac{\sin\varphi}{l_1\sin(\theta+\varphi)} \\ \dot{\varphi} = \dot{z}\dfrac{\sin\theta}{l_2\sin(\theta+\varphi)}. \end{cases} \qquad (14.137)$$

Equation (14.137) represents the inverse velocity relationship of the suspension mechanism. Further, we can obtain the inverse acceleration equations:

$$\begin{cases} \ddot{\theta} = \dfrac{\ddot{z}\sin\varphi}{l_1\sin(\theta+\varphi)} - \dot{\theta}^2\cot(\theta+\varphi) + \dfrac{\dot{\varphi}^2 l_2}{l_1\sin(\theta+\varphi)} \\ \ddot{\varphi} = \dfrac{\ddot{z}\sin\theta}{l_2\sin(\theta+\varphi)} + \dfrac{\dot{\theta}^2 l_1}{l_2\sin(\theta+\varphi)} - \dot{\varphi}^2\cot(\theta+\varphi) \end{cases} \qquad (14.138)$$

14.3.4 Dynamics Analysis and Numerical Simulations of the Suspension Mechanism

From the above analysis, we can find that the novel suspension mechanism shown in Figure 14.11 can guide the wheel to track an exact straight line during jounce and rebound. Therefore, this mechanism overcomes the major drawbacks of the traditional suspensions in motion characteristics. In this section, the dynamics problems of this suspension will be investigated with the Newton–Euler method.

14.3.4.1 Theoretical Dynamics of the Suspension Mechanism

For the sake of simplicity, we consider the case when the four kinematic chains are identical. The dynamics equations of one kinematic chain of the knuckle can be established in the coordinate frame shown in Figure 14.16.

First we immediately obtain the Euler equation for the eccentric crank A_1B_1:

$$m_1\boldsymbol{r}_1\times\boldsymbol{g}-\mathbf{J}_1\boldsymbol{\varepsilon}_1-\boldsymbol{\omega}_1\times\mathbf{J}_1\boldsymbol{\omega}_1-\boldsymbol{r}_{B_1}\times\boldsymbol{F}_{B_1}+\boldsymbol{M}_{A_1}-\boldsymbol{M}_{B_1}-c_1\boldsymbol{\omega}_1-c_2(\boldsymbol{\omega}_1-\boldsymbol{\omega}_2)=0 \tag{14.139}$$

where m_1 denotes the mass of the eccentric crank A_1B_1,$\mathbf{J}_1$ denotes the matrix of inertial moment of A_1B_1 about the fixed revolute joint, A_1,$\boldsymbol{r}_{B_1}$ denotes the vector of the point on which the force $\boldsymbol{F}_{B_1}$ act, $\boldsymbol{M}_{A_1}$ indicates the resistant moment of the fixed joint A_1,$\boldsymbol{M}_{B_1}$ indicates the moment exerting on the crank A_1B_1 by the link B_1C_1,c_1 indicates the viscous friction coefficient of moment of revolute joint A_1,$\omega_1 = \dot{\theta}$ and $\varepsilon_1 = \ddot{\theta}$ indicate the angular velocity and acceleration, respectively, $\omega_2 = \dot{\varphi}$ indicates the angular velocity of link B_1C_1 and c_2 represents the viscous friction coefficient of moment of revolute joint B_1.

Similarly, the Euler equation for the link B_1C_1 can be expressed as:

$$\begin{aligned}&-\mathbf{J}_2\boldsymbol{\varepsilon}_2-\boldsymbol{\omega}_2\times\mathbf{J}_2\boldsymbol{\omega}_2+\boldsymbol{r}_2\times\boldsymbol{F}_{B_1}-(\boldsymbol{r}_{B_1C_1}+\boldsymbol{r}_2)\times\mathbf{F}_{C_1}\\&+\boldsymbol{M}_{B_1}-\boldsymbol{M}_{C_1}-c_2(\boldsymbol{\omega}_2-\boldsymbol{\omega}_1)-c_3\boldsymbol{\omega}_2=0\end{aligned} \tag{14.140}$$

where $\mathbf{J}_2$ denotes the matrix of inertial moment of link B_1C_1 about its mass center, $\boldsymbol{r}_2$ denotes the vector from the mass center of link B_1C_1 to joint B_1,$\boldsymbol{r}_{B_1C_1}$ denotes the vector from joint B_1 to joint C_1,$\boldsymbol{F}_{C_1}$ represents the force acting on the revolute joint C_1 by the knuckle, $\boldsymbol{M}_{C_1}$ indicates the moment acting on the link B_1C_1 by the knuckle, and $\varepsilon_2 = \ddot{\varphi}$ indicates the angular acceleration of link B_1C_1 and c_3 represents the viscous friction coefficient of moment of revolute joint C_1.

Because the four kinematic chains of the end effector are assumed to be identical and $\pi_1 \perp \pi_4$, the Euler equation for the knuckle $C_1C_2C_3C_4$ can be expressed as:

$$\sum_{i=1}^{4}\boldsymbol{M}_{C_i}-\boldsymbol{M}_E+\sum_{i=1}^{4}(\boldsymbol{r}_{C_i}-\boldsymbol{r}_C)\times\boldsymbol{F}_{C_i}+2c_3\omega_2\boldsymbol{e}_1+2c_3\omega_2\boldsymbol{e}_2=0 \tag{14.141}$$

where $\boldsymbol{M}_E$ represents the external forces acting on knuckle, $\boldsymbol{r}_C$ indicates the position vector of the geometry of the knuckle, $\boldsymbol{e}_1$ and $\boldsymbol{e}_2$ denote the unit normals of π_1- and π_4-planes, respectively.

Next, we can consider the Newton equations of the system. Firstly, the Newton equation for the eccentric crank A_1B_1 can be expressed as:

$$\boldsymbol{F}_{A_1} - m_1\boldsymbol{g} - m_1\varepsilon_1 \times \boldsymbol{r}_1 - m_1\boldsymbol{\omega}_1 \times (\boldsymbol{\omega}_1 \times \boldsymbol{r}_1) - \boldsymbol{F}_{B_1} = 0 \tag{14.142}$$

where $\boldsymbol{F}_{A_1}$ indicates the resistance force of revolute joint $\boldsymbol{A}_1$.

The Newton equation for the link B_1C_1 is:

$$\boldsymbol{F}_{B_1} - m_2\boldsymbol{g} - m_2\boldsymbol{a}_2 - \boldsymbol{F}_{C_1} = 0 \tag{14.143}$$

where $\boldsymbol{a}_2 = \boldsymbol{\varepsilon}_1 \times \boldsymbol{r}_{B_1} + \boldsymbol{\omega}_1 \times (\boldsymbol{\omega}_1 \times \boldsymbol{r}_{B_1}) - \boldsymbol{\varepsilon}_2 \times \boldsymbol{r}_2 - \boldsymbol{\omega}_2 \times (\boldsymbol{\omega}_2 \times \boldsymbol{r}_2)$.

The Newton equation for the knuckle $C_1C_2C_3C_4$ is:

$$\sum_{i=1}^{4} \boldsymbol{F}_{C_i} - m\boldsymbol{g} - m\ddot{z} - \boldsymbol{F}_E = 0 \tag{14.144}$$

where m represents the mass of knuckle and $\boldsymbol{F}_E$ denotes the external forces exerted on the knuckle.

Substituting equation (14.139) into (14.140) yields:

$$\begin{aligned}\boldsymbol{M}_{C_1} &= \boldsymbol{M}_{A_1} - \mathbf{J}_1\boldsymbol{\varepsilon}_1 - \boldsymbol{\omega}_1 \times \mathbf{J}_1\boldsymbol{\omega}_1 - \mathbf{J}_2\boldsymbol{\varepsilon}_2 - \boldsymbol{\omega}_2 \times \mathbf{J}_2\boldsymbol{\omega}_2 + \boldsymbol{r}_2 \times \boldsymbol{F}_{B_1} \\ &\quad -(\boldsymbol{r}_{B_1C_1} + \boldsymbol{r}_2) \times \boldsymbol{F}_{C_1} + m_1\boldsymbol{r}_1 \times \boldsymbol{g} - \boldsymbol{r}_{B_1} \times \boldsymbol{F}_{B_1} - c_1\boldsymbol{\omega}_1 - c_3\boldsymbol{\omega}_2.\end{aligned} \tag{14.145}$$

14.3.4.2 Numerical Simulations of the Dynamics of the Suspension Mechanism

Considering the overconstraints analyzed above, we can expands equation (14.144) in the affine space $\boldsymbol{e}_1\boldsymbol{e}_2\boldsymbol{e}_3$:

$$\begin{bmatrix} 2N_{e_1} \\ 2N_{e_2} \\ 4F_{B_1C_1}\sin\varphi \end{bmatrix} - m\begin{bmatrix} 0 \\ 0 \\ g \end{bmatrix} - m\begin{bmatrix} 0 \\ 0 \\ \ddot{z} \end{bmatrix} + \begin{bmatrix} F_{e_1} \\ F_{e_2} \\ F_{e_3} - kz_3 \end{bmatrix} = 0 \tag{14.146}$$

where N_{e_1} represents the resultant normal forces of the planes π_1 and π_2, N_{e_2} represents the resultant normal forces of the planes π_3 and π_4, $F_{B_1C_1}$ represents the axial force of the link B_1C_1, F_{e_1}, F_{e_2} and F_{e_3} represent the external forces in $\boldsymbol{e}_1$-, $\boldsymbol{e}_2$- and $\boldsymbol{e}_3$- directions, k represents the elastic coefficient of the spring.

Immediately, we can find that

$$N_{e_1} = -\frac{1}{2}F_{e_1},\ N_{e_2} = -\frac{1}{2}F_{e_2},\ F_{B_1C_1} = \frac{-F_{e_3} + kz_3 + m\left(g + \ddot{z}\right)}{4\sin\varphi}. \tag{14.147}$$

Therefore, $F_{C_1}(i = 1,2,3,4)$ can be expressed as:

$$F_{C_1} = F_{C_2} = \begin{bmatrix} -\frac{1}{2}F_{e1} \\ \frac{-F_{e3} + kz + m\left(g + \ddot{z}\right)}{4}\cot\varphi \\ \frac{-F_{e3} + kz + m\left(g + \ddot{z}\right)}{4} \end{bmatrix},$$

$$F_{C_3} = F_{C_4} = \begin{bmatrix} \frac{-F_{e3} + kz + m\left(g + \ddot{z}\right)}{4}\cot\varphi \\ -\frac{1}{2}F_{e1} \\ \frac{-F_{e3} + kz + m\left(g + \ddot{z}\right)}{4} \end{bmatrix}. \quad (14.148)$$

Suppose that the coordinates of the four joints $C_1(0\ \ -p\ \ q)$, $C_2(0\ \ -p\ \ -q)$, $C_3(0\ \ p\ \ -q)$, and $C_4(0\ \ p\ \ q)$ in the local coordinate frame with the origin at point C and the $y_L o_L z_L$-coordinate plane perpendicular to the axis of the transfer shaft. The vector coordinates of $CC_i(i = 1,2,3,4)$ in the absolute coordinate frame can be expressed as:

$$r_{C_i} - r_C = \begin{bmatrix} \cos\frac{\pi}{4} & -\sin\frac{\pi}{4} & 0 \\ \sin\frac{\pi}{4} & \cos\frac{\pi}{4} & 0 \\ 0 & 0 & 1 \end{bmatrix}\begin{bmatrix} x_i \\ y_i \\ z_i \end{bmatrix} = \begin{bmatrix} -\frac{\sqrt{2}}{2}y_i \\ \frac{\sqrt{2}}{2}y_i \\ z_i \end{bmatrix}.$$

Considering the overconstraints analyzed above, we can expand equation (14.141) in the affine space $e_1e_2e_3$

$$\begin{bmatrix} 2M_1 \\ 2M_2 \\ 4M_3 \end{bmatrix} + \begin{bmatrix} M_{e1} \\ M_{e2} \\ M_{e3} \end{bmatrix} - \begin{bmatrix} 2c_3\omega_2 \\ 2c_3\omega_2 \\ 0 \end{bmatrix}$$

$$+ \begin{bmatrix} \sqrt{2}p \\ -\sqrt{2}p \\ 0 \end{bmatrix} \times \begin{bmatrix} -\frac{1}{2}F_{e1} \\ \frac{-F_{e3} + kz_3 + m\left(g + \ddot{z}\right)}{4}\cot\varphi \\ \frac{-F_{e3} + kz_3 + m\left(g + \ddot{z}\right)}{4} \end{bmatrix}$$

$$+ \begin{bmatrix} -\sqrt{2}p \\ \sqrt{2}p \\ 0 \end{bmatrix} \times \begin{bmatrix} \frac{-F_{e3} + kz_3 + m\left(g + \ddot{z}\right)}{4}\cot\varphi \\ -\frac{1}{2}F_{e1} \\ \frac{-F_{e3} + kz_3 + m\left(g + \ddot{z}\right)}{4} \end{bmatrix} = 0 \quad (14.149)$$

where M_{e_1}, M_{e_2} and M_{e_3} represent the external torques about the $\boldsymbol{e}_1$-, $\boldsymbol{e}_2$- and $\boldsymbol{e}_3$-directions, respectively.

From equation (14.149), we can find that

$$
\begin{aligned}
M_1 &= -\frac{1}{2}M_{e_1} + c_3\omega_2, \quad M_2 = -\frac{1}{2}M_{e_2} + c_3\omega_2, \\
M_3 &= -\frac{1}{4}M_{e_3} - \frac{\sqrt{2}}{8}p(F_{e_2} - F_{e_1}).
\end{aligned}
\tag{14.150}
$$

Therefore, $\boldsymbol{M}_{C_i}$ can be expressed as:

$$
\begin{aligned}
\boldsymbol{M}_{C_1} = \boldsymbol{M}_{C_2} &= \begin{bmatrix} 0 \\ -\frac{1}{2}M_{e_1} + c_3\omega_2 \\ -\frac{1}{4}M_{e_3} - \frac{\sqrt{2}}{8}p(F_{e_2} - F_{e_1}) \end{bmatrix}, \\
\boldsymbol{M}_{C_3} = \boldsymbol{M}_{C_4} &= \begin{bmatrix} -\frac{1}{2}M_{e_1} + c_3\omega_2 \\ 0 \\ -\frac{1}{4}M_{e_3} - \frac{\sqrt{2}}{8}p(F_{e_2} - F_{e_1}) \end{bmatrix}.
\end{aligned}
\tag{14.151}
$$

For the kinematic chain $A_1B_2C_3$, we can obtain the joint forces by the following steps. Substituting equation (14.148) into equations (14.143)

$$
\boldsymbol{F}_{B_1} = m_2\boldsymbol{g} + m_2\boldsymbol{a}_2 + \boldsymbol{F}_{C_1}. \tag{14.152}
$$

Substituting equation (14.152) into equation (14.142) yields

$$
\boldsymbol{F}_{A_1} = m_1\boldsymbol{g} + m_1\boldsymbol{\varepsilon}_1 \times \boldsymbol{r}_1 + m_1\boldsymbol{\omega}_1 \times (\boldsymbol{\omega}_1 \times \boldsymbol{r}_1) + \boldsymbol{F}_{B_1}. \tag{14.153}
$$

Substituting equations (14.148), (14.151), (14.152) into equation (14.140), we obtain

$$
\begin{aligned}
\boldsymbol{M}_{B_1} = \boldsymbol{M}_{C_1} &+ \mathbf{J}_2\boldsymbol{\varepsilon}_2 + \boldsymbol{\omega}_2 \times \mathbf{J}_2\boldsymbol{\omega}_2 - \boldsymbol{r}_2 \times \boldsymbol{F}_{B_1} + (\boldsymbol{r}_{B_1C_1} + \boldsymbol{r}_2) \\
&\times \boldsymbol{F}_{C_1} + c_2(\boldsymbol{\omega}_2 - \boldsymbol{\omega}_1) + c_3\boldsymbol{\omega}_2.
\end{aligned}
\tag{14.154}
$$

Substituting equations (14.151) and (14.154) into equation (14.139), we can gain

$$
\boldsymbol{M}_{A_1} = \boldsymbol{M}_{B_1} - m_1\boldsymbol{r}_1 \times \boldsymbol{g} + \mathbf{J}_1\boldsymbol{\varepsilon}_1 + \boldsymbol{\omega}_1 \times \mathbf{J}_1\boldsymbol{\omega}_1 + \boldsymbol{r}_{B_1} \times \boldsymbol{F}_{B_1} + c_1\boldsymbol{\omega}_1 + c_2(\boldsymbol{\omega}_1 - \boldsymbol{\omega}_2). \tag{14.155}
$$

Similarly, we can also obtain all the forces of the other joints. Next, a numerical simulation with the above analysis will be demonstrated with a set of real vehicle parameters.

Suppose the jounce and rebound function of the knuckle is:

$$
z = 100 \sin 2\pi t \ (mm). \tag{14.156}
$$

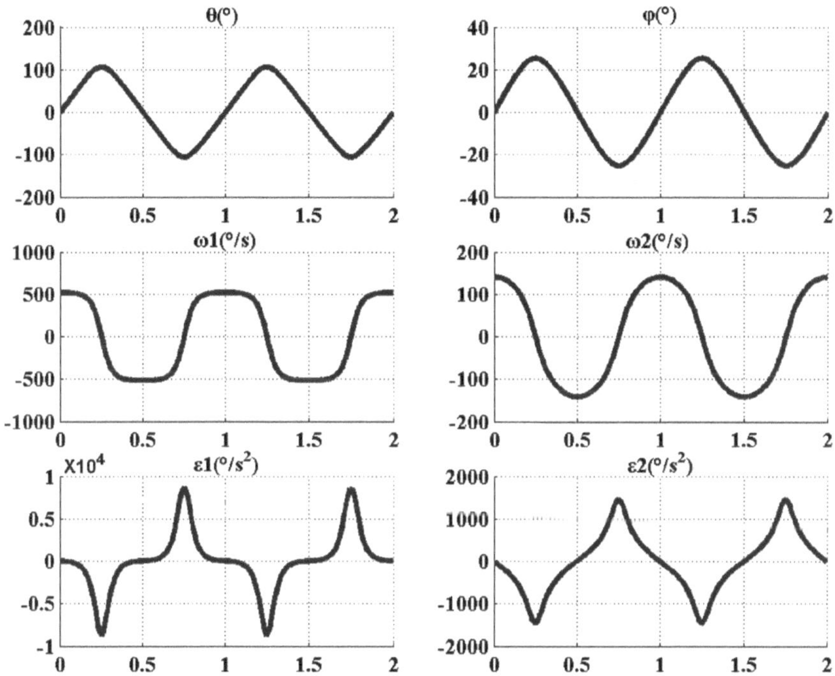

FIGURE 14.17 Kinematic curves of the joints A_i and C_i when the knuckle moves up and down in two circles.

From equations (14.137) and (14.138), we can immediately obtain the kinematics curves shown in Figure 14.17, the kinematic curves of the joints A_i and C_i when the knuckle moves up and down in two circles. The units of the abscissa are all in seconds and curves present the primary kinematic relationships of the joints A_i and C_i when the knuckle moves up and down in two circles.

Also suppose that the primary parameters of the knuckle are $l_1 = 15$ mm, $l_2 = 200$ mm, $d = l_2 - l_1 = 185$ mm, the distance from the revolute joint $A_i (i = 1,2,3,4)$ to the mass center of eccentric crank A_iB_i is $r_1 = 10$ mm and the mass of which is $m_1 = 0.5$ kg, the distance from the revolute joint B_i to the mass center of eccentric crank B_iC_i is $r_2 = 80$ mm and the mass of which is $m_2 = 4$ kg, the mass of the knuckle is $m = 35$ kg, the elastic coefficient of the spring is $k = 25$ N/mm, the viscous friction coefficients of moment of the revolute joints are identically equal to $c = 1$ Nms/rad. The external forces assumed are that $F_{e1} = 3200$ N, $F_{e2} = 3200$ N, $F_{e3} = 0$ N; $M_{e1} = 1000$ Nm, $M_{e2} = 1000$ Nm and $M_{e3} = 50$ Nm. With the above analysis, we can obtain the joint reaction forces of the kinematic chain $A_1B_1C_1$ shown in Figure 14.18 and the joint reaction torques shown in Figure 14.19. The abscissa coordinates are also all in seconds. The reaction forces and torques of the other joints can be similarly gained but these are omitted for the sake of brevity.

From Figures 14.17, 14.18, and 14.19, we can draw a conclusion that the joint resistant forces and torques are rational in value, and the merits that the

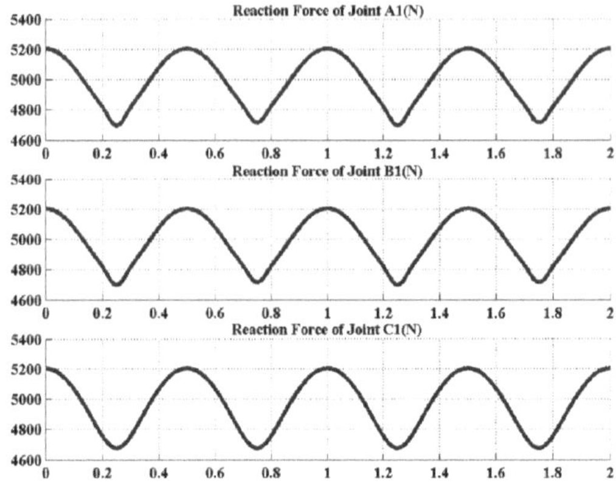

FIGURE 14.18 Reaction forces of the joints A_1, B_1, and C_1, when the knuckle moves up and down in two circles.

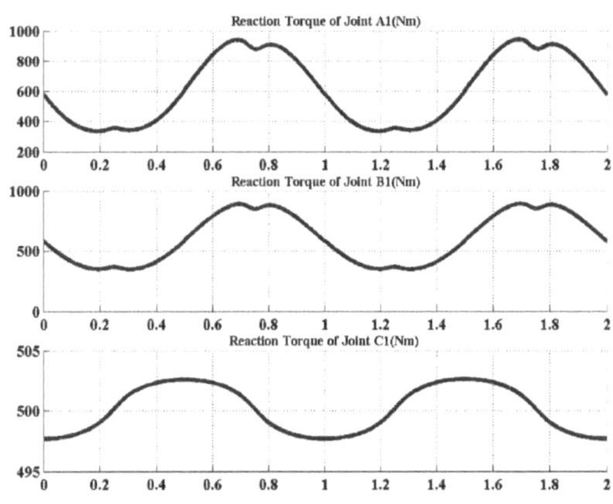

FIGURE 14.19 Reaction torques of the joints A_1, B_1, and C_1 when the knuckle moves up and down in two circles.

orientation parameters are invariant during jounce and rebound mean that the kind of suspension proposed in here has wide applications in advanced vehicles.

This section proposes a symmetric redundant suspension mechanism that has invariable orientation parameters. The mechanism mobility is firstly investigated with reciprocal screw theory, after which the displacement constraint equations of the suspension are established. Through differentiating the constraint equations with respect to time, the kinematic relationship is obtained and the dynamics equations of the suspension are built up with Newton–Euler method. Numerical simulations indicate that this kind of independent suspensions should not only eliminate the shambling shocks induced by the jumping of wheels because of the changing orientation parameters but also decrease the abrasion of the wheels. Therefore, this kind of independent suspensions can improve the ride and handling properties of advanced automobiles.

14.4 CONCLUSIONS

In this chapter, we adopt the natural coordinates to deal with the kinematics model for different spatial mechanisms. After obtaining the constraint equations in natural coordinates, we conduct the kinematics analysis of the mechanism, and then build the dynamics model with Newton–Euler's formulation. Regarding kinematics, many works have addressed this problem in the past. A fast kinematics model for the dynamics of a 2R1T spatial parallel manipulator with natural coordinates has been developed. The fast kinematics modeling of this method will surely satisfy the need for real-time analysis of dynamics. The actuators' driving forces and the constraints of the joints concerned are built up with Newton–Euler's formulation. The theoretical analysis and simulations are verified through experiments in engineering. The kinematic of suspension mechanism with invariable orientation parameters has been investigated, the novel suspension mechanism is designed to guide the wheel to track an exact straight line during jounce and rebound, and the dynamics problems of this suspension are analyzed with Newton–Euler method.

REFERENCES

[1] T. Szkodny, Dynamics of industrial robot manipulators, Mechanism and Machine Theory 30 (1995) 1057–1072.
[2] J.G. de Jalón, E. Bayo, Kinematic and Dynamic Simulation of Multibody Sytems: The Real-Time Challenge, Springer-Verlag, NewYork, 1994.
[3] M.D. Lioa, V. Cossalter, R. Lot, On the use of natural coordinates in optimal synthesis of mechanisms, Mechanism and Machine Theory 35 (2000) 1367–1389.
[4] R.V. Schwerin, Multibody System Simulation—Numerical Methods, Algorithms, and Software, Springer-Verlag, Berlin Heidelberg, 1999.

Index

A

B

G

H

P

Q

R

S

T

U

W

Kolibri 360 GmbH
Ferdinand-Jühlke-Straße 7
99095 Erfurt, Deutschland
produktsicherheit@kolibri360.de